Ecofriendly Insect Pest Management

Dr. S. Ignacimuthu, s.j.
Director
Entomology Research Institute
Loyola College, Chennai – 600 034, India

Dr. B.V. David
President
Sun Agro Biotech Research Centre
and
Executive Director
Sun Agro Biosystems Pvt. Ltd.
Chennai – 600 116, India

Elite Publishing House Pvt Ltd

302, JMD House, 4-B, Ansari Road, Daryaganj, New Delhi-110002 Phone: 011-41004299 E-mail: rg_elite@yahoo.com

ISBN: 81-88901-37-7

Published by R. Gangadharan for Elite Publishing House Pvt. Ltd.
302, JMD House, 4-B, Ansari Road, Daryaganj, New Delhi-110002
Phone : 41004299, 23074299
E-mail: rg_elite@yahoo.com

Graphic Design : Elite Studio, New Delhi-110 002
Printed at Chaman Offset, New Delhi-110002

PRINTED IN INDIA

IN MEMORY OF

LATE PROF. S. JAYARAJ
Former Vice Chancellor
Tamil Nadu Agricultural University
Coimbatore

Preface

The ever-increasing world population and consequent demands on agriculture products have led the agriculture to a critical situation in this country. Man has to increase the productivity of food grains from a limited area amongst many problems, of which pest and diseases come first. The intensification of farming practice has increased the problem of pest damage and hence control. The use of chemicals in pest management programmes has created many unwanted effects. Events during the past several decades have clearly shown that the use of chemical pesticides will not give adequate and lasting solutions to pest problems. The most important side effect created by synthetic chemicals is the development of resistance in pests to pesticides. Pest resurgence and secondary pest outbreak are the other two major problems. Besides these, many health hazards have been developing in human beings.

The farmers in developing countries are now renewing interest in approaches such as biological control, biopesticides and cultural control. However, these non-chemical methods are not popularized among farmers in developing countries. In future, the ecofriendly methods will play an important role in pest management, since they will bring about sustainable development. Applied research is required to develop and adapt the ecofriendly approaches for many crops and for today's production practices.

Ecofriendly insect pest management practices were discussed by eminent scientists and researchers from all over India in the two day national symposium on '*Ecofriendly Insect Pest Management*' held at the Entomology Research Institute, Loyola College, Chennai. The discussion focused on ecofriendly methods of insect pest management for sustainable agriculture with the following themes: 1. Botanical and microbial pesticides: Preparation and application methods of new formulations, 2. Biological control: Mass rearing techniques, field evaluation and conservation of biological agents, 3. Cultural control measures, 4. Vector control: Human vectors and vectors of household animals, 5. Genetic techniques and 6. Attractants and repellents

We are grateful to the scientists for their valuable contributions. We hope that this volume will serve as a useful resource material for scientists, research scholars, farmers and Industries. We are indebted to M/s Elite Publishing House, New Delhi for publishing this volume. We wish to thank Mr. C. Muthu for his assistance in bringing out this volume. We thank the Council of Scientific and Industrial Research (CSIR), New Delhi and Tamil Nadu State Council for Science & Technology (TNSCST), Chennai for partial financial assistance.

S. Ignacimuthu, s.j.
B.V. David

Contents

ENTOMOPHAGES– BIOLOGICAL CONTROL

BOTANICAL PESTICIDES

PHEROMONES, ATTRACTANTS AND REPELLENTS

VECTOR CONTROL

OTHERS

Alternate Strategies for Insect Pest Management in Tea in South India

A. BABU*

Tea, which is grown as a perennial crop is known to be infested by an array of insect and mite pests. Some of them are highly seasonal; many attack tea only during dry season while a few are prevalent during the wet weather; there are a few perennial pests too. As for as the south India is concerned, the major pests associated with the tea ecosystem include thrips, tea mosquito, red spider mite and shot hole borer. They cause considerable damage, resulting in substantial crop loss apart from escalating the cost on pest control. The bio-ecological and crop loss studies on these major pests revealed the fact that, they are accountable for an average crop loss of 15 – 20 per cent. Even though different pest control methods are being adopted over the past several decades, it is absolutely essential to have a closer look at the current status of important pests, in order to develop alternate strategies for implementing integrated pest management programme. The most important aspect one has to adopt for tackling the pests is to have a better understanding about the pest itself. Continuous use of pesticides and inappropriate execution of pest control measures will led to undesirable problems in tea gardens together with the incidence pesticide residues in made tea. At this point in time, switching over to non-chemical control strategies on large scale is indispensable, in order to race in global market for fetching excellent price for Indian tea. Pesticides intended at destructive pest populations have the potential to alter ecological balance of the tea ecosystem by destroying the natural enemies (non target organisms) such as predators and parasitoids. Incessant use of synthetic chemicals for the control, may also lead to development of resistance in pests. The enormous crop loss caused by pests requires execution of efficient, non-chemical and eco-friendly alternate strategies. Biopesticides are considered as environmentally safe, selective, biodegradable economical and renewable alternatives for use in IPM programme especially under organic farming systems. An attempt has been made in this paper to provide an in-depth knowledge about the important tea pests, their seasonal abundance, damage potential and integration of different biological control strategies as a package for implementing them for successful management of tea pests.

INTRODUCTION

Tea is grown as a perennial crop in the humid belt of the Western Ghats with varied agro-climatic conditions in south India. Majority of the tea growing areas in the North India are in flat land while in south India, it is cultivated in hill slopes at elevation ranging between 500 to 2400m. The weather conditions in the different planting districts vary distinctly. Being a perennial crop is known to be infested by a number of insect and mite pests (Muraleedharan and Chen 1997; Muraleedharan and Selvasundaram, 1996). Majority of the tea pests are highly seasonal; many attack tea only during dry season while a few are abundant in wet weather; there are a few perennial pests too. The major pests *viz.* thrips, tea mosquito bug, red spider mite and shot hole borer cause widespread damage, resulting in average crop loss (15 – 20%) apart from increasing the cost on pest control. Improper execution

* Division of Entomology, UPASI Tea Research Foundation, Valparai-642 127, Coimbatore District

of pest control measures using synthetic chemicals will lead to undesirable problems in tea gardens. Switching over to non-chemical control strategies to a larger extent is essential, in order to overcome the current situation and for fetching good price for our products in the international market. Pesticides aimed at destructive insect populations associated with the tea ecosystem may also destroy the natural enemies (non target organisms) such as predators and parasitoids (Muraleedharan, *et al.,* 2001), apart from the problems of development of resistance in pest itself. This requires implementation of efficient, eco-friendly alternate control strategies (Muraleedharan, 2005).

Several biological control methods including microbial control (use of fungus, bacteria), semio-chemicals (pheromones and kairomones) and other biological have been attempted (Selvasundaram *et al.*, 2001). Recently, Neem Kernel Aqueous Extract (NKAE) has been found effective against some of the pests like red spider, pink and purple mites, and catterpillars both in lab and field conditions. Although different pest control methods are adopted over the past several decades, there is a need for a closer look at the current status of important pests, to arrive at an efficient and eco-friendly control strategy for the management of major tea pests. Development of mass multiplication technology of the entomopathogens (microbes), natural enemies (major predators) and use of biologicals could definitely helpful in reducing the usage of synthetic insecticides for the control of tea pests. Degradation of pesticides applied on tea leaves takes place in different ways which is governed by weather factors and processing. Photolabile compounds undergo degradation due to sunlight in the field itself. The pesticide present on harvested shoots undergoes thermal degradation during manufacturing, at the high temperature, especially if the compound is thermolabile. In short, the level of residues of any pesticide on made tea depends on its response to temperature. Based on the degradation pattern and MRL, safe harvest intervals have been fixed for the more commonly used pesticides (see infra). Just as temperature influences the residues in made tea, the solubility of the compound decides the level of residues in the tea infusion. Regulations on pesticide residues in food and environment are becoming more the more stringent all over the world. Food industries, including tea industry will have to take note of the emerging scenario. The tea industry in India is in an advantageous position, since it had all along been adhering to the tolerance levels or MRLs declared by the EPA of USA or the EU in view of the large export market for Indian tea. The tea garden managements have to look at ways and means to further reduce the usage of pesticides. This, to a very large extent can be achieved by adopting the following measures against important tea pests.

Bio-Ecology of Major Pests

1 . Tea thrips *Scirtothrips bispinosus* Bagnall

Scirtothrips bispinosus (Bagnall) is an endemic species of peninsular India, causing severe damage to tea plantations **especially** in Vandiperiyar-Peermade area of Idukki District in Kerala. Tea fields recovering from pruning is more prone to the attack by this thrips. Leaf surface becomes uneven, curly and matty, exhibiting parallel lines of feeding marks on either side of the midrib. Heavily infested fields sometimes acquire a bronze color. Damaged terminal shoot may be discolored, stunted, and deformed. The threshold density is determined as three thrips per shoot, above which they cause economic damage leading to considerable crop loss.

The life cycle consists of an egg stage (incubation period of 3-4 days), two larval instars (larva I - 3 days, larva II- 5 days), a short pre pupal stage (< 1 day), a pupa (3 days). The larvae resemble the adults in general body form though they lack wings and are smaller. Larvae feed in groups, particularly along the leaf midrib and veins, and usually on young leaves and buds. Larval development time is determined principally by the suitability of temperature. Total duration of life cycle is about 14- 20 days depending on the weather conditions.

1.1 Symptoms of infestation

Lacerations of tissue and appear as streaks, leaf surface becomes uneven, curled and matty, appearance of parallel lines on either side of mid rib is due to feeding marks in buds which becomes prominently visible when the leaf unfolds. Yellowing of leaf margins is the typical symptom of thrips infestation. More damage in tea recovering from pruning and delayed tipping causes appreciable crop loss and also affects the quality of made tea.

1.2 Monitoring

Monitoring of thrips should have to be done, by inspecting the pluckable shoots (three leaves and a bud stage) and counting the number of larvae and adults separately. Shoots should be collected from the side branches, plucking table and below the plucking table. The shoots should be collected at random from different bushes plants in the field. Adults can be monitored by the use of yellow sticky traps even at the initial stage, thus could be prevented from the sudden flare-up by adopting control strategies.

1.3 Economic Threshold Level (ETL)

There are several recommendations for the economic threshold for different crops. One should remember that the economic threshold should be validated for each agro climatic conditions of different regions based on the thrips incidence, the anticipated crop loss, previous year's incidence and environmental conditions. Based on the research, the threshold is fixed as 3 thrips per shoot. Appropriate control measures have to be adopted to reduce economic loss due to thrips infestation if thrips population is around 3 thrips per shoot.

The monitoring of thrips population should be initiated from September onwards when they start their appearance in a small scale. The pest population should be monitored even after the preliminary rounds of spraying and the affected section should be marked to facilitate spot spraying instead of blanket spray, which is not necessary at that particular time

1.4 Control Strategies

(i) Chemical Control: Timely adoption of recommended chemicals having label clearance from the CIB is essential to contain the population of thrips. The follow-up round should be given at 7-10 days interval so as to have the pest under check. Care should be taken to avoid spraying chemicals either before plucking or without giving the stipulated safe harvest interval which may lead to taint/reside problems in made tea.

In view of pesticide residue problems general awareness of pesticide residues around the globe, it is very important to use pesticides as little as possible by integrating different control strategies by adopting integrated pest management (IPM) programs. Only the recommended pesticides with clearance from CIB have to be used in rotation at stipulated intervals. The selection of pesticides depending on the nature of the pest attack is very critical in achieving satisfactory control of pest.

The success of chemical control of thrips in tea is primarily depending on application method. It is necessary that the pesticide is reaches the target. Hence spraying has to be done in such a way that it covers both sides of the tea leaves and side shoots, where the majority of the thrips are located.

(ii) Biological control: There are several natural enemies that help in the control of thrips. However, none of them alone can reduce the thrips populations to a low, non-economical density. Furthermore, the intensive use of pesticides in the tea affected adversely on the ecosystem natural enemy's activity. There is a need for more study on the role of natural enemies in the tea ecosystem and to enhance natural enemy populations by planting a natural barrow row of flowering plants along the borders /

vacant places. The entomopathogenic fungus, *Verticillium lecanii* was also effective in controlling the thrips. Yellow sticky trap was found useful in monitoring the thrips population even at an early stage of infestation. It could be incorporated into the IPM schedule and the use of pesticides should be need based, and with sufficient safe-harvest intervals in order to keep residue levels of pesticide below stipulated maximum residue limit.

2. Red Spider Mite *Oligonychus coffeae* (Nietner)

The tea red spider mite, *Oligonychus coffeae* is the largest of all the mites and can be easily seen by the naked eye. It was considered as a "dry weather pest" until the last four to five years in south India, has gained importance in the last few years due to its sudden outbreak in the tea growing areas of south India (Muraleedharan *et al.,* 2005). It assumed the status of major pest causing severe damage to the tea plantations. Most of the tea estates are forced to spend a huge amount for the control of red spider mite.

The eggs are laid along the veins of leaves during the growing season. Red spider mite eggs are round and extremely large in proportion to the size of the adults. The maximum activity of spider mite peaks during the warmer months and can develop rapidly during this time, becoming full-grown in as little as a week after eggs hatch. At the same time, most of their natural enemies require more humid conditions and are stressed by arid conditions. Furthermore, plants stressed by drought can produce changes in their chemistry that make them more nutritious to mites.

After hatching, the egg shells remaining on the leaves, helps in diagnosing spider mite infestation. The fast development rate and high egg production can lead to extremely rapid increase in mite population. They feed more under dry conditions, as the lower humidity allows them to evaporate excess water they excrete.

2.1 Symptoms of infestation

They normally infest the upper surface of mature leaves, though in extreme cases they may invade the lower surface as well as young leaves and consequent to their feeding, the maintenance foliage turns ruddy bronze, making the infested fields distinct even from a distance. Adopting control measures in proper time will prevent the total defoliation of bushes.

2.2 Monitoring of Red spider mites

Sampling for red spider mite consists of collecting leaves and counting number of mite eggs, nymphs and adults. Large-scale sampling should be done by demarking 1 ha area into four or six plots and taking 50 leaves randomly from each plot and count the number of egg, nymphs and adults. Care should be taken to take into account even the presence of a single mite in a leaf as infested. If 30% of the leaves contain mites immediate spraying should be taken up.

2.3 Economic Threshold Level (ETL)

Based on the crop loss studies conducted, the economic threshold Level (ETL) for red spider mite is 4 mites per leaf.

2.4 Control Strategies

Lime sulphur is a potent acaricide useful in controlling red spider mite population, provided it is prepared in such a way that the Polysulphide-S content should be above 10%. It is rather inexpensive and it can be integrated with other chemical schedule. However, continuous usage of Lime Sulphur is discouraged because of its poor ovicidal activity and tainting nature.

The new herbal product Exodus derived from leguminous plant *Sophora favescens* was found effective against red spider mite. Mineral spray oils manufactured from Paraffin base has also found to be effective and is found safer to biological control agents also. New acaricides like Propargite 57EC, Fenpyroximate 5 %EC are found to be more effective and their efficacy lasted for more than 3 weeks. However, we have to go for an integrated approach by incorporating the various eco-friendly, non-chemical methods available, to achieve efficient control of this pest.

The entomopathogenic fungus, *Paecilomyces fumosoreus* has been isolated from the flushworm infesting tea and it has been formulated into a wettable powder (UPASI TRF Strain) and tested against the red spider mite It was found effective in controlling the red spider mite. The fungal formulation is now commercially available under the trade name "Mycomite". The formulation has to be applied @ 2000 g/ha in 450 litres of water, preferably using a hand operated/high volume sprayer. The spray solution should be filtered through muslin cloth to remove extraneous materials and to prevent clogging of the nozzles. Application of "Mycomite" has to be done early in the mite season when the red spider mite incidence is low. The ideal time for spraying will be early morning or late evening when the relative humidity is around 70-80%. All the life stages of the red spider mite are susceptible to the fungal pathogen and mortality could be noticed after 5-6 days of application. However, close monitoring of mite population may be necessary even after "Mycomite" spraying and spot application of Lime sulphur is recommended to achieve better control. It must be ensured to get minimum period of 10-12 days between fungicide application for the control of diseases and Mycomite spraying. Results obtained from the large scale field experiment on the bioefficacy study of NKAE on red spider mites indicated that application of NKAE @5.0 and 7.5% significantly decreased the number of red spider mites. It could be incorporated in to the integrated schedule for the control of red spider mite in the field.

The spray volume recommended is 350-400 l/ha (for power sprayers) 450-500 l/ha (for knapsack sprayers) and 600 l/ha (for Honda sprayers). Thorough drenching of maintenance foliage is essential for efficient control of mites, because large number of eggs, nymphs and adult mites are seen below the plucking table. Addition of no-ionic wetting agent @ 5 ml/10l is recommended for better control during wet weather. As in the case of any crop pests, integrated control measures are required to reduce the attack. The small-scattered pockets from where the infestation starts must be identified. Efforts should be made to keep these areas marked and to avoid spreading to other areas.

3. Tea Mosquito Bug *Helopeltis theivora* (Waterhouse)

The Tea Mosquito Bug *Helopeltis theivora* is yet another important pest causing serious threat to the tea plantations especially in Vandiperiyar-Peermade area of Idukki District in Kerala (Central Travancore region). They prefer dark, humid and shaded areas. They are very much active during the early hours of the day and late evenings. They hide themselves among the weeds and weed hosts during the day time. Incidence will be more during July – December with a peak during September/October. Initial infestation starts in the fields near jungle and swamps. The chinary and hybrid jats are more susceptible. Eggs are laid in tender stems, petioles. One can identify the egg laid spots by close observation of the two unequal respiratory horns of the eggs, exposed from the stem or petiole. The incubation period of the egg is about 5-7 days. It passes through 5 nymphal instars within a period of 9-12 days depending on the favorable weather conditions. Adults resemble mosquitoes and are black colour. Life cycle is completed within 19-29 days.

3.1 Symptoms of infestation

Adults and nymphs suck the sap from the buds, young leaves and tender stems by puncturing with needle like rostrum. Intensive feeding causes curling of leaves, badly deform, remain small and shoot dry up and crop loss is near total.

3.2 Monitoring of tea mosquito

Tea mosquito bug incidence may be monitored regularly during the monsoon season and suitable control measures should be taken up without any delay. Percentage of infestation has to be assessed by collecting 100 shoots (three leaves and a bud) at random from the plucking gang of different fields and counting the infested shoots from the hundred shoots. Even a single new puncture is counted as infested. Old punctures should not be included as infested.

3.3 Economic Threshold Level (ETL)

If the average infestation is above 5% immediate spraying should be done soon after plucking with a follow-up round at 7-9 days interval.

3.4 Control Strategies

Pesticide application using Quinalphos @ 750 ml/ ha may be taken up for the control of tea mosquito. Follow-up rounds may be done after 7-10 days in the severely affected fields.

Sex pheromones form an important component of IPM. Pheromone traps of *Spodoptera litura* (cutworms) infesting tea are commercially available and can be used for controlling the populations of the moths. The pheromones of tea leaf roller have been field tested and are used in Japan.. Similarly, the female sex attractants of the flushworm *(Cydia leucostoma)* have also been identified recently. Studies on the sex pheromones of tea mosquito are in progress. Kairomones from the partly dried stems of *Montanoa bipinnatifida* are found to attract shot hole borer beetles. At present these stems are used to attract shot hole borer beetles in tea fields.

3.5 Cultural Control

Keeping the fields free from weeds is the most important parameter to minimize tea mosquito infestation. Weeds like *Mikania cordata, Bidens biternata, Emillia sp., Polygonum chinense* and *Lantana camara* offer excellent hiding places and serve as alternate hosts for the tea mosquito. Growth of weeds and wild host plants in and around tea fields should be controlled and this will help to reduce the growth of *Helopeltis* population.

Removing the stubs intensively while plucking will remove eggs and thus we can reduce the next generation. Black plucking of affected bud can also be done to reduce the egg population.

(4) Shot- Hole Borer *Euvallacea fornicatus* (Eichhoff)

Shot-hole borer, *Euvallacea fornicatus* is one of the important pest of tea in the middle and low elevation tea growing districts. The females construct galleries in tea stems, leading to branch breakage and consequent indirect crop loss. They form the galleries in the primary branches formed after pruning. However, the attack may also see on the old wood, even at the collar region. Pruning helps in the removal of the affected stems and adult beetles. The first year fields will be free of shot hole borer infestation due the physiological changes taking place in the bushes immediately after pruning. We can notice the new infestation only from the end of the 2nd year. Being a perennial pest, it could be managed only by adopting system of pest management.

4.1 Symptoms of infestation

The shot hole borer infested fields does not show any visible symptoms like the foliage feeding insects. They attack the primary branches formed after pruning by forming both vertical and horizontal galleries. The attack may also found on the old wood, even at the collar region. The infested have to be divided into 2ha blocks and from each block, collect 200 cuttings of pencil thickness (1.5 cm diameter and 20 cm long) at random. The percentage of attack in each block has to be pooled and take the average infestation of all blocks.

4.2 Economic Threshold Level (ETL)

The economic threshold level is 15 per cent and control measures have to be taken up if the average infestation is above 15%.

4.3 Control Strategies

Integrated Pest management is the best way. It is better to avoid pruning during the wet weather. Removal of badly affected branches while pruning is another way of preventing the migration of adult beetles from one filed to the other. Chemicals like Quinalphos 25EC @ 750 ml + Dichlorvos 76 EC @ 250 ml/ha have to be sprayed on the pruned frames and prunings. Two applications are required during April and December for efficient management of this pest.

4.4 Mid Cycle Control

Assess the SHB infestation at the end of 2nd year after pruning, and it would be advantageous to commence mid cycle application in 3rd & 4th year fields, if the average infestation is above 15%.

4.5 Biological Control

Two rounds of the entomopathogen, *Beauveria bassiana* (Biopower) @ 2 kg/ha is recommended for spraying during end May and October, when the humidity levels are high addition of jaggery @ 1.5 kg/ found to be useful in enhancing the efficacy. Studies conducted at UPASI TRF revealed that partially dried cut stems of a jungle plant *Montanova bipinnatifida* .an exotic plant species introduced from Mexico, attract shot hole borer beetles in the field. Placing partially dried cut stems of *Montanova* has helped in trapping the beetles and reducing the incidence of this pest. Such dried stems could be used in SHB infested 3rd and 4th year fields during October to April to attract the adult beetles. Partially dried cut stems (3-4 weeks old), 90 cm long and 25-30 cm thickness may be used @ 4 cut stems in 100 m^2 and an average of 400 cut stems are required per ha. The stems have to be replaced once in three weeks.

4.6 Cultural control

Medium prune (60-70 cm) is best suited for shot - hole borer infested fields (except when other factors demand a different height of pruning). Longer pruning cycles will tend to increase the intensity of borer damage, especially in mid and low elevation areas.

4.7 Fertilizer application

Application of higher levels of potassium fertilizers is known to reduce the incidence of pests and disease in several crops. Application of higher levels of potassium fertilizer ($N:K_2O = 1:2$) in the pruned year had helped to decrease the incidence of shot-hole borer in the infested fields.

5. Other methods

Certain routine cultural operations such as plucking and manual removal, shade regulation, infilling the vacancies, heat treatment and soil solarisation, use of light traps, use of biocontrol using entomopathogens, use of multicellular organisms, use of plant products for pest control, use of inorganic compounds & hydrocarbon oils and host plant resistance can be manipulated to reduce the incidence of pests.

5.1 Plucking and manual removal

Populations of leaf folding caterpillars such as flushworms, leaf rollers and tea tortrix can be reduced by their manual removal while harvesting. Populations of foliage feeding caterpillars such as loopers, bunch caterpillars and faggot worms can be reduced to a great extent by manual removal of their larvae and pupae.

5.2 Shade regulation

Un-shaded areas are more prone to the attack of thrips and mites. The recommendations on shade management, if adopted, will help to prevent the excessive build up of thrips, mites and reduce the incidence of blister blight disease.

5.3 Infilling the vacancies in field

Weeds are problematic in young tea clearings, pruned fields and vacant patches in mature tea fields. These patches have to be infilled while pruning the field to avoid weed growth and to increase bush population and yield.

5.4 Host plant resistance

Use of pest resistant varieties is one of the important components of non chemical control strategies. Even low levels of resistance are important since the need for other control methods can be reduced. Being a perennial crop, research on clonal selection and breeding is primarily aimed at the production of high yielding and superior quality plants with practically no emphasis on resistance of pests or diseases. The emerging information could be successfully utilised in the management of pests.

5.5 Heat treatment and soil solarisation

Soil used in the nursery may be heated to 60-62°C for killing the infective juveniles of root knot nematodes. Care may be taken that the soil is not over heated since this will lead to phytotoxicity. Soil solarisation during summer is also found quite effective in reducing the *Meloidogyne javanica* populations in the infested soil.

5.6 Use of light/colour traps

Fluorescent light traps are useful in attracting the moths of caterpillars and other insects. These can be set up during the seasons of moth and beetle emergence and the attracted moths and beetles can be killed mechanically or by using insecticides. These traps are useful for monitoring the activity of the pests and also as a tool in their suppression. Yellow pan water traps have been found useful to trap thrips and aphids. Use of yellow sticky traps has been found to reduce thrips attack by trapping the adult thrips.

5.7 Biocontrol using multicellular organisms

The minor status of several pests such as aphids, scale insects, flushworms, leaf rollers and tea tortrix is due to the action of natural enemies. Often, the work of these beneficial arthropods goes unnoticed, especially when their hosts are minor pests. Efforts towards the conservation and augmentation of natural enemies in the tea ecosystem could offer significant advances biological control programme in tea in future.

5.8 Biocontrol using entomopathogens

Several microbes are pathogenic to tea pests. Formulations of the bacterial insecticide, *Bacillus thuringiensis* have been effectively used for the control of lepidopterous pests. A local strain of *Beauveria bassiana* has been found effective against the shot hole borer. This strain is now being marketed as a wettable powder formulation for the control of shot hole borer.

Certain commercial formulations of the entomopathogenic fungi, *Verticillium lecanii, Paecilomyces fumosoroseus* and *Hirsutella thompsonii* were evaluated and found effective against pink, purple and red spider mites. These formulations are recommended for the control of mites. *Trichoderma viride, T. harzianum* and *Gliocladium virens* are used to suppress the populations of soil borne pathogenic fungi causing root and stem diseases of tea.

Even though, a number of synthetic chemicals are available for pest control in tea, the choice is limited for widespread application owing to the Maximum Residue Limits (MRLs) declared by European Union (EU) and Codex Alimentarius Commission and EPA of USA. Hence need based, judicious and safe application of pesticides have become the vital aspect of chemical control under the integrated pest management strategy, which involves development of IPM skills. Monitoring the crop health, observing the ETL and identification and conservation of biological control agents including entomopathogens, evaluation of their field potential before taking a decision on the use of chemical pesticides as last resort.

The success of any control measure under IPM context depends on the various parameters such as :

(i) use of pesticides only when it is absolutely necessary
(ii) use of recommended chemical with label claim on tea
(iii) maintaining correct dosage and dilution rates
(iv) giving adequate pre-harvest interval
(v) change of pesticides in the subsequent rounds
(vi) incorporation of biologicals including entomopathogens.

Need based, judicious and safe application of approved pesticides is the most vital aspect of chemical control measures under IPM strategy. It involves developing IPM skills to play safe with environment by proper crop health monitoring, observing ETL and conserving the natural bio-control potential before deciding in favour of the use of chemical pesticides as a last resort. Adoption of such non-chemical control methods to evolve an integrated management programme against pests infesting tea will form an integral part of sustainable tea cultivation.

Acknowledgement

I am thankful to Dr.N. Mualeedharan, Adviser and Dr. P. Mohan Kumar, Director, UPASI Tea Research Foundation for their constant encouragement and support in the preparation of this paper.

References

Muraleedharan, N and Chen Z.M., 1997. Pests and diseases of tea and their management. *J. Plantation Crops,* 25(1), 15-43.

Muraleedharan, N., 2005. Sustainable cultivation of tea. *Planters' Chronicle,* 101(5), 5-17

Muraleedharan, N. and Selvasundaram, R., 1996. Control of tea pests in south India (Revised recommendations). *UPASI Hand Book of Tea Culture series,* 23, pp 33.

Muraleedharan, N. and Selvasundaram, R and Radhakrishnan, R., 2001. Parasitoids and predators of tea pests in India *J. Plantation Crops,* 29, 1-10.

Selvasundaram ,R., Muraleedharan, N. Sudhakaran R., and Sudarmani, D.N.P., 2001. New strategies for the management of shot hole borer of tea . *Bulletin of UPASI Tea Research Foundation,* 54, 82-87.

Recent Advances in Microbial Control of Vectors of Parasitic Diseases of Both Veterinary and Public Health Importance

K. NARAYANAN AND G.P. SHETTY*

Biological control of agricultural pests and weed plants by way of introduction of natural enemies like parasites, predators and pathogens is well known. However, traditional method of control of vectors of both veterinary and medical importance lies invariably on the sanitation of the environment combined with quarantine and immunization. To develop an efficient form of biological control of vectors of parasitic diseases of man and animal a thorough knowledge of various pathogens associated with vectors is very much important. In this review information mainly on various naturally occurring biological control agents viz., parasites, predators and pathogens associated with various vectors has been presented. Naturally occurring microbial control agents viz., entomopathogenic viruses like Nucleopolyhedroviruses (NPVs) and Granuloviruses (GVs) belonging to the family Baculoviridae; Entomopox viruses (EPVs) and Cytoplasmic viruses (CPVs) etc. has also been presented. Bacterial pathogens like *Bacillus sphaericus* and *B. israelensis* has been well studied for the control of vectors like *Culex, Anopheles* and *Aedes* spp. Advantage of *B. sphaericus* compared to *B. israeliensis*, having an increased duration of larvicidal activity against certain environments has been presented and discussed. Further, the pathogenecity potential of *Brevibacillus laterosporus* against house fly, *Musca domestica* and mosquito, *A. aegypti* has been reported very recently (Ruiu *et al.,* 2007). Also the toxicity of a mosquitocidal metabolite of *Pseudomonas fluorescence* on larvae and pupa of the housefly, *M. domestica* has also been reported (Padmanabhan *et al.,* 2005). Steers given *B. t* var *thuringiensis* containing exotoxin called "fly factor" as additives in their feed, produces faeces which prohibited normal imago development of 92% of the 48 h old *M. domestica* larvae reared in the faeces. Entomopathogenic fungi belonging to the family Deutromycetes *viz., Beuvaria bassiana* and *Metarhizium anisopliae* etc. as potential microbial control agents for the control of various vectors of public health importance like house fly, and mosquitoes *Culex, Anopheles, Aedes* and tsetse flies (*Glossina spp.*) which is responsible for the transmission of protozoan of genus *Trypnosoma*, which cause human and animal Trypanosomosis other wise referred as Sleeping sickness has been presented and discussed. The interaction between *Wolbachia*, a common cytoplasmic rickettsial alpha-proteobacterium and *Wucheria bancrofti*, which is a vector for causing river blindness and elephantiasis has been presented and discussed. Entomopathogenic nematodes (EPNs) belonging to Steinernematids, Heterorhabtidis and Mermithids, etc. for more effective control of vector borne diseases in the current concept of Integrated Vector Management (IVM) Programme has been presented. High pathogenicity of *Heterorhabditis spp.* against maggots, pupae and adults of house fly, *M. domestica* and ticks etc., has been recently reported (Narayanan *et al.* 2007, Unpublished). Information on certain promising parasities like pupal parasite of house fly, *Spalangia endius* and predatory fish like *Tilapia* and *Gambusia affinis* etc., for the control of mosquito vectors like *Anopheles stephensi* has also been briefed. Further, the existing information on various bio-technical

* Multiplex Bio-Tech Pvt. Ltd., Bangalore – 560 086

approaches made towards increasing the efficacy of certain vector pathogens like *B. sphaericus and B. israelents* for the control of mosquitoes through biological immobilization using ciliated protozoan like *Tetrahymena pyriformis* which biologically magnifies the spores of *B. sphaericus and B. isralensis* by way of engulfing not less than 6 to 8 spores in each of its 30 food vacules etc. and bio-technological means using recombinant (r-DNA) DNA technology using cyanobacteria like *Caulobacter crescentus* has also been presented and discussed.

Introduction

Many insect pests not only cause direct injury and annoyance to man and animals but also in quit some cases cause and transmit diseases and functional disorders in them by way of acting as vectors. The major vector borne diseases that affect human beings are the arthropod borne diseases like Typhus, Sleeping sickness, Malaria, Tick borne Relapsing fever, Tularemia, Filariasis, Japanese encephalitis (JE), Dengue, chicken guinea, Dengue haemorrhagic Fever (DHF) transmitted by mosquitoes, Leishmaniasis (dermal oriental sore and visceral-Kala Azar) transmitted by sand flies, Plague transmitted by rat fleas Kyasanur Forest Disease (KFD) transmitted by ticks, which results in high morbidity and mortality in humans. In the case of animals apart from transmitting various kinds of diseases like mosquito borne encephalitis of horses, tick borne babesiosis, theileriosis and rickettesial diseases, culicid fly borne blue tongue, onchocerciases and African horse sickness, Chagas disease etc., the productivity of affected animals is also reduced resulting in huge aggregate animal losses. The losses in livestock production is in terms of loss in milk, meat, hide and eggs. The role that insects play in the perpetuation and transmission of organisms pathogenic to man and animals is a complicated biological phenomenon involving interrelationships and mutual adaptations of the causative agent, vector, vertebrate host and the biotic and abiotic environment.

Traditionally, the most important aspect of insect vector control of practical value has been the sanitation of the environment combined with quarantine and immunization. All the other measures are supplementary. In medical and veterinary practice, chemical methods (chemical solutions, aerosols, gaseous substances) have been the most commonly employed control measures. Aerosol disinfection in large scale animal and poultry raising has become popular. However, in view of the high toxicity of the chemicals used for human and animal health, the development of insect resistance, energy crisis and the effect on non-target species, biological control appears to have a natural applicability in veterinary and zoonatic control operations.

The biological control of insects and weeds by introduction of natural enemies like parasites, predators and pathogens of pests is well known. The significance of microbial method lies in their narrow host range and selectivity of vector, innocuity to man, animals and other mammals, compatibility with the beneficial organisms and the ease with which such programmes can be integrated with zoonoses and animal diseases management. In this paper, the author intends to bring out information mainly on various naturally occurring microbial control of vector borne diseases in future. However, information on certain promising parasites and predators for the control of vectors of various parasitic diseases of vertebrates of both veterinary and medical importance has been presented. Further, the existing information on various biotechnological approaches made towards increasing the efficiency of certain vector pathogens through biological and biotechnological means using recombinant DNA technology has also been presented and discussed (Narayanan,1998).

Main Groups of Vector Pathogens

Viral Pathogens

Insect viruses belonging to the family Baculoviridae differ from that of plant and animal viruses wherein the infective virions are occluded in many sided occlusion bodies (OBs) which are more of proteinaceous-silicaceous framework which afford environmental stability to virions against certain external environmental factors. Based on the presence or absence of occlusions, site of multiplication (either at nucleus or cytoplasm of cell) and shape, insect viruses are generally classified as (i) Nuclear polyhedrovirus (NPV) (ii) Cytoplasmic polyhedrovirus (CPV) and (iii) Granulovirus (GV) as they are granular or capsule like in shape(Narayanan,2003a).

Nucleopolyhedrovirus (NPV)

Nucleopolyhedrovirus (NPV), baculoviruses, belonging to very efficient group destroying populations of lepidopteran and some sawflies. They form proteinic occlusion bodies in the nuclei of infected hosts, where the virus rods which are DNA in their genetic make up, can survive for long periods of time in natural habitats. All virus infections appearing in vector insects and known today are of minor importance in regulation of vector populations. Viral infections are recorded only from dipteran vectors (e.g. mosquito, crane fly and sciarid fly). The infection in the mosquito larva is similar to that in the sawfly larva because the infection is confined to the digestive tract. In mosquito larvae the virus attack cells of the gastric caeca and the posterior mid-gut. Infected cells are hypertrophic, with nucleus filling the cell. In its interior the virogenic stroma is filled with virus particles and these finally cluster together in occlusion bodies. In this stage the infection is evident and white parts of the gut with polyhedra visible through the cuticle. The infected larva is sluggish and gradually stops feeding. The milky- white area of the infected hypertrophied mid-gut cells is visible through the integument indicating a distinct sign of infection. Rupture of the infected cells results in larval death. Larvae of three mosquito species. *Aedes epactius*, *A. atropalpus*, and *A. scutellaris* were found to be susceptible to NPV. The virions are occluded in the nuclei of midgut cells. Unlike the polyhedra of other NPVs the small proteinic occlusions gradually coalesce and grow to large fusiform or ellipsoidal occlusions.

Cytoplasmicpolyhedrovirus (CPV)

The virions in the case of Cytoplasmicpolyhedrovirus (CPV) is RNA in their genetic make up, usually spherical in shape and they are invariably occluded in characteristic hexagonal shaped occlusion bodies and they invariably infect the cytoplasm of mid-gut cells, hence the name. By virtue of infection on mid-gut epithelium of digestive organs, they are usually considered to be slow acting and debilitative in their effect by way of regarding the normal growth of insect. Infection is common in black-flies, less so in mosquito larvae and is known from *Phlebotomus*. During moults at the end of larval development, the infected cells are removed and there is no mortality. CPV are stressing agents when in combination with NPV or with microsporidia. Moussa (1978) reported the first case of a frank virus infection in the case of house fly, *Musca domestica* the virus has been detected in adults and not in larvae. It infects the adult fly by ingestion. The mid-gut of heavily infected flies breakdown completely.

Iridescent Virus

The name Iridovirus is arrived from the characteristic iridescence observed in infected insect tissue and in centrifuged virus pellets. They are icosahedral in shape and they form the largest group of non- occluded viruses infecting insects. Mosquitoes iridescent virus (MIV) is the most common virus infection in larvae of boreal Aedine mosquitoes in their spring generations. The infected larvae are blue-green or violet, iridescent. The virus is localized in large masses in the cytoplasm of fat body cells, in muscles and the hypoderm. In histological sections, virogenic stroma is evident as rounded irregular mass. Infected larvae usually die before pupation, in-apparent infected animals bring the virus to pupae and adults and with the eggs to the new progeny. Usually 8-15% of the population are infected, infected animals pupate late or die 10-14 days after all those healthy pupated. At this period the relative incidence of the infection goes up. Artificial infections are not efficient and increase of virus frequency is not persistent. Differences in colour of the virus are connected with the size of virus particles. Infections in black fly larvae are less common, but sometimes cause local epizootics. Size of MIV is 0.1-0.08ì on an average.

Occurrence of iridescent virus, on the field populations of biting midge, *Culicoides variipennis sonorensis*, vector of blue tongue virus affecting mostly, sheep, cattle and other ruminants was reported. The virus infected larvae exhibited a striking opalescent, iridescent-blue colour consistent with iridescent virus infection. The poor natural infectivity of iridescent virus was enhanced when the midge was allowed to parasitise by a mermithid parasite, *Heledomermis magnapapula* by way killing both the host and mermithid before either can mature.

Denso-Nucleovirus

The term densonucleovirus is derived from the greatly hypertrophied nucleus (eosinophilic), which contains dense, voluminous, Feulgen-positive masses of virions which are single stranded DNA. Densonucleosis virus from the family of parvoviridae is less evident in fresh collections. It is common in some foci of *Aedes aegypti* in South-East Asia, not recorded from other parts. In water mounts infected cells of the fat body have giant nuclei with a very dense granulated material inside, the virus particles 0.1ì in diameter. Details of the distribution and the effect of the infection in mosquitoes are not known. The virus is infectious by oral administration. Infections are known from *Culex pipiens* and *Simulium vittatum* in the field.

Virus which do not form visible inclusion bodies and cause intestinal damage are known from tsetse flies in Kenya, but their roll in the field has not been studied in detail. Viruses can be stored in dead animals or in water the temperatures near 4 degree Celsius or in 50% glycerine for more than 10 months. The route of administration is *per os* with food. Recent study revealed that the tsetse DNA virus infection seriously affects the survival of the *Morsitans centralis* leading to early death.

Bacterial infections: Among the various spore forming bacterial pathogens *Bacillus thuringiensis* is a gram positive, rod shaped with a spore formed in one half of the cell and a protoxin crystal in the other half. These protein crystals are highly toxic to susceptible insects which ingest them. The protoxins dissolves in the alkaline pH of the insect mid-gut, where they are proteolytically activated and bind to epithelial cell membranes in the brush border. The cells are destroyed, and the larvae stop feedings and dies. Depending on the particular strain of *B. thuringiensis* the protein toxins may be specific for members of the Lepidoptera (butterflies, caterpillars), Coleoptera (beetles) or Diptera (mosquitoes and blackflies) while some are both Lepidoptera- Diptera specific. *B. thuringiensis* and its related species are of the most effectively and extensively used entomobacterial pathogens of control of almost all lepidopterous crop pests. Most of the insects infected by B.t. turns dark in colour their body tissue disintegrate with putrid colour (Narayanan, 1969;1991). Strains of bacteria which produce mosquitocidal toxins are restricted to *B. thuringiensis* subsp. *isrealensis*, a few strains in other *B. thuringiensis* serotypes, *B. sphaericus,* and *Clostridium bifermantans* serovar *malaysia.* while deposited. The main difference between B.t. and *B. cereus* which does not from the inclusion through the latter is also soil inhibiting and a spore former. The spore does not inflate the rod which remains tubular. All strains of B.t. where mortality appears after a longer period than 48 hours are not primarily of interest for vector control. Strains of *Bacillus sphaericus* which are not toxic for mosquito larvae are common. Therefore the test of activity is a primary requirement.

Spore forming bacteria are use full in insect control because there storage in dry preparations is not difficult, they have a long shelf-life and can been grown on a relatively cheap nutrient media. There are real prospects of recovery of strains with specific activity on snails on flies and other disease vectors. Non-spore forming microorganisms some time cause mortalities in mosquito larvae in cultivation. Their study will be usefull but only if their activity equals or is higher than the known set of spore forming bacteria. The *B. thuringiensis* subsp, *sphaericus* is widespread in soil and aquatic environments. Isolation, pathogenicity and small scale field trails of *B. sphaericus* in different formulations have been carried out against *C. quinquefasciatus* or *C. pipiens* mosquitoes. Similarly *B. thuringiensis* subsp. *isrealensis* has been advocated for use as mosquito larvicide by the World Health Organisation though the different formulations *B. sphaericus* and *B. israelensis* generally cause high rates of mortality in susceptible species, the rapid settling of the bacterial spores lead to a very relatively short duration of control, which has been addressed under immobilization of mosquitocital toxins both by biological and biotechnological means later in this paper.

Entomotoxicity of recently isolated *B. thuringiensis* strains (VCRC-17) from soil sample, near Pondicherry, against fourth instar larvae of *Cx. quinquefasciatus, C. tritaeniorhynchus, A. culicifacies and A. stephensi* (vectors of filariasis, Japanese encephalitis and Malaria, respectively) revealed its high virulent nature by way of killing them within 24hr, and was found to be more virulent than *B. t.* var. *isrealensis.* Onchocerciasis (river blindness) is a chronic disease, caused by a nematode transmitted by black flies, which in the past has blinded some one million individuals. Over the past two decades, however, aerial spraying of *B. thuringiensis* to attack blackfly larvae has interrupted transmission of the disease over some 60,000 square kilometers with no cases of blindness reported among the nine million children within that area.

During the recent years there has been evidence of some Bacilli with molluscicidal activity. E.g. *Bacillus brevis*-active against vector snail-*Biomphalaria glabrata. Bacillus laterospores* an aerobic spore forming isolated from water samples produces during sporulation on the side of the central spore a canoe-shaped parasporal mass firmly adhere to the spore. Some strains display pathogenicity to mosquito larvae *Culex quinquefasciatus* and *Aedes aegypti* and for black fly larvae, *Simulium vittatum.* Laboratory studies showed that steers given *Bacillus thuringiensis* var. *kurstaki* as an additive in their feed, produced faeces which prohibited normal imago development of 92 percent of the 48-hour-old *Musca domestica* larvae reared in the faeces. Most of the mortality occurred in the pupa rather than the larval stage. Faeces from control animals in the same experiment allowed development of all but 19 percent of the fly larvae which were added to them. It is suggested that a toxin rather than a septicemia is the cause of mortality. In India the thermo stable exo-toxin containing *B. thuringenesis* is available from International Biotech, New Delhi. The exotoxin containing formulation of Bt is extensively used in foreign countries like Central America, Finland, New Zealand etc. (Sharma 1998 Personal communication). Larvicidal effects of diffusible insecticidal toxin (exotoxins) from the culture supernatants of a novel *Pseudomonas fluorescens* (strain MSS-1) originally isolated from mosquito larvae to *Culex quinquefasciatus, Anopheles stephensi* and *Aedes aegypti* has also been reported

Protozoan infections

Protozoans have been recognized as important factors in the natural regulations of the population of certain insects. However, they have been given little consideration as applied microbial control agents because entomophillic species in general cause chronic or debilitative infections in a narrow range of hosts. From a control stand point, they could not compete with more virulent pathogens such as bacteria, viruses and certain fungi. Protozoa enter insect bodies with food and can infect the host or pass with food into the gut and are voided with faeces. Among the groups of protozoa which have this impact on vector insects are amoeba, flagellates, gregarines, microsporidia and ciliates. Amoeba invading the body tissue of *Biomphalaria* snails was reported (Weiser, 1991).

Flagellates:

Only a limited number of insect flagellates is known from vectors. Flagellats which do not form cysts are destroyed during desiccation. *Blastocrithidia triatomae* infects T*riatomae infestans* causing mortality up to 75%.

Gregarines: Gregarines are most common in mid-gut of insects with their trophozoites fixed in the epithelial layer of the gut, but some species cause disease mortality because they close up the surface of some cells, and impede their functions. The gregarines are isolated from *Culex pipeins, Aedes aegypti, A. geniculatus, A. and A. albopictus.* Further gregarine *Lamkesteria mackiei* has been reported from *Phlebotomus argentipes*. The red flour beetle *Tribolium castaneum* usually found in flour and other starch material occasionally acts in the transmission of parasitic diseases in poultry cestodes *Raillietina cesticillius* was invariably affected with Schizo-gregarines, *Farinocystis tribolli.*

Microsporidia: Microsporiadians are very frequent in mosquitoes, and blackflies, but not reported in reduviid bugs, sandflies and tsetse flies where the sucking sterile blood makes infections difficult. Some microsporidia infect snails and others infect nematode larvae inside the snail host. The important microsporidians are *Nosema lagerae* infects *A. steptensi, An. gambiae and A. aegypti, Culex pipiens and Amblyospora and Prarathelohania* infecting *Aedes* and *Anophelinis.*

Ciliates: In a very limited number of larvae of mosquitoes and blackflies are infected with ciliates. E.g. *Lambornella stegomyiae* on the larvae of *A. scutellaris*. Another ciliated protozoan, Tetrahymena pyriformis, a bacteria feeder, present mainly in mosquito larvae killed with *B. thuringiensis.* Active invasion of *Tetrahymena* with mosquito larvae was not observed.

Fungus infections: Entomogenous fungi have long been recognized as potentially important naturally mortality factors affecting insect population. Fungi have been successfully incorporated into multifaceted insect suppression strategies. All the four classes of viz. Phycomycetes, Ascomycetes, Basidiomycetes and Deuteromycetes (fungi imperfecti include organisms causing mycosis in insects). Among these classes Phycomycetes and Deuteromycetes are very important. Mostly fungus infected insects are characterized by the presence of fungal mat, ramifying the entire body of the insect.

Records of fungi attacking vector insects are only a fragment of existing interactions of vectors insects with fungi only three groups are important viz. Mastigomycolina, Entomophthorales and Deutromycetes. They are the most common pathogens killing vector insects. Mastigmycolina are mainly aquatic, with a common characteristic that their zoospores are searching for suitable hosts by swimming in their aquatic habitats. Exception is *Helicosporidium* which does not have motile zoospores but has common aquatic habitat. The important species are E*ntomophthora destruns* which kills mosquitoes in shelters, caves and *E.culicis* appear on floating dead mosquitoes and *E. aquatica* ane E. *larai* on aquatic stages of mosquitoes and blackflies and *E.muscae* for housefl*y Musca domestica. Laginidium giganteum* cause high infection rates of C*ulex tarsalis.* Because *Lagenidium* asexually recycles throughout the season as well as persist over winter, considerable interest and further assessment of its potential as biological control agent against dipteran vectors has been stressed recently.

Deutromycetes: These fungi represent imperfect stages of other fungi, eventually known as Ascomycetes. (*Cordyceps, Torrubiella etc.)* in their perfect (sexual from). The important genera are *Aspergillus, Metarzhizium, Poecilomyces, Culicinomyces, Hirsutella, Beauveria, Verticillium* and *Tolypocladium* etc. Deutromycetes fungi are not host specific, but have some preference in *Beauveria and Aspergillus* are common. Close to water habitats are *Culicinomyces* or *Tolypocladium. Verticillium* is present under damp condition and rather high temperature. *Metarhizium, Poecilomyces* or *Beauveria* in fungi are directly infectious to mosquito larvae. Recent laboratory study suggest that *M. anisipliae* has good potential for tick, *Boophilus micropilus* control which is the most important ecto parasite affecting Eureopean and hybrid (European x Zebu) cattle. Barson et al. (1994) evaluated six species of entomopathogenic fungi for the control of housefly, *B. bassiana, B. brongniarti, M. anisopliae, Paecilomyces farinosus, Tolyplocadium cylindrosporum and Verticillium lecanii* and shown high effectiveness of *M. anisopliae* and *T. cylindrosporum* at 10^8 condia /ml by way of bringing cent percent kill of both larvae and adults.

Both *Beauveria bassiana* and *Metarhizium anisopliae* induced approximately 30% mortalities in adult *Rhipicephalus appendiculatus* feeding on rabbits while *M.anisopliae* induced a mortality of 37% in adult *Amblyomma variegatum.* Both fugal species induced reductions in engorgement weights, fecundity, and egg hatchability in adult *A. veriegatum. M. anisoplae* reduced fecundity by 94% in *A. veriegatum.* Furthermore, *B. bassiana* and *M. anisopliae* induced high mortalities ranging from 76-85%, a remarkable reduction in fecundity (85-99%), and a significant reductiin in egg hatchability (94-100). Both *B. bassiana* and *M. anisoplae* induced a mortality of approximately 100%, 76-95% and 36-64% in larvae, nymphs, and adults respectively of *R. appendiculatus* seeded in grass in the field.

Nematode Pathogens

Steinernematids/ heterorhabditis

Insect parasite nematodes, particularly, the Neoaplectanid and Heterorhabditid groups which produce infective larval stages capable of entering hosts and releasing associated pathogenic bacteria, *Xenorhabdus spp.* and *Photorhabdus spp.* are potential biocontrol agents. Steinernematidae are efficient parasites of terrestrial insects in and on soil with invading larvae which enter the host through the mouth, spiracular openings and eventually wounds. Studies on the effect of two species of entomophilic nematodes. *Heterorhabditis heliothidis* and *Steirnema feltea* against both larvae and adult stages of *Musca domestica* in intensive animal units especially poultry houses and piggeries revealed 90% and 93.3% of parasitation of both larval and adult stages with either species of entomophilic nematodes. Future potential of Steinernematids *(Neoplectana bibionis* and *N. carpocapsae)* for the control of stable fly, *Cromoxys calcifrans,* which is the major live stock infesting insects has also been reported

Mermithid Parasites of Vectors

Mermithidae are parasite of aquatic as well as terrestrial vector insects. Terrestrial hosts such as testse flies or reduvid bugs are less common hosts mermithids. The cultivation of mermithids from blackfly material has been performed only to a limited extent on *Mesomermis flumenalis* with the use of natural *Similium venustein* maintained in laboratory rearing cascades. After the release of 1.5 million pararsites in a stream resulted in the infection of 71.4% of blackfly larvae immedietly below the point of release. *Isomermis,* a mermithid infecting *Similium* damvosum in the ivory coast is a very common parasite in natural habitats. 80% of the nulliparous females and 35% of the whole population during the rainy season contain mermithids: 98% of the infected females were sterilized by the mermithid. Maintaining a colony of terrestrial mermithids is possible under laboratory conditions using laboratory produced tsetse pupae or blackflies.

Mermithid parasites of mosquitoes

Mermithid infecting mosquitoes are divided into several groups. One group include parasites of boreal mosquitoes, mainly *Aedes* species in the snow pools of Northern Canada. Another group of mermithids is present in mosquitoes living in tree holes and *Octomyomermis muspratti* is a typical representative. Third group infects a wide range of Anopheline, Culicine and Aedine hosts in the subtropic and tropics represented by the genus *Romanomeris.* Further development they need clean water.

They are well adapted to the life cycle of mosquitoes including the dry period they are host specific and kill the host before pupation. Several species of *Romanomermis (R. culicivorax, R. iyengari, R. nielseni)* were studied but only *R. culicivorax* was developed to a perfect tool for laboratory and brand common commercial production and use. *Romanomermis culicivorax* was tested at least 87 species of mosquitoes from 13 genera including *Culex quinquifasciatus, C. pipiens* and *Anophelis* mosquitoes and high susceptibility of most them was recorded.

Insect parasitisation by mermithid nematodes are often abundant and common in crickets and grasshoppers. Helminth parasites having indirect life cycle require intermediate host (IH) for development of larval stages to the ineffective cystic stage and transmission to the definitive hosts. This mode of transmission is known as biological transmission. Many insects act as IH. Among them, various species of ants, lice, grasshoppers and dragon files serve as IH of helminth parasitic of the class Trematoda, Cestoda and Nematoda. In the case of Cestode, *Choanotaenia infunlibulum,* parasitising definite host like fowl and turkey. Some of the grasshoppers of the genera, *Attractomorpha, Hieroglyphus, Oxya, Melanoplus* etc. act as a intermediate host. Cephalogregarines are gaining important component of IPM in recent years since they show promise as efficient biocontrol agents of grasshoppers. Two Cephalogregarines parasites viz. *Leidyana subramanii* and *Retraetocephalus dhawanii* parasitising effectively on the tobacco grasshopper *Attractomorpha crenulata*

The biting midge, *Culicoides variipennis sonorensis* is regarded as the primary vector of blue tongue virus, a pathogen of sheep, cattle and other ruminants. Recently a mermithid parasite. *Heleidomermis magnapapula* as a principal natural enemy of *Culicoides spp,* parasitising 69% of the host larvae in the field has been reported.

Predators and Vectors

Insect predator

Among the insects which reduce vector populations, the predators play a more important, but also more difficult to estimate their role. They are not very specific and their regulations is very difficult. The role of terrestrial insects attacking adult vector insects is not well documented by adequate evaluation, but there is no doubt that mosquitoes in the shrub and plant growth shelters are victims of attacks of dragon flies in the same way as are adult black flies. In aquatic habitats, larvae of *Trichoptera* are the most important consumers of early instars of blackfly larvae, especially larvae belonging in the genera *Plectrnemia* (*P.* canspus) and *polycentropus (P. flavomaculatus).*

Fish Predators

Of the several biological control agents, larvivorous fishes have their greatest potential as biocontrol agents against vectors and against stream inhibiting blackfly larvae and pupae. The *Ponnilia reticulata* commonly known as guppy is a bottom feeder, can tolerate high degree of pollution and used effectively for the control of *Culex* breeding whereas, the top minnow, *Gambusia affinis* continues to occupy a unique place in worldwide distribution as a tool for control of mosquitoes. *G.affinis* is a surface feeder and prefers to breed in clean oxygenated water and is suitable for the control of Anophiline breedings. Studies on the control of mosquito breeding is wells by way of using *G. affinis* and *Aplocheisus blochii* revealed that extent of Anophiline breeding was brought down from 32.8% of the wells breeding before fish introduction to less than one percent one month later. Successful control of *An. Stephensi*, the principle vector in the transmission of malaria by way of using *G. affinis* as an adjunct. Methodology at urban malaria scheme area of Vellore in Tamil Nadu which was carried from 1995 onwards with 85% reduction in the indigenous malaria cases. Its negative influence is the competition with the local fish species. Therefore, a search for local mosquito predatory fish in natural habitats adopted to local conditions is recommended.

Parasites of Vectors

Insect parasitoids or entomophagous insects play only a minor role in vector control, but because they develop inside the host body, the activity of parasitoids is limited mainly to terrestrial stages of vectors, therefore tsetse flies and reduviid bugs have their specific parasitoids but to some extent parasitoids are known from aquatic animals such as blackflies and snails. Mosquitoes do not have parasitoids but in blackflies planedial larve are recorded from adult flies and at least two pupal parasites have been recorded in the literature. Experiments conducted in India have shown that sustained release of *Spalangia endius,* a parasite of housefly, *M. domestica* resulted in enhanced parasitism ranging 84% to 100% of housefly, *M. domestica* and stable fly *Stomoxys calcitrans* breeding in poultry, bovine and pig manures not only in United States but also in India. In India, these parasites are available from Bio-Control Research Laboratories (BCRL), Bangalore, India.

Blue Green Algae (Cyanobacteria):

Planktonic blue green algae are part of the food filtered by mosquitoes or blackfly larvae because the larvae are not able to eliminate any material suspended in the aquatic habitat during their filtration activity. Some blue green algae are present in masses in fish ponds and lakes during the summer, especially in water with a high content of nutrients and they form what is known as "waterblooms". The members of waterblooms are one or two species in low quantities. They produce oxygen when growing in numbers , but at the end of the bloom, they are concentrated, blown by the wind to the shores of the lake where they die and decay. They are present in the outer water of these habitats and become the food of blackfly larvae. Several species appearing in large quantities have endotoxins, cycle polypeptides, which are absorbed by aquatic animals with the food and cause intoxications of vertebrates. Their impact on vector insects is not very clear, but in several cases dead mosquitoes harbour blue green algae in their gut contents. The important species which are occasionally toxic are *Microcysti* and *Anabaena sp. E*ecophysiological investigations have indicated the differential sensitive of different mosquito species towards the tannins and other phenolic compounds released from fallen leaves decomposing in water (for e.g, insensitive insects such as *Aedes Anopheles*, can colonise waters associated with abundant vegetation rich in tannins, wheras. *Culex pipiens* populations are confined to open habitats without leaf litter.

Recently in laboratory experiments some laboratory reared solitary blue green algae were used for incorporation of the toxin-bearing plasmid of *Bacillus thuringiensis, B. israelensis.* The species involved grows well in artificial media and in nature, remains floating without sedimentation in the experimental suspension and is toxic for mosquito larvae when ingested. This method is far from being ready for field application and for this the survival of the manipulated blue green algae and its participation in the nutrition of larval mosquitoes must be further investigated under different conditions of mosquito.

Water-fern: *Azolla* spp.

Of the different mosquito habitats rice fields are the only type where the blue green algae appear in large quantities together with mosquito larvae. *Azolla,* a small aquatic fern, lives in symbiotic association with blue green algae, *Anabaena,azollae.* The combination of these two species makes *Azolla* a valuable source of organic nitrogen, of particular interest to rice cultivation. *Azolla* multiplies with paddy fields very quickly and covers the entire field in the form of a thick mat within a period of two weeks. A significant reduction in mosquito larvae breeding in paddy fields in which paddy was cultivated in association with *Azolla* was observed in China. The use of a combination of the fern *Azolla* together with *Anabaena azollae* as a nitrogen fixing algae fertilizer for rice fields brings mosquitoes living in rice fields much closer to a possible toxic agent than anywhere else in natural mosquito habitats, but this algae does not produce any known toxin. Preliminary laboratory evaluation of *Azolla pinnata* revealed that the complete coverage of water surface, with this fern drastically reduced the oviposition and adult emergence of both the species of mosquitoes viz. *Culex quinquefaciatus* and *Anophels culcifacies,* However, egg hatchability was partially affected only in *A. culcifacies.*

Immobilization of Mosquitocidal Bacterial Toxin

i. Biological Immobilazation Technique Using Ciliate Protozoa

With advancement in biological science, especially after the advent of recombinant DNA (r-DNA) technique an opportunity has emerged in alleviating certain commercial short comings of pathogens viz. low persistence of commercial preparations of biopesticides in the field and fostering the creation of new generations/formulations of biopesticides with more persistent preparations. Formulation of insect vector pathogens especially through genetic engineering is anticipated to modify delivery system effectively thereby delivering the active agent to the target organism so as to increase their effectiveness greatly. In this part the author intends to bring out the existing information and the recent advances made on the impact of biological and biotechnological immobilization/encapsulation of various microbial control agents of insect vectors especially, bacteria in increasing their effectiveness by way of increasing their persistence and residual activity by means of effective immobilization/ encapsulation of either pathogens *per se* or of their toxin products by biological means using ciliated protozoan, *Tetrahymena pyriformis,* by way of encapsulating some of the bacterial spores and toxins crystals in their food vacuoles-a process known as "bioencapsulation" and biotechnological means using recombinant DNA techniques (r-DNA).

Mosquitoes and blackflies are vectors of many human infectious diseases as mentioned earlier. One of the best biocontrol agents against mosquitoes and blackfly larvae is the bacterium *Bacillus thuringiensis subsp. israelensis (Bti)* and its larvicidal activity is included in polypeptides of a para-sporal crystalline body (ä-endotoxin) that is produced during sporulation. The use of Bti is however limited by the low efficacy of current preparations under field conditions. The major reasons for the low efficacy are (1) rapid sinking of spore and crystals to the bottom of the water body, (2) adsorption onto silt particles and organic matter (3) consumption by other organisms to which it is non-toxic and (4) inactivation by sunlight. Since most mosquito larvae feed at or near the water surface, the effective larvicidal activity of the toxin is limited to the shortest period prior to sedimentation or sinking.

Recently a novel system has been described by Zariksky *et al.*.(1991) in which the ciliate protozoan *Tetrahymena pyriformis* is used to bioencapsulate the larvicidal activity of *Bacillus thuringiensis* subsp .*israelensis* and to deliver it to the target organism viz. the mosquito larvae. Death of larvae preying on *T. pyriformis* loaded with active powder of *B. thuringiensis subsp.israelensis* clearly demonstrated that the protozoan delivered toxicity to the predator. To increase the persistence of Bti larvicidal activity, a commercial powder was usually incubated with *T.pyriformis*. Each cell of an exponentially growing *T. pyriformis* can form up to about 30 food vacuoles and it has been estimated that each food vacuole can accommodate 6-8 spores of Bti with their toxins. Thus each cell has been shown to concentrate between 180-240 Bti spores and their toxins and its food vacuoles and keep the spores at the water surface. The motile protozoan cells would be then ingested by mosquito larvae thus delivering the toxin to the larvae. This active process of concentrating and delivering intact ä-endotoxin to the target organism was termed by Zaritsky as "Bioencapsulation". It is based on the following (1) the toxic crystal is stable at pH below 9 (2) the pH of food vacuoles in *Tetrahymena* is substantially lower than 9 (3) *T. pyriformis* feeds on bacteria (4) *T pyriformis* shares the natural habitat of mosquito larvae i.e. the upper layer of fresh water (5) mosquito larvae prey on microorganisms.

Advantages of Bioencapsulation: Utilisation of Bti encapsulated in *T.pyriformis* as a biological control agent of mosquitoes would be advantageous over treatments with commercial formulations of Bti because protozoan cells. (1) prevent fast sinking of spores and toxic crystals and their adsorption to soil (2) concentrate large quantities of spores and toxins that cause rapid larval mortality and increase the spectrum of affected population such as late fourth instar larvae and (3) retain the toxic principle in the feeding zone of surface feeding species such as Anopheles due to the motile behaviour and attraction to oxygen of the protozoan. Additional advantage can be achieved by way of recycling of Bti spores in the excreted food vacuoles.

Future possibilities of bioencapsulation

Ciliated protozoa as *Tetrahymena* exhibit a substantial biotechnological potential , but fermentation strategies yielding biomass comparable to those of bacteria and yeast cultures have been lacking for these organisms. An efficient method of growing ciliated protozoa in a 2-litre perfused bioreactor which routinely achieved in standard batch fermentations. However, this method is yet to be exploited for biotechnological purposes. The tactics of raising large amounts of *Tetrahymena* and loading them with Bti for efficient delivery in nature are the subjects of future studies. The programme using bioencapsulation techniques will go a long way both for the control pests and mosquitoes harbouring in a common rice ecosystem by this biotechnological approach, will also from the subject of future studies IPM in rice in India especially surface feeder species such as *Anopheles* rather than bottom feeder.

ii. Biotechnological Immobilization Using Cyanobacteria

One of the major factors limiting the duration of mosquito control following the application of *B. sphaericus* or *B. thuringiensis* strains as mentioned earlier is the rapid sedimentation of bacterial spore. Since most mosquito larvae feed at or near the water surface, the effective larvicidal activity of the toxins is limited to the short period prior to spore sedimentation. Other workers have recognized this problem and have transferred insecticidal toxin genes from *B. sphaericus* or *B. thuringiensis* subsp. *israelensis* into *Cyanobacteria*

Various species of Cyanobacteria have been investigated as vehicles for prolonged delivery of insecticidal toxins to the larval feeding zone. Like *Caulobacter* species, the ubiquitous photosynthetic Cyanobacteria exist at or near the water surface. These bacteria are adaptable to both fresh water and salt water environments, they have a wide temperature tolerance and being protoautotrophs, they have limited nutritional requirements. The first attempt at was successful. There was a delay in the onset of toxicity of the recombinant *Agmenellum quadruplicatum* cells from 1 to 3 days and 100% mortality occurred 5.5 days after the cells were first fed to the larvae.

Increasing Larvicidal Activity and Host Range of the Mosquitocidal Bacilli-Through Recombinant Technology

Recombinant "r-DNA" Technique

Recombinant DNA technology provides the tools for developing safe, efficient and cost effective microbial control agents. It is now possible to combine the best traits of several different organisms into a single strain, expressing ä-endotoxins that exhibit enhanced insecticidal activity, longer residual activity and broader host range. Thus the feasibility of using Bt for insect vector control has been increased by advances of toxin genes and their expression in other organisms. As a result of the enormous potential that recombinant DNA technology provides for advanced in insect vector control.

Mosquitocidal Toxin Expression in Prokaryotic System

Caulobacter crescentus

In the quest for effective control of mosquitoes, attention has turned increasingly to strains of the bacteria *Bacillus sphaericus* and *Bacillus thuringiensis subsp. israelensis* which produce potent toxins with specific mosquitocidal activities. However, as mentioned earlier, sedimentation of the bacterial spores limits the duration of effective control after field application of these bacilli. Cloning of genes encoding the 51.4 and 41.9 kDa toxins from *B. sphaericus2297,* the 100 kDa toxin from *B. sphaericus* SSII-I and the 130 kDa toxin from *B. thuringiensis subsp. israelensis* into the broad-host-range plasmid pRK248 and the transfer of these genes for expression in *Caulobacter crescentus* CB15 has been reported. Bacteria belonging to the genus *Caulobacter* were chosen for the following reasons. *Caulobacter* species are found in almost every aquatic habitat, and within these habitats they are found predominantly in the regions at or close to the water surface. In the flagellate swarmer stage, *Caulobacter* species are motile, thus allowing distribution through the habitat. In both the swarmer and the stalked stages, they are capable of attachment to solid particles at or near the water surface. *Caulobacter* species are able to persist and grow in environments with low nutrient concentration Mosquito larvicidal activity of intact recombinant *C. crescentus* has been demonstrated The recombinant *C. crescentus* cells were shown to be toxic to mosquito larvae since *Caulobacter* species are ubiquitous microorganisms residing in the upper regions of aquatic environments and therefore provide the potential for prolonged control by maintaining mosquitocidal toxins in larval feeding zones.

Development of novel cloning vectors for Bt has made it possible to construct improved Bt strains for use as microbial insecticide for the control of insect vectors. A number of convenient shuttle vectors functional in *E. coli* and *Bacillus* species have been constructed using replication origins form resident Bt plasmid. The shuttle vectors which exhibit good segregational stability are employed to introduce new toxin Bt genes to Bt strains. All the major crystal toxins of *B. thuringiensis* subsp. *israelensis* have been expressed with various efficiencies in *E. coli* or *B. subtilis* or *B. sphaericus* etc. Toxin gene CryIIA has also been cloned into *B. megaterium* and the transformed bacterial cells were highly toxic to mosquito larvae.

Expression in Bacillus Strains

Among the new generation of biological control agents which are currently being deployed for control of mosquitoes are strains of *Bacillus sphaericus* and *Bacillus thuringiensis* subsp *.israelensis.* The highly toxic strains of *B. sphaericus* used in the field produce two proteins with masses of 51.4 and 41.9 kDa which are jointly responsible for toxicity. These toxins are produced at the onset of sporulation and accumulate in parasporal crystalline inclusions. Binding of the two proteins occurs in the gastric caecum and posterior midgut of *Culex quinquefasciatus* larvae causing disruption of the gut cells and larval death. In addition to the 51.4 and 41.9 kDa toxins, toxic *B. sphaericus* strains may produce a 100 kDa toxin during vegetative growth. *B. sphaericus* is generally most toxic for mosquitoes of the genus *Culex*, with less activity against *Anopheles* species and least activity against *Aedes* species. This mosquitocidal target spectrum is complemented by that of the 130 kDa toxin of subsp. *israelensis.* This protein is also produced during *B. thuringiensis* sporulation is deposited as an endosporal crystals and is most active against *Aedes* mosquitoes, with less toxicity to *Anopheles* species and least to *Culex* species. The site of the action of the 130 kDa toxin, like that of the *B. sphaericus* toxin causes cell lysis by a process of colloid osmotic lysis. The 130 kDa protein of *B. thuringiensis* subsp. *israelensis* is also toxic to the blackfly, a vector of onchocerciasis. The toxins produced by each of these microorganisms have different host range specificities. *B. sphaericus* toxin is toxic mainly to the larvae of *Culex* and *Anopheles* species, while the *B. thuringiensis* subsp. *israelensis* inclusions are more active against *Culex* and *Aedes* species *B. sphaericus* also has the benefit of longer persistence in polluted aquatic ponds.

It is therefore of great advantage for mosquito control to combine, by genetic engineering, the toxic genes from the two microorganisms in one microorganism in order to widen the host range spectrum. The genes encoding the toxic determinants of *Bacillus sphaericus* have been expressed in a nontoxic and a toxic strain of *Bacillus thuringiensis* subsp. *israelensis.* In both cases, the *B. sphaericus* toxin proteins were produced at a high level during sporulation of *B. thuringiensis* and accumulated as crystalline structures, *B. thuringiensis* transformants expressing *B. sphaericus* and *B. thuringiensis* subsp. *israelensis* toxins did not show significant enhancement of toxicity against *Aedes aegypti*, *Anopheles stephensi* and *Culex pipiens* larvae.

Bt Mosquitocidal Toxin Expression In Eukaryotic System:

Baculovirus expression: Several baculoviruses belonging to lepidopteran insects are already in use as microbial insecticides. They have particular advantage over chemical insecticides, such as lack of toxicity towards non-pest insects just like insecticidal bacilli. A good deal of the molecular biology of baculoviruses has been known and their development as expression vectors. The recombinant virus resulted in the synthesis of a polyhedrin –Cry IV D protoxin fusion protein, which is crystallized into cuboidal inclusions in the cytoplasm. The infected cells and purified inclusions were toxic to mosquito larvae (LC50 – 250mg/ml). The suitability of the various bacteria as host for the delivery of engineered toxin to mosquito larvae was summarized and presented in table.1.

Table 1. Suitability of various bacteria as host for the delivery of engineered toxins to mosquito larvae.

Property	Suitability of					
	a	*b*	*c*	*d*	*e*	*f*
Good capacity to express foreign proteins	Y	Y?	Y?	Y	?	?
Good genetic maps and mutants available	Y	N	N	Y	N	N
Structurally and segregationally stable plasmids	Y	?	?	Y	Y	N
Persists at larva feeding zone	?	N	N	N	Y	Y
Known food source for larvae	Y?	Y	Y	Y	Y	Y
Survival in diverse environments	?	N	Y	Y	Y	Y
Experience in production	Y	Y	Y	Y	N	N
Formulation	N	Y	Y	N	N	N
Mammalian toxicity	Y	N	N	N	?	?

a- E. Coli; b- B. thuringiensis subsp. *Israelensis, c- B. sphaericus, d- B. subtilis, e- Caulobacter spp., f- Cyanobacteria*; Y: yes ; N: no.Conclusion

With the intensive use of DDT in the 1950, tropical diseases were reduced to minimum but not eradicated. They returned having developed increased multiple resistance of the vector insects to most chemical insecticides. For e.g. not only the malarial vector *Anopheline* mosquito had developed resistance to DDT but the human malarial parasite *P. falcifparum* has also developed resistance against drug chloroquine. The high doses of chemicals caused high levels of environmental contamination, dangerous for man and animals. The use of environment friendly insect pathogens for vector control initiated an intensive search for new entomopathogens, which can enable to support or replace some of the chemical insecticides. As a result there should be an intensifying search for new pathogens in nature, which may be more efficient than those known is the important task. Efficient biological factors are rare and disappear soon after they have destroyed all available susceptible hosts. Therefore an intensive search for new active organisms is most important. Further it is so important since people started using not only for the control of pest of veterinary importance but also for the control of microbial infections and cancer in humans for the control of pathogenic and spoilage microorganisms in food. Hence the future for using vector pathogen information for more effective control of vector borne diseases is promising.

With the advent of recent recombinant DNA technology not only the certain of new generation of biopesticides is possible but also for the control of insect vectors or parasitic diseases by way of stable integration and expression of bacterial gene in the mosquito *Anopheles gamibiae(*Narayanan,2003b*)* Further it is now possible to modify a genome and create transgenic insect with an objective of debilitating the individuals by way of reduced reproductive potential or vector competence and increased vector susceptibility to existing measures. In the case of tsetse fly control, alternative methods of control include the use of insect attractants and pheromones and the application of sterile insect techniques (SIT) as a supplement to biological control since the *Glossina* spp. are to known to be affected by a number of microorganisms leading to high mortality.

Insect antibacterial proteins such as cecropin, defensin etc., are the best characterized invertebrate antibacterial proteins (Narayanan, 2003c) Contrary to earlier belief that these proteins were present only in insects and can act only against bacteria/bacterial disease the recent finding had shown presence of homologous counterparts in vertebrates including humans and more versatile in their function by way of acting against parasites that cause malaria, chagas disease and leishmaniasis. With the recent knowledge of insect immunity in general and induction of cecropin and attacin like antibacterial factors in the haemolymph of *Glossina morsitans* cloning and expression of some of these anti-bacterial protein genes and understanding the molecular structure and relating to it to biological function for the future designing and management of certain insect vectors which causes human and animal diseases, hold much promise. Similar such approach has to be made not only in mammalian arthropod disease vectors like mosquitoes, but also other vectors of both man and animal parasitic diseases, since transgenesis techniques have already been developed for a variety of animals such as *Drosophilla,* to mammalian species and aquatic organisms or amphibians. These later activities have opened up new possibilities for future vector borne disease control through microbial products in the current concept of Integrated Vector Management (IVM) programmes. Attempts to control vector-borne diseases either by drug or vector control through chemical pesticides alone is like to empty a reservoir without closing the inlet. Therefore, the approach of vector control should be changed slowly from a state of overkill by pesticides to ecofriendly Integrated Vector Management (IVM).

References

Barson, G., Renn, N. and Bywater, A.,1994. *J.invertebr.pathol*, 64, 104-113.

Chandler, D., Davidson, G., Pell, J.K., Ball, B.V., Shaw, K. and Sunderland, K.D., 2000. Fungi biocontrol of Acari., *Science and Technology* 10, 357-384.

Govindarajan, R., Jayaraj,S. and Narayanan, K., 1975. Observations on the nature of resistance in *Spodoptera litura* (F) *Bacillus thuringiensis* Berliner. *Indian J. Exp. Biol.*, 13, 548-550.

Moussa, A.Y., 1978. *J. Iinvertebr.pathol*, 31, 204-216.

Narayanan, K., 1969. Pathogenicity of *Bacillus thuringiensis* Berliner on some lepidopteran pests with particular reference to *Plutella maculipennis* (Curtis) (Plutellidae: Lepidoptera) on cruciferous crops. *M.Sc (Ag). Thesis*, T.N.A.U., Coimbatore.

Narayanan, K., 1990. Past, present and future prospects of *Bacillus thuringiensis* Berliner in pest management in India. Paper presented at the Dept. of Biotechnology during the task force committee meeting on *Bacillus thuringiensis* held at DBT, New Delhi. Dec.6, 1989.

Narayanan, K., 1998. Microbial control of vectors of parasitic diseases. Paper presented during the fourth national training programme on vectors and vector borne parasitic diseases held from 25th Jan to 8th Feb. 1999 at Centre for Advanced Studies, Department of parasitology, Veterinary College, University of Agricultural Sciences, Bangalore.

Narayanan, K., 2003a. Mass production and utilization of microbial agents with special reference to insect pathogens. In *"Biopesticides and Bioagents in integrated pest management of agricultural crops"*. (Ed., .R.P. Srinivastava) International Book Distributing Co., Lucknow, U.P. pp 493-524.

Narayanan, K., 2003b. *Genetic engineering of insect viruses: Novelty or necessity?* "Biopesticides and pest management, (Ed. By Opender Koul, G.S. Dhaliwal, S.S Marwaha and Jatinder K.Arora) Pub. Society of Biopesticides Sciences, India. Vol. 1, 153-178.

Narayanan, K., 2003c. Insect resistance : Its impact on microbial control of insect pests. Paper to be sent to workshop on *Biotechnology Awareness and Education* held at Thiruvananthapuram, 1-2, Feb.2003.

Samish, M, and Rehacek,J., 1999. Pathogens and predators of ticks and their potential in biological control. *Ann.Rev Entomol* 44, 159-181

Weiser, J., 1991. Biological control of vectors. John Wiley and Sons, England , 189 pp.

Zaritsky, A., Zalkinder, V., Ben-Dov, E. and Barak, Z.J., 1991. *J. invetebr.pathol*, 58, 455-457.

Bio-efficacy of nucleopolyhedrovirus (*Amal*NPV) against red hairy caterpillar, *Amsacta albistriga* (Walker) (Lepidoptera: Arctiidae), on groundnut in Karnataka

P. C. Ganiger,* V.T. Sannaveerappanavar* and A.R.V. Kumar*

The bio-efficacy of *Amsacta albistriga* nucleopolyhedrovirus (*Amal*NPV) was evaluated against red-hairy caterpillar through two field trials during *Kharif*, 2004. In the first trial, at higher doses (6.5×10^{11} and 7.5×10^{12} POB's/ha), *Amal*NPV was as effective as fenvalerate dust (25 kg/ha) against early instar larvae, ten days after its application. At lower doses also ($1.25 \times 10^{11} - 2.5 \times 10^{11}$ POBs/ ha) *Amal*NPV was fairly effective in suppressing the pest (0.67 – 0.87 larvae/m^2 compared to 1.87 larvae/m^2 in untreated control plots). While in second trial, plots dusted with fenvalerate recorded the lowest population of the pest (0.27 larvae/m^2) followed by plots treated with *Amal*NPV at higher doses ($1.5 \times 10^{12} - 7.5 \times 10^{12}$ POBs/ha) with a population of 0.93 – 1.67 larvae per square meter compared to 7.07 larvae, ten days after treatment in untreated control plots. These treatments were statistically on par in suppressing RHC populations. In second trial also, at lower doses ($1.25 \times 10^{11} - 2.5 \times 10^{11}$ POBs/ ha) *Amal*NPV was moderately effective in suppressing the pest.

Introduction

The red hairy caterpillar (RHC), *Amsacta albistriga* (Walker) (Lepidoptera: Arctiidae), is one of the major pests causing severe damage to the crop in south India (Ghewande and Nandagopal, 1997). In Karnataka, *Amsacta albistriga* has been a major pest in red loamy soils in the districts of Raichur, Bellary, Gulbarga, Belgaum, Bijapur, Chitradurga, Chikkamagalur and Kolar (Puttarudriah, 1956; Thontadarya *et al.*, 1976). Recently, the outbreak was also reported from some parts of Bidar and Bellary districts.

The pest completes two generations per annum in Southern Karnataka (Ganiger and Sannaveerappanavar, 2007). In endemic areas, groundnut and other crops such as sesamum, cowpea, green gram, sunflower, etc. suffer heavy defoliation from large populations of RHC and the farmers often resort to resowing of crops. The severity of the pest in endemic areas can be gauged from the fact that the Karnataka State Department of Agriculture spent Rs. 23 lakhs for the manual collection and destruction of 65,728 kg of grown up larvae in Pavagada Taluk of Tumkur district alone to suppress the pest during the year 2000 (Personal communication, Department of Agriculture, Pavagada). Use of biocontrol agents such as baculoviruses, fungi and bacteria could be a alternative method to manage this pest. NPVs are known for high epizootic levels and are naturally occurring, self perpetuating, safer due to host specificity and environment friendly. The potential of NPV infected cadavers in epizootic spread of the disease was assessed by taking the POB counts of diseased larvae of *A. albistriga* infected by *Amsacta albistriga* nucleopolyhedrovirus (*Amal*NPV) in groundnut ecosystem. Though the potential of *Aa*NPV has been reported by earlier workers (Rabindra and Jayaraj, 1975; Jayaraj *et al.*, 1977), very little information is available on the utilization of the virus in the suppression of the pest on different crops. In the present study, the efficacy of the virus against *A. albistriga*

* Department of Entomology, College of Agriculture, UAS, Bangalore – 560 065

in groundnut ecosystem was evaluated through two field trials during *Kharif*, 2004.

Material and Methods

The field trials were conducted in Pavagada taluk of Tumkur district to assess the bio-efficacy of *Amal*NPV against second and third instar larvae in two generations of RHC in the early and later stages of groundnut crop. *Amal*NPV was tested at six doses *viz*., 1.25×10^{11}, 2.5×10^{11}, 6.5×10^{11}, 1.5×10^{12}, 3.75×10^{12}, 7.5×10^{12} POB's/ha) with fenvalerate 0.4 D (25 kg/ha) as the standard check.

Both the experiments were laid out in randomized block design (RCBD) with three replications. Each experimental plot measured 3 x 3 m. The groundnut crop (var. TMV-2) was grown as per the recommended practices. The polyethylene sheet barriers of one-meter height were erected all-round the individual plots to prevent the migration of larvae from one plot to another. The NPV sprays were given using a hydraulic high volume sprayer. Jaggery (1g/l) and cloth whitener blue (Ujjala(R)) (0.5 ml/l) were added to spray fluid as phagostimulant and UV protectant, respectively. The pre-treatment counts of the larvae were recorded one day before the spray. The post treatment observations were made on fourth and tenth day after application of *Amal*NPV. The population of the pest was recorded as number of larvae per square meter area at five spots in each experimental plot. The data were subjected to square root ($\sqrt{x + 0.5}$) transformation and the treatment means were compared by Duncan's Multiple Range Test (DMRT).

Results and Discussion

In the first trial, the population of RHC before treatment ranged from 2.34 to 4.26 larvae per square meter (Table 1). Five days after spraying, *Amal*NPV reduced the pest population significantly (1.60 – 1.87 larvae/m^2) compared to untreated control (2.33 larvae/ m^2) at all the doses tested. However,

Table 1. Bio-efficacy of *Aa*NPV against early larval instars of *A. albistriga* on groundnut (First generation ; *Kharif* 2004)

S. No.	Dose (No. of POB's / ha)	Pre-treatment count	No. of Larvae/ m^2		
			5th DAS**	7th DAS	10th DAS
1	1.25×10^{11}	4.26 (2.18)^{d}	1.73 (1.49)^{b}	1.53 (1.42)^{bc}	0.87 (1.17)^{b}
2	2.5×10^{11}	2.68 (1.78)^{ab}	1.87 (1.54)^{b}	1.47 (1.40)^{bc}	0.67 (1.08)^{b}
3	6.5×10^{11}	2.90 (1.84)^{b}	1.87 (1.54)^{b}	1.20 (1.30)^{ab}	0.33 (0.91)^{ab}
4	1.5×10^{12}	2.67 (1.78)^{ab}	1.60 (1.45)^{b}	1.13 (1.28)^{ab}	0.27 (0.88)^{ab}
5	3.75×10^{12}	2.87 (1.84)^{b}	1.80 (1.52)^{b}	0.73 (1.11)^{bc}	0.20 (0.84)^{ab}
6	7.5×10^{12}	3.02 (1.88)^{b}	1.60 (1.45)^{b}	1.27 (1.33)^{bc}	0.53 (1.02)^{ab}
7	Fenvalerate 0.4 D	3.73 (2.05)^{c}	0.07 (0.75)^{a}	0.03 (0.73)^{a}	0.00 (0.71)^{a}
8	Control	2.34 (1.69)^{a}	2.33 (1.68)^{b}	2.07 (1.60)^{c}	1.87 (1.54)^{c}

* Values in parentheses are $\sqrt{x + 0.5}$ transformed values ; Means followed by same letters along the columns are not significantly different by DMRT; ** DAS – Days After Spraying

the plots treated with fenvalerate dust recorded the lowest pest population density (0.07 larvae/m^2) at this stage and the chemical was superior to *Amal*NPV in suppressing the pest. Same trend was observed at seven days after treatment with fenvalerate application being the most effective treatment. Ten days after treatment, plots dusted with fenvalerate recorded zero population of the pest. At this stage, application of *Amal*NPV at higher doses (6.5 × 10^{11} – 7.5 × 10^{12} POB's/ ha) was as effective (0.20 – 0.53 larvae/m^2) as dusting of fenvalerate. At lower doses also (1.25 × 10^{11} – 2.5 × 10^{11} POB's/ ha) *Amal*NPV was fairly effective in suppressing the pest (0.67 – 0.87 larvae/m^2 compared to 1.87 larvae/m^2 in untreated control plots).

In the second trial, the population density of the pest before imposition of the treatments was fairly higher (7.93 – 9.47 larvae/m^2) compared to the population level in the first trial (Table 2). Five days after spraying, the standard check, fenvalerate dust was highly effective (0.53 larvae/ m^2) followed by *Amal*NPV at higher doses (6.5 × 10^{11} – 3.75 × 10^{12} POB's /ha). The reduction in population of the pest at lower doses of the virus was not significant compared to the population in untreated control plots. Seven days after treatment, the virus suppressed the RHC population significantly at all the doses tested, but dusting of fenvalerate was the most effective treatment. There was further reduction in the pest population 10 days after spraying. Plots dusted with fenvalerate recorded the lowest population of the pest (0.27 larvae/m2) followed by plots treated with *Amal*NPV at higher doses (1.5 × 10^{12} – 7.5 × 10^{12} POB's/ha) with a population of 0.93 – 1.67 larvae per square meter compared to 7.07 larvae in untreated control plots. These treatments were statistically on par in suppressing RHC populations. Application of the virus at lower doses (1.25 × 10^{11} – 2.5 × 10^{11} POB's/ ha) was moderately effective in suppressing the pest.

Table 2. Bio-efficacy of *Aa*NPV against early instars of *A. albistriga* on groundnut (Second generation ; *Kharif* 2004)

S l . No.	Dose (No. of POB's / ha)	Number of larvae /m^2			
		Pre-treatment count	5th DAS**	7th DAS	10th DAS
1	1.25 x 10^{11}	8.20 (2.95)^{a}	6.27 (2.60)^{bc}	5.07 (2.36)^{d}	3.53 (2.01)^{e}
2	2.5 x 10^{11}	9.40 (3.15)^{c}	6.33 (2.61)^{bc}	4.87 (2.32)^{d}	3.67 (2.04)^{e}
3	6.5 x 10^{11}	8.20 (2.95)^{bcd}	5.93 (2.54)^{b}	4.20 (2.17)^{d}	2.80 (1.82)^{d}
4	1.5 x 10^{12}	7.93 (2.90)^{b}	4.87 (2.32)^{b}	2.93 (1.85)^{c}	1.53 (1.43)^{c}
5	3.75 x 10^{12}	9.47 (3.16)^{c}	4.80 (2.30)^{b}	1.53 (1.43)^{b}	0.93 (1.20)^{b}
6	7.5 x 10^{12}	9.00 (3.08)^{bc}	5.67 (2.48)^{b}	2.67 (1.78)^{c}	1.67 (1.47)^{c}
7	Fenvalerate 0.4 D	9.33 (2.05)^{cd}	0.53 (1.02)^{a}	0.40 (0.95)^{a}	0.27 (0.87)^{a}
8	Control	8.07 (2.93)^{bc}	7.87 (2.89)^{c}	7.53 (2.83)^{e}	7.07 (2.75)^{f}

* Values in parentheses are $\sqrt{x + 0.5}$ transformed values; Means followed by same letters along the columns are not significantly different by DMRT; ** DAS – Days After Spraying

Even though the virus was moderately effective at lower doses, the inoculum will build up in the soil with time and hence may help in suppressing the pest forever in the long run. Different levels of suppression of RHC by spraying *Amal*NPV has also been reported by Jayaraj *et al.* (1977), Rabindra and Jayaraj (1975) and Veenakumari *et al.* (2005). As observed in the present studies, the efficacy of the NPV against the RHC on groundnut was also established through field experiments in Tamil Nadu by Chandramohan and Kumaraswami (1979) and Rabindra and Balasubramanian (1980).

References

Chandramohan, N., and Kumaraswami, T., 1979. Comparative efficacy of chemical and nuclear polyhedrosis virus in the control of groundnut red hairy caterpillar, *Amsacta albistriga* (Walker). *Science and Culture*, 45(5), 202-204.

Ganiger, P.C. and Sannaveerappanavar, V.T., 2007. Field biology of red-headed hairy caterpillar, *Amsacta albistriga* (Walker) on groundnut. *Int. J. Tropical Agri.*, 25(1-2), 271-278.

Ghewande, M. P. and Nandagopal, V., 1997. Integrated pest management in groundnut (*Arachis hypogea* L.) in India. *Integrated Pest Mng. Reviews*. 2, 1-15.

Jayaraj, S., Sundaramurthy, V.T., Mahadevan, N.R. and Swamiappan, M., 1977. Relative efficacy of Nucleopolyhedrosis virus and *Bacillus thuringiensis* in the control of groundnut red-headed hairy caterpillar,

Biodegradation of endosulfan by phyllosphere bacteria

J.S. KENNEDY,* G. PRASAD** AND V. KARTHIKA**

Phyllosphere bacteria were enumerated from endosulfan treated leaves at 0 h, 24 h and 72 h after application. At recommended level initially there was no effect on microbes. The populations were found to decrease at 24 h and slight increase was observed population at 72 h after application of the pesticide, which indicated their adaptability. The bacterial isolates obtained from endosulfan treated leaves were mostly gram-negative rods, forming smooth, circular, raised and glistening colonies on nutrient agar medium. Based on the morphological and biochemical characters isolates were tentatively identified as *Pseudomonas* sp. Endosulfan was effectively degraded in the mineral medium having endosulfan as carbon source. Application of bacterial isolates on endosulfan treated plants resulted in the significant reduction of pesticide residue levels. The rate of endosulfan residues degradation was much slower in the plants that received no bacterial treatments.

INTRODUCTION

Endosulfan is a cyclodein insecticide possessing a relatively broad spectrum of activity, used throughout the world for the control of numerous insects to a wide variety of food and non food crops. Because of its abundant usage and potential transport, endosulfan is frequently found in the environment at considerable distances from the point of its original applications (Mansingh and Wilson, 1995; Miles and Pfeuffer, 1997). Endosulfan has been detected in the atmosphere, soils, sediments, surface and rain water and food stuffs (United States Department of Health and Human Services, 1990). It is extremely toxic to fish and aquatic invertebrates (Paul and Balasubramanium, 1997, Sundaram *et al*., 1992) and has been implicated in mammalian gonadal toxicity (Sinha *et al*., 1997). Recently in Kasorgod district of Kerala in India, causation of high morbidity and greater prevalence of congenital malformation were reported (Saiyed *et al*., 2003). Though the use of organochlorine pesticides has been banned in several developed countries, due to various socioeconomic factors and cheapness they are still widely used in developing countries including India. These health, environmental and socioeconomic concerns have led to an interest in detoxification of endosulfan in the environment.

Among the biologicals, microorganisms play a major role in pesticide degradation because they are the source of xenobiotics degrading enzymes (Chen and Mulchandani, 1998). Many works have been carried out on degradation of endosulfan by soil microorganisms but the present research work is on degradation of endosulfan by phyllosphere microbes. The role of phyllosphere microbe in degrading the pesticides is still an unexplored domain. Since most of the pesticides are applied on the surface of leaves, isolating a pesticide degrading phyllosphere microbe and its application will surely ensure a reduction in the residual effect of these pesticides. Hence, the present study was conducted.

MATERIALS AND METHODS

Isolation of endosulfan degrading bacteria

Endosulfan treated leaf samples from bhendi plants that received endosulfan were collected at 0 h, 24 h, and 48 h after application in sterile polythene bags and brought to the laboratory. One gram of

* Department of Entomology, **Department of Microbiology, Tamil Nadu Agricultural University, Coimbatore - 641 003

leaf sample was transferred to 100 ml sterile water blank and shaken for 10 min to dislodge microbes from phyllosphere. Ten ml of suspension was taken and transferred to 250 ml conical flask containing 100 ml sterile water blank. One ml aliquots from first and second dilutions were transferred to sterile peteridishes and plated with nutrient agar medium and incubated at room temperature. Numbers of bacterial colonies were counted after 24 h. The single colonies were further purified by streak plate method.

Apart from the above, endosulfan treated leaves were collected and brought to the laboratory. A moderate sized single leaf was taken and made an imprint with the dorsal side on the surface of plated nutrient agar medium. The same way ventral surface was also imprinted in a separate petridish containing nutrient agar medium. The plates were incubated at room temperature for 24 h and observed for microbial growth. Bacterial colonies were isolated and purified.

Enrichment of endosulfan degrading cultures

The isolated cultures were grown in 50 ml Luria Bertani broth (LB) supplemented with 150 ppm endosulfan at 27°C for five days. Five ml of culture thus grown was transferred to 50 ml of fresh medium containing 150 ppm endosulfan, incubated for 5 days at 27°C. Ten ml was taken, centrifuged at 10,000 rpm for 10 min, discarded supernatant, washed the pellets with mineral medium and suspended in 10 ml of sterile distilled water. Five ml of the cell suspension was transferred to mineral medium containing 100 ppm of endosulfan, and incubated at 30 °C for 7 days on a rotary shaker (50 rpm). The culture thus obtained was streaked on solid, endosulfan mineral salt medium for isolation of single colonies.

Tolerance of the isolates to higher concentrations of endosulfan

From the filter sterilized stock solutions of pesticides appropriate volumes were pipetted out into nutrient agar medium so as to get the pesticides at various final concentrations like 0, 100, 250, 500, 750 and 1000 ppm. Twenty ml of each medium was poured into sterile petriplates. After solidification, the isolates were streaked on the plates. After 72 h of incubation, tolerance level of the isolates was determined by their growth.

Effect of selected bacterial isolates on Endosulfan

Rate of degradation of endosulfan by bacterial isolates in mineral broth: Mineral medium containing endosulfan at 10 ppm, was prepared and dispensed in 50 ml quantities in 250 ml side arm flask, sterilized at 15 lb for 20 min. The bacterial cells grown in endosulfan (200 ppm) impregnated LB broth were harvested by centrifuging at 10,000 rpm for 10 min. The pellet was washed and resuspended in sterile water such that a final cell concentration of 10^6 cells ml^{-1} was achieved. Transferred 1ml of the cell suspension into endosulfan impregnated sodium succinate sodium glutamate medium depleted of sodium succinate and incubated at 30°C with orbital shaker (130 rpm) for 3 days. The cells were harvested by centrifuging at 1000 rpm for 20 min and the supernatant was taken for residue analysis.

Fifty ml of sample was transferred to 1 litre separating funnel. To the sample 50 ml of acetone was added shaken for 2-3 min. To that 50 ml saturated sodium chloride and 50 ml hexane were added. Then it was shaken for about 2-3 min and the layers were allowed to separate by keeping for a period of 10 minutes. The organic layer was collected separately and aqueous was once again extracted with 50 ml hexane. The organic layers were pooled and dried by passing through a anhydrous sodium sulfate matrix. Then the extract was condensed to 3-5 ml using rotary vacuum evaporator. The content was dissolved to a desired volume with hexane and finally the content of pesticide was estimated using GC fitted with Electron Capture Detector (ECD) and with capillary column.

Pot culture experiment: A pot culture experiment to test the bacterial strains for their ability to degrade endosulfan on bhendi, *Abelmoschus esculentus*. L (variety: *Arka anamica*) was conducted. Pots were filled with soil collected from garden land and the experiment was performed in completely randomized block design (CRBD) with three replications for each treatment as follows:

T_1 - Endosulfan + BP1 culture ; T_2 - Endosulfan + BP6 culture
T_3 - Endosulfan + BP7 culture ; T_4 - Endosulfan + BP8 culture
T_5 - Endosulfan + BP10 culture ; T_6 - Control ; T_7 - Without Endosulfan

Dislodgeable residue analysis: Endosulfan was sprayed on bhendi plants during preflowering stage at recommended level of 2 ml. After the application of endosulfan the respective bacterial cultures were sprayed on the plants. The bacterial suspension for spray contained 10^6 cells ml^{-1} and 0.05 per cent Tween 20. Leaf samples were collected at 0, 1 and 3 days after application of samples using sterile forceps and polythene bags. Leaf discs (12.56 cm^2) were cut with the help of sterile cork borer, transferred to conical flask containing sterile distilled water (50 ml), shaken well for 20 min to extract dislodgeable residues. The leaf washing was further processed for detecting residues after removing the leaf disc as per Gunther *et al.*, 1974. Endosulfan was estimated by gaschromatography - Chemito model 2865 – equipped with electron capture detector (ECD) Ni ^{63} employing the following operating parameter: Temperature - Oven: 220 °C, injector: 250 °C, detector: 280 °C.; Column: Packed glass column – 1.5% O. V 17 + 1.95% Q. F1 on gaschrom Q; Carrier gas: Nitrogen flow rate - 20ml min^{-1} Volume injected: 1 μl in split less mode.

Results

Enumeration of phyllosphere bacteria on endosulfan treated Bhendi leaves

Enumerated phyllosphere bacteria by leaf impression and serial dilution plate count method at 0 h, 24 h and 72 h. The results are presented in tables 1 and 2.

Table 1. Enumeration of phyllosphere bacteria on endosulfan treated bhendi leaves (Leaf impression method)

Leaf surface	Population after treatment (Colonies/cm^2)			
	0 h	24 h	72 h	Untreated control
Dorsal	12.45 (1.09)	4.12 (0.61)	5.29 (0.72)	13.6 (1.13)
Ventral	8.67 (0.94)	2.45 (0.39)	4.00 (0.60)	9.3 (0.97)

* Values in parentheses are log $_{10}$ transformed populations

Table 2. Enumeration of phyllosphere bacteria on endosulfan treated bhendi leaves (Serial dilution method)

Population after treatment (X 10^2 Cells ml^{-1})			
0 h	24 h	72 h	Control
36 (3.56)	3.3 (2.52)	4.6 (2.66)	42 (2.62)

* Values in parenthesis are log $_{10}$ transformed populations

At recommended doses, the results clearly indicated that, there was no immediate effect of the pesticide endosulfan at 0 h, as the population 12.45 colonies cm^{-2} was almost similar to the untreated leaf sample. The population was found to decrease at 24 h after application to 4.12 cells cm^{-2}. There was a slight increase in the population at 72 h after application of the pesticide (5.29 cells cm^{-2}), which indicated the adaptability of the microbes.

On the leaves, the population was found to be more on the dorsal surface compared to that of ventral surface. On the dorsal surface, the population was 12.45, 4.12, 5.29, and 13.6 colonies cm^{-2} at 0, 24, and 72 h after application of the pesticide and untreated leaves respectively. In the case of ventral surface the population was 8.67, 2.45, 4.0 and 9.3 cells cm^{-2} for the respective treatments.

The population was also enumerated by serial dilution plate count method. The results are presented in table 2. In the case of 0 h, it showed the maximum population of 36.0 X 10^{2} cells ml^{-1} where as, 24 h after application the population was slightly declined showed 3.3 X 10^{2} cells ml^{-2}. There was a slight increase in the population at 72 h after application of the pesticides (4.6 X 10^{2} cells ml^{-1}), which indicated the adaptability of the microbes.

Tolerance level of the selected isolates to endosulfan

The tolerance of the selected isolates to endosulfan was examined *in vitro* by incorporating various concentrations of the insecticides into the culture medium. The results are presented in table 3. The results indicated that, all the isolates were able to tolerate 500ppm of endosulfan. The isolates BP8 and BP9 tolerated up to 750ppm where as, BP6, BP7 and BP10 tolerated the presence of endosulfan up to 1000 ppm.

Table 3. Maximum tolerance level of the selected bacterial isolates to Endosulfan

S. No	Isolate	Concentration (ppm)					
		0	100	250	500	750	1000
1.	BP1	+	+	+	+	-	-
2.	BP2	+	+	+	+	-	-
3.	BP6	+	+	+	+	+	+
4.	BP7	+	+	+	+	+	+
5.	BP8	+	+	+	+	+	-
6.	BP9	+	+	+	+	+	-
7.	BP10	+	+	+	+	+	+

+ = Growth; - = No growth; Degradation of endosulfan by the selected isolates in mineral broth

The rate of degradation of endosulfan by the microbial isolates was studied by analyzing the residue in the mineral broth, initially impregnated with 10 ppm of endosulfan, after 72 h of incubation. The results are presented in table 4.

Table 4. Estimation of endosulfan degradation in culture medium by gas chromatographic analysis

Isolate	Residues in µg ml^{-1}			Per cent degradation after 3 days of incubation
	α Endosulfan	ß Endosulfan	Total Endosulfan	
BP1	0.242	BDL	0.242	53.50
BP2	0.190	BDL	0.19	63.50
BP6	BDL	BDL	BDL	100.00
BP7	0.014	0.010	0.024	95.39
BP8	0.158	BDL	0.158	69.60
BP9	0.069	0.027	0.096	81.50
BP10	0.058	0.027	0.085	83.60
Control	0.388	0.133	0.521	

BDL - Below detectable limit

Table. 5. Effect of selected bacterial isolates on the endosulfan sprayed bhendi plants (*Abelmoschus esculentus*)

Treatments	Residues in μg cm^{-2}									Percentage of degradation		
	0 h			24 h			72 h					
	α E	ß E	Total Endosulfan	α E	ß E	Total Endosulfan	α E	ß E	Total Endosulfan	0 h	24 h	72 h
T1	0.220	0.152	0.372	0.210	0.090	0.300	BDL	0.070	0.070	12.6	24.6	71.0
T2	0.220	0.169	0.389	0.120	0.076	0.196	0.060	0.079	0.139	8.68	50.7	42.5
T3	0.209	0.160	0.369	0.116	0.048	0.164	0.042	0.092	0.134	13.3	58.7	44.6
T4	0.210	0.179	0.389	0.197	0.114	0.311	0.081	0.076	0.157	8.06	21.8	35.1
T5	0.242	0.168	0.410	0.132	BDL	0.132	BDL	0.024	0.024	3.75	66.8	90.0
T6	BDL	BDL	BDL	BDL	BDL	BDL	BDL	BDL	BDL	BDL	BDL	BDL
T7	0.247	0.179	0.426	0.228	0.170	0.398	0.144	0.098	0.242			

T_1 - Endosulfan + BP_1 culture; T_2 - Endosulfan + BP_6 culture; T_3 - Endosulfan + BP_7 culture; T_4 -Endosulfan + BP_8 culture; T_5 - Endosulfan + BP_{10} culture; T_6 - Control without endosulfan and bacterial isolates; T_7 - Endosulfan; BDL - Below detectable limit; α E - α endosulfan; ß E - ß endosulfan

The concentration of endosulfan residues in mineral broth was found greatly reduced in different treatments within 3 days of incubation. The residual endosulfan was not detected in the case of medium inoculated with BP6 isolate indicating 100 per cent of degradation. Medium inoculated with isolate BP7 degraded endosulfan up to 95.39 per cent and the isolate BP10 degraded up to 83.60 per cent. Endosulfan sulfate was not produced in the medium by these isolates, except BP1 inoculated broth during degradation. The final pH in all the treatments was around 6.8 – 7.5. Among the seven isolates, BP1 and BP2 were found to be poor degraders.

Effect of selected bacterial isolates on the endosulfan sprayed to bhendi plants (*Abelmoschus esculentus. L*)

The experiment was conducted to find out whether the selected bacterial isolates were potential candidates for the degradation of endosulfan. The results are presented in the table 5. The residue levels assessed by employing GC-ECD indicated a significant reduction in all the treatments that received endosulfan and different bacterial isolates. The percentage degradation effected by BP10 was the maximum (90.01) followed by BP1 (71.0). The treatments that received the isolates BP6, BP7 and BP8 were found to be moderate as far as the endosulfan degradation concerned. The isolate BP10 was the most potent of all the six isolates tested for the ability to degrade endosulfan. There was no residue found in the control samples which received neither pesticide nor bacterial isolates. The rate of degradation was very slow in the treatment that received only endosulfan.

Discussion

Naiter and Amal (1977) reported the endosulfan metabolism on tobacco leaves. He reported that α and ß-endosulfan are interconvertible and endosulfan sulfate is converted into α-endosulfan but not into ß-endosulfan. Alpha endosulfan, ß-endosulfan and endosulfan sulfate can be directly hydrolyzed to endosulfan diol. Maier -Bode (1968) also reported the levels of endosulfan residues in fodder grass diminished from 0.3 ppm to 0.1 ppm, in 7 days of application. In the present study, only the dislodgeable residues on treated leaves were analyzed. This is one reason for the low values of the residues of endosulfan.

The results revealed that significantly reduced levels of residues in the sample were due to microbial degradation, as the leaves without bacterial treatments showed no or very less degradation. In endosulfan treated Bermuda grass and potato tuber the principal metabolite found was endosulfan sulfate (Naiter and Amal, 1977). In the present study endosulfan sulfate was not observed. This result is in accordance with that of Gi-Seok *et al.* (2002) who reported that *Klebsiella pneumoniae* degraded endosulfan without formation of the toxic metabolite endosulfan sulfate. The present study revealed that more ß-endosulfan than α- endosulfan was estimated after 3 days, which is due to inter conversions of α into ß-endosulfan as elucidated by Naiter and Amal (1977). Evans *et al.* (1971) reported that the degradation activity of *Pseudomonas* spp. on endosulfan was due to the presence of enzymes such as hydroxylases and monooxygenases.

The results of this study revealed that plants may enhance microbial degradation by providing specific environment for pollutant degrading commensal such as pseudomonads. (Johnson *et al.*, 2003). Plants may support a microflora in the phyllosphere with much greater adaptability for growth on different carbon sources including pollutants and helps in enhanced biodegradation. Essential are the studies on molecular characterization of the degradation pathway *i.e.*, identification of plasmids, genes and enzymes involved in the degradation. Recombinant DNA techniques and cloning of desired genes into the plants can be taken up as future plan of work to enhance biodegradation.

References

Chen, W. and Mulchandani, A., 1998. The use of live biocatalysts for pesticide detoxification. *TIBTECH.*, 16, 71-76.

Evans, W.C., Smith, B.S., Moss, P. and Ferney, H.N., 1971. 2,4-D-oxidation at 6 position by *Pseudomonas. Biochem. J.*, 122, 509-543.

Gi-seok, K., Kim, J., Kim, T.K., Sohn, H.Y., Koh, S.C., Shin, K.S. and Kim, D.G., 2002. Klebsiella pneumonia KE-1 degrades endosulfan without formation of the toxic metabolite, endosulfan sulfate. *FEMS Microbial Lett.*, 215, 255-259.

Gunther, F.A., Barkely, J.H. and Westlake, W.E., 1974. Worker Environment Research II. Sampling and processing techniques for determining dislodgeable pesticide residues in leaf surface. *Bull. Environ. Contam. Toxicol.*, 12, 641-644.

Johnson, D.L., Maguire, K.L., Anderson, D.R. and McGruth, S.P., 2003. Enhanced dissipation of chrysene in planted soil the impact of a rhizobial inoculum. *Soil Biol. Biochem.*, 36(1), 33-38.

Maier-bode, H., 1968. Properties, effect, residues and analytical of the insecticide endosulfan. *Res. Rev.*, 22, 1-44.

Mansingh, A. and Wilson, A., 1995. Insecticide contamination of Jamanican environment. Baseline studies on the status of insecticidal pollution of Kingstom Harbour. Mar. *Pollut. Bull.*, 30, 640-645.

Miles. C.J. and Pfeuffer, R.J., 1997. Pestides in canals of South Florida. *Arch. Environ. Contamin. Toxicol.*, 32, 337-345.

Naiter, M.C. and Amal, M.M., 1977. Metabolism of endosulfan I, endosulfan II, and endosulfan sulfate in tobacco. *J. Agric. Food. Chem.*, 25, 32-36.

Paul, V. and Balasubramanium, E., 1997. Effect of single and repeated administration of endosulfan on behaviour and its interaction with centrally acting drugs in experimental animals: a mini review. *Environ. Toxicol. Pharmacol.*, 3, 151-157.

Saiyed, H., Dewan., A., Bhatnagar V., Shenoy, U., Shenoy, R. and Rajmohan, H., 2003. Effect of endosulfan on male reproductive development. *Environ. Health perspect.*, 111, 1958-1962.

Sinha, N., Narayan, R. and Saxena, D.K., 1997. Effect of endosulfan on testis of growing rats. *Bull. Environ. Contam. Toxicol.*, 58, 79-86.

Sundaram, R.I..M., Cheng, D.M.H. and Thompson, G.B., 1992. Toxicity of endosulfan to native and introduced fish in Australia. *Environ. Toxicol. Chem.*, 11, 1469-1476.

United States Department of Health and Human Services, 1990. Toxicological profile for endosulfan. Agency for toxic substances and disease registry, Atlanta, GA.

Entomopathogenic Nematodes: Ecofriendly Tool for Insect Pest Management

NETHI SOMASEKHAR*

Entomopathogenic nematodes (EPN) in the genera *Steinernema* and *Heterorhabditis* together with their symbiotic bacteria are lethal insect parasites. Their impressive list of attributes including high virulence, ease of mass production, broad host range, host search ability, safety to non target organisms and exemption from registration in many countries has generated a great deal of scientific and commercial interest in these insect-killing nematodes as biological control agents in recent years. Currently, they are the second largest selling biological control products after the *Bacillus thuringiensis* (Bt) in the world market. Predictability of control, a key objective in the sound pest management strategy can be achieved with EPN in protected agricultural systems and they have successfully replaced chemical insecticides in some greenhouse crops. However, implementation of EPN under field conditions still remains hampered by the lack of predictability in efficacy of control. This underscores the need for increasing our understanding of bioecology of these nematodes. Sensitivity to environmental stresses (heat, UV, and desiccation) is one of the key factors attributed to the inconsistencies in the field performance. Nevertheless, solutions to this problem could be found in the high degree of genetic diversity present in the natural populations of these nematodes for many desirable traits. Genetic improvement of environmental tolerance of EPN has shown to improve their field performance. No significant adverse effects of EPN were observed on non-target soil fauna, a few field studies have shown that these nematodes selectively suppress the population of plant parasitic nematodes. In India, research on these beneficial worms is receiving considerable attention in recent years. There is great potential for developing these nematodes as biological alternatives to hazardous pesticides, particularly in organic farming and export oriented crops. This paper discusses the prospects and problems in implementing EPN in insect pest management in the light of recent developments in EPN research.

INTRODUCTION

Control of insect pests using chemical pesticides often leads to complex problems such as insecticide resistance, pest resurgence, elimination of natural enemies, toxic hazards to non target organisms and environmental pollution. Increasing concern about these ill effects of chemical pesticides provides strong impetus for developing newer technologies that are safe to the environment and well being of mankind.

Entomopathogenic nematodes (EPN) in the genera *Steinernema* Travassos and *Heterorhabditis* Poinar (Nematoda: Steinernematidae and Heterorhabditidae) are obligate and lethal parasites of insects (Gaugler, 2002; Grewal *et al.,* 2005). These nematodes have a unique symbiotic relationship with enteric bacteria. Bacteria of the genera *Xenorhabdus* and *Photorhabdus* (Enterobacteriaceae), with the exception of *Photorhabdus asymbiotica* associate mutualistically with nematodes of the

Central Potato Research Station, Muthorai, Ooty – 643 004

* Present address: Entomology Department, Directorate of Rice Research, Hyderabad – 500 030, A.P.

families Steinernematidae and Heterorhabditidae, respectively. The bacteria produce toxins for killing insect, protect cadavers from secondary infections and provide congenial conditions for nematode survival and reproduction. Nematodes in turn protect bacteria by harboring them in their gut and act as a vector for transporting bacteria in to insect host. EPN together with their symbiotic bacteria show great promise as biological alternatives to chemicals for the control of economically important insect pests (Grewal and Georgis, 1999; Shapiro-Ilan *et al.,* 2002; Shapiro-Ilan, 2004; Grewal *et al.,* 2005). EPN are currently marketed world wide for the biological control of insect pests and currently, they are the second largest selling biological control products after the *Bacillus thuringiensis* (Bt) in the global market. (Grewal and Georgis, 1999). Species of *Steinernema* and *Heterorhabditis* have shown to be effective in biological control of several important insect pests. Their impressive list of positive attributes including high virulence (ability to kill host within 48 h), availability of scalable mass production techniques, broad host range, host search ability, safety to non-target organisms, and exemption from registration in many countries has generated a great deal of scientific and commercial interest in these insect-killing nematodes. Species and isolates of EPN exhibit considerable differences in host range, virulence, tolerance to environmental stresses, foraging behavior, longevity and amenability to mass production and formulation (Gaugler *et al*., 1988; Grewal *et al*., 1994; Somasekhar *et al.* 2002a). This has given impetus for surveys seeking efficient native strains for biocontrol programs in several countries. This has resulted in a spurt in number of research publications on this group of nematodes and description of several new species in recent years.

Biogeography and biodiversity of *Steinernema* and *Heterorhabditis*

The species of *Steinernema* and *Heterorhabditis* have worldwide distribution. Steinernematids have been reported from all the continents except Antarctica. Among the species of *Steinernema*, some species like *S. carpocapsae* and *S. feltiae* have a global distribution while other species show restricted geographic distribution (Hominick *et al.,* 1996). However, with increasing number of surveys in many countries, the geographical distribution range of many known species is expanding besides discovery many new species. The distribution of heterorhabditids is relatively narrow compared to the steinernematids because fewer number of species. Among *Heterorhabditis* species, *H. bacteriophora* has widest geographical distribution and is found in North and South America, Europe, Australia and Asia. H. indica which was originally described from India also has wide distribution including several countries in tropics and subtropics (Hominick *et al.,* 1996). However, *H. zealandica* and *H. marelatus* have a limited distribution and are found only in New Zealand (Akhurst, 1987) and USA (Liu and Berry, 1996), respectively.

The family Steinernematidae is composed of two genera *viz. Steinernema* Travassos, 1927 and *Neosteinernema* Nguyen and Smart, 1994. The genus *Steinernema* is more diverse with more than 45 described species while the genus *Neosteinernema* has a lone species, *N. longicurvicaudata* (Table 1). The family Heterorhabditidae Poinar, 1976 has only one genus, *Heterorhabditis* which has about 11 described species (Table 2). EPN species have been described using Linnaean and biological species concept mainly based on morphometric analyses and cross-breeding studies (Poinar, 1990, Dix et al., 1994, Nguyen and Smart, 1996). Later, molecular techniques based on protein and isozyme profiles (Akhurst, 1987; Poinar and Kozodoi, 1988), RFLP (Joyce *et al.,* 1994), RAPD (Gardner *et al.,* 1994), Satellite DNA (Greiner *et al,* 1996), genomic DNA sequences (Adams *et al,* 1998) and RFLP analyses of rDNA ITS sequences (Joyce *et al.,* 1994) have been used to over come the limitations of traditional morphological studies. Using *Heterorhabditis* as an example, Adams *et al* (1998) demonstrated that most suitable species concept for nematodes is a combination of the phylogenetic and evolutionary species concept.

Considerable diversity has also been observed among symbiotic bacteria associated with EPN

(Adams *et al.,* 2006). Nine species of *Xenorhabdus* have been described (Table 3). Among them, seven species are associated with a single species of *Steinernema* while *X. bovienii* is associated with four (Akhurst and Boemare, 1988) and *X. poinarii* with two *Steinernema* species (Fischer-Le Saux *et al.,* 1999). The genus *Photorhabdus* consists mostly the bacterial symbionts of the nematode genus *Heterorhabditis* (*P. luminescens and P. temperate*) besides a non symbiotic species *P. asymbiotica* (Fischer-Le Saux *et al.*, 1999). Further, *P. luminescens* is split in to five subspecies and *P. temperate* split in to two subspecies (Fischer-Le Saux *et al.,* 1999; Hazir *et al.,* 2004).

Table 1. The genera and species of the family Steinernematidae

Family: Steinernematidae Chitwood and Chitwood, 1937
= Neoaplectanidae Sbolev, 1953
Type genus: *Steinernema* Travassos, 1927
Type species: *S. kraussei* (Steiner, 1923) Travassos, 1927
Syn. *Aplectana krausseei* Steiner, 1923

Other species:

S. abbasi Elawad *et al.*, 1997
S. aciari Qiu *et al.*, 2005
S. affine (Bovien, 1937) Wouts *et al.*, 1982
S. anatoliense Hazir *et al.,* 2003
S. arenarium (Artyukhovsky, 1967) Wouts *et al.*, 1982
S. asiaticum Anis *et al.,* 2002
S. apuliae Triggiani *et al.,* 2004
S. bicornutum Tallosi *et al.,* 1995
S. carpocapsae (Weiser, 1955) Wouts et al., 1982
S. caudatum Xu *et al.*, 1991
S. ceratophorum Jian *et al.,* 1997
S. cubanum Mrácek *et al.,* 1994
S. diaprepesi Nguyen and Duncan, 2002
S. feltiae (Filipjev, 1934) Wouts *et al.,* 1982
S. glaseri (Steiner, 1929) Wouts *et al.,* 1982
S. guangdongense Qiu *et al.,* 2004
S. intermedium (Poinar, 1985) Mamiya, 1988
S. hermaphroditum Stock, Griffin and Chaenari, 2004
S. jollieti Spiridonov *et al.,* 2004
S. karii Waturu *et al.,* 1997
S. kushidai Mamiya, 1988
S. loci Phan *et al.,* 2001a
S. longicaudum Shen and Wang, 1991
Syn. *S. serratum* Liu, 1992
S. monticolum Stock *et al.,* 1997
S. neocurtillae Nguyen and Smart, 1992
S. oregonense Liu and Berry, 1996b
S. pakistanense Shahina *et al.,* 2001
S. puertoricense Román and Figueroa, 1994
S. rarum (de Doucet, 1986) Mamiya, 1988
S. riobrave Cabanillas *et al.,* 1994
S. ritteri de Doucet and de Doucet, 1990
S. robustispiculum Phan *et al.,* 2005
S. sangi Phan *et al.,* 2001b
S. scapterisci Nguyen and Smart, 1990
S. scarabaei Stock and Koppenhöfer, 2003
S. siamkayai Stock *et al.,* 1998
S. tami Van Luc *et al.,* 2000
S. thanhi Phan *et al.,* 2001a
S. thermophilum Ganguly and Singh, 2000
S. websteri Cutler and Stock, 2003
S. weiseri Mrácek *et al.,* 2003
Genus: *Neosteinernema* Nguyen and Smart, 1994

Type and only species: *Neosteinernema longicurvicauda* Nguyen and Smart, 1994

Table 2. The genera and species of the family Heterorhabditidae

Family: Heterorhabditidae Poinar, 1976
Type and only genus: *Heterorhabditis* Poinar, 1976
Syn. *Chromonema* Khan *et al.,* 1976
Type species: *H. bacteriophora* Poinar, 1976
Syn. *Chromonema heliothidis* Khan *et al.,* 1976
H. heliothidis (Khan *et al.,* 1976) Poinar *et al.,* 1977
H. argentinensis Stock, 1993

Other Species
H. baujardi Phan *et al.*, 2003
H. brevicaudis Liu, 1994a
H. downesi Stock *et al.*, 2002
H. indica Poinar *et al.*, 1992
Syn. *H. hawaiiensis* Gardner *et al.*, 1994
H. marelatus Liu and Berry, 1996a,b,c
Syn. *H. hepialius* Stock *et al.*, 1996
H. megidis Poinar *et al.*, 1987
H. mexicana Nguyen *et al.*, 2004
H. poinari Kakulia and Mikaia, 1997a
H. taysearae Shamseldean *et al.*, 1996
H. zealandica Poinar, 1990

Table 3. The genera and species of bacterial symbionts of entomopathogenic nematodes

Bacteria	Nematode
Genus: Xenorhabdus	
X. nematophila (Poinar and Thomas, 1965) Thomas and Poinar, 1979	Steinernema carpocapsae
X. bovienii (Akhurst, 1983) Akhurst and Boemare (1993)	*S. affine*, *S. feltiae*, *S. intermedium*, *S. Kraussei*
X. poinarii (Akhurst, 1983) Akhurst and Boemare (1993)	*S. glaseri*, *S. cubanum*
X. beddingii (Akhurst, 1986) Akhurst and Boemare (1993)	S. longicaudum
X. japonica Nishimura *et al.*, 1994	S. kushidai
X. budapestensis Lengyel *et al.*, 2005	S. bicornutum
X. ehlersii Lengyel *et al.*, 2005	S. serratum
X. innexi Lengyel *et al.*, 2005	S. scapterisci
X. szentirmaii Lengyel *et al.*, 2005	*S. rarum*
Genus: *Photorhabdus*	
P. luminescens (Thomas and Poinar, 1979) Boemare *et al.*, 1993; subsp. *Luminescens* Fischer-Le Saux *et al.*, 1999	*H. bacteriophora* Brecon
P. luminescens subsp. *akhurstii* Fischer-Le Saux *et al.*, 1999	H. indica
P. luminescens subsp. *laumondii* Fischer-Le Saux *et al.*, 1999	*H. bacteriophora* HP88
P. temperata Fischer-Le Saux *et al.*, 1999	*H. zealandica*, *H. bacteriophora* NC1,*H. megidis* (Nearctic strains)
P. temperata subsp. *temperata* Fischer-Le Saux *et al.*, 1999	*H. megidis* (Palearctic strains)
P. luminescens subsp. *kayaii* Hazir *et al.*, 2004	*H. bacteriophora* (grassland/clover fields, Turkey)
P. luminescens subsp. *thracensis* Hazir *et al.*, 2004	*H. bacteriophora* (sunflower/fallow field or pine forest, Turkey)

Biology and life cycle of *Steinernema* and *Heterorhabditis*

The species of *Steinernema* have a simple life cycle that includes the egg, four juvenile stages and the adult. The invasive stage is a special third-stage juvenile (J_3) also called as infective juvenile (IJ) or dauer juvenile, which is resistant to environmental conditions. Infective juveniles carry live cells of the mutualistic bacterium *Xenorhabdus* in their digestive tract. The nematodes, together with their symbiotic bacteria form a unique biocontrol system. EPN locate their insect hosts in soil by detecting chemical cues such as excretory products and carbon dioxide gradient. The infective juveniles enter into the haemocoel of host insect through natural openings and release symbiotic bacterium. Toxins produced by the developing nematodes (Burman, 1982) and bacteria (Akhurst and Boemare, 1990) cause septicemia and kill the insect host usually within 48 h. The immature nematodes feed on bacterial cells and host tissues and undergo successive molts to develop into adult males and females. Steinernematids are amphimitic in all generations. Large females and smaller males mate and produce progeny. The nematode reproduction continues until resources in the host cadaver are depleted and usually nematodes complete 2-3 generations. When the host cadaver is exhausted of resources, next batch of infective juveniles is produced. The infective juveniles that retain the cuticle of second stage as a protective sheath and carry symbiotic bacteria in their intestine emerge from the cadaver in search of new hosts in soil. Infective juveniles do not feed and can survive in the soil for several months. The cycle of entry of infective juveniles in to the insect host to emergence of new batch of infective juveniles usually requires 10 to 14 days.

The life cycle of *Neosteinernema* is similar to that of Steinernema but it completes only one generation inside host and females emerge from out from host cadaver and the infective juveniles are formed in side the female body. The bacterial symbiont of *Neosteinernema* is yet to be characterized. The life cycle of *Heterorhabditis* is also similar to *Steinernema,* but in this case the infective juveniles develop in to hermaphrodites in the first generation while second generation juveniles develop in to adult males and females inside the insect host. *Heterorhabditis* carries a bacterial symbiont *Photorhabdus* instead of *Xenorhabdus*. Further, *Heterorhabditis* can enter the insect haemocoel directly through the insect's integument (Wang and Gaugler, 1998). The phenomenon of endotokia matricida (hatching of eggs within female body is not uncommon among species of *Heterorhabditis* and *Steinernema.*

Biology and life cycle of *Xenorhabdus* and *Photorhabdus*

Bacteria of the genera *Xenorhabdus* and *Photorhabdus*, with the exception of *Photorhabdus asymbiotica* associate mutualistically with nematodes of the families Steinernematidae and Heterorhabditidae, respectively. In nature, these bacteria cannot exist in soil without their nematode associates. *Steinernema* species carry the bacteria inside a specialized vesicle located in the anterior part of the intestine (Bird and Akhurst, 1983; Forst and Clarke, 2002), whereas in *Heterorhabditis* species the bacteria colonize the entire intestine (Ciche and Ensign, 2003). The bacteria remain in a quiescent state within the nematode gut. Upon entry into the host, the infective juveniles release their bacterial symbionts through mouth (*Photorhabdus*) or defecation (*Xenorhabdus*) (Martens et al., 2003; Wouts, 1991; Ciche and Ensign, 2003). The bacteria and nematodes cooperate to overcome the host's immune response which allows proliferation of the bacterial cells. *Steinernema* species are able to suppress the host's immune response by periodically releasing enzymes, which may facilitate the release of their symbionts (Boemare and Akhurst, 1999; Wang and Gaugler, 1998). The bacteria multiply by feeding on host tissues and produce toxins that result in septicemia and bioconversion of the insect cadaver (Forst and Clarke, 2002). *Photorhabdus* produces a toxin that destroys the insect's midgut and hemocytes and this toxin encoded by the *mcf* (makes caterpillars floppy) gene (Daborn *et al.,* 2002). The *Xenorhabdus* releases endotoxins that are lipopolysaccharide (LPS) in nature. (Brillard *et al.,* 2001; Dunphy and Thurston, 1990). The *Xenorhabdus* and *Photorhabdus* produce a variety of antimicrobial compounds (bacteriocins, xenorhabdicin, lumicins, etc) that protect the cadaver from secondary infections (Akhurst, 1982; Boemare *et al.,* 1997) The developing nematodes feed on the bacteria and bioconverted host tissue. Once the food resources in the cadaver are exhausted, the nematodes produce a new batch of infective that acquire bacterial cells and emerge from the host cadaver in search of a new host. Both *Photorhabdus* and *Xenorhabdus* occur in two forms i.e. primary (phase I) and secondary (phase II) form. Phase I cells are larger than phase II cells and produce greater amounts of antibiotics and toxins. Infective juveniles carry only phase I cells. The role of phase II cells is yet to be understood.

Host range

The EPN as a group have a broad host range. Insects from over 17 orders and 135 families have been reported to be susceptible to EPN. *Steinernema carpocapsae* alone infected more than 250 species of insects from 11 orders in laboratory tests (Poinar, 1979). The EPN-bacterium complex kills insects rapidly so there is not much scope for development of intimate host parasite relationship as in case of other parasites like mermithids (Grewal and Georgis, 1999). In laboratory studies, where host contact is assured and no ecological and behavioral barriers exists nematodes infect and kill a wide range of insects. However, their efficacy under field conditions is influenced by wide range of biotic (foraging strategies of nematodes, behavioral responses of insects, etc) and abiotic factors (UV radiation, temperature, relative humidity, etc) (Gaugler *et a.l,* 1988).

Mass production

EPN can be mass produced by *in vivo* or *in vitro* methods. Insect host serve as small biological reactor for mass production of EPN in the *in vivo* production method. The larvae of wax moth *Galleria mellonella* are commonly used for *in vivo* nematode production in laboratory. The basic *in vivo* production technique was first described by Dutky *et al* (1964). Later some improvements were suggested by other workers (Woodring and Kaya, 1988; Lindgren *et al*., 1993). The yield of infective juveniles varies from 0.5 x 105 to 4 x 105 IJs/larva. Several lepidopterous insect larvae have also been used for mass production of *Steinernema* and *Heterorhabditis* species. The yield of infective juveniles varies with nematode species/isolate, inoculum level, insect host species, environmental conditions and nutrient status of host. The *in vivo* technique is very simple, requires minimum equipment and investment. However, this method is not suitable for large scale commercial production as it lacks economy of scale and is sensitive to biological variations.

EPN are cultured *in vitro* on a variety of solid or liquid substrates such as dog biscuit-agar, potato mash, chicken offal, synthetic liquid media, etc. The *in vitro* culture method was first described by Glaser as early as in 1931. However, the first successful large scale *in vitro* production method using solid media was developed by Bedding (Bedding, 1981). This method is now widely used. In this method, polyurethane sponge pieces are impregnated with artificial media along with symbiotic bacterium. *In vitro* culture using solid media generally yields 6 x 10^5 to 10 x10^5 infective juveniles/g media. Friedman (1990) developed a liquid fermentation technique for large scale mass production of EPN. Several steinernematids (*S. carpocapsae, S. riobravae, S. scapteresci, S. feltiae* and *S. glaseri*) and heterorhabditids (*H. bacteriophora, H. megidis* and *H. indica*) were efficiently produced using monoxenic liquid fermentation technique and it is commonly used in large-scale commercial mass production of EPN in industries with fermentors of 80000-300000 l capacity. *In vitro* culture using liquid media generally yields 50 x 10^{12} infective juveniles /g. This method is very economical for large scale production and gives consistent out put and quality. Nematode species and isolates exhibit considerable differences in suitability for *in vitro* production. Therefore this technique needs to be fine-tuned to suit the nematode species for getting optimum yield.

Formulations

Formulations of Entomopathogenic nematodes are developed basically to extend their shelf life and deliver live nematodes to the customers. Initially, simple formulations were developed using carriers like polyurethane sponge, peat, vermiculite, etc which have sufficient interstial space to facilitate gas exchange required for nematode survival. In these formulations carrier substances are impregnated with nematodes. Nematodes from the sponge formulations are extracted by squeezing them in to water before applications. Although these formulations are simple and easy to produce, they are labor intensive and needs constant refrigeration thus useful for only for small scale applications. Later on newer formulations with improved shelf life and scalability were developed. These formulations are aimed at extending the shelf life by conserving storage reserves of infective juveniles by restricting their movement (physical immobilization) or oxygen consumption by inducing a state of partial anhydrobiosis (physiological immobilization). The formulations based on principle of physical immobilization include activated charcoal, alginate gels, flowable gels, etc. (Yukawa and Pitt, 1985; Kaya and Nielsen, 1985; Georgis and Manweiler, 1994). Nematodes are extracted by dissolving these gels in water containing sodium citrate before application. The alginate gel formulations of *S. carpocapsae* possessed the shelf life of 3-4 months even at room temperature. However, preparation of these formulations and disposal of consumables used is a cumbersome process and may not be suitable for large scale use. A significant advance in formulation of EPN was made with the development of novel water dispersible granule (WDG) formulations based on the principle of physiological immobilization. In these formulations IJs are encapsulated in 10-20 mm diameter granules made up of a complex mixture of silica, clays, cellulose, lignin and starch. The granular matrix allows supply of oxygen to nematodes in the center of the granule in a partially desiccated state (Silver *et al*, 1995). The WDG formulations have a shelf life of several months at room temperature. Nematodes can be extracted by dissolving these granules in water before application. The shelf life of a formulation varies depending on nematode species, storage temperature and type of formulation (Grewal and Georgis, 1998) (Table-4).

Application technology

No matter how efficient a biological control agent may be against the target pest, the application will fail if the agent is not delivered in a manner that facilitates the infection of the host. EPN can be applied using a wide range of application equipment including ground and aerial spray equipment like pressurized sprayers, mist blowers, and electrostatic sprayers (Georgis, 1990) and irrigation systems like drip and sprinkler irrigation systems. The efficacy of application influenced by the nematode species, type of application equipment and the manner in which the nematodes are applied. A general recommendation for entomopathogenic nematodes has been common nozzle type sprayers with openings larger than 50 u and operating pressures less than 2000 kPa (290 psi) (Georgis, 1990). However, some nematode species can tolerate operating pressures up to 2000 kPa while other species like *H. megidis* can not withstand operating pressure above 1380 kPa. EPN can be used as soil application, foliar sprays in the field (Grewal and Georgis, 1999); they can be applied to trap crops (Purcell *et al.,* 1992) or propagation material like cuttings and root stalks (Bedding and Miller, 1981; Bari, 1992) or they can be used in insect traps (Parkman and Frank, 1993; Georgis *et al.,* 1999).

Field efficacy

Under optimum environmental conditions that exist in protected agriculture systems (e.g. glasshouses or polyhouses) (Richardson, 1990) and intensive systems (e.g. orchards with drip irrigation or golf courses with sprinkler irrigation systems) (Duncan and MacCoy, 1996; Grewal and Georgis, 1999) the level and consistency of pest control achieved with EPN was comparable to that of best insecticides and they have successfully replaced insecticides in these high value systems. EPN were also found effective against several insect pests in the field. However, the predictability of control the key objective of any pest control program is difficult achieve under field conditions because of their sensitivity to environmental factors like UV radiation, moisture and temperature (Richardson, 1990; Georgis and Gaugler, 1991). Tremendous progress made in research and development of EPN during the past two decades has lead to their successful implementation for pest management in many sectors. Some examples of successful use of EPN are presented in Table 5.

Table 4. Shelf life of some commercially available formulations of entomopathogenic nematodes

Formulation	Nematode species	Shelf Life (months)	
		Room	Refrigerated
Alginate gels	*S. carpocapsae*	3.0 – 4.0	6.0 – 9.0
	S. feltiae	0.5 – 1.0	4.0 - 5.0
Flowable gels	*S. carpocapsae*	1.0 – 1.5	3.0 - 5.0
Attapulgite Clay chips	*H. bacteriophora*	1. 5 – 2.0	0
	S. feltiae	1.0 – 1.5	4.0 - 6.0
	S. carpocapsae		
Water dispersible granules	*S. carpocapsae*	4.0 - 5.0	9.0 - 12.0
	S. feltiae	1. 5 – 2.0	5.0 – 7.0
	S. riobrave	2.0– 3.0	4.0 - 5.0

(Source: Grewal and Georgis, 1999)

Table 5. Examples of successful use of entomopathogenic nematodes in pest control

Crop/ sector	Nematode	Insect Pests	% Control achieved
Greenhouse crops	*S. carpocapsae*	Black vine weevil,	87-100 %
	S. feltiae	Sciarid flies, Stem borers, weevil	
	H. bacteriophora	Mint weevil, root borer	
Mushrooms	*S. feltiae*	Sciarid flies	47 – 90 %
Citrus	*S. riobrave*	Citrus root weevil (*Diaprepes abbreviatus*)	77 – 90 %
Berries	*S. carpocapsae*	Black vine weevil, strawberry root weevil, cranberry girdler	70 –100 %
	H. Bacteriophora	Black vine weevil, white grubs	
Mint	*S. carpocapsae*	Mint weevil, root borer, flea beetle, cutworms	67-94 %
Turf grass	*S. carpocapsae*	Billbugs, cutworms, army worms	70-100 %
	S. Riobrave	Mole crickets	
	S. Scapteresci	Mole crickets	
Pet/Vet Sector	*S. carpocapsae*	Cat fleas of Dogs and Cats	91-97 %

Ecological considerations

The field efficacy of EPN is influenced by several biotic and abiotic factors. The foraging behavior in entomopathogenic nematodes varies with the species and influences the distribution and host preference of the nematode. Based on foraging strategy they could be either ambushers (sit and wait) or cruisers (seek and kill). In order to ambush the prey, some *Steinernema* species nictate, or raise their bodies off the soil surface and attach to passing insects. Other species adopt a cruising strategy where they roam through the soil searching for potential hosts. Ambushers such as *S. carpocapsae* are effective against mobile insects that move on the surface, while cruisers like *Heterorhabditis bacteriophora* are effective against sedentary insects that live deep in the soil (Gaugler *et al.,* 1988; Campbell and Gaugler, 1993). Behavioral barriers restrict nematode efficacy to a few selected hosts. Therefore, it is important to match the proper nematode species with target insect pest for achieving successful and predictable insect control under field conditions. Since EPN are highly sensitive to environmental stresses like heat, UV, hypoxia, and desiccation, their successful use under field conditions is highly influenced by these factors (Kaya, 1990). They should be normally applied in the late evening hours to avoid exposure to intense sunlight and UV radiation with suitable adjuvants like glycerol, liquid paraffin, etc. Entomopathogenic nematodes require adequate soil moisture for survival and movement, but too much moisture may cause oxygen deprivation and restrict movement (Kaya, 1990; Koppenhöfer *et al*., 1995; Womersley, 1993). Optimum moisture levels will vary by nematode species and soil type. A light irrigation before and after application of nematodes provides optimum soil moisture for nematode activity in case of soil applications. Isolation and identification of native nematode strains with high level of tolerance to environmental stresses and development of stress tolerant strains through genetic improvement are necessary to increase the efficacy of these nematodes under field conditions.

Safety and regulatory procedures

Traditionally, biocontrol agents are projected as ecologically safer products. However, recent studies on ecological consequences of the introduction of biocontrol agents suggest that biological control carries benefits as well as risks. To ensure continuing safety and positive public image of biological control, many countries are requiring risk assessment of all biocontrol agents. The broad spectrum of potential host range observed in laboratory studies and large scale inundative releases of EPN species and strains warrants critical assessment of their safety to non-target organisms and development of regulatory protocols to ensure their safe usage. Studies on safety of EPN to non target organisms and regulatory protocols in place in different countries are recently reviewed by Akhurst and Smith (2002). The available evidence indicates that EPN had no adverse long term impact on non target insects and other invertebrates (Bathon, 1996). Although EPN could infect wide range of insect in laboratory studies with high dosages, their impact on field populations of non target invertebrates was found to be negligible (Bathon, 1996).

Investigations on safety of EPN and their symbiotic bacteria to vertebrates and humans also showed no negative effects despite widespread use during the past 20 years. Some isolates of *Photorhabdus luminiscens* were isolated from human wounds (Farmer *et al.,* 1989). However, detailed studies based on DNA analyses clearly showed that these clinical isolates are different from isolates having symbiotic association with nematodes and later they were grouped under a new species *P. asymbiotica* (Fischer-Le Saux *et al*., 1999). The existence of bacterial species with and without pathogenic effects on humans within one genus is common (e.g. *Bacillus*).

Inundative applications of EPN showed no adverse effects on plants. In some cases, plant vigor increased following soil application of EPN. EPN were found to suppress populations of plant parasitic nematodes that cause serious damage to plants (Smitley et al., 1992; Grewal *et al.,* 1997; Jagdale *et al.,* 2002). Further, they were found to suppress plant parasitic nematodes selectively i.e. without adversely effecting the populations of free-living microbivorus nematodes that are beneficial to plant growth and hence this phenomenon is referred to as beneficial non-target effect (Somasekhar *et al.,* 2002b).

A joint workshop supported by the EU COST Action 819 "Entomopathogenic Nematodes" and the OECD (Organization for Economic Co-operation and Development) Research Programme "Biological Resource Management for Sustainable Agriculture Systems", which met in 1995 to discuss potential risks related with the use of EPN in biological control, concluded that EPN are safe to production and application personnel and the consumers of agriculture products treated with EPN (Ehlers and Hokkanen, 1996). Based on available information it can be concluded that there are no safety considerations that preclude continued use of EPN as biological control agents. Akhurst and Smith (2002) reviewed the regulatory procedures for EPN in different countries. EPN products are completely exempted from registration requirements in most of the countries. Even in countries like

Japan, New Zealand, Sweden and Norway where registration is required the procedures for EPN products are relatively a less stringent compared to the insecticides in view of their safety to non target organisms and humans. Environmental Protection Agency of USA has exempted even the genetically modified EPN also from registration (Gaugler, *et al.,* 1997). However, the introduction of exotic strains of EPN is regulated in many countries to avoid biological pollution.

Entomopathogenic Nematode Research in India

Research on EPN in India was started in 1960s with an exotic strain (DD136) of S. carpocapsae. Several important insect pests of rice, sugarcane and apple were found susceptible to this nematode in laboratory (Rao and Manjunath, 1966). However, attempts to expand their use under field conditions were not successful due to problems in adaptability of this exotic strain to local conditions and lack of mass production and application techniques. Rahman *et al* (2000) reviewed the research done on EPN in India. EPN research in India got slowed down during 1970s-80s due to greater emphasis given for use of chemical pesticides and fertilizers to support high yielding varieties released during green revolution period.

Organized research on EPN with special emphasis on isolation and identification of indigenous strains was initiated in late 1980s at Sugarcane Breeding Institute (SBI), Coimbatore and Tamilnadu Agricultural University, Coimbatore. Several indigenous isolates of EPN were collected from tropical and subtropical sugarcane growing regions of India in surveys carried out by SBI, Coimbatore. A new isolate of *Heterorhabditis* was recovered from sugarcane top borer, *Scirpophaga excerptalis* (Pyralidae: Lepidoptera) in the vicinity of Coimbatore, India in one of these surveys. This isolate was later described as a new species *Heterorhabditis indica* Poinar, Karunakar and Dävid, 1992 (Poinar *et al.,* 1992). Several insect pests of sugarcane including borers (*Scirpophaga excerptalis, Chilo infuscatellus and Chilo sacchariphagus indicus*) and white grubs (*Holotrichia serrata*) were found susceptible to EPN (Karunakar *et al.,* 1999 and 2000; Sankarnarayanan *et al.,* 2003 and 2006). Research on biological control of white grubs in sugarcane using indigenous EPN strains was carried out at SBI, Coimbatore under an international collaborative research project involving leading research laboratories from Ireland and Germany. During this period detailed studies on molecular characterization of isolates of *Heterorhabditis indica* isolates from India and other countries(Stack *et al.,* 2000), development of large scale mass production technique using fermentors for Indian isolate of *H. indica* (Ehlers *et al.,* 2000) and field evaluation of EPN for biological control of white grub *H. serrata* in sugarcane were carried out. A new species *Steinernema thermophilum* was described by Ganguly and Singh from Indian Agricultural Research Institute, New Delhi, India (Ganguly and Singh, 2000). Occurrence of several isolates of known species of EPN were also reported from India including *S. carpocapsae* and *S. bicornutum* (Hussaini *et al.*, 2001), *S. riobrave* (Ganguly *et al.,* 2002), *H. bacteriophora* (Sivakumar *et al.,* 1989). Recently, *S. asiaticum* and a rhabditid nematode *Oscheius sp* were reported to be infecting rice yellow stem borer (*Scirpophaga incertulas*) and leaf folder *(Cnaphalocrosis medinalis*) (Katti *et al.*, 2006; Prasad *et al.*, 2006; Padmakumari *et al.,* 2007).

Progress made in mass production and formulation technology coupled with increasing demand to reduce pesticide usage in recent years gave boost to EPN research in India. Currently several other laboratories in India are working on EPN including Project Directorate of Biological Control (PDBC, Bangalore), Gujarat Agricultural University (GAU), Anand and Tamil Nadu Agricultural, University (TNAU), Coimbatore, Directorate of Rice Research, Hyderabad. Commercial formulations of EPN (Soil commander and Green commander) were released in market in India by Ecomax company during 1980s but they were not successful. The failure of these formulations may be due to the poor adaptability of exotic strains used in these formulations or production or formulation problems and lack of awareness about proper use of nematodes in field at that time. Recently, Multiplex company from Bangalore started producing commercial formulations of *Steinernema* and *Heterorhabditis*. Information on field efficacy of these formulations is yet to be available.

Conclusions and Future Prospects

Entomopathogenic nematodes together with their symbiotic bacteria represent a unique biocontrol system. EPN are exceptionally safe biological control agents. They are certainly more specific and are less of a threat to the environment than chemical insecticides. Progress has been made in mass production, formulation and application technology that EPN are now available for commercial large scale field applications in many countries. These nematodes have shown promise as biological alternatives to hazardous insecticides in protected agriculture systems and gave consistent results. However, their use in field crops is currently limited and needs attention. Sensitivity of EPN to environmental stresses (UV radiation, temperature, moisture, etc) and lack of awareness about proper use of nematodes are the major limitations in expanding their use in field. Future research efforts should be focused on discovery of native strains of EPN with high virulence and environmental stress tolerance, genetic improvement of native strains using conventional or genetic engineering methods, improvement of formulations with public- private partnership model involving research laboratories and industry. Further, creating awareness about these beneficial worms among farmers and extension agencies and promoting them as an ecofriendly component of integrated pest management (IPM) programs is vital for successful use of entomopathogenic nematodes.

References

Adams, B.J., Burnell, A.M., Powers, T.O., 1998. A phylogenetic analysis of *Heterorhabditis* (Nemata: Rhabditidae) based on internal transcribed spacer 1 DNA sequence data. *J. Nematol.*, 30, 22–39.

Adams, B.J., Fodor, A., Koppenhöfer, H.S., Stackebrandt, E., Stock, S.P., and Klein, M.G. 2006. Biodiversity and systematics of nematode–bacterium entomopathogens. *Biol. Control*, 37, 32–49.

Akhurst, R. and K. Smith. 2002. Regulation and safety. In: *Entomopathogenic Nematology.* (ed.) R. Gaugler. CABI Publishing, Oxon, UK. p. 311-332.

Akhurst, R.J., 1982. Antibiotic activity of *Xenorhabdus* spp., bacteria symbiotically associated with insect pathogenic nematodes of the families Heterorhabditidae and Steinernematidae. *J. Gen. Microbiol.* 128, 3061–3065.

Akhurst, R.J., 1987. Use of starch gel electrophoresis in the taxonomy of the genus *Heterorhabditis* (Nematoda: Heterorhabditidae). *Nematologica,* 33, 1–9.

Akhurst, R.J., Boemare, N.E., 1988. A numerical taxonomic study of the genus *Xenorhabdus* (Enterobacteriaceae) and proposed elevation of the subspecies of *X. nematophilus* to species. *J. Gen. Microbiol.* 134, 1835–1845.

Akhurst, R.J., Boemare, N.E., 1990. Biology and taxonomy of *Xenorhabdus*. In: *Entomopathogenic Nematodes in Biological Control.* (Eds.) Gaugler, R., Kaya, H.K. CRC Press, Boca Raton, FL, pp. 75–90.

Bari, M.A. 1992. Disinfestation of artichoke stumps with entomopathogenic nematode against the artichoke plume moth. In: *Proceedings XIX International congress of Entomology*, Abstracts. Beijing, China, P.319.

Bathon, H. 1996. Impact of entomopathogenic nematodes on non-target hosts. *Biocontr. Sci. Technol.* 6, 421-434.

Bedding, R.A. 1981. Low cost in vitro mass production of *Neoaplectana* and *Heterorhabditis* species (Nematoda) for field control of insect pests. *Nematologica,* 27, 109-114.

Bedding, R.A. and Miller, L.A. 1981. Use of a nematode *Heterorhabditis heliothidis*, to control black vine weevil, *Otiorhynchus sulcatus*, in potted plants. *Ann. Appl. Biol.* 99, 211-216.*Biocontrol Science and Technology,* 5, 607-616.

Bird, A.F., Akhurst, R.J., 1983. The nature of the intestinal vesicle in nematodes of the family Steinernematidae. *Int. J. Parasitol.*, 13, 599–606.

Boemare, N., Akhurst, R., 1999. Genera *Photorhabdus* and *Xenorhabdus*. In: *The Prokaryotes: An Evolving Electronic Resource for the Microbiological Community*, third ed. (Eds.) Dworkin, M., Falkow, S., Rosenberg, E., Schleifer, K.-H., Stackebrandt, E. Springer, NY. release 3.4.

Boemare, N., Givaudan, A., Brehelin, M., Laumond, C., 1997. Symbiosis and pathogenicity of nematode–bacterium complexes. *Symbiosis,* 22, 21–45.

Brillard, J., Ribeiro, C., Boemare, N., Brehelin, M., Givaudan, A., 2001. Two distinct hemolytic activities in *Xenorhabdus nematophila* are active against immunocompetent insect cells. *Appl. Environ. Microbiol.* 67, 2515–2525.

Burman, M. 1982. *Neoplectana carpocapse*: toxin production by axenic insect parasitic nematodes. *Nematologica,* 28, 62-70.

Campbell, J.F. and Gaugler, R. 1993. Nictation behaviour and its ecological implications in the host search strategies of entomopathogenic nematodes (Heterorhabditidae and Steinernematidae). *Behaviour,* 126, 3-14.

Ciche, T.A., Ensign, J.C., 2003. For the insect pathogen *Photorhabdus luminescens*, which end of a nematode is out? *Appl. Environ. Microbiol.* 69, 1890–1897.

Daborn, P.J., WaterWeld, N., Silva, C.P., Au, C.P.Y., Sharma, S., Vrench-Constant, R.H., 2002. A single *Photorhabdus* gene, makes caterpillars floppy (*mcf*), allows *Escherichia coli* to persist within and kill insects. *Proc. Natl. Acad. Sci. USA,* 99, 10742–10747.

Dix, I., Burnell, A.M., Griffin, C.T., Joyce, S.A., Nugent, M.J. and Downes, M.J. 1992. The identification of biological species in the genus *Heterorhabditis* (Nematoda: Heterorhabditidae) by cross breeding second generation amphimictic adults. *Parasitology*, 104, 509-518.

Duncan, L. and McCoy, C.W. 1996. Vertical distribution in soil, persistence, and efficacy against citrus root weevil (Coleoptera: Curcilionidae) of two species of entomopathogenic nematodes (Rahbditida: Steinernematidae) in citrus. *Environ. Entomol.*, 25, 174-178.

Dunphy, G.B., Thurston, G.S., 1990. Insect immunity. In: *Entomopathogenic Nematodes in Biological Control.* (Eds.) Gaugler, R.,Kaya, H.K. CRC Press, Boca Raton, FL, pp. 301–323.

Dutky, S.R. Thompston, J.V. and Cantwell, G.E. 1964. A technique for the mass production of the DD-136 nematode. *J. Insect Pathol.* 6, 417-422.

Ehlers RU; Niemann-I; Hollmer-S; Strauch-O; Jende-D; Shanmugasundaram-M; Mehta-UK; Easwaramoorthy-SK; Burnell, A. M. 2000. A Mass production potential of the bacto-helminthic biocontrol complex *Heterorhabditis indica-Photorhabdus luminescens. Biocontrol Science and Technology.,* 5, 607-616.

Ehlers, R.U. and Hokkanen, H.M.T. 1996. Insect Biocontrol with non-endemic entomopathogenic nematodes (*Steinernema* and *Heterorhabditis* spp.): Conclusions and recommendations of a combined OECD and COST workshop on scientific and regulatory policy issues. *Biocontrol Science and Technology,* 6, 295-302.

Farmer III, J. J., J. H. Jörgensen, P. A. D. Grimont, R. J. Akhurst, G. O. Poinar Jr., E. Ageron, G. V. Pierce, J. A. Smith, G. P. Carter, K. L. Wilson and F. W. Hickman-Brenner. 1989. *Xenorhabdus luminescens* DNA hybridization group 5 from human clinical specimens. *J. Clin. Microbiol.* 27, 1594-1600.

Fischer-Le Saux, M., Viallard, V., Brunel, B., Normand, P., Boemare, N.E., 1999. Polyphasic classification of the genus *Photorhabdus* and proposal of new taxa: *P. luminescens* subsp. *luminescens* subsp. nov., *P. luminescens* subsp. *akhurstii* subsp. nov., *P. luminescens* subsp. *Laumondii* subsp. nov., *P. temperata* sp. nov., *P. temperata* subsp. *temperate* subsp. nov. and *P. asymbiotica* sp. nov. *Int. J. Syst. Bacteriol.*, 1645–1656.

Forst, S., Clarke, D., 2002. Bacteria–nematode symbioses. In: *Entomopathogenic Nematology.* (Ed.) Gaugler, R. CABI Publishing, Wallingford, UK, pp. 57–77.

Friedman, M.J. 1990. Commercial production and development. In: *Entomopathogenic nematodes in biological control* (Eds) Gaugler, R. and Kaya, H.K., Boca Raton, Fl, USA, pp 153-172.

Ganguly, S., Singh, L.K., 2000. *Steinernema thermophilum* (Rhabditida: Steinernematidae) from India. *Int. J. Nematol.* 10, 183–191.

Ganguly, S., Singh, M., Lal, M., Singh, L.K., Vyas, R.V., Patel, D.J., 2002. New record of an entomopathogenic nematode, *Steinernema riobrave* Cabanillas, Poinar and Raulston, 1994 from Gujarat. *India Indian J. Nematol.* 32, 223.

Gardner, S.L., Stock, S.P., Kaya, H.K., 1994. A new species of *Heterorhabditis* from the Hawaiian islands. *J. Parasitol.*, 80, 100–106.

Gaugler, R., 2002. *Entomopathogenic Nematology*. CABI Publishing, Wallingford, UK,

Gaugler, R., Wilson, M. and Shearer, P. 1997. Field release and environmental fate of a transgenic entomopathogenic nematode. *Biological control*, 9, 75-80.

Gaugler, R., 1988. Ecological considerations in the biological control of soil inhibitory insects with entomopathogenic nematodes. *Agric. Ecosys. Environ*, 24, 351

Georgis, R. and Manweiler, S.A. 1994. Entomopathogenic nematodes: a developing biological control technology. *Agric. Zool. Rev.* 6, 63-64.

Georgis, R., 1990. Formulation and application technology. In: *Entomopathogenic Nematodes in Biological Control.* (Eds.) Gaugler, R., Kaya, H.K., Boca Raton, FL, USA, CRC Press, pp. 173–194.

Georgis, R., Gaugler, R., 1991. Predictability in biological control using entomopathogenic nematodes. *J. Econ. Entomol.* 84, 713–720.

Georgis, R., Wojciech, W.F., Shetlar, D.J., 1989. Use of *Steinernema feltiae* in a bait for the control of black cutworms (*Agrotis ipsilon*) and tawny mole crickets (*Scapteriscus vicinus*). *Florida Entomol.* 72, 203–204.

Grenier, E., Bonifassi, E., Abad, P. and Laumond, C. 1996. Use of species specific Satellite DNAs as diagnostic probes in the identification of Steinernematidae and Heterorhabditidae entomopathogenic nematodes. *Parasitology,* 114, 497-501.

Grewal, P., Ehlers, R.-U., Shapiro-Ilan, D.I. (Eds.), 2005. *Nematodes as Biocontrol Agents*. CABI Publishing, Wallingford, UK.

Grewal, P.S., Georgis, R., 1999. *Entomopathogenic nematodes*. In: Hall, F.R., Menn, J.J. (Eds.), Methods in Biotechnology, vol. 5: Biopesticides: Use and Delivery. Humana Press, Totowa, NJ, pp. 271–299.

Grewal, P.S., Martin, W.R., Miller, R.W. and Lewis, E.E. 1997. Suppression of plant parasitic nematode populations in turfgrass by application of entomopathogenic nematodes. *Biocontrol Science and Technology,* 7, 393-399.

Grewal, P.S., Selvan, S., and Gaugler, R., 1994. Thermal adaptation of entomopathogenic nematodes: niche breadth for infection, establishment, and reproduction. *J. Therm. Biol.* 19, 245-253.

Katti, G.R., Prasad, J.S., Padmakumari, A.P. and Sankar, M. 2006. Effect of storage period on survival and infectivity of indigenous entomopathogenic nematodes of insect pests of rice. *Nematologia Mediterranea*, 34, 37-41.

Hazir, S., Stackebrandt, E., Lang, E., Schumann, P., Ehlers, R.U. .and Keskin, N., 2004. Two new subspecies of *Photorhabdus luminescens*, isolated from *Heterorhabditis bacteriophora* (Nematoda: Heterorhabditidae): *Photorhabdus luminescens* subsp *kayaii* subsp nov and *Photorhabdus luminescens* subsp *thracensis* subsp nov. Syst. *Appl. Microbiol.* 27, 36–42.

Hominick, W.M., Reid, A.P., Bohan, D.A. and Briscoe, B.R., 1996. Entomopathogenic nematodes: biodiversity, geographical distribution, and convention on biological diversity. *Biocont, Sci. Technol.*, 6, 317–331.

Hussaini, S.S., Ansari, M.A., Ahmad, W. and Subbotin, S.A., 2001. Identification of some Indian populations of *Steinernema* species (Nematoda) by RFLP analysis of ITS region of rDNA. *Int. J. Nematol.* 11, 73–76.

Jagdale, G.B., Somasekhar, N., Grewal, P.S. and Klein, M.G., 2002. Suppression of plant parasitic nematodes by application of live and dead entomopathogenic nematodes on boxwood (*Buxus* spp). *Biological control* 24, 42-49.

Joyce, S.A., Burnell, A.M. and Powers, T.O., 1994. Characterization of Heterorhabditis isolates by PCR amplification of mtDNA and rDNA genes. *J. Nematol*, 26, 260-270.

Karunakar, G., Easwaramoorthy, S. and David, H., 1999. Susceptibility of nine lepidopteron insects to *Steinernema glaseri, S. feltiae* and *Heterorhabditis indicus* infection. *International Journal of Nematology*, 9, 68-71.

Karunakar, G., Easwaramoorthy, S. and David, H., 2000. Pathogenicity of steinernematid and heterorhabditid nematodes to whitegrubs infesting sugarcane in India. *International Journal of Nematology*, 10, 19-26.

Kaya, H.K., 1990. Soil ecology. In: *Entomopathogenic Nematodes in Biological Control.* (Eds.) Gaugler, R., Kaya, H.K., Boca Raton, FL, USA, CRC Press, pp. 93–116.

Kaya, H.K. and Nelsen, C.E., 1985. Encapsulation of steinernematid and heterorhabditid nematodes with calcium alginate: a new approach for insect control and other applications. *Environ. Entomol.* 14, 572–574.

Koppenhöfer, A.M., Kaya, H.K. and Taormino, S.P., 1995. Infectivity of entomopathogenic nematodes (Rhabditida: Steinernematidae) at different soil depths and moistures. *J. Invertebr. Pathol.* 65, 193–199.

Lindgren, J.E., Valero, K.A. and MacKey, B.E. 1993. Simple in vivo production and storage methods for *Steinernema carpocapsae* infective juveniles. *J. Nematol.* 25, 193-197.

Liu, J. and Berry, R.E., 1996. *Heterorhabditis marelatus* n. sp. (Rhabditida: Heterorhabditidae) from Oregon. J. Invertebr. Pathol. 67, 48–54.

Martens, E.C., Heungens, K. and Goodrich-Blair, H., 2003. Early colonization events in the mutualistic association between *Steinernema carpocapsae* nematodes and *Xenorhabdus nematophila* bacteria. J. Bacteriol. 185, 3147–3154.

Nguyen, K.B. and Smart, G.C., 1996. Identification of entomopathogenic nematodes in the Steinernematidae and Heterorhabditidae (Nematoda: Rhabditida). *J. Nematol,* 28, 286-300

Padmakumari, A.P., Prasad, J.S., Katti, G.R. and Sankar, M. 2007. Efficacy of Rhabditis (Oscheius) sp. against rice yellow stem borer, *Scirpophaga incertulas* (Walker). *Indian Journal of Plant Protection,* 35, 255-258.

Parkman, J.P. and Frank, J.H., 1993. Use of a sound trap to inoculate *Steinernema scapterisci* (Rhabditida: Steinernematidae) into pest mole cricket populations (Orthoptera: Gryllotalpidae). *FL. Entomol.* 76, 75-82.

Poinar Jr., G.O. and Kozodoi, E.M., 1988. *Neoaplectana glaseri and N. anomali*: sibling species or parallelism, *Rev. Nematol.* 11, 13-19.

Poinar Jr., G.O., 1979. Nematodes for the biological control of insects., Boca Raton, FL, USA,CRC Press. 227 pp.

Poinar Jr., G.O., 1990. Biology and taxonomy of Steinernematidae and Heterorhabditidae. In: *Entomopathogenic Nematodes in Biological Control.* (Eds.) Gaugler, R., Kaya, H.K., CRC Press, Boca Raton, FL, pp. 23–62.

Poinar Jr., G.O., Karunakar, G.K. and David, H., 1992. *Heterorhabditis indicus* n. sp. (Rhabditida, Nematoda) from India: separation of *Heterorhabditis* spp. by infective juveniles. *Fundam. Appl. Nematol.* 15, 467–472.

Prasad, J.S., Gururaj Katti, Padmakumari, A.P. and Sankar, M. 2006. Bioefficacy of indigenous entomopathogenic nematodes against leaf folder, *Cnaphalocrocis medinalis*

Guenee in rice. *Abstracts of International Rice Congress*. October 9-13, 2006, New Delhi

Purcell, M., Johnson, M.W. LeBeck, L.M. and Hara, A.H. 1992. Biological control of *Helicoverpa zea* with *Steinernema carpocapsae*(Rhabditida: Steinernematidae) in corn used as trap crop. *Environ Entomol.* 21, 1441-1447.

Rahman, P.F., Sharma, S.B. and Wightman, J.F., 2000. A review of insect-parasitic nematodes research in India: 1927–1997. *Int. J. Pest Manag.* 1, 19–28.

Rao, V.P. and Manjunath, T.M., 1966. DD-136 nematode that can kill many insect pests. *Indian Farming,* 16, 43–44.

Richardson, P.N., 1990. Uses for parasitic nematodes in insect control strategies in protected crops. *Aspects Appl. Biol.* 24, 205-210.

Sankaranarayanan, C., Somasekhar, N. and Singaravelu, B., 2006. Biocontrol potential of entomopathogenic nematodes *Heterorhabditis* and *Steinernema* against pupae and adults of white grub *Holotrichia serrata* F. *Sugar Tech*, 8, 268-271.

Sankaranarayanan, C., Easwaramoorthy, S. and Somasekhar, N., 2003. Infectivity of entomopathogenic Nematodes *Heterorhabditis* and *Steinernema* spp. to Sugarcane Shoot Borer, *Chilo infuscatellus* Snellen at Two Temperatures. *Sugar Tech* 5, 167-171.

Shapiro-Ilan, D.I., 2004. Entomopathogenic nematodes and insect management. In: E*ncyclopedia of Entomology,* vol.1 (Ed.), Capinera, J.L., Kluwer Academic Publishing, Dordrecht, The Netherlands, pp. 781– 784.

Shapiro-Ilan, D.I., Gouge, D.H. and Koppenhöfer, A.M., 2002. Factors affecting commercial success: case studies in cotton, turf and citrus. In: *Entomopathogenic Nematology.* (Ed.) Gaugler, R., CABI Publishing, Wallingford, UK, pp. 333–356.

Silver, S.C., Dunlop, D.B., and Grove, D.I., 1995. Granular formulation of biological entities with improved storage stability. *Int. Patent. WO 95/05077.*

Sivakumar, C.V., Jayaraj, S. and Subramanian, S., 1989. Observations on Indian population of the entomopathogenic nematode *Heterorhabditis bacteriophora. J. Biol. Control,* 2, 112–113.

Smitley, D.R., Warner, F.W. and Bird, G.W., 1992. Influence of irrigation and *Heterorhabdtis bacteriophora* on plant parasitic nematodes in turf. *Journal of Nematology,* 24, 637-641.

Somasekhar, N., Grewal, P.S., and Klein, M.G., 2002a. Genetic diversity in stress tolerance and fitness among natural populations of *Steinernema carpocapsae. Biological Control* 23, 303-310.

Somasekhar, N., Grewal. P.S., DeNardo, E.AB., and Stinner, B.R., 2002b. Non-target effects of entomopathogenic nematodes on soil nematode community. *Journal of Applied Ecolog,* 39, 735-744.

Stack, C.M., Easwaramoorthy, S., Mehta, U.K., Downes, M.J., Griffin, C.T. and Burnell, A.M., 2000. Molecular characterization of *Heterorhabditis indica* isolates from India, Kenya, Indonesia and Cuba. *Nematology.* 2, 477-487.

Wang, Y. and Gaugler, R., 1998. Host and penetration site location by entomopathogenic nematodes against Japanese beetle larvae. *J. Invertebr. Pathol.,* 72, 313–318.

Womersley, C.Z., 1993. Factors affecting physiological fitness and modes of survival employed by dauer juveniles and their relationship to pathogenicity. In: *Nematodes for the Biological Control of Insects.* (Eds.) Bedding, R., Akhurst, R., Kaya, H. CSIRO Press, East Melbourne, Australia, pp. 79–88.

Woodring, J.L. and Kaya, H.K., 1998. *Steinernematid and Heterorhabditid nematodes: A handbook of biology and techniques.* Southern Cooperative Series Bulletin 331, Arkansas Agricultural experiment Station, Fayettesville, AR. USA.

Wouts, W.M., 1991. *Steinernema* (Neoplectana) and *Heterorhabditis* species. In: *Plant and Insect Nematodes*, (Ed.) W.R. Nickle, Marcel Dekker, NY, pp. 855–897.

Yukawa, T. and Pitt, J., 1985. Nematode storage and transport. *Int. Patent WO85/03412.*

Evaluation of Mycohit (*Hirsutella thompsonii* Fisher) against eriophyid mite *Aceria guerreronis* Keifer in coconut

S. JEYARANI,* E.V. BHASKARAN,* J. KAVITHA,* K. RAMARAJU,*
P. SREERAMA KUMAR** AND R.J. RABINDRA**

Field trials were conducted to evaluate the efficacy of Mycohit (a commercial formulation of *Hirsutella thompsonii* Fisher) against coconut eriophyid mite. In the preliminary trial conducted at Vedapatti, the populations of eriophyid mites in Mycohit treated palms were 31.92 and 26.72 mites/4 sq.mm in 4th and 5th bunch, respectively. The Mean Grade Index (MGI) recorded at the time of harvest in tagged bunch 1 was found to be 2.3 in Mycohit treated palms and 2.1 in untreated check and in bunch II, there was no variation in both treatment and control (2.2). In the second trial conducted at Chitraichavadi, the pre-treatment population of mites ranged from 22.36 to 27.42 and 22.65 to 28.08 mites/4sq.mm in 4th and 5th bunches, respectively. Among the treatments, Mycohit (1%) spray recorded the lowest mean mite population (15.15 and 4.04 mites/4 sq. mm.) followed by Mycohit + adjuvant 1 (glycerol) (16.13 and 5.57 mites/4 sq. mm.) as against control (38.81 and 23.67 mites/4 sq. mm) in 4th and 5th bunches respectively, after three rounds of spray. The MGI was the lowest in Mycohit + adjuvant 2 (2.3) treated palms followed by Mycohit + adjuvant 1 (2.4), Mycohit (1 %) alone (2.6) and Mycohit + adjuvant 3 (2.7) as against the control (2.9) in the tagged bunch 1. The same trend was also observed in the tagged bunch 2.

INTRODUCTION

In recent years, the non-insect pest, coconut eriophyid mite, *Aceria guerreronis* Keifer (Acari: Eriophyidae) has attained the status of a major pest. The estimated yield loss of copra due to the attack of coconut eriophyid mite in Tamil Nadu was reported to be 27.5 per cent (Ramaraju *et al.*, 2000). Several chemical and biological methods have been evaluated for the management of coconut eriophyid mite, but, chemical control has not been viewed as a good long term control method because of potential problems of residues in coconuts, hazards to workers, environment and the potential development of resistant strains of mite (Moore, 2000). The environmental impact and safety of chemical applications caused public concern forcing us to seek research on safer, alternative and sustainable methods for mite management, emphasizing more on biological control. Among the biological control agents, Acaropathogenic fungi which invade their host through the cuticle, is well suited against eriophyid mites since they are mobile, soft-bodied organisms with no cuticular barrier. In addition, the habitat of these mites which is mostly dark and moist may favour the germination of infective propagules of fungal pathogens. Among the acaro-pathogenic fungal genera infecting coconut mite, *Hirsutella thompsonii* Fisher is reported to be the best regulator of pest and was first described by Fischer (1950). *H. thompsonii* was reported as the causative agent of epizootics in the citrus rust mite and its pathogenicity was confirmed by McCoy and Kanavel (1969). In India, on-site

* Department of Agricultural Entomology, Tamil Nadu Agricultural Entomology, Coimbatore - 641 003
** Project Directorate of Biological Control, ICAR, Bangalore – 580 024

observation in both fallen and harvested nuts indicated the association of *H. thompsonii* with the mite in 10 - 25% of the samples (Kumar, 2001). Hence, the present investigation was made to bring out the scope of using Mycohit, a commercial formulation of *H. thompsonii,* as a successful biocontrol agent in combating the menace of coconut eriophyid mite.

Materials and Methods

Field trials were conducted to evaluate Mycohit, a commercial formulation of *H. thomsonii* supplied by the Project Directorate of Biological Control (PDBC), Bangalore through frequent sprays against coconut eriophyid mite in 10 year old local hybrid coconut palms. In the preliminary trial, Mycohit 1 % (20 g/2 lit/tree) with out any adjuvant was evaluated against control with twenty replications per treatment (@ one tree per replication) at Vedapatti village near Coimbatore. In the subsequent field trial, the efficacy of Mycohit along with adjuvant was evaluated at Chitraichavadi village near Coimbatore. The treatments *viz*., T_1- Mycohit + adjuvant 1 (Glycerol @ 10 ml/2 lit/tree), T_2- Mycohit + adjuvant 2 (Yeast extract powder @ 10 g/2 lit/tree), T_3- Mycohit + adjuvant 3 (Malt extract powder @ 10 g/2 lit/tree), T_4- Mycohit alone (1 %), T_5- Triazophos @ 10 ml/2 lit. and T_6- untreated check, were evaluated for its efficacy with twelve replications per treatment (@ one tree per replication). Both the experiments were laid out in a randomized block design. Three sprays were given at 15 days interval.

In both the experiments, the number of live mites (both nymphs and adults) were assessed in an area of 4 sq. mm (2 x 2 mm) on the inner side of the inner most bracts *viz*., 4th, 5th and 6th bract and on the nut surface (at three places) in every sample. The observations were made in the region wherein maximum populations were noticed. The mite population was assessed using a binocular stereo zoom microscope (Nikon SMZ 800 Model). The bunch next to the fully opened inflorescence was counted as 1st bunch and the preceding older bunches were numbered as 2, 3…n. The 2nd and 3rd bunches were tagged for further observations. Pre–treatment population of mite was recorded in 4th and 5th bunches @ 1 nut/bunch. Post treatment population count was recorded on 45th day after three rounds of spraying from the tagged bunches @ 1 nut/bunch. The 3rd button from the bottom of the bunch was sampled each time. The damage grading of the tagged bunches were recorded at the time of harvest following standard scale.

Grade		Damage/ symptom category
Grade 1	-	Nuts with no damage
Grade 2	-	Nuts with 1-10% of mite damage
Grade 3	-	Nuts with 11-25% of mite damage
Grade 4	-	Nuts with 26-50% of mite damage
Grade 5	-	Nuts with > 50% of mite damage with reduced size and greatly distorted

Results and Discussion

In the preliminary trial conducted at Vedapatti village, pre-treatment observations on live mite population in 4th and 5th bunch in Mycohit and control were 33.04, 28.44 and 35.44, 31.48 mites/4 sq.mm. The mean damage grade index (MGI) recorded in the harvested nuts prior to treatment were 4.0 and 4.2 respectively. Post treatment observation after 45 days of treatment in Mycohit treated palms were 31.92 and 26.72 mites/4 sq.mm and in control trees, 32.48 and 27.92 mites/4 sq. mm, respectively, in tagged bunches I and II (Table 1). Damage grading done at the time of harvest in tagged bunch I in Mycohit treated and control were 2.3 and 2.1, respectively (Table 2). In the tagged bunch II the overall MGI was 2.2 in both the treatments and control and were on par with each other.

The mean per cent reduction in MGI over pre treatment, in Mycohit and control palms was found to be 50.0 and 42.5 per cent in tagged bunch I. In tagged bunch II, the per cent reduction was found to be 47.6 and 45.0 per cent. Naseema Beevi *et al.* (2003) reported that the mortality of mites due to *H. thompsonii* was 20 to 40 per cent in Kerala. However, in our preliminary trial, the mite population in both treated and control palms were on par with each other and this may be due to the variation in the climatic conditions both in Kerala and Tamil Nadu.

Table 1. Efficacy of Mycohit against coconut eriophyid mite at Vedapatti village, Coimbatore District

Treatments	Pretreatment (Live mites no./4 sq. mm)		Post treatment (Live mites no./4 sq.mm) ± SE	
	4th bunch	5th bunch	Tagged bunch I	Tagged bunch II
Mycohit	33.04	28.44	31.92 ± 0.38	26.72 ± .039
Control	35.44	31.48	32.48 ± 0.47	27.92 ± 0.37

Means followed by the common letter(s) are not significantly different at 5 % level by LSD

Table 2. Efficacy of Mycohit on damage grade index of nuts at the time of harvest at Vedapatti village, Coimbatore District

Treatments	Pre-treatment MGI	Mean Grade Index (MGI)			
		Tagged bunch 1*	% red. over pre-treatment MGI	Tagged bunch II*	% red. over pre-treatment MGI
Mycohit	4.2	2.3	50.0	2.2	47.6
Control	4.0	2.1	42.5	2.2	45.0

* The means are not statistically significant

In the subsequent field trial conducted at Chitraichavadi village, the pre-treatment population of mites ranged from 22.24 to 27.42, 22.65 to 28.08 mites/4 sq.mm. The MGI recorded in the harvested nuts prior to treatment ranged from 3.8 to 4.1. Among the treatments, Mycohit (1 %) spray recorded the highest percent reduction over control (60.96 and 82.93 %) with a mean mite population of 15.15 and 4.04 mites/4 sq. mm. followed by Mycohit + adjuvant 1, Mycohit + adjuvant 2 and were on par with each after 45 days of spraying, respectively in tagged bunches I and II (Table 3). The MGI revealed similar trend as that of preliminary trial. It was the lowest in Mycohit + adjuvant 2 (2.3 and 2.0) followed by Mycohit + adjuvant 1 (2.4 and 2.6) and Mycohit (1 %) alone (2.6 and 2.6) as against control (2.9 and 3.3) at harvest (Table 4). The MGI recorded at the time of harvest in both the experiments were found to be on par with control. This may be due to the unpredictable population dynamics of eriophyid mites in relation to weather parameters (Bhaskaran and Ramaraju, 2007). However, the mean per cent reduction in MGI over pre treatment in Mycohit + adjuvant 2 treated palms was significantly high (39.4 and 47.3 %) in tagged bunches I and II. Chinnamadegowda *et al.* (2003) reported that the combination of Mycohit with monocrotophos or econeem was better than Mycohit alone. The present investigation also indicates that the efficacy of Mycohit may be enhanced by adjuvants. However, the choice of adjuvant needs in-depth study. Apart from this a high relative humidity (higher than 90%) is essential to favour the infection of fungi on mites (Brandemburg and Kennedy, 1982; McCoy, 1978). The present investigations were carried out during July to February 2004 and December 2006 to September 2007. The Relative humidity prevailed in Coimbatore during this trial period ranged between 75 and 89 per cent. Multilocation trials conducted with Mycohit in different states of South India also revealed that the reduction in damage is not significant under field conditions (Nair *et al.,* 2005). In the environment, various factors interact which determine the efficacy of the fungus. Hence, the unpredictable population dynamics of mite coupled with non-availability of suitable climate (high relative humidity) for the establishment of the fungi may be the probable reason for varying results. In addition, further studies are needed to find out the efficacy of *H. thompsonii* by repeated sprayings under field conditions.

Table 3. Efficacy of Mycohit against coconut eriophyid mite at Chitrachavadi village, Coimbatore District

Treatments	Pre- treatment (Live mites/4sq.mm)		Post- treatment (45 DAT) (Live mites /4sq.mm)			
	4th bunch	5th bunch	Tagged bunch I	% red. over control	Tagged bunch II	% red. over control
Mycohit + adjuvant 1	26.68	27.03	16.13[a]	58.44	5.57[a]	76.47
Mycohit + adjuvant 2	22.36	24.72	18.07[a]	53.44	7.06[a]	70.17
Mycohit + adjuvant 3	24.28	28.08	18.68[a]	51.87	9.94[a]	58.01
Mycohit alone (1 %)	27.42	22.65	15.15[a]	60.96	4.04[a]	82.93
Triazophos @ 5 ml/lit.	23.55	26.00	14.52[a]	62.59	1.33[a]	94.38
Untreated control	22.24	26.29	38.81[b]	--	23.67[b]	--

Means followed by the common letter(s) are not significantly different at 5 % level by LSD

Table 4. Efficacy of Mycohit on damage grade index of nuts at the time of harvest at Chitrachavadi village, Coimbatore District

Treatments	Pre-treatment MGI	Mean Grade Index (MGI)			
		Tagged bunch 1*	% red. over pre-treatment MGI	Tagged bunch II*	% red. over pre-treatment MGI
Mycohit + adjuvant 1	4.0	2.4	40.0[b]	2.6	35.0[b]
Mycohit + adjuvant 2	3.8	2.3	39.4[b]	2.0	47.3[a]
Mycohit + adjuvant 3	3.9	2.7	30.7[bc]	2.8	28.2[b]
Mycohit alone (1 %)	4.1	2.6	36.5[b]	2.6	36.5[b]
Triazophos @ 5 ml/lit.	4.0	1.7	57.5[a]	1.8	55.0[a]
Untreated control	3.8	2.9	23.6[c]	3.3	13.1[c]

* The means are not statistically significant; Means followed by the common letter(s) are not significantly different at 5 % level by LSD

References

Bhaskaran, E.V. and Ramaraju, K., 2007. Population dynamics of coconut eriophyid mite, *Aceria guerreronis* Keifer. *Journal of Acarology*, 16(1&2), 72-75.

Brandemburg, R.L and Kennedy, G.G., 1982. Relationship of Neozygites floridana (Entomophthorales: Entomophthoraceae) to two-spotted spider mite (Acari: Tetranychidae) populations in corn field. *Journal of Economic Entomology*, 75, 691-694.

Chinnamadegowda, C. Guruprasad, H., Jayappa, J., Onkarappa, S. and Mallik, B., 2003. Comparative performance of chemicals, plant products and bioagents against coconut eriophyid mite. In: *Proceedings of National*

Symposium on Frontier Areas of Entomological Research, Division of Entomology, Indian Agricultural Research Insititute, New Delhi, 5-7 November, 2003, pp. 277-278.

Fisher, F.E., 1950. Two new species of *Hirsutella* Patouillard. *Mycologia*, 42, 13–16.

Kumar, P.S., Singh, S.P. and Gopal, T.S., 2001. Natural regulation of the coconut eriophyid mite, *Aceria guerreronis* Keifer, by the acarine parasite, *Hirsutella thompsonii* Fisher, in certain districts of Karnataka and Tamil Nadu States in India. *Journal of Biological Control*, 15(2), 151-156.

McCoy, C.W. and Kanavel, R.F., 1969. Isolation of *Hirsutella thompsonii* from the citrus rust mite, *Phyllocoptruta oleivora* and its cultivation on various synthetic media. *Journal of Invertebrate Pathology*, 14, 386-390.

McCoy,C.W., 1978. Entomopathogens in arthropod pest control programs for citrus. In. Future Strategies in Pest management Systems. G.E.Allen, C.N.Igmoffo and R.P.Jaques .pp 211-224 NSF-USDA. University of Florida, Gainesville.

Moore, D., 2000. Non-chemical control of *Aceria guerreronis* on coconuts. *Biocontrol News and Information,* 21, 83-87.

Nair, C.P.R., Rajan, P. and Chandrika Mohan, 2005. Coconut Eriophyid mite *Aceria guerreronis* Keifer – An overview. *Indian Journal of Plant Protection*, 33, 1-10.

Naseema Beevi, S., Thomas Biju Mathew, Hebsy Bai and Saradamma, K., 2003. Status of Eriophyid mite in Kerala – Resume of work done. In: Coconut Eriophyid Mite – Issues and Strategies. In: *Proceedings of the International workshop on coconut mite* held at Bangalore. (Eds.) H.P. Singh and P. Rethinam. Coconut Development Board, Kochi, pp. 64-75.

Ramaraju, K., Natarajan, K., Sundarababu, P.C. and Palaniswamy, S., 2000. Studies on coconut eriophyid mite *A.guerreronis* K. in Tamil Nadu, India. In: *proceedings of the international workshop on coconut eriophyid mite,* Coconut Research Institute, Sri Lanka, pp. 13-31.

Thomas Mathew, M., 2004. Coconut Industry in India: An overview. *Indian Coconut Journal*, 35(7), 3-15.

Pathogenicity of *Pieris brassicae* granulosis virus infecting the caterpillars of *Pieris brassicae* (Lepidoptera : Pieridae)

W. Sangeeta Devi,* M. Ingobi Singh,* K. Dhanapati,*
R.R. Pandey* and R. Varatharajan*

Bioassay of *Pieris brassicae* granulosis virus (*Pb*GV) was carried out against III instar larvae of *P. brassicae* and the results indicated that LC_{50} was 7.7×10^4 OBs/ml and LT_{50} was 125.72 hours. To augment the virus through *in vivo* method, rearing 200 larvae per culture in a wooden cage (30×30×30 cm) was found to be optimum and to rear the virus inoculated larvae, cabbage and cauliflower leaves were found to be better than the mustard foliage, since the mean percentage mortality was 78, 75 and 43% respectively on 6.5 days after post infection.

Introduction

The cabbage white, *Pieris brassicae* (L.) [Lepidoptera: Pieridae] is an oligophagous pest. It is known to infest 83 species of food plants belonging to the families such as Cruciferae, Torpaeolaceae, Capparaceae, Resedaceae and Papilionaceae (Feltwell, 1982). It is cosmopolitan in distribution and found wherever cruciferous vegetables are grown (Atwal and Dhaliwal, 1999). In India, it is distributed in the Himalayas from Chittal to Bhuttan up to 3030 m MSL (Bhalla *et al.*, 1997). The pest causes heavy damage to leaves and inflorescence of rapseed and cruciferous vegetables in the entire Himalayan ranges including foothills and north eastern states of India (Sachan and Gangawar, 1990).

Management of this pest using parasitoids is not a viable option under the agro climatic conditions of Manipur, since they parasitise only up to 40% (Lata, 2007). They also attack the pest invariably at the later stage of the crop and have little impact on the outbreak of pest population. Control of this pest using chemical insecticide is discouraged because it entailed recurring cost and long term residual effect to the environment. Therefore, to check the population, it becomes imperative to use ecofriendly biocontrol agents like microbials. In this context, baculovirus can be considered as an ideal option due to many advantages (Rabindra, 2003). Baculovirus can be applied at the early stage of infestation and causes significant mortality of the larva. They have a very restricted host range and pose no threat to beneficial organisms unlike other microbials like *Bt*. Occurrence of granulosis virus of *Pieris brassicae* (*Pb*GV) was reported from India by Ingobi Singh *et al.* (2006). However detailed studies about *Pb*GV have not been made so far. Therefore, in the present investigation, an attempt has been made to know the pathogenicity of *Pb*GV on *Pieris brassicae.*

* Department of Life Sciences, Manipur University, Canchipur-795 003

Materials and Methods

Insect rearing and virus inoculation

Colonies of *P. brassicae* (II instar larvae) were collected from pesticide free cabbage field. The larvae were collected along with the leaves so that there is minimum disturbance to the larvae. Virus suspension of 5×10^5 OBs/ml was prepared as per the method adopted by Narayanan, 2003. The prepared viral suspension was applied on cabbage leaves. Subsequently the leaves were air dried and the petioles along with cotton were inserted in a conical flask containing water. The caterpillars (III instar) which were starved for 2 hours were released on the air dried virus contaminated foliage and this culture was kept inside a wooden cage (30×30×30 cm). On feeding the contaminated leaves, fresh foliage was provided daily till the death of all the larvae. The culture was cleaned daily and the data on larval mortality was recorded.

Purification of OBs

Purification of occlusion bodies (OBs) from the GV infected dead larvae was done following the methods of Sudhakar *et al.* (1997) and Narayanan (2003). The virus infected larvae showed the symptoms such as gradual turning of whitish brown body colour, loss of appetite, loose integument, oozing of haemolymph and eventually hang in an inverted 'V' shaped fashion by their middle pair of legs. The infected dead larvae were collected and macerated in distilled water and filtered through fine muslin cloth so as to remove debris and centrifuged at 500 rpm for 5 mins. The supernatant was centrifuged again at 15000 rpm for 25 mins to pellet the virus. Further purification was done by washing the pellet with 0.1% SDS solution. This exercise was repeated 3 or 4 times until it formed a clear white coloured pellet. The pellet was suspended in sterile distilled water and stored in an amber coloured bottle at 4°C. This method gives fairly good quality of OBs (Gopalakrishnan, 2001).

Bioassay (LC_{50} and LT_{50})

Different concentrations of the viral suspension ranging from 10^3 to 10^7 OBs/ml were prepared. Each concentration was applied individually on the cabbage leaf by following the leaf disc method (Rabindra *et al.*, 1997). Fresh and clean cabbage leaf was cut into circular disc (6 cm diam.). The leaf disc was sprayed individually with 5µl of the viral suspension of different concentrations and then air dried. Third instar larvae of uniform age, size and starved for 2 hours were released on the treated leaves individually in a petridish of 9 cm diameter. On feeding the virus treated leaf, all the larvae that were inoculated with the same concentration were pooled together and reared in a netted wooden cage (30×30×30 cm) with 30 individuals per replication. Fresh foliage was provided daily till death. Data on larval mortality was collected individually for each treatment and the recorded data were processed for probit analysis namely concentration factor versus mortality rate (Finney, 1977).

Similar exercise was also carried out for LT_{50}, wherein the *Pb*GV viral suspension @ 7.7×10^4 OBs/ml [=LC_{50} of III instar] was applied on the leaf disc @ 5µl per disc. Totally 150 larvae [@ 30 larvae/replication] were bioassayed. The data on larval mortality were plotted against time factor and later on processed for Probit analysis (Finney, 1977).

Larval Density

To study the optimum larval density for rearing the larvae, the leaves were sprayed with viral suspension @ 10^6 OBs/ml. The petioles of the contaminated leaves were taken in 150 ml conical flask containing tap water. The virus contaminated leaves were shade dried and kept within the cage. In each cage different density (50, 100, 150, 250 larvae) of III instar larvae were released after starving them for 2 hours. The cage size and rearing conditions were kept constant but only the larval density varied between the treatments i.e. there were totally 5 treatments and each treatment was replicated 3 times. Larval mortality was noted daily in each cage.

Influence of plant host on larval mortality

To study the influence of plant host on larval mortality, virus inoculated III instar larvae were reared on the leaves of cabbage, cauliflower and mustard @50 larvae per plant host with 3 replications each. The caterpillars were reared till death or pupation. The larval mortality was recorded and compared between the host plants.

Result and Discussion

Bioassay (LC_{50} and LT_{50})

Experiment on concentration versus mortality rate of the III instar larvae of *P. brassicae* indicated that the lethal concentration to kill 50% of the population [LC_{50}] was 7.7×10^4 OBs/ml and the fiducial limit ranged from 7.3×10^4 - 8.06×10^4 OBs/ml (Table 1). Earlier studies on the efficacy of *Pb*GV showed that application of *Pb*GV @10^6 OBs/ml against *P. brassicae* reduced upto 72-95% pest population in USSR. Similarly application of *Agrotis segetum* granulosis virus @ 10^7 OBs/ml effectively checked the population of *A. segetum* upto 65-85%. Almost the same trend was observed for *Cydia pomonella* granulosis virus where application @ 9×10^6 – 2.9×10^{11} OBs/ml could reduce 74-88% pest population (Lipa, 1998). Again in the Indian context, use of *Chilo infuscatellus* granulosis virus @1×10^7 -1×10^9 OBs/ml reduced the incidence of *C. infuscatellus* below the economic injury level of 20% (Easwaramoorthy, 1998). While comparing the above examples, it revealed that although there appeared to be slight variation in terms of concentration of granulosis virus of different species of insects, their dose and relative efficacy seem to be close to each other and such minor variation could be due to certain factors such as climatic condition, geographical location, age and stage of test insect etc. Further in the present study, inoculation of *Pb*GV with 7.7×10^4 OBs/ml against III instar larvae yielded an average of 5.5×10^9 OBs per larva.

Again experiment on time factor versus mortality rate of III instar larvae of *P. brassicae* indicated that the lethal time to kill 50 percent of their population was 125.72 hrs. (5.24 days). The fiducial limit ranged from124.89-127.30 hrs (Table 1). However LT_{50} value was found to vary from instar to instar. For instance, LT_{50} value of II and III instar larvae was 124 and 187 h, espectively. Thus, the present experiment has unequivocally showed that the efficacy of *Pb*GV was appreciably better at the early stages than the later stages of caterpillar indicating the view that spraying the OBs at the early stages of caterpillars would lead to good control of the pest especially under the field condition. Similar range of LT_{50} value was observed in China with 5.5-6.7 days when *Pr*GV was sprayed against the larvae of *P. rapae* (Entwistle, 1998). Further studies on the LT_{50} value of granulosis virus of other insect species also revealed similar trends of LT_{50} value. For example, LT_{50} values of *Px*GV were 4.9-5.4 days when it was inoculated to the different larval stages of *Plutella xylostella* (Jones *et al.*, 1998).

Table 1. Bioassay of PbGV on III instar larvae of *P. brassicae.*

	χ^2	Regression Equation	LC_{50}/LT_{50}	Fiducial limit
LC_{50}	1.99*	Y=12.87x-57.91	7.7×10^4 OBs/ml	7.3×10^4-8.06×10^4 OBS/ml
LT_{50}	3.96*	Y=30.15x-58.29	125.72hrs	124.89-127.30hrs

*Data showed homogeneity, Indoor rearing temperature = 22-29°C; Inoculum dose = 7.7×10^4 OBs/ml (=LC_{50} value) for LT_{50} expt; 5 treatments each replicated 3 times with 30 larvae each (LC_{50} expt.); 5 replications each with 30 larvae (LT_{50} expt.)

Optimum Larval density for rearing

When virus inoculated larvae were reared in individual groups of 50, 100, 150, 200 & 250 per culture, the mean percentage death due to viral infection was 31, 40, 42, 63 and 66 percent respectively (Table 2). This result indicated that rearing 200 larvae per culture were found more suitable than 250 larvae. It is because higher the larval density greater the problem for rearing due to congested condition, difficulties in cleaning and harvesting the infected dead larvae etc. But rearing the larvae with optimum density enables to maintain the culture in a compact manner that helps in spreading viral infection. Therefore, this system is advantageous for large scale viral production by *in vivo* method. Moreover rearing the larvae inside the netted wooden cage helps in aeration and movement of larvae, which promote healthy growth of larvae particularly during early stage of infection.

Table 2. Relation between larval density and mortality due to viral infection.

Treatment (No. of larvae/ Culture	Average mortality %
50	31c
100	40bc
150	42b
200	63a
250	66a
CD at 5% level (ANOVA)	9.09

Figures followed by the same alphabet are not different from each other at 5% level (ANOVA); Each value is mean of 4 replication & inoculum dose $=1\times10^{6}$ OBs/ml

Host foliage Versus larval mortality

Test on host foliage versus larval mortality revealed that the larval mortality at 6.5 day after treatment (DAT) was found to be more on cabbage with 78% followed by cauliflower with 75%. But a moderate larval mortality of 43% was obtained when reared on mustard leaf (Table 2). Although cent percent mortality was noticed at the end of the trial, variation in percentage of death was observed on 6th day after virus treatment. This difference in virus induced larval mortality on various host foliage can be attributed to variation in phenol content of the leaves of different plant species. It is known that higher the phenol content lesser the performance of insect (Ananthakrishnan, 1992). Earlier observations revealed the presence of more quantity of phenol in mustard foliage than that of cabbage leaf. Extrapolating this view in the present context, the low mortality of larvae on mustard leaf could be due to more quantity of phenol in mustard foliage. Further, presence of thioglucosides in mustard plant acts as a phagodeterrent, thereby reduces feeding propensity of the insect (Thorsteinson, 1953). Similar variation on susceptibility to virus and larval mortality due to food plant was reported in other insects such as *Amsacta albistriga* (Jayaraj *et al.*, 1976), *Bombyx mori* (Sosa Gomez *et al.*, 1991), *Helicoverpa armigera* (Rabindra *et al.*, 1994), *Spodoptera litura* (Muralibaskaran *et al.*, 1996) and *Spilarctia obliqua* (Singh and Varatharajan, 2001).

Table 3. Influence of host plant on larval mortality due to viral infection.

Treatment	Average mortality %
Mustard	43a
Cauliflower	75b
Cabbage	78b
CD at 1% level (ANOVA)	21.71

Figures followed by the same alphabet are not significantly different at 5% level (ANOVA); DAT = 6.5 days (156 hrs); Test insect = III instar *P. brassicae* larvae; Concentration = 1×10^6 OBs/ml.

In the present study, rearing the larvae on cabbage leaf has not only resulted in quick mortality but the cabbage leaf could also withstand in culture for 2 or 3 days without withering/drying and no need to replace them daily like that of mustard leaf. Therefore analyzing from different angles, it is clearly evident that cabbage leaf is better for rearing the virus inoculated larvae for mass production of *Pb*GV under *in vivo* system.

Acknowledgement

The authors thank DBT, New Delhi for financial support and Head, Department of Life Sciences for encouragement.

References

Ananthakrishnan, T.N., 1992. *Dimensions of insect plant Interaction;* Oxford & IBH publishing co. Pvt. Ltd. New Delhi, 184 pp.

Atwal, A.S. and Dhaliwal, G.S. 1999. Pest of vegetables; in *Agricultural pests of South Asia and their management.* Kalyani Publishers, New Delhi, pp. 253-273.

Bhalla, S., Kapur, M.L., Verma, B.L. and Singh, R., 1997. An unusual abundance of cabbage butterfly *Pieris brassicae* (Linnaeus) on various *Brassica* species in the environs of Delhi for locating sources of resistance among different cultivars; *J. Ent. Res.*, 21, 147-151.

Entwistle, P.F., 1998. Peoples Republic of China; in *Insect virus and pest management* (eds) Frances R Hunter Fujita, Philip F Entwistle, Hugh F Evans, Norman E Crook (John Wiley & Sons, New York) 258-268 pp.

Easwaramoorthy, S., 1998. Indian subcontinent; in *Insect virus and pest management* (eds) Frances R Hunter-Fujita, Philip F Entwistle, Hugh F Evans, Norman E Crook (John Wiley & Sons, New York) 232-243 pp.

Feltwell, J., 1982. *Large butterfly, the biology, biochemistry and physiology of Pieris brassicae* (Linnaeus). Dr. W Junk Publishers, The Hague, Bostan-London 533pp.

Finney, D., 1977. *Probit Analyis,* Cambridge University Press London 333p.

Gopalakrishnan, C., 2001. Mass production and utilization of microbial agents with special reference to insect pathogen; in *Microbials in insect pest management* (eds) Ignacimuthu S and Sen A (Oxford and IBH publishing Co Pvt. Ltd Kolkata) 19-29 pp.

Ingobi Singh, M., Sangeeta, W. and Varatharajan R., 2006. On the occurrence of viral infection in *Pieris brassicae; Ann. Pl. protect. Sci.,.* 14, 241-242.

Jayaraj, S., Sundaramurthy, V.T. and Mahadevan, N.R., 1976. Laboratory studies on the nuclear polyhedrosis of red hairy caterpillar of groundnut *Amsacta albistriga (Walker); Madras Agri. J.*, 63, 567-569.

Jones, A.K., Zelazny, B., Ketunuti, U., Cherry, A. and Grzywacy, D., 1998. South east Asia and the Western Pacific; in *Insect virus and pest management* (eds) Frances R Hunter-Fujita, Philip F Entwistle, Hugh F Evans, Norman E Crook (Hohn Wiley & Son, New York) 244-257pp.

Lata Devi O., 2007. *Parasitoids of certain Lepidopteran insects from valley region of Manipur* Ph. D. Dissertation, Manipur University.

Lipa, J.J., 1998. Eastern Europe and the former Soviet Union; in *Insect virus and pest management* (eds) Frances R Hunter-Fujita, Philip F Entwistle, Hugh F Evans, Norman E Crook (John Wiley & Sons, New York) 216-231pp.

Muralibaskaran, R.K., Venugopal, M.S. and Mahadevan, N.R., 1996. Effect of host plant on the infectivity and yield of nuclear polyhedrosis virus on *Spodoptera litura* (Fabricus); *J. Biol. Control.,* 10, 73-78.

Narayanan, K., 2003. Occurrence of granulosis virus of *Spodoptera exigua* (Hubner) (Lepidoptera: Noctuidae); *Proceedings of the symposium on Biological control of Lepidopteran Pests*, Bangalore, July 18-19, p 139-142.

Rabindra, R.J., Rajasekaran, B. and Jayaraj, S., 1997. Combined action of nuclear polyhedrosis virus and neem bitter against *Spodoptera litura* (Fabricius) larvae: *J. Biol. Control,* 11, 5-9.

Rabindra, R.J., Muthuswami, M. and Jayaraj, S., 1994. Influence of host plant surface environment on the virulence of nuclear polyhedrosis virus against *Helicoverpa armigera* (Hub) (Lepidoptera: Noctuidae) *J. Appl. Ent.*, 118, 453-460.

Rabindra, R.J., 2003. Microbial control of lepidopteran pest; *Proceedings of the Symposium on Biological control of Lepidopteran pests*, Bangolore, July 17-18, p 103-132.

Sachan, J.N. and Gangawar, S.K., 1990. Seasonal incidence of insect pest of cole crops; *Indian. J. Ento*., 52, 111-123.

Singh, Y.R. and Varatharajan, R., 2001. Certain aspects of NPV infecting the larvae of *Spilosoma obliqua* Walker (Arctiidae); in *Microbials in insect pest management* (eds) S Ignacimuthu, A Sen (Oxford & IBH Publishing Co. Pvt. Ltd.) 133-140 pp.

Sosa-Gomez, D.R., Alves, S.B. and Marchini, L.C., 1991. Variation in the susceptibility of *Bombyx mori* L. to nuclear polyhedrosis when reared on different mulberry genotypes; *J Appl. Ent*., 111, 316-320.

Sudhakar, S., Varatharajan, R. and Mathavan, S., 1997. Simple method to purify inclusion bodies form *Nosema* (Microspora: Nosematidae) contamination; *Entomon,* 22, 89-93.

Thorsteinson, A.J., 1953. The chemotactic responses that determine host specificity in an oligophagous insect (*Plutella maculipennis*: Lepidoptera); *Can. J. Zoo*., 31, 52-72.

Dynamics of *Beauveria bassiana* Infecting *Hypothenemus hampei* and Its Management

Stephen D. Samuel,* S. Irulandi,* G. Antony Raj,* P.K. Vinod Kumar* and A. Ravikumar**

Studies were carried out at the Regional Coffee Research Station, Thandigudi, Tamil Nadu, India during 2005-06 to assess the natural monthly infection levels and effect of the white muscardine fungus, *Beauveria bassiana* (Balsamo) on the coffee berry borer, *Hypothenemus hampei* (Ferrari). During the survey, the level of natural incidence of *B. bassiana*, a potential fungal pathogen on berry borer was recorded. It was high between November and January and low from March to June. The mean infection of *B. bassiana* on berry borer adults in the left-over berries was high (56.70%) in July and low (9.69%) in March. In field experiments, the cumulative mean per cent mycosis of coffee berry borer by *B. bassiana* recorded throughout the period of observation, revealed that the spray dosage of 1 X 10^9 spores/ml recorded the highest per cent mycosis of 96.01, closely followed by 1 X 10^8 spores/ml (95.88) and 1 X 10^7 spores/ml (93.56).

Introduction

The coffee berry borer, *Hypothenemus hampei* (Ferrari) (Scolytidae: Coleoptera), a serious pest of coffee in many of the worlds chief coffee producing countries, has caused great losses (Le-Pelley, 1968). Though, it was recorded in India during 1990 from Gudalur in Nilgiris District of Tamil Nadu, it made its first appearance in the Pulney Hills at Pethuparai village, an isolated pocket in Perumalmalai liaison zone (Anonymous, 1996). It spread to the main coffee areas of Pulneys during 1997 and by 2000 to the entire coffee areas in Pulneys. The coffee berry borer is known to feed and reproduce only in the seeds of coffee species. The female beetle enters the coffee berry by cutting a circular hole, generally at the tip of the berry. Occasional attempts by the coffee berry borer to penetrate into the immature endosperm cause decaying of endosperm by secondary infection resulting in premature fruit drop. Any delay in harvesting will aggravate the damage as rate of reproduction is faster near to harvest (Baker, 1999; Sreedharan *et al.*, 2001). A range of natural enemies including predators, parasitoids and pathogens attack the coffee berry borer. Among these, white muscardine fungus, *Beauveria bassiana* is widely used in the biocontrol of coffee berry borer. The pathogen is present in nature itself, and under favourable conditions of temperature and humidity it causes mortality of the beetles. Once viable spores of this fungus fall on the berry borer, they germinate in about 12 hours. Within three days the insect body is colonized by the mycelium, eventually leading to death. The fungus is most effective at high humidity and a temperature range of 23 to 28 ⁰C (Alves, 1986). At present, *B. bassiana* is naturally infecting coffee berry borer, *H. hampei* in coffee ecosystem of Pulney hills. The present paper deals with the seasonal incidence and effect of *B. bassiana* on coffee berry borer.

* Regional Coffee Research Station, Thandigudi - 624 216, Dindigul District
** Agricultural College and Research Institute, Madurai – 625 104

Materials and Methods

a. Natural incidence of *B. bassiana*

The level of infection by *B. bassiana* on coffee berry borer was recorded at fortnightly intervals from January 2005 to December 2006 at the farm of the Regional Coffee Research Station situated at Thandigudi in the Pulney hills. Observations were made on 10 randomly selected coffee plants and on each plant, three branches (one top, one middle and one bottom) were chosen. The per cent mycosis on berry borer by *B. bassiana* was calculated based on the record of the number of diseased borer at the entry point or inside the bean and the uninfected borers. The data were corrected using Abbot's formula. The means were compared using Duncan's multiple range test (DMRT) Erwin *et al.* (1962).

b. Mycosis of *B. bassiana* on berry borer in left over berries

The left-over berries were collected at three different locations in the lower Pulney hills from March to July of 2005 and 2006. One hundred left-over berries were collected randomly from each location. The fruits were sliced open and the total number of beetles and infected beetles recorded. The mean percentage of beetle infection due to *B. bassiana* was computed from this data.

c. Field efficacy of *B. bassiana* on berry borer

A field experiment was conducted at RCRS, Thandigudi, to test the field efficacy of *B. bassiana* on *C. canephora* (25 year old plantation) the serially diluted spore concentration of $1x10^5$, $1x10^6$, $1x 10^7$, $1x10^8$ and $1x10^9$ / ml was tested. The treatments were replicated four times. Five branches per replication. These plants with sufficient infestation only were selected at random and the infested berries were counted. The conidial suspensions of *B. bassiana* were sprayed on the berry clusters. The post treatment counts on mortality of coffee berry borer due to *B. bassiana* infection were recorded at weekly intervals 7, 14, 21 and 28 days after treatment and percentage infection was worked out.

Results and Discussion

a. Incidence of *B. bassiana*

The data obtained on the natural infection of coffee berry borer by *B. bassiana* indicated that the incidence increased from the month of July onwards, during both the years of study and continued to do so till December-January (Table 1). After this period there was a slight drop in the infection level as the summer approached. The lowest incidence was during the peak summer months of April, May and June. This is also a period where very less fruits are available on the coffee plants, especially in Arabica coffee. On robusta plants fruits are available during this period and the berry borer is also active. The prevailing high temperature and dry condition are probably not conducive for the growth of the pathogen. One characteristic feature of the coffee tracts on the eastern hill ranges are the receipt of some quantum of rainfall during all the months; February 2006 being an exception with no rainfall (Tables 2 & 3). The frequent rainfall assists in the sustenance of the fungal pathogen which is not seen in the coffee growing tracts of the Western Ghats. Hence, there is appreciable suppression of this coffee berry borer by this natural pathogen at every stage of the infection cycle. When the berry borer attacks the coffee fruit, it usually enters from the tip region of the fruit and remains just below the surface till the endosperm of the fruit is mature enough to be tunneled. This is a vulnerable stage in the attack cycle and the spores of the pathogen has easy access to the body surface of the berry borer.

Table 1. Natural incidence of the entomopathogenic fungus, *B. bassiana* on coffee berry borer, *H. hampei*

Month /Period	2005		2006		Pooled mean
	Per cent infection of *B. bassiana* Range	Mean per cent infection of *B. bassiana**	Per cent infection of *B. bassiana* Range	Mean per cent infection of *B. bassiana**	
January	12.71 - 88.12	74.00 (61.97)[b]	12.50 - 82.00	73.00 (60.88)[a]	73.50
February	13.00 - 86.00	70.86 (61.85)[d]	8.00 - 78.00	68.89 (62.18)[b]	69.61
March	4.41 - 68.00	56.28 (64.51)[g]	6.00 - 69.00	61.65 (52.62)[e]	58.97
April	4.22 - 65.20	39.32 (38.91)[k]	5.55 - 70.36	47.33 (37.06)[j]	43.33
May	5.61 - 67.42	42.66 (40.78)[j]	4.00 - 58.00	36.55 (20.20)[l]	39.61
June	6.21 - 68.31	44.81 (42.02)[i]	3.25 - 64.51	43.96 (38.33)[i]	44.39
July	7.62 - 71.00	51.39 (45.79)[h]	4.00 - 66.17	52.61 (39.81)[g]	52.00
August	7.81 - 72.25	57.27 (49.18)[f]	5.10 - 70.00	58.90 (29.17)[k]	58.09
September	8.52 - 74.12	63.20 (52.65)[e]	4.16 - 85.00	66.20 (39.97)[h]	64.70
October	8.65 - 78.25	70.66 (57.20)[d]	8.89 - 82.46	70.69 (46.39)[f]	70.68
November	9.15 - 80.16	72.17 (62.16)[c]	8.68 - 86.57	75.48 (54.65)[d]	73.83
December	11.16 - 88.25	79.16 (62.83)[a]	10.55 - 89.84	80.25 (61.94)[c]	79.71

* Each value is the mean of three replications; Figures in parentheses are arc sine transformed values Means followed by a common letter (s) are not significantly different by DMRT (P=0.05)

The ambient temperature and humidity play a definite role at this crucial instance. Heavy and continuous rain can have a wash down effect, while high humidity and reasonable high temperature is ideal for fungal growth. According to Pascalet (1939) the temperature ranging from 20 – 30 ^{0}C, sufficient rain to produce humidity required for sporulation, followed by one or two sunny days for an even distribution of spores and then by mist or light rains to induce the development of spores on the beetles is ideal for the growth of this fungus. This fungus is most effective at at humidity > 90% with temperature regime of 23 to 28 ^{0}C (Alves, 1986; Baker, 1999). When conditions like temperature and relative humidity are ideal there will be natural increase in the entomopathogenic activity. This was also confirmed from the earlier report of Balakrishnan *et al.* (1994) who noticed an epizootic incidence of *B. bassiana* in some plantations at Kilkotagiri areas. Bustillo and Posado (1996) who reported that *B. bassiana* and *Metarhizium anisopliae* had potential to control berry borer populations emerging from fallen infested coffee berries and the efficacy of fungi in the field varies depending on the relative humidity, radiation and timing of sprays to borer attack. In the conditions prevailing across the coffee tracats of Pulney hills, *B. bassiana* plays a key role in the natural suppression of the berry borer. Enhancement of the activity is possible by taking up sprays of *B.bassiana* spore suspension.

Table 2. Weather parameters – RCRS, Thandigudi - 2005

Sl. No	Months/Year	Temperature		Relative humidity		Rainfall (mm)
		Av. Maximum	Av. Minimum	Av. Maximum	Av. Minimum	
1	January	23.0	10.26	96	57	56.0
2	February	25.2	9.46	96	39	54.0
3	March	26.09	11.97	99	50	63.0
4	April	25.86	15.23	100	67	203.0
5	May	27.06	15.35	100	66	117.0
6	June	25.3	15.73	100	71	85.0
7	July	24.35	15.94	99	71	192.0
8	August	25.42	16.06	94	63	105.0
9	September	24.3	14.8	99	73	124.0
10	October	23.93	15.58	100	75	302.0
11	November	21.63	14.6	99	78	417.0
12	December	22	13.19	98	68	164.0

Table 3. Weather parameters – RCRS, Thandigudi - 2006

Sl. No	Months/Year	Temperature		Relative humidity		Rainfall (mm)
		Av. Maximum	Av. Minimum	Av. Maximum	Av. Minimum	
1	January	22.65	10.96	98	59	26.5
2	February	24.04	8.71	96	43	0
3	March	25.55	12.61	100	58	220.0
4	April	27.2	13.7	100	55	35.2
5	May	26.3	15.7	99	65	145.8
6	June	25.1	14.5	100	74	139.3
7	July	25.4	15.6	97	61	9.8
8	August	24.8	15	97	68	205.0
9	September	24	15.3	99	76	177.0
10	October	23	15	100	79	239.7
11	November	22.1	15.4	100	84	199.2
12	December	21.6	11.6	99	72	16.3

b. Mycosis of *B. bassiana* on berry borer in left over berries.

The fruits left over on the plants after the main harvest are the source of inoculation for carry over the borer population to the next season's crop. This becomes a key issue in areas where both arabica and robusta coffee varieties are grown. This is because of the difference in the fruit maturation period of the two varieties. While arabica ripens at a faster rate (8-10 months) robusta takes more time and hence the berry borer can have multiple generations during this time and increase in population, damaging more fruits. Hence, it is very important to reduce this residual population of berry borer in the left over fruits. The present study on robusta coffee indicated that the natural activity of *B. bassiana* on coffee berry borer in the left over fruits also, but the highest percent mycosis was only 56.70 (Table 4). This is because of the dry condition. Yet, there is suppression of the population to some extent.

Table 4. *B. bassiana* infection on berry borer in left over berries (Robusta)

Month	Mean per cent infection on berry borer adults*			
	2005	2006	2007	Pooled mean
March	12.00	7.50	9.56	9.69
April	18.16	15.00	17.65	16.94
May	20.33	25.10	27.55	24.30
June	37.47	41.16	42.17	40.26
July	56.55	52.16	61.39	56.70

*Mean of 100 fruits per observation

c. Field efficacy of *B. bassiana* on berry borer

If intervention is required, it is necessary to know the dosage at which maximum mycosis of the berry borer can be achieved using *B. bassiana* spore suspension. Among the various doses tested, application of *B. bassiana* at 1x10^9 spores / ml was found superior at all the periods of observations and the mycosis ranged from 62.30 to 97.14 per cent. Next to this in descending order of efficacy noticed in the treatment with 1x10^8 spores / ml (90.26 to 96.64%) and 1x10^7 spores / ml (87.07 to 96.28%) (Table 5). This trend has also been observed by the earlier workers (Rosa *et al.,* 1997; Samuels *et al.,* 2002). The observations indicated that a dosage of 1 X 10^8 spore/ml may be ideal as that range of mean per cent mycosis much narrower.

Table 5. Bio-efficacy of *B. bassiana* against coffee berry borer under field condition

Treatments	Corrected mean per cent mycosis at weekly interval*				Cumulative mean
	7th day	14th day	21st day	28th day	
T1 - *B. Bassiana* (1x10^5 spores/ml)	58.56 e	86.62 c	92.12b	93.36 b	83.16 d
T2 - *B. bassiana* (1x10^6 spores/ml)	82.43 d	92.23 bc	95.25 ab	95.18 b	91.27 c
T3 - *B. bassiana* (1x10^7 spores/ml)	87.07 c	94.64 b	96.31. ab	96.28 a	93.56 bc
T4 - *B. bassiana* (1x10^8 spores/ml)	90.26 b	96.46 b	97.01 a	96.64 a	95.88 ab
T5 - *B. bassiana* (1x10^9 spores/ml)	92.30 a	97.13 a	97.48 a	97.14 a	96.01 a

* Each value is the mean of four replications; Figures in parentheses are arc sine transformed values
In a column, means followed by a common letter (s) are not significantly different by DMRT (P=0.05)

Mostly from this study, a better understanding of the natural infection of coffee berry borer by *B. bassiana* is achieved. It is possible to develop this study further correlating the infection cycle with ambient conditions to develop prediction model.

References

Anonymous, 1996. Coffee Board, Research Department, Forty ninth Annual Report, Central Coffee Research Institute, Coffee Research Station, Chikmagalur, Karnataka. . 141 pp.

Alves, S. B., 1986. Fungus entomopathogenicos. *In:* Alves, S. B. (Ed.) Controle microbiano de insectos. Sao Paulo, Editora Mande, pp.73-126.

Baker, P.S., 1999. The Coffee Berry Borer in Colombia: Final report of the DFID- Cenicafe-CABI Bioscience IPM for the coffee Project (CNTR 93 / 1536 A). CABI Bioscience, Silwood Park, Ascot SL5 7TA, U. K: 144pp.

Balakrishnan, M.M., Sreedharan, K. and Krishnamoorthy Bhat, P., 1994. Occurrence of the Entomopathegenic fungus, *Beauveria bassiana* on certain coffee pests in India. *J. Coffee Res.,* 24(2), 33-35.

Bustillo, P. and Posada, F., 1996. Use of entomopathogens to control coffee berry borer in Colombia. *Manejo Integrado de plagas,* 42, 1 – 13.

Erwin L.L., Leonard, W.H. and Clark, A.G., 1962. Field plot technique. Burgess Publishing Company, Minneapolis 23, Minnesota, USA. 373 pp.

Le-Pelley, R.H., 1968. *Pests of Coffee.* Longman Green and Co Ltd, London, 590pp.

Pascalet, P., 1939. La lutte biologique contre *Stephanoderis hampei* ou scolyte du cafeier au Cameroun. *Revue Bot. app. Agric. Trop.,* 19 (219), 753-764.

Rosa, W., Alatorre, R., Trujillo J. and Barrera, J.F., 1997. Virulence of *Beauveria bassiana* (Deuteromycetes) strain against the coffee berry borer (Coleoptera: Scolytidae). *J. Econ. Entomol.*, 90, 1534-1538.

Samuels, R.I., Pereira, R.C. and Gava, C.A.T., 2002. Infection of the coffee berry borer, *Hypothenemus hampei* (Coleoptera: Scolytidae) by Brazilian isolates of the entomopathogenic fungi *Beauveria bassiana* and *Metarhizium anisopliae* (Deuteromycotina: Hyphomycetgs). *Biocontrol Sci. Tech.,* 12, 631 – 635.

Sreedharan, K., Vinod Kumar, P.K. and Prakasan, C.B. (Eds.), 2001. *Coffee berry borer in India.* Central Coffee Research Institute, Coffee Research Station, India. 122pp.

In vitro studies on the efficacy of *Beauveria bassiana* (Balsomo) Vuillumin and *Metarhizium anisopliae* (Metschnikoff) Sorokin (Deuteromycotina: Hyphomycetes) against the coffee berry borer, *Hypothenemus hampei* (Ferrari) (Coleoptera: Scolytidae).

C. Balasubramanian* and S. Sundaresan*

The efficacy of *Beauveria bassiana* and *Metarhizium anisopliae* was evaluated against the coffee berry borer, *Hypothenemus hampei.* Suitable medium for the growth of the experimental fungal species was also studied under *in vitro* condition. The pathogenic fungi were isolated from infected borers by culturing them in PDA medium. *B. bassiana* was found to be the dominant fungal species followed by *M. anisopliae*. Bioassay was performed with the fungal spore dilutions (103-108 spores ml-1) in sterile distilled water with 0.02% Tween 80 against the adult coffee berry borer under *in vitro* condition (27±2°C, 75±5%RH). The observations were made at 5th, 8th 12th and 17th day after treatment. The percent mortality of the beetles was calculated for each fungal treatment. DMRT analysis was performed in order to group the treatments. The bio-assay revealed 100% mortality of the berry borer with *B. bassiana* treatment at 1 X 108 spores ml-1 whereas 65.4% was observed with *M. anisopliae* on 17th day after treatment. Both the rice water and carrot broth were found to be effective substrates in the promotion of growth and spore production of *M. anisopliae* but these did not support the growth of *B. bassiana.*

Introduction

Coffee berry borer, *Hypothenemus hampei* (Ferrari) (Coleoptera: Scolytidae), is prevalent in about 35% of the total coffee growing areas in the country. But it is not a native of our country. It is a native of Central Africa, which is considered to be the most dreaded primary pest of coffee in many of the growing countries. It is an endemic pest causing extensive loss (Le Pelley, 1968). It was first noticed in February, 1990 in a few plantations in Gudalur in Nilgiri district of Tamilnadu (Kumar *et al.,* 1990). Infested coffee berries that fall to the soil are the main source of reinfestation of the coffee plantations at the end of the harvest period (Chamorro *et al.,* 1995).

Routine spraying of synthetic pesticides is becoming less acceptable because of increasing concerns about their detrimental effects on human health / animal health and the environment, in particular their negative impact on natural enemies. Hence, biological control methods are recently involved in the control of coffee berry borer in many countries. Among the many biocontrol methods, being explored for use against this insect, the fungal entomopathogen, *B. bassiana* (Balsamo) Vuillemin (Ascomycota: Hypocreales) remains one of the most promising. Pascalet (1939) was the first to record *B. bassiana* infections in *H. hampei* in Cameroon. *Metarhizium anisopliae* is also a potential entomopathogen that could infect the coffee berry borer in the soil (Bernal *et al.,* 1994).

Hence in the present study, the work was framed in-such- a way as to evaluate the efficacy of native strain of *Beauveria bassiana* and *Metarhizium anisopliae* against the coffee berry borer,

* Department of Zoology and Microbiology, Thiagarajar College (Autonomous), Madurai-625 009

Hypothenemus hampei under *in vitro* condition. Apart from this bioassay, a suitable medium for the growth of the experimental fungal species was also studied.

Materials and Methods

Isolation of natural pathogen against *Hypothenemus hampei*

A survey was conducted in the coffee plantations at Horticultural Research Station, Yercaud" Tamil Nadu, India during July-November, 2006. The samples such as soil, infested berries, leaf materials, coffee twigs and fallen coffee berries were collected and screened in the laboratory for the presence of infected and dead Coffee Berry Borer (CBB) (Plate 1). Dead CBB were washed with sterile distilled water, homogenized using a mortar and pestle in sterile distilled water and plated on Potato dextrose agar (PDA). The cultured plates were incubated at 27±2°C for seven days Haraprasad *et al.* (2001).

Identification of the isolated fungal pathogen

After seven days of incubation, fungal colony was abundantly found. *Beauveria bassiana* and *Metarhizium anisopliae* were identified.

Subculture of the isolated fungal colonies

The identified fungal colonies were then sub-cultured in Sabouraud Dextrose Agar (SDA) (40 g dextrose, 10 g peptone, 15 g agar, 10 g yeast extract, 40 mg tetracycline, 1000 ml distilled water). The sterilized medium was transferred into sterile petridishes and test tubes that were then inoculated with conidia by the streak plate method or by placing the mycelia mat of 8 mm diameter using sterilized cork borer, in a laminar flow cabinet. The inoculated medium was maintained at 27±2°C, 80±5% RH. After 12-15 days of incubation, the conidia were removed with a sterile bacteriological loop and stored in sterilized distilled water at 4°C.

Bioassay of fungal pathogen against CBB

The coffee berry borers were collected from the infested coffee berries by cutting the berries carefully and the CBB were removed. The adults of CBB were selected for the bioassay.

Efficacy of *Beauveria bassiana* and *Metarhizium anisopliae* on CBB

The isolated *B. bassiana and M anisopliae* were tested to determine the pathogenicity of fungal culture towards CBB. Conidial suspension of the fungal culture was used. Pure cultures of the test fungal species were grown on SDA at 27±2°C until a dense sporulation mat was produced (> 14 days). The conidiospores of each fungal species were harvested by flooding the fungal plates with sterile distilled water containing 0.02% Tween-20. The conidial suspension of 1 x 10^8 conidia ml^{-1}) was prepared by counting the spores in improved Neubauer counting chamber. The conidial suspension of up to 10^3 conidia ml^{-1} (10^8 to 10^3) was prepared for both the fungal species.

Bioassays with these two different fungal spore suspensions were carried out by dipping 10 adult CBB in conidial suspensions of each concentration for 30 seconds to 1 minute. After a minute, the CBB were transferred to sterile filter paper and then placed in individual sterilized containers having single coffee grain, which was already surface sterilized with sodium hypochlorite (1 %) and dried. The setup was kept in controlled condition of 27±2°C at 95±5% RH. Triplicates were maintained for each concentration of both fungal species. One group of borers was maintained as a control treatment consisting of distilled water plus 0.02% Tween-20. The observations were carried out at regular intervals viz., 5th, 8th, 12th and 17th day after inoculation of CBB with conidial suspension. The dead borers were placed in a controlled humidity to stimulate the development of fungal mycelia and confirm that the deaths were caused by infection of the particular fungus that was used.

Suitability of certain culture media for- mass production of the fungus

For large scale field application, sufficient quantum of fungal inoculum would be required. Hence studies were conducted to identify the suitable culture media for the growth and sporulation of *B. bassiana* and *M anisopliae.* The media inoculated for the study was carrot and rice water (i.e. water filtered during cooking).

Preparation of media

100 g of carrot was washed, crushed, boiled along with distilled water and filtered.100 ml of the pure filtrate was taken in a 250 ml conical flask and another 100 ml with agar. The media were autoclaved and the streptomycin was added to each medium after the medium gets hand bearable temperature and the agar added medium was poured in to the sterilized petrlplates. Similarly, two sets of 100ml of the rice washed water were taken in separate 250ml conical flask. In one of the conical flask, agar was added. Both the conical flask containing medium was autoclaved and streptomycin was added after cooling and the agar medium was poured in to sterilized petrlplates.

Biomass production

After 14 days of the fungal growth in SDAY medium, the mycelial mat from the edges were cored out using 8mm diameter cork borer and placed in broth culture which was then incubated for 27±2°C for 14 days. After 14 days incubation, the broth culture was filtered through a pre-weighed Whatman No.1 filter paper with 0.001 mg accuracy digital balance, which was dried at 80°C for 3-4 hours and reweighed. The difference in the weight can be calculated for biomass produced.

Results

Isolation and identification of fungal species in natural infection of CBB:

Two different fungal strains were isolated from the dead CBB from natural habitat. In the laboratory, after seven days of incubation, a fungus with white fluffy cottony growth was abundantly expressed on PDA plates inoculated with homogenized suspension of CBB cadavers and was identified as *B. bassiana.* Another green coloured powdery colony was also observed along with *B. bassiana,* which was identified as *M anisopliae.* The pure cultures of the isolated fungal colonies were maintained in Sabourad dextrose agar (SDA) (Plate 1).

Efficacy of *B. bassiana* and *M. anisopliae* on CBB

The spores formed over the SDA medium were harvested by flooding the plate with sterile distilled water containing 0.02% Tween-20 and made up to 10 ml then different spore dilutions ranging from 10^3 to 10^8 spores ml^{-1} was prepared for both the fungus, using Neubauer counting chamber.

The mean percent mycosis of *B. bassiana* against CBB at various spore dilutions on5th, 8th, 12th and 17th day of post inoculation was represented in Table 1. It showed 100% mortality on 17th day at 1 x 10^8 spores ml^{-1}). It was then followed by 92.6% mortality on the same day of observation at 1 x 10^7 spores ml^{-1}). Whereas in control noticed a least mortality on 17th day of CBB (Table 1). The mortality of CBB was found to be very low with the treatment of 1 x 10^3 spores ml^{-1}) in all the days of observation when compared to other spore concentrations. The observations made on other days of post treatment (i.e. 5th, 8th and 12th day) represents a steady increase in the mortality of CBB at each concentration from 10^3 to 10^8 spores ml^{-1}). The mean mortality of CBB for all the days ranges from 15.97% (1 x 10^3 spores ml^{-1}) to 58.77% (1x 10^8 spores ml^{-1}) and 3.45% for the control treatment.

In the case of *M anisopliae* treatment, on the 5th day of observation, 9.7% was the maximum mortality obtained in 10^8 spore ml-1) (Table 2). The 8th day observation showed the mortality range between 2.9% (1 x 10^3 spores ml^{-1}) and 18.1% (1 x 10^8 spores ml^{-1}). The percentage of morality caused by 1 x 10^8 spores ml^{-1}) on the 12 th day of post treatment was 42.5% and on 17 th day it was 65.4% which was the maximum mortality observed during the laboratory study when compared to all other concentrations on all days of observation. Nearly fifty percent mortality of CBB (52.2%) was observed in the treatment of 1 x 10^7 spores ml^{-1}) on 17th day of post treatment. When compared to treatment with *B. bassiana, M anisopliae* has a least mean mortality (5.0%) at 1 x 10^3 spores ml^{-1}) and a maximum mean mortality (33.92%) at 1 x 10^8 spores ml^{-1}) (Table 2). The similarity correlation was analysed between *B. bassiana* and *M anisopliae* at various days after treatment represented a positive correlation (Table 3).

Table 1. Efficacy of *Beauveria bassiana* against Coffee Berry Borer

Treatment	Mean Mortality of CBB (Days after treatment)				mean
	5	8	12	17	
T1 (1×10^3 spore/ml)	0.8 (5.13)f	2.9 (9.80)f	6.11 (14.30)f	10.2 (18.63)f	5.0 (12.92)f
T2(1×10^4 spore/ml)	1.9 (7.92)e	3.2 (10.30)e	7.9 (16.32)e	12.8 (20.96)e	6.45 (14.71)e
T3(1×10^5 spore/ml)	3.4 (10.62)d	5.8 (13.93)d	12.1 (20.36)d	25.4 (30.26)d	11.67 (19.97)d
T4(1×10^6 spore/ml)	5.3 (13.31)c	8.6 (17.05)c	15.3 (23.03)c	34.1 (35.73)c	15.82 (23.44)c
T5(1×10^7spore/ml)	7.1 (15.45)b	11.7 (20.00)b	21.6 (27.69)b	52.2 (46.26)b	23.15 (28.76)b
T6(1×10^8 spore/ml)	9.7 (18.15)a	18.1 (25.18)a	42.5 (40.69)a	65.4 (53.97)a	33.92 (35.62)a
Control	0.00 (1.08)g	0.00 (1.08)g	1.5 (7.03)g	3.8 (11.24)g	1.32 (6.60)g

Biomass Production

The fungal cultures that were isolated were inoculated in the carrot broth and rice water broth respectively. After 14 days of incubation, the broth was filtered through pre-weighed Whatman No.1 filter paper (1.058 g), and dried in oven at 60°C. Then the filter paper was weighed and the difference between the initial and final weight of the weighed sample were, *B. bassiana* in carrot broth was 0.252% (w/v), whereas *M. anisopliae* was 0.457% (w/v). In rice water broth the calculated biomass of *M. anisopliae* was 0.618% (w/v). But *B. bassiana* showed no growth in rice water broth (Plate 1).

Discussion

The results proved that both the fungi act as entomopathogens against coffee berry borer. In the present study, treatment with *B. bassiana* at 1 x 10^8 spores ml^{-1}) caused 100% mortality on 17th day of post treatment. A similar experiment with 18 strains of *B. bassiana* against *H. Hampei* conducted by De La Rosa *et al.* (1997) showed that all the 18 strains tested were found to be pathogenic towards *H. hampei.* In their experiment the percentage of mortality ranged from 20.1% to 100%. Mendez (1990) observed that the strains isolated from the coffee berry borer demonstrated greater virulence (LC_{50}=5.0 x 10^6 conidia ml^{-1}) than strains isolated from other hosts. The work of De La Rosa *et al.* (1997) revealed that *B. bassiana* is a potential biological control agent, suitable for incorporation in to a program for the integrated control of the coffee berry borer.

Plate 1. Mass production of *B. bassiana* and *M. anisopliae* and its natural iinfection on Coffee Berry Borer

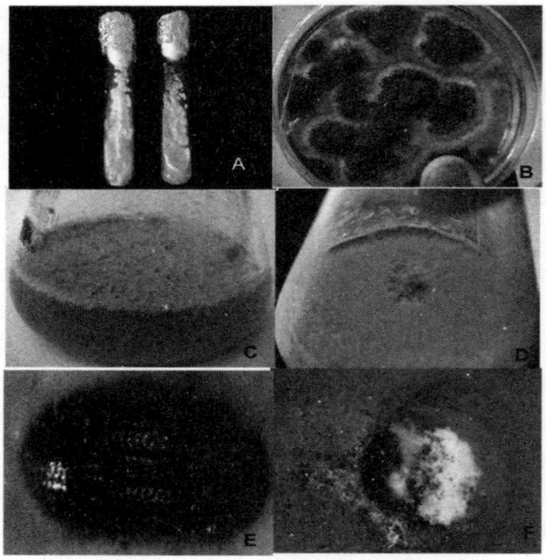

A -Pure culture of *B. bassiana* B-*M. anisopliae*.
C and D- Mass production of Entomopathogenic fungi
E- Cofee Berry Borer *Hypothenemus hampei*
F- Natural infection of B. bassiana on CBB

Table 2. Efficacy of *Metarhizium anisopliae* against Coffee Berry Borer

Treatment	Mean Mortality of CBB (Days after treatment)				mean
	5	8	12	17	
T1 (1×10^{3} spore/ml)	0.8 (5.13)[f]	2.9 (9.80)[f]	6.11 (14.30)[f]	10.2 (18.63)[f]	5.0 (12.92)[f]
T2(1×10^{4} spore/ml)	1.9 (7.92)[e]	3.2 (10.30)[e]	7.9 (16.32)[e]	12.8 (20.96)[e]	6.45 (14.71)[e]
T3(1×10^{5} spore/ml)	3.4 (10.62)[d]	5.8 (13.93)[d]	12.1 (20.36)[d]	25.4 (30.26)[d]	11.67 (19.97)[d]
T4(1×10^{6} spore/ml)	5.3 (13.31)[c]	8.6 (17.05)[c]	15.3 (23.03)[c]	34.1 (35.73)[c]	15.82 (23.44)[c]
T5(1×10^{7}spore/ml)	7.1 (15.45)[b]	11.7 (20.00)[b]	21.6 (27.69)[b]	52.2 (46.26)[b]	23.15 (28.76)[b]
T6(1×10^{8} spore/ml)	9.7 (18.15)[a]	18.1 (25.18)[a]	42.5 (40.69)[a]	65.4 (53.97)[a]	33.92 (35.62)[a]
Control	0.00 (1.08)[g]	0.00 (1.08)[g]	1.5 (7.03)[g]	3.8 (11.24)[g]	1.32 (6.60)[g]

Table 3. Similarity correlation between *Beauveria bassiana* and *Metarhizium anisopliae* at various days after treatment

M. anisopliae \ *B. bassiana*	5	8	12	17
5	0.9802	0.9802	0.9411	0.9832
8	0.9788	0.9841	0.9618	0.9847
12	0.9709	0.9750	0.9478	0.9831
17	0.9733	0.9655	0.9150	0.9819

The results revealed that the emergence of the pest population was reduced to 80% by the fungus in the highest conidia concentration of 2.52 x 10^{10}. When this was compared to with the present investigation with regard to coffee berry borer, it showed a relative resemblance (i.e.) *M anisopliae* produced 65.4% mortality of *H. hampei* at 1 x 10^8 spores ml^{-1}. If the conidial concentration was increased to 10^{10}, then it may create 80-90% of mortality of *H. hampei.* But the conidial concentration was maintained up to 10^8 because the highest mortality (100%) was caused by *B. bassiana* at 1 x 10^8 spores ml^{-1}. So that it is easy to compare the effect of both the fungal species at the same concentrations.

As a whole, from the experimental study under laboratory condition, it was evident that *B. bassiana* at a spore concentration of 1 x 10^8 per milliliter was effective in the control of *H. hampei* when compared to *M anisopliae.* With reference to the work undertaken to identify the suitability of certain culture media for the mass production of the isolated fungus, the results of the study revealed that in both the carrot and rice water substrates, *M anisopliae* showed a better growth. Of which, in rice water, the biomass was found to be high (0.610%) when compared to that of carrot broth (0.457%). But in the case of *B. bassiana,* it showed a limited growth in carrot broth (0.252%), whereas there was no growth in the case of rice water broth. Hence that rice water and carrot can be used as a substrate in the mass production of *M anisopliae,* but not of *B. bassiana.*

References

Bernal, M.G., Bustillo, A. E. and Posada, Y.F.J., 1994, Virulencia de aislamientos de *Metarhizium anisopliae* y su eficacia en campo sobre *Hypothenemus hampei.*Revista *Colombiana de Entomologia,* 20, 225-228

Chamorro, G.R, Cardenas, R. and Herrera, Y.A., 1995. Evaluacion economica y de la calidad en taza del cafe proveniente de diferentes sistemas de recoleccion manual,utilizables como control en cafetales infestados de *Hypothenemus hampei. Revista Cenicafe,* 46, 164-175.

De la Rosa, W., Alatorre, R., Trujillo, J. and Berrera, J.F., 1997. Virulence of *Beauveria bassiana* (Deuteromycetes) Strains against the Coffee Berry Borer (Coleoptera: Scolytidae). *Journal of Economic Entomology,* 90, 1534-1538.

Haraprasad, N., Niranjana, S.R., Prakash, H. S., Shetty, H.S. and Seema Wahab, 2001. *Beauveria bassiana-A* Potential Mycopesticide for the Efficient Control of Coffee Berry Borer, *Hypothenemus hampei* (Ferrari) in India. *Biocontrol Science and Technology,* 11(2), 251 - 260.

Kumar, P.K.V., Prakasan, C.B. and Vijayalakshmi C.K., 1990. Coffee berry borer, *Hypothenemus hampei* (Coleoptera: *Scolytidae)* first record from India. *Journal of Coffee Research,* 20(2), 160-164.

Le-Pelley, R.H., 1968. Pests of Coffee. Longman Green and Co Ltd, London, p. 590.

Mendez, L.I., 1990. Control microbiana de la broca cafeto *Hypothenemus hampei* (Coleoptera: Scolytidae), con el honga *Beauveria bassiana* (Bals.) Yuill. (Deuteromycetes) on el Soconusco, Chiapas, Mexico. M. S. thesis, Colegio de Posgraduados, Chapingo, Mexico.

Pascalet, P., 1939. La lutte biologique contre *Stephanoderis hampei* ou scolyte du cafeier au Cameroun. *Revue Bot. App. Agric. Trop.,* 19 (219), 753-764.

Effect of Abamectin 1.9 EC - Microbial Insecticide to Coccinellids of Cotton Ecosystem

R. Sheeba Jasmine *, S. Kuttalam,** P. Sivasubramaniam,** and J. Stanley**

Abamectin a spectrum microbial insecticide derived from the soil actinomycete, *Streptomyces avermitilis* and is effective against several pests in number of crops. Two field experiments in cotton were conducted from June 2003 - April 2004 to evaluate the toxicity of abamectin on coccinellids in cotton ecosystem. Abamectin 1.9 EC @ 8.1, 10.8, 14.5, 18.5 and 22.5 g a.i. ha^{-1} was tested in comparison with spinosad 45 SC @ 70 g a.i. ha^{-1} and untreated check. Observations on the population of coccinellids were made prior to spraying and on 3, 7, 10 and 14 days after spraying from 10 randomly tagged plants per plot and the mean was worked out. Results clearly showed that abamectin 1.9 EC @ 8.1 and 10.8 g a.i.ha^{-1} was found to be least toxic to coccinellids.

Introduction

Cotton *Gossypium* spp., the king of fibre crop, is widely grown in India and plays a key role in shaping the economy of the country. It is attacked by several insect pests from sowing to maturity. Among the pests, bollworms and sucking pests are very important. To control these pests, insecticides like carbaryl, endosulfan, quinalphos and many formulations of synthetic pyrethroids were used. The indiscriminate use of insecticides has affected the population of bio control agents as all the recommended insecticides are highly toxic to predatore and parasitoids (Dhawan *et al*., 1992, 1994; Singh 1994). The population of predators has declined by 68.4 % during the last two decades and many parasitoids have been eliminated from cotton ecosystem (Dhawan and Simwat, 1996). To a large extent, problems of environmental and human risk have been overcome through the development of newer compounds that can be handled safely and that do not persist as environmental contaminants. Abamectin is a spectrum microbial insecticide derived from the soil actinomycetes, *Streptomyces avermitilis* effective against several pests in a number of crops, alternate to existing formulation and also ecologically sound for the effective management of cotton pests. Keeping this in view, the present work was taken up to study the impact of abamectin on coccinellids.

Materials and Methods

Two field experiments were conducted one each at Rudhiriampalayam and TNAU from June 2003 - April 2004 1.9 EC to assess the toxicity of abamectin in cotton. The experiments were carried out in plots of 4 x 5 m size in a randomized block design with seven treatments, each replicated four times. The treatments imposed were abamectin 1.9 EC @ 8.1, 10.8, 14.5, 18.5 and 22.5 g a.i. ha^{-1}, spinosad 45 SC @ 70 g a.i. ha^{-1} and untreated check. The treatments were imposed four times at 14 days interval commencing from 60th day after sowing with pneumatic Knapsack sprayer using 1000 litres of spray fluid per hectare. Applications were made during morning hours to avoid photo oxidation of the insecticides. Observations on the population of coccinellids (grubs and adults), a day before each spraying and on 3, 7, 10 and 14 days after each spraying from 10 randomly tagged plants per

*Regional Plant Quarantine Station, Chennai – 600 027

**Department of Agricultural Entomology, Tamil Nadu Agricultural University, Coimbatore – 641 003

plot were made and the mean worked out. The statistical analysis was carried out using IRRISTAT ver 3.1. ANOVA. The data were transformed into $\sqrt{x+0.5}$. The mean values of treatments were separated using Duncan's Multiple Range Test (DMRT) (Gomez and Gomez, 1976).

Results and Discussion

The population of coccinellids ranged from 8.0 to 9.5 per 10 plants before imposing the treatments in the first field experiment. (Table 1). Abamectin at the lowest dose recorded the higher mean coccinellid population of 9.7 per 10 plants next to untreated check (10.0per 10 plants). Abamectin at 10.8 g a.i.ha^{-1} recorded 9.0 coccinellids per 10 plants followed by abamectin at 14.5 g a.i.ha^{-1} (8.8 per 10 plants) and spinosad 75 g a.i.ha^{-1} (8.6 per 10 plants). It is interesting to note that the population increased significantly three days after each spraying in all treatments (Tables 1 and 2). After the second round of spray, spinosad 45 SC 75 g a.i.ha^{-1} recorded 10.4 coccinellids per 10 plants followed by abamectin at 18.5 g a.i.ha^{-1} (10.2 per 10 plants) and abamectin at 14.5 g a.i.ha^{-1} (10.1 per 10 plants). The same trend was noticed throughout the study period. All the treatments were found to have little effect on coccinellids when compared to untreated check throughout the investigation period (Table 2).

In the second field experiment, the pretreatment population of coccinellids ranged from 21.3 to 22.8 per ten plants in various treatments. After the first round, abamectin 1.9 EC at 8.1 and 10.8 g a.i. ha^{-1} recorded a mean population of 23.8 and 22.9 coccinellids per ten plants respectively, next to untreated check (25.2 per ten plants). Spinosad 45 SC at 75 g a.i.ha^{-1} and abamectin 1.9 EC at 14.5 g a.i. ha^{-1} were on par with each other on 7 DAT (24.5 per 10 plants). Abamectin 1.9 EC at 18.5 g a.i. ha^{-1} recorded 21.5 per ten plants followed by abamectin 22.5 g a.i. ha^{-1} (21.1 per ten plants). After second spraying, the mean population reduction percentage was 21.9 and 23.3 for abamectin 1.9 EC at 18.5 and 22.5 g a.i. ha^{-1}, while spinosad 45 SC at 75 g a.i. ha^{-1} recorded a mean of 20.3 per cent reduction over untreated check. The same trend was observed throughout the experimental period (Tables 3 and 4).

The effect of abamectin 1.9 EC on coccinellids revealed that after first spray, abamectin at all doses reduced the population significantly on 3 DAT. The observations on 7, 10 and 14 showed the recolonization of coccinellids in all the treatments irrespective of concentrations.

The present finding is in accordance with the observations of Badawy and Arnaouty (1999) who reported that abamectin caused 7 per cent mortality of third instar larvae of *Chrysoperla carnea* Stephens and safer than all the tested conventional insecticides to egg at the recommended rate of application. Acharya *et al.* (2002) stated that abamectin was safer to lady bird beetles. Avermectins were safe to non target organisms *viz*., *Dolycoris bauarum* (L.), *Pentatoma rufipes* (L.), *Adalia bi-punctata* (L.) and *Coccinella septempunctata* (L.) (Chizhov *et al*., 2000)

Spinosad recorded a mean of 8.6 - 13.5 and 21.8 - 36.8 coccinellids per 10 plants in Rudhiriam-palayam and TNAU experiments, respectively. The present finding is in accordance with the observations of Sansone and Minzenmayer (2000) who reported that spinosad had the least impact on spiders and *Scymnus* sp as compared to indoxacarb (Steward®) and emamectin benzoate (Denim®). Medina *et al.* (2001) concluded that spinosad caused a significant reduction in the adult life span and fecundity of *C. carnea.*

Although an abamectin reservoir with the mesophyll layer of leaf tissues is accessible to phytophagous insects, the parasitic and predatory arthropods continue to proliferate because of the short lived surface residues. Therefore, the application of abamectin is less harmful to the important natural enemies in cotton fields.

Table 1. Effect of abamectin 1.9 EC on the population of coccinellids in cotton ecosystem after first and second round of application – Field Experiment I (Location- Rudhiriampalayam) (Mean of four observations)

Treatments	2 PTC	Number of coccinellids per 10 plants: Days after first treatment 3	7	10	14	Mean	Days after second treatment 3	7	10	14	Mean
T1 - Abamectin 1.9 EC @ 8.1g a.i.ha^{-1}	9.0	8.8^{a} (3.04)	9.0^{a} (3.08)	10.0^{a} (3.24)	11.0^{ab} (3.39)	9.7	11.0^{b} (3.39)	11.5^{b} (3.46)	13.0^{b} (3.67)	14.5^{b} (3.87)	12.5
T2 - Abamectin 1.9 EC @ 10.8 g a.i.ha^{-1}	8.5	8.3^{ab} (2.95)	8.5^{ab} (2.99)	9.0^{b} (3.08)	10.3^{bc} (3.28)	9.0	9.8^{c} (3.20)	10.3^{c} (3.28)	11.0^{c} (3.39)	13.8^{c} (3.77)	11.2
T3 - Abamectin 1.9 EC @ 14.5 g a.i.ha^{-1}	9.0	8.0^{bc} (2.91)	8.3^{abc} (2.95)	8.8^{bc} (3.04)	10.0^{bc} (3.24)	8.8	9.0^{cd} (3.08)	9.5^{d} (3.16)	10.5^{c} (3.31)	13.5^{cd} (3.74)	10.6
T4 - Abamectin 1.9 EC @ 18.5 g a.i.ha^{-1}	9.0	7.5^{cd} (2.82)	7.8^{bc} (2.87)	8.3^{bc} (2.95)	9.8^{c} (3.20)	8.4	8.5^{d} (2.99)	9.0^{d} (3.08)	10.0^{c} (3.24)	13.3^{cd} (3.71)	10.2
T5 - Abamectin 1.9 EC @ 22.5 g a.i.ha^{-1}	8.8	7.3^{d} (2.78)	7.5^{c} (2.82)	8.0^{c} (2.91)	9.8^{c} (3.20)	8.2	8.5^{d} (2.99)	9.0^{d} (3.08)	10.0^{c} (3.24)	13.0^{d} (3.67)	10.1
T6 - Spinosad 45 SC @ 75 g a.i.ha^{-1}	9.5	8.0^{bc} (2.91)	8.0^{bc} (2.91)	8.5^{bc} (2.99)	10.0^{bc} (3.24)	8.6	8.8^{d} (3.04)	9.3^{d} (3.12)	10.3^{c} (3.28)	13.3^{cd} (3.71)	10.4
T7 - Untreated check	8.0	8.5^{ab} (2.99)	9.0^{a} (3.08)	10.5^{a} (3.31)	12.0^{a} (3.53)	10.0	12.5^{a} (3.60)	13.0^{a} (3.67)	14.5^{a} (3.87)	16.0^{a} (4.06)	14.0

PTC – Pre treatment count; Figures in parentheses are $\sqrt{x + 0.5}$ transformed values

In a column, means followed by a common letter(s) are not significantly different by DMRT(P=0.05)

Table 2. Effect of abamectin 1.9 EC on the population of coccinellids in cotton ecosystem after third and fourth round of application – Field Experiment I (Location- Rudhiriampalayam) (Mean of four observations)

Treatments	2	Number of coccinellids per 10 plants									
		Days after third treatment					Days after fourth treatment				
	PTC	3	7	10	14	Mean	3	7	10	14	Mean
T1 - Abamectin 1.9 EC @ 8.1g a.i.ha^{-1}	15.0	14.5^{b} (3.87)	15.0^{b} (3.94)	15.3^{b} (3.97)	15.5^{b} (4.00)	15.1	15.0^{b} (3.94)	15.8^{b} (4.03)	15.8^{b} (4.03)	16.0^{b} (4.06)	15.7
T2 - Abamectin 1.9 EC @ 10.8 g a.i.ha^{-1}	14.8	13.5^{c} (3.74)	14.0^{c} (3.81)	14.3^{c} (3.84)	14.5^{c} (3.87)	14.1	13.5^{c} (3.74)	14.3^{c} (3.84)	14.5^{c} (3.87)	15.0^{c} (3.94)	14.3
T3 - Abamectin 1.9 EC @ 14.5 g a.i.ha^{-1}	14.8	13.0^{cd} (3.67)	13.5^{cd} (3.74)	13.8^{cd} (3.77)	14.0^{cd} (3.81)	13.6	13.0^{de} (3.67)	13.5^{cd} (3.74)	14.0^{c} (3.81)	14.5^{cd} (3.87)	13.8
T4 - Abamectin 1.9 EC @ 18.5 g a.i.ha^{-1}	14.8	12.3^{de} (3.6)	13.0^{d} (3.67)	13.0^{de} (3.67)	13.5^{d} (3.74)	13.0	12.0^{e} (3.53)	12.5^{e} (3.60)	13.3^{d} (3.71)	13.8^{de} (3.77)	12.9
T5 - Abamectin 1.9 EC @ 22.5 g a.i.ha^{-1}	14.0	11.5^{e} (3.46)	12.0^{e} (3.53)	12.5^{e} (3.60)	12.8^{e} (3.64)	12.2	11.0^{f} (3.39)	11.5^{f} (3.46)	12.5^{e} (3.60)	13.0^{e} (3.67)	11.8
T6 - Spinosad 45 SC @ 75 g a.i.ha^{-1}	14.3	12.8^{cd} (3.64)	13.5^{cd} (3.74)	13.8^{cd} (3.77)	14.0^{cd} (3.81)	13.5	12.5^{de} (3.60)	13.0^{de} (3.67)	14.0^{c} (3.81)	14.5^{cd} (3.87)	13.5
T7 - Untreated Check	16.5	16.5^{a} (4.12)	17.0^{a} (4.18)	16.5^{a} (4.12)	17.0^{a} (4.18)	16.8	17.0^{a} (4.18)	18.0^{a} (4.30)	18.3^{a} (4.33)	18.5^{a} (4.36)	18.0

PTC – Pre treatment count; DAIIT- Days after second treatment ; Figures in parentheses are $\sqrt{x + 0.5}$ transformed values
In a column, means followed by a common letter(s) are not significantly different by DMRT(P=0.05)

Table 3. Effect of abamectin 1.9 EC on the population of coccinellids in cotton ecosystem after first and second round of application– Field Experiment II (Location-Eastern Block, TNAU) (Mean of four observations)

Treatments	2	Number of coccinellids per 10 plants									
		Days after first treatment					Days after second treatment				
	PTC	3	7	10	14	Mean	3	7	10	14	Mean
T1 - Abamectin 1.9 EC @ 8.1g a.i.ha^{-1}	22.0	21.5^{b} (4.69)	22.5^{b} (4.79)	24.0^{b} (4.95)	27.0^{ab} (5.24)	23.8	26.0^{b} (5.15)	27.0^{b} (5.24)	28.0^{b} (5.34)	31.0^{a} (5.61)	28.0
T2 - Abamectin 1.9 EC @ 10.8 g a.i.ha^{-1}	21.5	21.0^{bc} (4.64)	22.0^{bc} (4.74)	23.0^{bc} (4.85)	25.5^{bc} (5.10)	22.9	24.0^{c} (4.95)	25.0^{c} (5.05)	26.5^{c} (5.20)	29.0^{b} (5.43)	26.1
T3 - Abamectin 1.9 EC @ 14.5 g a.i.ha^{-1}	21.3	20.5^{bcd} (4.58)	21.8^{bc} (4.72)	22.5^{cd} (4.79)	24.5^{cd} (5.00)	22.2	23.5^{c} (4.90)	24.0^{c} (4.95)	25.0^{d} (5.05)	27.8^{bc} (5.31)	25.1
T4 - Abamectin 1.9 EC @ 18.5 g a.i.ha^{-1}	22.5	20.0^{cd} (4.53)	21.0^{c} (4.64)	21.5^{de} (4.69)	23.5^{d} (4.90)	21.5	21.5^{d} (4.69)	22.5^{d} (4.79)	24.0^{de} (4.95)	25.8^{d} (5.12)	23.5
T5 - Abamectin 1.9 EC @ 22.5 g a.i.ha^{-1}	22.0	19.5^{d} (4.47)	20.8^{c} (4.61)	21.0^{e} (4.64)	23.0^{d} (4.85)	21.1	21.0^{e} (4.64)	22.3^{d} (4.77)	23.8^{e} (4.92)	25.3^{d} (5.07)	23.1
T6 - Spinosad 45 SC @ 75 g a.i.ha^{-1}	22.8	20.3^{cd} (4.55)	21.0^{c} (4.64)	21.5^{de} (4.69)	24.5^{cd} (5.00)	21.8	22.3^{de} (4.77)	22.5^{d} (4.79)	24.5^{de} (5.00)	26.5^{cd} (5.20)	24.0
T7 - Untreated check	21.3	23.0^{a} (4.85)	24.0^{a} (4.95)	25.8^{a} (5.12)	28.0^{a} (5.34)	25.2	28.5^{a} (5.38)	29.5^{a} (5.48)	30.0^{a} (5.52)	32.3^{a} (5.72)	30.1

PTC – Pre treatment count; Figures in parentheses are $\sqrt{x + 0.5}$ transformed values

In a column, means followed by a common letter(s) are not significantly different by DMRT(P=0.05)

Table 4. Effect of abamectin 1.9 EC on the population of coccinellids in cotton ecosystem after third and fourth round of application– Field Experiment II (Location- Eastern Block, TNAU) (Mean of four observations)

Treatments 2	Number of coccinellids per 10 plants										
		Days after third treatment					Days after fourth treatment				
	PTC	3	7	10	14	Mean	3	7	10	14	Mean
T1 - Abamectin 1.9 EC @ 8.1g a.i.ha^{-1}	31.0	30.0[b] (5.52)	32.3[b] (5.72)	35.8[b] (6.02)	38.0[b] (6.20)	34.0	37.5[b] (6.16)	40.8[b] (6.42)	43.0[a] (6.60)	45.8[b] (6.80)	41.8
T2 - Abamectin 1.9 EC @ 10.8 g a.i.ha^{-1}	29.0	27.5[c] (5.29)	30.0[c] (5.52)	34.5[b] (5.92)	36.5[c] (6.08)	32.1	35.5[c] (6.00)	38.0[c] (6.20)	41.3[b] (6.46)	44.3[c] (6.69)	39.7
T3 - Abamectin 1.9 EC @ 14.5 g a.i.ha^{-1}	27.8	26.5[c] (5.20)	28.0[d] (5.34)	33.0[c] (5.79)	35.5[c] (6.00)	30.8	34.0[d] (5.87)	36.5[d] (6.08)	39.8[c] (6.34)	42.5[d] (6.56)	38.2
T4 - Abamectin 1.9 EC @ 18.5 g a.i.ha^{-1}	25.8	24.0[d] (4.95)	26.5[ef] (5.20)	31.0[d] (5.61)	34.3[d] (5.89)	29.0	32.3[e] (5.72)	34.0[e] (5.87)	38.3[d] (6.22)	41.0[e] (6.44)	36.4
T5 - Abamectin 1.9 EC @ 22.5 g a.i.ha^{-1}	25.3	23.5[d] (4.90)	26.0[f] (5.15)	31.0[d] (5.61)	33.0[e] (5.79)	28.4	31.0[e] (5.61)	33.5[e] (5.83)	37.5[d] (6.16)	41.0[e] (6.44)	35.8
T6 - Spinosad 45 SC @ 75 g a.i.ha^{-1}	26.5	24.0[d] (4.95)	27.5[de] (5.29)	32.5[c] (5.74)	34.0[d] (5.87)	29.5	32.0[e] (5.70)	34.5[e] (5.92)	38.0[d] (6.20)	42.5[d] (6.56)	36.8
T7 - Untreated check	32.3	33.0[a] (5.79)	35.3[a] (5.98)	38.0[a] (6.20)	40.3[a] (6.38)	36.7	41.5[a] (6.48)	42.3[a] (6.54)	44.5[a] (6.71)	48.0[a] (6.96)	44.1

PTC – Pre treatment count; DAIIT- Days after second treatment; Figures in parentheses are $\sqrt{x + 0.5}$ transformed values
In a column, means followed by a common letter(s) are not significantly different by DMRT(P=0.05)

References

Acharya, S., Mishra, H.P. and Dash, D., 2002. Efficacy of insecticides against okra jassid, *Amrasca biguttula biguttula* Ishida. *Ann. Pl. Prot. Sci.,* 10(2), 230-232.

Badawy, H.M.A. and Arnaouty, S.A.E., 1999. Direct and indirect effects of some insecticides on *Chrysoperla carnea* (Stephens) (Neuroptera: Chrysopidae). *J. Neuropterol.,* 2, 67-74.

Chizhov, V.N., Shukina, E.V. and Yurkin, V.A., 2000. The effect of avermectin preparations on arthropods. *Zashchita I Karantin Rasteni*, 8, 14-15.

Dhawan, A.K. and Simwat, G.S., 1996. Status of natural enemy complex in cotton agro eco system and its impact on present pest scenario in Punjab. In: *First "Indian Ecological Congress"*, National Institute of Ecology, New Delhi, December, 27-31.1996, p77.

Dhawan, A.K. and Simwat, G.S. and Madan, V.K., 1994. Impact of synthetic pyrethroids on the arthropod diversity and productivity of upland cotton, *Gossypium hirsutum J. Cotton Res. Dev*., 8(1), 81-99.

Dhawan, A.K. and Simwat, G.S. and Makwana, D.N., 1992. Impact of bollworm management with different insecticides on target and non target insects, some plant characters and fibre quality of up land cotton variety F 286, *J. Cotton Res. Dev.,* 6 (2), 171-179.

Gomez, K.A. and Gomez. A.A., 1976. *Statistical Procedures for agricultural Research with Special Emphasis on rice*. International Rice Research Institute, Los Banos, Phillippines, 368 p.

Medina, P., Budia, F., Tirry, L., Smagghe, G. and Vinuela, E., 2001. Compatibility of spinosad, tebufenozide and azadiractin with eggs and pupae of the predator *Chrysoperla carnea* (Stephens) under laboratory conditions. *Biocontrol Sci. Tech.*, 11(5), 597-610.

Sansone, C.G. and Minzenmayer., R.R., 2000. Impact of new bollworm insecticides on natural enemies in the southern rolling plains of Texas. In: *Proc. Beltwide Cotton Conf.,* P. Dugger and D. Richter (eds.), San Antonio, USA, 4-8 January, Volume-2, 1104-1108.

Singh, S.P., 1994. "Fifty years of AICRP on biological control", Project Directorate of Biological Control, Bangalore.

Biocontrol of Insect Pest: Present Scenario and Future Strategies

HEM SAXENA*

Biocontrol has been viewed as a sound alternative to the chemicalcontrol. During the past several years, some noticeable success of biocontrol has been achieved such as classical bio-control of cottony cushion scale *Icerya purchasia* by the introduction of *Rodalia cardinalis*. A large number of parasitoids have been reported to be very successful such as *Isotima javensis* – larval parasitoid of sugarcane top borer, *Epiricarnia melanoleuca* – adult and nymphal parasitoid of *Pyrilla purpusilla*, *Campoletis chlorideae* – larval parasitoid of *Helicoverpa armigera* and many more. During 2004, heavy parasitisation of *Copidosoma floridanum* – a larval parasite of *H. armigera* was reported for the first time at Kanpur. The most successful predators are coccinellid beetles, reduviid bugs, pentotomid bugs, carabids, araneids and formicids. A large number of arthropods and insectivorous birds have also been reported as successful predators.

Of the different types of viruses, the nuclear polyhedrosis virus (NPV) has the greatest potential since they are more virulent, killing the insects much faster than granulosis virus (GV) and cytoplasmic virus (CPV). The NPV of *H. armigera* is effectively controlling it on chickpea, pigeonpea, groundnut, sunflower, cotton, tomato and sorghum. The NPV of *Spodoptera litura* is also very effective in cotton and tobacco. *Bacillus thuringiensis* has been used in the management of several insect pests such as *Plutella xylostella* in vegetables, gram pod borer *H. armigera* in chickpea and pigeonpea, *S. litura* in cabbage and fruit borer of okra. A number of commercial formulations of Bt is available. Fungal pathogens particularly *Beauveria bassiana, Metarhizium anisopliae, Vertcillium lecanii* and *Nomuraea riley* have been found to be promising in the control of several agricultural pests. *Nomuraea rileyi* occurs in epizootic form on *H. armigera*, *S. litura* and *S. exigu* in field conditions. Recently, *Aspergillus flavus* and *A. niger* are reported from the larvae of *H. armigera* in pulse eco-system. *Nosema* sp. and *Vairiomorpha* sp. two protozoans are reported on *H. armigera.* A large number of entomopathogenic nematodes are also reported on *H. armigera*. Recently two new species of entomopathogenic nematodes (EPN) *i.e., Steinernema masoodi* and *S. seemae* have been reported from the larvae of *H. armigera* causing upto 67% mortality of *H. armigera* larvae in laboratory. Many bio-agents operate unnoticed and it is amazing to note how complex bio-ecological interaction is going on. For greater use of bio-agents, their intensive commercial production and formulation with strict enforcement of quality control measures are necessary.

INTRODUCTION

Modernization of agriculture has lead to dependence on various agrochemicals which are highly hazardous and resulted not only in the mortality of the pests, but also lead to development of resistance, residue in harvested produce and adverse effect on other non-target organisms in the eco-system including human-beings. There is considerable pressure from environmentalist for the use of bio-control agents against insect pests. Biocontrol has been viewed as a sound alternative to

* Crop Protection Division, Indian Institute of Pulses Research, Kanpur – 208 024 (U.P.)

the chemical-control (Ali, *et. al.*, 2007). During past several years, some noticeable success of bio-control has been achieved such as classical biocontrol of cottony cushion scale *Icerya purchasia* by the introduction of *Rodalia cardinalis*.

Bioagents are parasites, predators and pathogens (virus, bacteria, fungi, protozoan and entomopathogenic nematodes), which offer great scope in biocontrol of insect-pests. A large number of parasites have been reported very successful such as *(-Isotima) javensis*, larval parasite of sugarcane top borer, *Epiricarnia melanoleuca*, adult and nymphal parasitoid of *Pyrilla purpusilla*, *Campoletis chlorideae,* larval parasitoid of *Helicoverpa armigera* and many more. During 2004, heavy parasitisation of *Copidosoma floridanum,* a larval parasite of *H. armigera* was reported for the first time at IIPR, Kanpur (Saxena, 2007). The most successful predators are coccinellid beetles, reduviid bugs, pentomid bugs, carabids, araneids and formicids. A large number of arthropods and insectivorous birds have been also reported as a successful predators (Rabindra *et al.*, 2005).

Of the different types of viruses, the nuclear polyhedrosis virus (NPV) has the greatest potential since they are more virulent, killing the insects much faster than granulosis virus (GV) and cytoplasmic virus (CPV). The NPV of *Helicoverpa armigera* is effectively controlling *H. armigera* in chickpea, pigeonpea, groundnut, sunflower, cotton, tomato and sorghum. The NPV of *Spodoptera litura* is also very effective in cotton and tobacco.

Bacillus thuringiensis (Bt.) is one of the earliest microbial insecticides to be commercially produced. In India, this bacterial insecticide has been used in the management of several insect pests such as *Plutella xylostella* in vegetables, gram pod borer *H. armigera* in chickpea and pigeonpea, *S. litura* in cabbage and fruit borer of okra. A number of commercial formulations of Bt are available in the market.

Fungal pathogen particularly *Beauveria bassiana, Metarhizium anisopliae, Verticillium lecani* and *Nomuraea rileyi* have been found to be promising in the control of several agricultural pests. *Nomuraea rileyi* is highly active against a number of insect pests and occur in epizootic form on *H. armigera*, *S. litura* and *Spodoptera exigua* in field conditions. Recently, *Aspergillus flavus* and *Aspergillus niger* are reported from the larvae of *H. armigera* in pulse eco-system at IIPR, Kanpur. Both the fungus are very effective against *H. armigera* (Saxena, 2007).

Nosema sp. and *Vairiomorpha* sp. two protozoans are reported on *H. armigera.* A large number of entomopathogenic nematodes were also reported on *H. armigera*. Recently two new species of entomopathogenic nematodes (EPN) *i.e., Steinernema masoodi* and *S. seemae* have been reported from the larvae of *H. armigera* at IIPR, Kanpur, causing upto 67% mortality of *H. armigera* larvae in laboratory.

Present Status of bio-control in India

(i) Pulses

Helicoverpa armigera is a major insect pest of chickpea and pigeonpea, causing maximum damage at podding stage of crops.

Bio-agents

A large number of bio-agents have been recorded on gram pod borer *H. armigera* infesting chickpea. However, in spite of a huge list of bio agents, *H. armigera* is causing severe crop losses by defying the suppressive role of natural enemies.

1. Parasite

A comprehensive list of parasites and predators of *H. armigera* has been published by entomologist from time to time (Lingappa *et al.* 2005). Egg parasitism by *Trichogramma* spp. is negligible on chickpea due to the acid exudates of chickpea foliage, which is considered to repel the egg parasites.

About half a dozen of larval parasites known to occur on *H. armigera* on chickpea in India. *Campoletis chlorideae* Uchida is one of the most important and potent larval parasite, which parasitizes 20-76% larvae of *H. armigera* in the field. Saxena and Ahmad (2001) reported that endosulfan and match are safer insecticides to *C. chlorideae*.

In North India the pupa of gram pod borer, *H. armigera* undergoes the facultative diapause, resulting into a prolonged pupal period of more than 100 days. This results into poor availability of host to the *Campoletis chlorideae* during November to February, resulted into significant decrease in parasite population in the absence of sufficient food. During post-winter months, *H. armigera* moths emerge out from hibernating pupae and breed freely till the parasite is in position to put any significant check on its multiplication. It is the population of *H. armigera* during March-April which causes, heavy crop losses.

H. armigera has been recorded to occur around the year but in north India it is usually not a serious pest except during March-April. Thus, absence or almost negligible parasitism has been an important factor in the rapid build up of *H. armigera* population along with favourable climatic condition. This coincide with the podding stage of chickpea and pigeonpea in north India. During this period *i.e.* March-April in North India the inundate release of potent parasites should be done.

The inundative release of egg-larval parasite, *Chelonus heliopea* and larval parasites such as *C. chlorideae* and *Eribours* spp. are not economical as they are not amenable for mass rearing. On the other hand, the prospects of using classical biological control, involving introduction of exotic natural enemies are not considered to be bright, in view of the migratory behaviour of moths and the sudden influx of populations to which resident natural enemies may not be able to respond readily. Larval parasitism of *H. armigera* has been found to be higher on susceptible cultivars than on resistant cultivars in chickpea, suggesting that host-plant resistance influences the incidence of larval parasitism. There are reports that in pulses ecosystem (chickpea & pigeonpea) egg parasitism to some extent has been found in other countries (Kenya, Newzealand *etc.*) and effort are on to import these egg parasite, evaluate, colonize and release. Six species of parasitoids have been recorded from field collected pupae.

2. Predator

A total of 33 species of insects and many spiders have been reported form *Helicoverpa* spp. in India. The wasps *Delta* spp. have been observed to predate on *H. armigera* larva by placing them in the cells of their mud nests before oviposition. The most common predator of *Helicoverpa* include *Chrysoperla* spp., *Nabis* spp. *Geocoris* spp. *Orius* spp; *Polistes* spp. and various Pentatomidae, Reduviidae, Coccinellidae, Carabidae, Formicidae and Araneidae. Some predators have been used in augmentative release studies, notably *Chrysoperla carnea.* Although effective in large numbers, the high cost of large -scale production precludes its economic use in biological control of *H. armigera.*

Clever and Ambrose (2002) studied the functional response of the predator *Rhynocoris fuscipes* a reduviid bug against *H. armigera* in laboratory found that the prey density was negatively correlated with the handling time of the predator. More than 60 species of arthropod predators have been recorded preying upon eggs and larvae of *H. armigera* (Romies and Shanower, 1996). Predation of pre-pupae and pupae by ants may be significant. Physical factors impeding adult emergence may also account for 10 to 20 % pupal mortality.

Insectivorous birds: A number of predatory (insectivorous) birds have been reported from different parts of India such *Passer domesticus* L., *Sterna aurantia* and *Bubulcus ibis* etc. Ojha *et. al.* (2001) studied the relative abundance of seven insectivorous birds in Rajasthan in chickpea. Installation of bird perches @ 30-40/ha is now being recommended as an augmentative method to increase predation of *H. armigera* larvae. Gopali and Lingappa (2001) reported the use of the black drongo, *Dicrurus adsimilis* (Bechstein) an insectivorous bird against *H. armigera* in pigeonpea in Gulbarga (Karnataka) is effective. Tall sorghum perches was the most effective and exhibited high crop yields. Besides predation, birds have quite helpful in dispersing HaNPV though fecal matter.

3. Entomophagous pathogens:

There has been a considerable interest in classical biological control as a tactic for the management of *Helicoverpa armigera* and exploitation of entomophagous pathogens such as *Bacillus thuringiensis, H. armigera* NPV and entomogenous fungi etc.

Amongst the microbial pathogens, 3 species of virus, viz; Nuclear Polyhedrosis Virus (NPV), Cytoplasmic Polyhedrosis Virus (CPV) and Granulosis virus (GV), 1 spp. of bacteria, *Bacillus thruingiensis* Berliner (Bt), 3 species of fungi *Metarhizium anisopliae, Beauveria bassiana* and *Nomuraea rileyi*, 2 species of protozoa *Nosema* spp. and *Vairimorpha* spp. and 2 species of nematodes *Hexamermis* sp. and *Steinernema* sp. have been recorded from the larvae of gram pod borer. Practical utility of nuclear polyhedrosis virus (*Ha*NPV), *B. bassiana* and *Bacillus thuringiensis* has been demonstrated. Results of field trials have revealed that application of *Ha*NPV 250 LE (Larval equivalent/ha) plus endosulfan 0.035% gave better control of *H. armigera* than *Ha*NPV 250 LE/ha alone. Addition of adjuvants like crude sugar, groundnut oil cake and ULV retardants such as tinopal, etc., has also been considered useful to make NPV use more effective.

Incidence of cytoplasmic polyhedrosis virus (CPV) and grandulosis virus has also been reported on the larvae of the pest. However, the use of *Ha*NPV for suppressing larval population of *H. armigera* under field conditions is considered most promising. Two sprays with *Heliothis* nuclear polyhedrosis virus @ 250 LE/ha are as effective as two sprays of endosulfan 0.07% in reducing the infestation of *Heliothis* larvae and pod damage, resulting an increase in the seed yield, alternating sprays of *Ha*NPV + endosulfan *Ha*NPV have also been advocated.

Saxena and Ahmad (1977a) reported *Beauveria bassiana* is effective against *H. armigera* infesting chickpea. Saxena and Chaudhary (2002) recorded two new entomogenous fungi infesting larvae of *H. armigera* in fields. They are identified as *Aspergillus flavus* and *Aspergillus niger*. *A. flavus* was found 80% pathogenic to 3rd and 4th instar larvae of *H. armigera* in laboratory (Saxena, *et. al.*, 2005).

Nomuraea rileyi has been established as a viable pathogen of *H. armigera* which is suitable for commercialization, as it is readily amenable to the mass production. Commercial formulations of *Bacillus thuringiensis* (Bt.) such as thuricide, dipel, delfin etc., have provided high mortality of *H. armigera* in laboratory as well as in field conditions (Saxena and Ahmad, 1998).

Recently two new species of entomopathogenic nematodes (EPN) *i.e., Steinernema masoodi* and *S. seemae* have been reported causing 67% mortality of the larvae of *H.armigera* in the laboratory. There has been some success in the use of pathogens such as *Bacillus thuringiensis, N. rileyi* and *Helicoverpa armigera* nuclear polyhedrosis virus (*Ha*NPV). However, the relatively high cost and rapid inactivation by ultraviolet light often lead to poor performance under field conditions. In case of *Ha*NPV, the difficulty in obtaining consistently the high levels of purity and virulence necessary to achieve satisfactory control has limited its usefulness.

(ii) Sugarcane

Sugarcane is one of the most important cash crop of India. About 20% yield is lost annually due to insect pests and diseases. Major insect pests are borers, sucking pests and white grubs. Sugarcane provides stable agro eco-system ideal for the colonization of natural enemies and bio- control has emerged as one of the sustainable method of control of top borer, stalk borer, shoot borer, internode borer, scale insect and *Pyrilla*, etc.

Sugarcane leaf hopper *Pyrilla perpusilla* was successfully controlled by *Epiricania melancleuca* in Rajasthan, Karnataka, Madhya Pradesh, Maharashtra, Orissa, West Bengal and Gujarat (David,

1986; Joshi and Sharma, 1989; Balasubramaniam, 1993; Verma and Tiwari, 1994 and Sharma 2004) and by *Metarhizium arisopliae* (Singh, 1996).

Sugar Top borer *Scipophga excerptalis* was effectively controlled by inundative releases of *Isotima javensis* in Pugalur in Tamil Nadu giving 70-81% parasitization (Easwaramorthy and David, 1979)

Sugarcane Stalk borer *Chilo auricillus* by egg parasite *Trichogramma chilonis* in Bihar, by larval parasite *Stumiposis infernes* in Haryana and by exotic larval parasite *Allorhoghus pyralophagus* in U.P., Haryana was effectively controlled (Balasubramaniam, 1993 and Nigam and Varma, 1989).

Sugarcane Shoot borer *chilo infuscatellus* and Intermode borer chilo *sacchariphagus inddicus* were effectively controlled by egg parasite *Trichogramma chilonis* in Coimbatore and Karanataka (Singh, 1996)

(iii) Cotton

Cotton is one of the most important cash crop, damaged by more than 12 major insect pests which causes economic loss. Balasubramanian (1993) reported that *Chrysopa scelestes, Trichograma chilonis* and *Eucelatoria bryani* are effective against *Helicoverpa armigera* in cotton in Tamil Nadu. While Dhaliwal and Vikas (2005) reported that *Trichograma chilonis* , HaNPV and neem are increasing yield of cotton about 30% in Punjab.

(iv) Oil Seeds

A number of parasitoids, predator and pathogens were tried against important pest of oil seeds such as *Coccinella septempunctata* against mustard aphid *Lipaphis erysini, Bacillus popilae* against white grub of groundnut, Bt. and AjNPV against castor semilooper *Achaea janata* and HaNPV, Bt. against *Helicoverpa armigera* and SlNPV against *Spodoptera litura* in sunflower are giving good control(PDBC- RH -1993-94, 1995-96, and Bassapa,2005).

(v) Paddy

Paddy stem borer *Scirpophaga incertulas,* yellow stem borer and leaf folder *Chaphalocrocis medinalis* are effectively controlled by egg parasite *Trichograma japonicum* in Tamil Nadu (Balasubramanian, 1993).

(vi) Vegetable Crops

Vegetables are short duration crops, pest activity is observed for a limit period only. That's why inundative releases of parasites were done. In tomato against *Helicoverpa armigera* inundative release of *Trichogramma brasiliensis, T. pretiosum & T. chilonis* were effective (Sharma, 2004). Besides this HaNPV were also effective against *H. armigera* (Gopal Krishnan, 1999) in tomato.

In Cabbage *Plutella xylostella* was effectively controlled by *Cotesia plutellae* in Bangalore in (Krishna Moorthy, 1999). In Brinjal *Leucinodes orbonalis* by *Trathala flovoorbitalis* in Bihar was effectively control (Krishna Moorthy, 1999).

(vii) Fruit crops

Most of the fruit crops provide a stable environment offering good opportunities for biological control. Fruit crop pests are more amenable for biological control because many of them are sessile in nature and are easily available for the attack of the natural enemies throughout the year.

In Citrus Cottony cushionscale *lcerya purchasi* by *Rodolia cardinalis* in Niligiris was effectively controlled. In grapes Pink Mealy bug *Maconellicoccus hirsutus by Cryptolaemus montrouzien* is effectively controlled (Mani, 1999). In mango Mealy bug *Rastrococous iceryoides by Cryptolaemus montrouzien* is effectively controlled (PDB+C-RH 1995-96). In apple San Jose Scale *Quadriaspidiotus perniciosus by Encarsia pemiciosi* in Kashmir reduced (PDBC-RH 1993-94).

(viii) Plantation crops

Coconut has received greater attention of bio- control researchers than other plantation crops like coffee, cocoa, etc. Important pests of coconut may be controlled by the application of indigenous parasitoids .

In Coconut Black Headed caterpillar by *Goniozus nephantidis* + *Bracon brevicornis* giving good control (Dhaliwal and Arora, 1998). The *Rhinoceros Beetle Oryctes rhinoceros* by *Metarhizium anisopliae* was effectively controlled (Singh, 1996).

In India tremendous progress has been made in all aspects of biological control. India has a very rich biodiversity of biotic agents. Atleast 26 biotic agents of India origin have been introduced to other countries and successfully established. The credit for the first classical biological control success goes to India, as way back in 1762 the common India myna bird, *Acridotheres tristis* was exported to Mauritius, and it could successfully control the red locust, *Nomadaris septemfasciata.*

For the conservation of natural enemies minimum spray of pesticides, retaining or planting of weeds or pollen and nectar bearing plants on the bunds for providing shelter and food to the biotic agents and providing suitable nesting and perching site for predatory birds, cultural lowering of the pest population by physically removing the off season flushes on citrus and other crops, regular removal of the fallen fruits collecting and killing of the pest, manipulation of sowing dates, spacing, avoidance of excessive use of fertilizers, irrigation, growing of pest tolerant varieties, growing trap crops for attracting pests and killing them with biotic agents have been recommended.

In 1981 there was one private insectory supplying natural enemies to the farmers on demand. The current information indicates that there are 128 units in the country of which 80 are in private sector, which supply predators, parasitoids, entomopathogens, plant disease antagonists, *etc.* to the farmers. This growth of private sector is an indicator of the bright prospects of biological control.

Farmer's psychology about bio-agents and pesticides – (Yadav, 2001)

- Ninety per cent farmers still rely on chemical pesticides.
- Only four per cent of the farmers used bio-agents available commercially.
- Twenty per cent of the farmers showed awareness about usefulness of parasites and predators.
- Twenty-three per cent of farmers approach Agriculture College to seek guidance.
- Fifty per cent of the farmers rush to nearby pesticides dealers for counseling.
- Fifteen per cent of farmers approach Gram sevaks.
- Twelve per cent of the farmers follow what their neighbor says/does.

Future strategies

Transfer of technology

1. Intensification of education amongst the farmers on all the aspects of bio-control.
2. Integration of bio-control technology with information technology.
3. Establishing counseling cells at each village.

Commercial production of bio-agents:

1. Establishing commercial laboratories for mass production of proven bio-agents.
2. Quality control laboratories to be setup as early as possible.
3. Registration procedure should be simplified.
4. Each SAU should establish parasites/predator production unit and supply bio-agents to the farmers.
5. Government may give soft loans to unemployed Agricultural Graduates to take up commercial production.

Future needs

1. **Biosystematic and bioecological studies**
   For the successful multiplication of natural enemies it is essential to study the bioecology of the host as well as natural enemies.
2. **Importation of potential natural enemies**
   To improve co-operation among research workers, Institutes and International organisations for exchange of natural enemies.
3. **Improving the performance of natural enemies**
   1. In nature there are number of geographical ecotypes/biotypes available, it is possible to collect and hybridise with a view to enhance the key attributes *i.e.* potency and pesticide and climatic tolerance.
   2. Artificial structures and their placements in the field may be done so that predators like wasps and birds utilise these for building up their nests etc.
   3. Providing additional food for adults of predators and parasitoids increases their fecundity as well as searching ability. The need for planting or retention of certain plant species in and around cropped area, which could serve as a source of nectar or pollen.
4. **Development of mass production technology**
   The first and foremost task in bio-control is to mass-produce the host and bio-control agents.

Conclusion

A good amount of research has been done on various aspects of biological control. Now the farmers are aware about bio- agents. Sustained efforts/collaboration with other laboratories state department of agriculture and NGOs will be long way in promoting this noble method of pest suppression.

With the globalization and liberalization of agriculture the bio-control of the insect-pests has attained immense importance and has very bright future. A number of strategies and techniques are involved. Many bio-agents operate unnoticed and it is amazing how complex bio-ecological interaction is going on. For the promotion of bio-agents their intensive commercial production and formulation with strict enforcement of quality control measures are necessary.

References

Ali, S. S., Saxena, H. and Mohan, R.P., 2007 . Prospects of biopesticides in safe environment. page 123. In Abstracts: Natational Symposium on Legumes for Ecological Sustainability : Emerging Challenges and Opportunities, November 3-5 2007, IIPR, Kanpur.

Balasubramanian, G., 1993. Field application of parasites and predators. In *National Training on Mass Multiplication of Bio-control Agents*, conducted by Training Division, Directorate of Extension Education, TNAU, Coimbatore – 641 003. pp. 49-56.

Basappa, 2005. Role of biopesticides in the Biointensive IPM of oilseed crops. Proceedings of Biopesticide conference, November 11-13, 2005, Palampur, Himachal Pradesh.pp.55-57.

Claver, M.A. and Ambrose, D.P., 2002. Functional response of the predator, *Rynocoris fuscipes* (Heterodera : Reduviidae) to three pest of pigeonpea. (*Cajanus cajan*). *Shashpa*, **9** (1), 47-51.

Dhaliwal and Vikas, J., 2005. Biointensive IPM with trichogramma, *Ha*NPV, NSKE in cotton. Proceedings of Biopesticide conference, November 11-13, 2005, Palampur, Himachal Pradesh.pp.63-65.

Dhaliwal, G.S. and Ramesh Arrora, 1998. Parasitoids and Predators. *In Principles of Insect Pest Management*, Kalyani Publishers, New Delhi. pp. 112-127.

Easwaramoorthy, S. and David, H., 1979. A granulosis virus of sugarcane shoot borer, *Chilo infuscatellus*(Lep idoptera : Crambidae). *Curr. Sci.*, 48, 685-686.

Gopalakrishnan, C., 1999. Microbial control of vegetable crops. Integrated pest Management in Horticultural crops – Compendium released by Trainers Training Centre, IIHR, Bangalore, pp.99-108.

Gopali, J.B. and Lingappa, S., 2001. Utilization of black drongo, *Dicrurus adsimilis* (Bechstein) an insectivorous bird in pigeonpea ecosystem. *Karnataka Journal of Agricultural Sciences* **14,** 1078-1082.

Joshi, R.K. and Sharma, S.K., 1989. Augmentation and conservation of *Epiricania melanoleuca*, for the population management of sugarcane leaf hopper, *Pyrilla perpusilla*, under arid conditions of Rajasthan. *Indian Sugar* **39,** 625-628.

Krishanamoorthy, A., 1999. Biological control of pest of vegetable crops. Integrated pest Management in Horticultural crops – Compendium released by Trainers Training Centre, IIHR, Bangalore, pp92-98.

Lingappa, S., Saxena, Hem and Devi Vimla, P.S., 2005 Role of biocontrol agents in management of *Helicoverpa armigera* (Hubner). Pages 159 -184. In: *Recent Advances in Helicoverpa Management* (Hem Saxena , A. B. Rai, R. Ahmad and Sanjeev Gupta (Eds.)) Indian Society of Pulses Research and Development , IIPR, Kanpur.

Mani, M., 1999. Biological control of Tropical Fruits. Integrated pest Management in Horticultural crops – compendium released by Trainers Training Centre, IIHR, Bangalore, pp.81-91.

Nigam, Hem and Varma, A., 1989. Field releases of an exotic parasite, *Allorphogas pyralophagus* Marsh against sugarcane stalk borer, *Chilo auricilius* Ddgn. at two bio-ecological zones of the pest habitat and its recoveries. *Indian Journal of Entomology* **51**(2), 136-138.

Rabindra, R. J., Singh, A, Saxena, H. and Ballal, R.C., 2005. Biological control of insect pests and diseases in food legumes. Pages 25-26. In Abstracts *–4 International Food Legumes Research Conference* October 18 -22., New – Delhi.

Research Highlights 1993 – 1998. Project Directorate of Biological Control, Bangalore.

Romeis, J. and Shanower, T.G., 1996. Arthropod natural nemies of *Helicoverpa armigera* (Hubner) (Lepidoptera : Noctuidae) in India. *Biocontrol Sciences and Technology* **6,** 481-508.

Saxena, H., 2007. Record of new natural enemies of *Helicoverpa armigera* in pulses. page 134 In Abstracts: Natational Symposium on Legumes for Ecological Sustainablity : Emerging Challenges and Opportunities, November 3-5, 2007 , IIPR, Kanpur.

Saxena, H., 2007. Microbial management of crop pest. page 26 Invited paper *In* : Souvenir of Biopesticide International Conference " Biocicon -2007" , St. Xavier's College (Autonomous) , Palayamkottai (Tamil Nadu), 28 -30 November, 2007.

Saxena, Hem and Ahmad, R., 1998. Comparative performance of *Bacillus thuringiensis* (Bt) formulations against *Helicoverpa armigera* infesting chickpea. *Proceedings of National Symposium on Management of Biotic and Abiotic Stresses in Pulse* Crops held at IIPR, Kanpur on 26th to 28th June, 1998. p. 61.

Saxena, Hem and Ahmad, R., 1997b. Field evaluation of *Beauveria bassiana* (Bals.) Vuillemin against *Helicoverpa armigera* (Hubner) infesting chickpea. *Journal of Biological Control* (1-2), 93-96.

Saxena, Hem and Ahmad, R., 2001. Search for safer insecticides to *Campoletis chlorideae* Uchida a larval parasite of *Helicoverpa armigera* (Hubner). *Proceedings of Symposium on biocontrol based pest management for quality crop protection in current millennium.* July 1`8-19, 2001. Punjab Agricultural University, Ludhiaan . pp. 13.

Saxena, Hem and Chaudhary, R.G., 2002. Record of new entomogenous fungi from the larvae of *Helicoverpa armigera. IIPR News Letter* **13** (2), 3.

Saxena, Hem, Chaudhary, R.G. and Dar, M.H., 2005. Bio- efficiency of two entomogenous fungi against *Helicoverpa armigera* pages 342. In Abstracts- *4 International Food Legumes Research Conference* October 18 -22. New – Delhi.

Sharma, A.K., 2004. A hand book of organic farming. Agrobios India, Jodhpur.pp.627.

Singh, S.P., 1996. Biological control of crop pests. In *Second International Congress of Entomological Sciences*, National Agricultural Research Centre, Islamabad, Pakistan (March 19-21, 1996), pp.21.

Varma, A and Tiwari, N.K., 1994. Recent concepts in Insect Pest Management in Sugarcane. *In* : G.S. Dhaliwal and Ramesh Arora (eds.). *Trends in Agricultural Insect pest Management.* Common wealth publishers, New Delhi, India, pp.238-264.

Yadav, D.N., 2001. Present day scenario of biocontrol agents in integrated pest management with special reference to parasites and predators. In: *Plant protection - New Horizons* (Eds: S.C. Bhardwaj, R.C. Saxena and R. Swaminathan). p.53-60.

Biological Control of Thrips of Horticultural Importance in India: Present Status and Future Scope

P.N. Ganga Visalakshy* and A. Krishnamoorthy*

Thrips are important insects causing substantial yiled losses as pests and vectors of viral diseases of several horticultral crops in India. The rising demand for organic produces in the internaitonal and national markets in the recent years seeks alternates to conventional chemical pesticides. Biologicla control is high on the list of petential alternate techniques. Natural enemies such as entomopathogens, predators and parasitoids have shown promise. The present paper discusses on the procedures involoved to include potential natural enemies as a component in biological control based integrated management strategy with special reference to *Scirtothrips dorsalis* and *Thrips tabaci*.

Introduction

There has been increasing interest in recent years in adopting more of eco-friendly control methods as part of the Integrated Pest Management (IPM) approach in our country. Factors such as awareness by the consumers on the health problems associated with the pesticide sprayed produces, resurgence of pests resulting in poor results by chemical pesticides application coupled with the profit incurred by organic farming by the farmers are some of the contributing factors that have led for the increasing demand for pesticide free agricultural produces. In addition, there is support at the national level for promoting export oriented agriculture as well as organic farming, as means of capturing niche markets and so enhance the incomes and livelihood of the rural and farming communities in India. These market-linked farming systems necessitate development of eco-friendly management practices such as biological control.

In the recent decades, sucking pests such as thrips, mites, whiteflies, are gaining prominence as economically important pests of many horticultural crops. The present paper gives a review of status of research work carried out on biological control of thrips of horticultural importance in India and future research thrusts for improved use of the bio-control agents for sustainable management of thrips with reference to *Scirtothrips dorsalis* and *Thrips tabaci* of horticultural importance.

Thrips species *Scirtothrips dorsalis*, *Thrips tabaci* and *T. palmi* are considered as serious pests of many horticultural crops in India (chilli, capsicum, cucurbits, pomegranate, grapes, rose, gerbera etc.). Larvae and adults pierce and suck the plant content that under severe infestation affects the yield and marketability of the produce. In addition, they play an important role as vectors of many viral diseases (Ananthakrishnan, 1984; 1993).

1. Crop loss in relation to thrips infestation in horticulture

The information on the crop loss estimate in relation to thrips damage in horticultural crops is scanty. The available estimates of horticultural crop losses due to thrips is about 30 – 50% loss in chilies due to *S. dorsalis* (Kandasamy *et al.*, 1990) and 46-87% yield loss in onion by *Thrips tabaci* (Srinivas and Lawande, 2004 ; Mohite and Moholker, 1992). Quantitative yield reduction due to *S. dorsalis*

* Indian Institute of Horticultural Research, Hessarghatta Farm P.O., Bangalore – 560 089

has also been found to be more than 90% on chilli pepper, compared to 11- 32 % in sweet pepper, while quality loss of 88-92% was observed in sweet pepper (Krishna kumar, 1995).

Though thrips are reported to be serious pests of corps such as rose, gerbera, cucumber, and watermelon etc. information on the crop loss due to their infestation is many crops is lacking. There is a need to develop baseline data on crop losses caused by thrips in nationally important horticultural crop representing the different agro-eco-zones. This would provide convincing data on crop loss and also the basis for developing appropriate bio-control strategies besides.

2. Effect of ecological and climatic factors on thrips incidence

Ecological factors such as temperature, humidity, rainfall are reported to be limiting factors in the infestation /incidence of thrips. Incidence levels of *S. dorsalis* on chilli has been reported to be greater during dry periods, when the temperature is 30° C and above, with no rainfall (Lingeri *et al.*, 1998; Manjunatha *et al.,* 2001). A positive correlation of *S. dorsalis* incidence has been found with maximum temperature and negative association with rainfall, minimum temperature, mean relative humidity and mean vapour pressure in chilli ecosystem (Panickar and Patel, 2001; Varadarajan and Veeravel, 1995).

Thrips incidence was found to subsequently reverse and with the ageing of crop, they became more abundant at the wetter end of the gradient. Incidence of *S. dorsalis* was highest during March – April, with the emergence of new flush (Wheatly *et al.*, 1989). However, temperature, relative humidity, rainfall and sunshine were not found to significantly affect infestation levels of *S. dorsalis* on mango in Gujarat (Kumar *et al.*, 1994). In onion, *T. tabaci* showed higher incidence level (78%) in *Rabi* season crops in Maharashtra (Srinivas and Lawande, 2004).

3 Status of biological control of thrips In India

3.1 Pathogens as biological control agents of thrips

3.1.1 Pathogens isolated from thrips in India

A list of the mycopathogens recorded so far in India on thrips is furnished in Table 1. The pathogens reported showing promising effect for their control include the Microsporadian parasite - *Mrazekia* sp. infecting the larval forms of *Scirtothrips oligochaetus* (Raizada, 1976), besides the nematode, *Anguillulina aptini* attacking the flower inhabiting thrips namely *Frankliniella schultzei* and *Microcephalothrips abdominalis* occurring on sunflower (Varatharajan, 1985).

Certain fungal pathogens namely, species of *Cephalosporium* and *Aspergillus*, *Cladosporium cladosporioides* and *Entomophthora thripidium* are also reported to be effective against thrips species (Ananthakrishnan, 1984).

Table 1: List of mycopathogens recorded as natural enemies on Thrips in India*

Mycopathogen	***Thrips species***
Cladosporium cladosporioides	*Thrips flavus*
Trichothecium roseum	*Thrips tabaci*
Entomophthora thripidium	*Thrips tabaci*
Cephalosporium sp.	*Thrips tabaci*
Aspergillus sp.	*Thrips tabaci, Frankliniella* sp.
Paecilomyces fumosoroseus	*Thrips tabaci*
Hirsutella thompsonii	Thrips in general
Verticillium sp.	*Scirtothrips dorsalis, Thrips palmi*

Taking into view the importance of thrips in horticultural crops, a biological control programme with emphasis on *S. dorsalis* and *T. tabaci* was initiated at IIHR, Bangalore, with the below mentioned objectives

i. Isolation of entomopathogens from thrips and assessing their infectivity
ii. Evaluation of strains of the entomopathogens against different stages of *S. dorsalis from* laboratory to field conditions
iii. Prolonging the viability of the entomoapthogens by identifying oils, stickers and emulsifiers that act as synergists
iv. Determining the effectiveness of the isolated entomopathogens to thrips in relation to different temperature and humidity.
v. Compatibility of chemical pesticides and fungicides against potential entomopathogens to *S. dorsalis* and *T. tabaci*
vi. Development of a biocontrol based management strategy for the management of *S. dorsalis* and *T. tabaci* to selective horticultural crops.

3.1.2 Pathogenecity of entomopathogens against S. dorsalis

In view of developing a biological control management for thrips, a strain of *V. lecanii* isolated from *T. palmi* infesting cucumber denoted as IIHR strain (Ganga Visalakshy *et al.*, 2004), *Metarhizium anisopliae* a strain isolated from a sucking pest and *Beauveria brongniartti*, isolated from the soil were evaluated under laboratory conditions against *S .dorsalis.* Preliminary tests indicated that the fungal pathogens are infective to *S. dorsalis.*

Further bioassay studies to determine the host pathogen relationship were carried out under laboratory conditions on *S. dorsalis* Results indicated that all the three entomopathogens caused significant mortality to adults and larvae of *S. dorsalis* at dosages varying from $1x10^4$ to $1x10^9$. However, the mortality was positively correlated with the stage of the insect, dose and period of time of inoculation. Among the different dosages of the three fungus tested, application of spore suspension @ $1x10^9$ spores/ml recorded highest mortality (Ganga Visalakshy, unpublished report).

3.1.3. Effect of relative humidity and temperature on infectivity of V. lecanii to S. dorsalis

Temperature and relative humidity are important factors that influences growth, infection and viability of entomopathogens against the target pest. Increased interest in the use of entomopathogens within IPM programmes of thrips has necessitated the selection of fungal pathogens tolerant to wide range of temperature and humidity.

The present strain of *V. lecanii* (IIHR strain) was found to cause significant mortality to the adults and larvae of *S. dorsalis* at even low RH of 50%. Similarly, the entomopathogen was observed to be virulent even at temperatures of 30-32°C indicating that the strain of *V. lecanii* could prove to be an effective biological control agent of *S. dorsalis* in the field even dry periods and high temperatures (IIHR, Annual report, 2007)

3. 1.4 Evaluation studies

Evaluation studies carried out on potted plants, preliminary field trails against *S. dorsalis* on chilli and capsicum under protected conditions respectively indicated that the entomopathogens *M. anisopliae, V. lecanii* and a combination *of M. anisopliae, V. lecanii* and + *B. brongniartti* @ $1x10^9$ spores/ml, significantly reduced the thrips population (Ganga Visalakshy, unpublished report).

3.1.5 Factors that enhance effectiveness of entomopathogens

A complex set of interacting processes, both environmental and biotic is necessary for or inhibitory to development of epizootics caused by entomopathogenic fungi. These include the following factors such as:

1. Persistence and infectivity of inoculums
2. Compatible schedule with other agricultural practices (linked with irrigation, chemicals sprayed, microbial antagonists)
3. Presence of susceptible stage of the host
4. Host behavior
5. Host age
6. Sensitivity to solar radiation

Persistence and infectivity: Formulation of fungi in oil is reported to increase their effectiveness by prolonged persistence, preventing conidial desiccation, increasing adhesion, and spreading the inoculums evenly on the host body thereby interfering with the defensive nature of cuticle. Similarly addition of additives such as stickers, binders etc with the spray suspension also helps in increasing the infectivity by way of enhancing the adherence of spores to the insect body. Screening of the effect of oils, stickers, binders are carried out to determine the compatible one with *V. lecanii, M. anisoplaie and B. brongniartti* that enhances the viability of the entomopathogens. While sunflower oil, pongamia and neem oil were reported to be compatible with *V.lecanii* (Visalakshy *et al*., 2005), coconut oil, sunflower, castor oils caused good germination of *M. anisopliae* spores.

Presence of pesticides and their safety to entomopathogens: One of the important factors for developing BIPM strategy for the target pest is scheduling of other agricultural practices such as linking with irrigation, chemical pesticides and microbial antagonists with biological control. This needs identification of chemical pesticides that are non-toxic to germination, growth and sporulation of applied entomo-pathogen.

Studies were made on the compatibility of various chemical pesticides (insecticides and fungicides) and microbial antagonists with on germination, growth and sporulation of *V. lecanii*. The susceptibility of entomopathogenic fungus to different pesticides and antagonist is varied. Germination of conidial spores of V. *lecanii* was observed in pesticides such as endosulfan, acephate, abamectin, ethion, pongamia oil, and Dinocap and *T. harzianum.* Except carbendazim and iprodion + carbendazim all the treatments were found to be compatible with mycelia growth (Krishnammorthy.*et al.*2007). In vitro studies have indicated that the microbial plant antagonists *Trichoderma harzianum and T. viridae* inhibited the entomopathogens *(*Ganga Visalakshy, *et al.* 2007*)*. Hence in instances, where these two biological control agents are to be incorporated for management of pests and diseases, further studies are to be carried out on their compatibility before application.

3.2.Current and recent research on predators as bio-control agents:

3.2.1. Predators recorded from thrips in India

Predatory insects such as mirid bugs, larvae of syrphids, lace wings, coccinellids, predatory mites, spiders, sphecid wasps and pseudoscorpion are also listed among the common predators of thrips on various crop plants (Ananthakrishnan, 1984; Varatharajan and James Keisa, 2000). Among the different predators reported anthocorids of *Orius* species are the common ones. In Karnataka, *Orius tantillus* and *O. maxidentex* are the most common species occurring on maize, marigold, rose, sunflower, pulses, grams and cotton.

Table 2: Potential predators of thrips of horticultural crops in India

Predators	Host plant	Target thrips	Source
Orius albidipennis (Reut.)	Onion	*Thrips tabaci*	Saxena, (1977; 1981)
Orius maxidentex Ghauri and *O. tantillus* Motsch.	Onion and garlic Chilli,	*Thrips tabaci* *Scirtothrips dorsalis*	Muraleedharan and Ananthakrishnan (1978)
O. insidiosus. O.maxidentex, Aeolothrips *collaris, Scymnus nubilus, Laius externenotatus, Chrysopa* sp, *Amblyseius ovalis, Chrysoperla carnea*	Chrysanthemum, dahlia and Marigold Onion	*Haplothrips ganglbaueri* *Thrips tabaci*	Ananthakrishnan (1984) Manjunatha *et al.* (2001)
Orius minutus (L.)	*Gliricidia sepium* (*maculata*) (Jacq.)	*Megalurothrips distalis* *Frankliniella schultzei* *Haplothrips ganglbaueri*	Viswanathan and Ananthakrishnan (1974)
Carayonocoris indicus Muraleedharan	mango	*Frankliniella schultzei*	Muraleedharan and Ananthakrishnan (1978) and Suresh Kumar and Ananthakrishnan (1984)
Orius spp.	Egg plant	*Thrips palmi*	Nagai (1991)
Orius indicus (nymphs and adults)	*Cajanus cajan* (Flowers)	*Megelurothrips nigricornis*	Rajasekhara and Chatterji (1970)
Holothrips anacardii Hood	Rose	*Rhipiphorothrips cruentatus* Hood (Nymphs)	Dhaliwal (1975)

3.2.2. Extent of natural control/impact and scope for conservation

Scymnus nubilus Muls., *Laius externenotatus* Pic, *Orius albidipennis* (Reut.), *Chrysopa* sp. and *Aeolothrips collaris* Priesner were found preying on *Thrips tabaci* Lind on onion. The predators consumed 23-96 thrips larvae per day, but their incidence was very low in comparison with that of their prey (Saxena, 1981).

Field studies carried out in apple orchards in the Shimla district of Himachal Pradesh, indicated the importance of conservation of biotic agents in apple orchards. Insecticidal sprays appeared to be detrimental to predatory activity. Monocrotophos 0.05% alone did not give an effective level of control and was detrimental to predator activity. In various orchards, populations of *Thrips flavus* remained below the economic injury level and the high predator activity of the anthocorid *Orius* sp. appeared to keep the pest population at low levels (Gupta and Bhalla, 1993).

3.2.3. Attempts on biological control of thrips

Manjunatha *et al.* (2001) evaluated the efficiency of the predatory mite, *Amblyseius ovalis* for the control of yellow mite and thrips on chilli under field conditions and recorded significant control of thrips (0.81 per 3 leaves) in the released plots.

Other predators for thrips biocontrol: An experiment carried out on the effectiveness of controlling thrips in *C. annum* indicated that releasing of the first and second instar larvae of *Chrysoperla carnea* at 3 per plant, resulted in significant reduction of the pest as compared to control treatments (Wadaskar *et al.*, 2004).

3.2.4. Work at IIHR. on biological control of thrips by predators

Among the various species of *Orius* species reported (Table 3) basic studies on the biology and predatory potential of *Orius tantillus*, a predator of *S. dorsalis* was carried out. . The bug completed its life cycle within 17-20 days, with an adult longevity of 12-17 days and consumed an average of 164 thrips during its life period (ranging 43-336). Preliminary evaluation studies under field conditions (under protected cultivation) indicated that *O. tantillus* could effectively suppress S. *dorsalis* population on capsicum. Further trails are being initiated.

Table 3. Predators recorded from various species of thrips at IIHR.

S.no	Predator species collected	Host plant /thrips species
1	*Orius maxidentex*	from thrips infesting marigold
2	*Oirus tantillus*	from thrips infesting maize
3	*Orius sp*	from thrips infesting crossandra
4	*Orius sp*	from thrips infesting niger

3.3. Status of research on parasitoids as bio-control agents

The information on parasitoids as natural enemies of thrips so far in India is rather limited. The list of parasitoids of thrips recorded in India (Ananthakrishnan, 1984) is furnished below.

Parasitoid species	**Host thrips species**
Ceranius sp.	Onion thrips - Thrips *tabaci*
Ceranisus maculatus	Grape wine thrips - *Rhipiphorothrips cruentatus*
Megaphragma longiciliatus	Common flower thrips - *Frankliniella sp.*
Tetrastichus rhipiphorothripsidis	Grape wine thrips - *Rhipiphorothrips cruentatus*
Tetrastichus thripophonus	Gall thrips - *Gynaikothrips uzeli*
Thripobius semiluteus	Turmeric thrips – *Panchaetothrips indicus*

There is very limited information on the seasonality of parasitoids, their impact and scope for conservation in India Parasitism of *T. tabaci* on onion by *Ceranisus sp*. has been estimated to range from 2 to 18%, but the incidence of the parasite was generally low (Saxena, 1981). Parasitism of *T. tabaci* by *Thirophobis sp* sp was reported for the first time from India, It was observed that about 80 % of the *T.tabaci* in onion were parasitised during the months of September to December. (Krishna Moorthy and Ganga Visalakshy, 2006).

3.4 Progress and success in thrips biocontrol elsewhere in tropics

3.4.1. Use of pathogens for biocontrol of thrips

In many countries of the world like Japan, France, Kenya, Germany, UK entomopathogens such as

M. anisoplaie and *V.lecanii* have been used in glass house and field for the management of thrips. Aerial or soil application of M.anisoplaie is reported to effectively reduce the population and damage by western flower thrips *F.occidentalis* in chrysanthemum, cucumber, lettuce, gerbera cultivated under glass house in Israel, Kenya, Brazil, Denmark respectively (Maniania *et al.,* 2001, Azaizeh *et al.,* 2002, Lopes *et al,.*2000)

Incorporation of the fungus with peat mixtures was found to effectively suppress the infestation of F.occidentlais up to 6th week after planting of gerbera (Vestergaard *et al.,* 1996.)

Application of M. *anisoplaie* was reported to effectively suppress the infestation of *T.tabaci* on Onion in Kenya (Maniania *et al.,* 2003.). It was reported that the beneficial agents were more in entomopathogen-applied plots than to insecticides treated plots (Maniania *et al.,* 2003)

Similarly application of *V.lecanii* alone or in combination with predatory mite *A. cucmeris* is reported to give promising results in the management of *F. occidentalis* in Chrysanthemum, melons, cucumber, tomato and bush beans cultivated in glass house and open field in countries such as Japan, Australia, UK and Germany respectively (Heyler *et al.*1992, Saito, 1992)

3.4.2.Use of predators for biocontrol of thrips elsewhere

Successful attempts have been made to mass rear *Orius* spp. on various lepidopteron eggs and artificial diets such as *Orius insidiosus* on eggs of *Dacus dorsalis* (Takara and Nishida, 1978); *Orius minutus* (L.) on *Sitotroga cerealella* (Niemczyk, 1970; 1978); *Orius albidipennis* and *Orius laevigatus* on paralysed larvae of *E. kuehniella, Phthorimaea operculella* and *T. confusum* (Zaki, 1989), *Orius sauteri* on *Melanaphis sacchari* and eggs of *E. kuehniella* or acarid mites or pollen and *Aphis gossypii* (Wang *et al.,* 1996; Yano and Van Lenteren, 1996); *O. majusculus* on eggs of *E. kuehniella* (Alauzet *et al.*, 1992) or acarid mite *Tyrophagus putrescentiae* (Hussaini *et al.*, 1993); *O. insidiosus* on eggs of *Heliothis virescens* (Bush *et. al.*, 1993) etc. artifical diet is standardized for rearing *O.* sauteri (Zhou and Wang, 1989). Castane and Zalom (1994) found that an artificial ovipositional medium made of carrageenan salt of potassium chloride and covered with paraffin wax was suitable for rearing *O. insidiosus*. Martin *et al.* (1978) observed that some developmental stages of *O. insidiosus* could feed on the diet used for rearing *Chrysoperla carnea*

The anthocorid species, which have been commonly used for field releases, are *Anthocoris nemoralis*, and *Orius* spp. Anthocorids are now being commercially produced in several countries. In Ontario, the cost of production of *O. insidiosus* was 0.03 dollar per bug and up to 100,000 bugs could be reared per week and 200 to 600 bugs could be reared in one zip - lock plastic bag (Schmidt, 1994).

In countries like France, UK, Netherlands and Germany, anthocorid predators have been multiplied and released in the green houses and fields for management of thrips. *O. sauteri* is known to be an important predator of thrips, effective in suppressing *Thrips palmi* in egg plants Japan, China, Korea and Russian Far East, *Orius majusculus* could reduce populations of thrips on sweet pepper and cucumber (Fischer *et al.*, 1992). In Germany, good results were achieved with 2 releases of *Orius* at fortnight intervals (0.5-1 insect/m^2) against *F. occidentalis* in cucumbers (Buhl and Bassler, 1992). In France, the efficiency of *Orius majusculus* for the control of *F. occidentalis* on cucumbers was studied in greenhouses and the predator was shown to have great potential (Trottin Caudal *et. al.,* 1991).

Orius majusculus was introduced for the control of *F. occidentalis* on AYR chrysanthemums in a greenhouse in the UK and appeared to reduce thrips numbers for at least the first 2 plantings. In Netherlands, *Orius insidiosus* was released on chrysanthemums for the control of *F. occidentalis* in greenhouses and pest numbers were reduced to a great extent (Fransen *et al.,* 1993). In Sweden, the predators *Amblyseius cucumeris* (*Neoseiulus cucumeris*) and *Orius insidiosus* were very effective when they were released together on ornamental plants (*Saintpaulia impatiens*, *Gerbera* and *Brachyscome multifida*) infested with *F. occidentalis* in greenhouses (Sorensson and Nedstam, 1993).

3.5. Complimentary control methods for development BIPM of thrips

There are several promising control methods for thrips that can be complimentary to bio-control and could be targeted for integration. These include the following:

i) *Field sanitation:* Removal of weed hosts from the cropping areas, as weeds are often the reservoirs for thrips (Ananthakrishnan, 1984). This is applicable in general for vegetable crops.

ii) *Cropping practices*: netting the nursery, intercropping, traps (water traps and sticky traps). Diraviam and Uthamasamy (1992) found that yellow sticky traps are effective not only to monitor thrips density, but also to trap the thrips as a means of population reduction. Infestations of *Scirtothrips dorsalis* on *Capsicum annum* have been found significantly reduced due to yellow pan water trap (James Keisa *et al.*, 1996).

iii) *Botanicals:* Neem pesticides were also reported to be effective against chilli thrips (James Keisa and Varatharajan, 1995) and onion thrips (Sattar Shah *et al.* 2005).

iv) *Host plant resistance: Thrips palmi* is a common pest in Latin America, infesting legumes, cucurbits, solanaceous and ornamental plants; after repeated screening trials, five genotypes of *Phaseoles vulgaris* were identified as resistant to *T. palmi* (Cardona *et al.*, 2002).

v) Effect of volatiles on biological control incidence*: volatiles such as jasmonate are reported to cause increased parasitism of thrips*

3.6. Major thrust areas for future research and development in thrips biocontrol of horticultural importance

1. Studies on crop loss assessment in relation to thrips infestation| damage
2. Natural suppression of thrips by biocontrol agents in different crops
3. Effect of abiotic and biotic factors on natural enemies
4. Mechanization of mass rearing \culturing of biological control agents
5. Developing conservation of biological control agents in IPM of thrips
6. Influence of physical and chemical attributes of cultivars in the field impact studies of bioagents - an aid for developing pest resistance varieties
7. Effect of plant volatiles on pest and its natural enemies.
8. Optimization of culture methods, formulation, application methods – management of solar sensitivity, desiccation, viability etc.
9. Role of alternate host in pest and natural enemies incidence
10. Community involvement approach for production, management of thrips by biological control and providing market linkages for price incentives for the crop produced by biopesticides.

References

Alauzet, C., Dargagnon, D. and Hatte, M., 1992. Production of the heteropteran predator: *Orius majusculus* (Heteroptera: Anthocoridae). *Entomophaga,* 37, 249-252.

Ananthakrishnan, T.N. ,1993. Bionomics of thrips. *Annual Review of Entomology*, 38, 71-92.

Ananthakrishnan, T.N., 1984. *Bioecology of Thrips*. India Publishing House, Michigan, USA, pp. 233.

Azaizeh, H., Gindin, G., Said, O. and Barash, I., 2002, Biological control of the western flower thrips *Frankliniella occidentalis* in cucmber using the entomopathogenic fungus *Metarhizium anisoplaie,* Phytoparasitica, 30,18-24.

Bush, L., Kring, T.J. and Ruberson, J.R., 1993. Suitability of green bugs, cotton aphids and *Heliothis virescens* eggs for development and reproduction of *Orius insidiosus*. *Entomologia Experimentalis et Applicata,* 67, 217-222.

Cardona, C., Frei, A., Bueno, J. M., Diaz, J., Gu, H. and Dorn, S., 2002. Resistance to *Thrips palmi* (Thripidae: Thysanoptera) in beans. *Journal of Economic Entomology*, 95,1066 – 1073.

Castane, C. and Zalom, F.G. ,1994. Artificial oviposition substrate for rearing *Orius insidiosus* (Hemiptera: Anthocoridae). *Biological Control,* 4, 88-91.

Diraviam, J. and Uthamasamy, S., 1992. A new sampling technique involving yellow sticky trap for monitoring of thrips infesting different crops. *Journal of Entomological Research,* 16, 78-81.

Fischer, S., Linder, C. and Freuler, J., 1992. Biology and utilization of *Orius majusculus* Reuter (Heteroptera: Anthocoridae) for the control of the thrips *Frankliniella occidentalis* Perg. and *Thrips tabaci* Lind. in greenhouses. *Revue Suisse de Viticulture d'Arboriculture et d'Horticulture,* 24, 119-127.

Fransen J.J., Boogaard, M. and Tolsma, J., 1993. Minute pirate bug, *Orius insidiosus* (Say) (Hemiptera: Anthocoridae), as a predator of western flower thrips, *Frankliniella occidentalis* (Pergande), in chrysanthemum, rose and saintpaulia. *Bulletin, OILB-SROP,* 16, 73-77.

Ganga Visalakshy, P.N., Manoj Kumar, A. and Krishnamoorthy, A., 2004. Epizootics of a Fungal Pathogen, *Verticillium lecanii* Zimmermann on *Thrips palmi* Karny. *Insect Environment*. 10, 134-135

Gupta, P. R. and Bhalla, O. P., 1993. Conservation of biotic agents in apple orchards of Himachal Pradesh. *Journal of Insect Science,* 6, 204-206.

Helyer, N, Gill, G and Bywatrer, A., 1992 control of Chrysanthemum pests with *Verticillium lecanii.* Phytoparasitica, 20, 5-9

James Keisa, T. and Varatharajan, R., 1995. Efficacy of two neem products in the field control of *Scirtothrips dorsalis* Hood on *Capsicum annum. Indian Journal of Plant Protection*, 23, 166-168.

James Keisa, T., Singh, O.D., Singh, S. A. and Varatharajan, R., 1996. Impact of water pan colour trap and application of neem pesticides on the abundance of *Scirtothrips dorsalis* Hood (Thysanoptera) on *Capsicum annum* L. (Solanaceae). *Proceedings of the National Symposium on IPM and Sustainable Agriculture- an Entomological Approach*, Muzaffernagar, U.P., 6,105-107.

Kandasamy, C., Mohanasundaram, M. and Karuppuchamy, P., 1990. Evaluation of insecticides for the control of thrips, *Scirtothrips dorsalis* Hood on chillies (*Capsicum annum* L.). *Madras Agricultural Journal,* 77,169 – 172.

Krishna Kumar, N.K., 1995. Yield loss in chilli and sweet pepper due to *Scirtothrips dorsal is* Hood (Thysanoptera: Thripidae). *Pest Management in Horticultural Ecosystem,* 1, 61-69.

Krishnamoorthy, A., Ganga Visalakshy, P..N, and Mani, M., 2006. A new record of *Thirophobis* sp a parasitoid of onion thrips, *Thrips tabaci* Lindermann in India Presented in National Symposium on Biological control of Sucking pests in India, IAT, Bangalore, 28 and 29th May 2006

Krishnamoorthy, A., Ganga Visalakshy, P.N., Manoj Kumar, A, and Mani, M., 2007 Influevence of some pesticides on entomoapthogenic fungus *Lecanicillium (verticillium) lecanii* (Zimm) ZAFE and GAMS, *J. Hort. Sciences*, 2,53-57

Kumar S. P., Patel, C.B., Bhatt, R.I. and Rai, A.B., 1994. Population dynamics and insecticidal management of the mango thrips. *Scirtothrips dorsalis* Hood (Thysanoptera: Thripidae) in South Gujarat, *Pest Management and Economic Entomology*, 2, 59-62.

Lingeri., M.S., Awaknavar, J.S., Lingappa, S. and Kulkarni, K.A., 1998. Seasonal occurrence of chilli mites (*Polyphagotarsonemus latus* Banks) and thrips *Scirtothrips dorsalis* Hood), *Karnataka Journal of Agricultural Sciences*, 11, 380-385.

Lopes, R.B., Alves, S.B. and Tamai, M.A., 2000. Control of *Frankliniella occidentalis* in hydroponic lettuce by *Metarhizium anisopliae*, Scientia Agricola, 57,239-243

Maniania, N. K., Sithanantham, S., Ekesi, S., Ampong-Nyarko, K., Baumgärtner, J., Löhr B. and Matoka, C.M., 2003. A field trial of the entomogenous fungus *Metarhizium anisopliae* for control of onion thrips, *Thrips tabaci*. *Crop Protection,* 22, 553-559.

Maniania, N.K., Ekesi, S., Löhr. B. and Mwangi, F., 2001. Prospects for biological control of the western flower thrips, *Frankliniella occidentalis* with the entomopathogenic fungus, *Metarhizium anisopliae*, on chrysanthemum. *Mycopathologia,* 155, 229-235.

Manjunatha, M., Hanchinal, S.G. and Kulkarni, S.V., 2001. Interaction between *Amblyseius ovalis* and *Polyphagotarsonemus latus* and efficacy of *A. ovalis* on chilli mite and thrips. *Karnataka Journal of Agricultural Sciences,* 14, 506-509.

Martin, P. B., Ridgway, R.L. and Schuetze, C.E, 1978. Physical and biological evaluations of an encapsulated diet for rearing *Chrysopa carnea. Florida Entomologist,* 61,145-152.

Mohite, P.B., Teli, V.S. and Moholkar, P.R., 1992. Efficacy of organic synthetic pesticides against onion thrips, *Thrips tabaci* Lind. *Pestology*, 16, 8 – 10.

Nair, M.R.G.K., 1986. Insect and mites of crops in India. *ICAR Publications. New Delhi*, 408 pp.

Niemczyk, E., 1970. The development and fecundity of the bark bug - *Anthocoris nemorum* (L.) (Heteroptera: Anthocoridae) reared on the eggs of the Angoumois grain moth - *Sitotroga cerealella* Oliv. (Lepidoptera: Gelechiidae). *Polskie Pismo Enotomologiczne,* 40, 857-865.

Panickar, B. K. and Patel, J.R, 2001. Population dynamics of different species of thrips on chilli, cotton and pigeon pea, *Indian Journal of Entomology,* 63, 170-175.

Raizada, U., 1976. On the occurrence of *Mrazekia* sp., a microsporadian parasite infecting some thrips larvae. *Current Science*, 45, 627-628.

Saito, T., 1992. Control of *Thrips palmi* and *Bemisia tabaci* by a mycoinsecticidal preparation of *Verticillium lecanii.* In Proc.of the kanto Tosan plant prot. Society, 209-210

Sattar Shah, Md. A., Singh, H. C., and Varatharajan, R., 2005. Effect of neemazal on Onion thrips, *Thrips tabaci* Lindmann., (Thripidae: Thysanoptera). *Annals of Plant Protection Science*, 13, 470-471.

Saxena, R.C., 1981. Observations on some predators and parasites of *Thrips tabaci* Lind. *Bulletin of Entomology,* 22, 97-100.

Schmidt, J., 1994. Pirate bugs plunder greenhouse pests. *Agricultural Food Research in Ontario,* 17, 12-15.

Sorensson, A. and Nedstam, B., 1993. Effect of *Amblysieus cucumeris* and *Orius insidiosus* on *Frankliniella occidentalis* in ornamentals. *Bulletin-OILB/SROP,* 16,129-132.

Srinivas, P.S., and Lawande, K.E., 2004. Impact of planting dates on *Thrips tabaci* Lindeman infestation and yield loss in onion (*Allium cepa* L.). *Pest Management in Horticultural Ecosystems,* 10,11-18.

Takara, J. and Nishida, T., 1978. Eggs of the oriental fruit fly for rearing the predaceous anthocorid *Orius insidiosus* (Say). *Proceedings of the Hawaiian Entomological Society*, 23, 441-445.

Thaler, J. S., 1999. Jasmonate-inducible plant defences cause-increased parasitism of herbivores. *Nature,* 399, 686-688.

Trottin Caudal, Y., Grasselly, D., Trapateau, M., Dobelin, H, and Millot, P., 1991. Biological Control of *Frankliniella occidentalis* with *Orius majusculus* on cucumber. *Bulletin SROP,* 14,: 50-56.

Varadarajan, S, and Veeravel, R, 1995. Population dynamics of chilli thrips *S. dorsalis* in Annamalainagar, *Indian Journal of Ecology*, 22, 27-30.

Varatharajan, R, 1985. Parasite-host interactions in relation to the nematode *Anguillulina aptini* (Sharga) – a parasite on *Microcephalothrips abdominalis* (Crawford) and *Frankliniella schultzei* (Trybom). *Current Science*, 54, 396-398.

Varatharajan, R. and James Keisa, T., 2000. Bioȩcology and Management of Thrips. In: *IPM System in Agriculture*, (ed. Rajeev K. Upadhyay, Mukerji, K.G and Dubey, O.P.), Aditya Books Private Limited, New Delhi, and *Key Animal Pests,* 7, 219-234.

Vestergaard, S., Gillespie, A.T. and Elienberg, J996. Control of western flower thrips *Frankliniella occidentalis* (Thysanoptera: thripidae) in gerbera incorporating the entomopathogenic fungus *Metarhizium anisoplaie* into the growth medium, In Insect pathogens and insect parasitic nematodes, Proc. Of the first meeting, bulletin –OLB-SROP.240-246

Visalakshy, P. N. G., Krishnamoorthy, A, and Manoj Kumar, A. 2005. Effect of plant oils and adhesive stickers for the mycelia growth and sporulation of *Verticillium lecanii*, a potential entomopathogen. *Phytoparasitica*, 33, 367-369.

Wadaskar, R. M. , Deotale, V. Y., Sharnagat, B.K., 2004, Evaluation of *Chrysoperla* releases along with insecticides against pests of chilli. *Journal of Soils and Crops,* 14, 62-65.

Wang, F. H., Zhou, W. R., and Wang, R., 1996. Studies on the method of rearing *Orius sauteri*. *Chinese Journal of Biological Control,* 12, 49-51.

Wheatly, A. R. D. , Wightman, J. A., Williams, J. H, and Whaetly, S.J., 1989. The influence of drought on the distribution of insects on four groundnut genotypes grown near Hyderabad, India. *Bulletin of Entomological Research*, 79, 566-577.

Yano, E., and Van Lenteren, J.C., 1996. Biology of *Orius sauteri* (Poppius) and its potential as a biocontrol agent for *Thrips palmi* Karny. *Proceedings of the meeting on Integrated control in glasshouses, held in Vienna, Austria,* 20-25 May, 1996. *Bulletin OILB SROP,* 19, 203-206.

Zaki, F. N., 1989. Rearing of two predators, *Orius albidipennis* (Reut.) and *Orius laevigatus* (Fieber) (Hemiptera: Anthocoridae) on some insect larvae. *Journal of Applied Entomology*, 107, 107-109.

Zhou, W. and Wang, R., 1989. Rearing of *Orius sauteri* (Hemiptera : Anthocoridae) with natural and artificial diets. *Chinese Journal of Biological Control,* 5: 9-12.

Pests of mulberry and silkworm, *Bombyx mori* and their management, with emphasis on biocontrol

M. Geetha Bai*

Mulberry, *Morus alba* L. being a perennial plant is prone to attack by a large number of pests. Mulberry silkworm, *Bombyx mori* L. (Lepidoptera: Bombycidae), a highly domesticated insect is also infested by several pests. *B. mori* is sensitive to chemical pesticides. Therefore, it is essential to develop eco-friendly pest management strategies for pests of mulberry and mulberry silkworm. Biological control using natural enemies of pests of mulberry and silkworm has tremendous scope. Many natural enemies have been recorded on these pests and detailed studies on their bio-ecology and potential to contain the pests have been conducted. However, only some of them are being used for management of these pests in Integrated Pest Management Programmes. Scope for integrating the use of natural enemies and botanicals in management of these pests is discussed.

Introduction

Sericulture is one of the most lucrative industries in the rural scenario in India. Sericulture, being an agro-based industry, is known for its capacity to generate high income per unit area, high employment potential, frequent periodical income and values related to preservation of our cultural heritage.

Mulberry, *Morus alba* L. (Moraceae), being a perennial plant is prone for attack by a large number of pests. Mulberry silkworm, *Bombyx mori* L. (Lepidoptera: Bombycidae), a highly domesticated insect is also infested by several pests. *B. mori* is sensitive to chemical pesticides. Therefore it is essential to develop eco-friendly pest management strategies for pests of mulberry and mulberry silkworm.

Mulberry pests

Pests of mulberry include defoliators, sap suckers and stem borers. Earlier little importance was given for pests of mulberry. Sericulturists in irrigated areas were of the opinion that they can grow mulberry leaves easily, but management of silkworm diseases is the challenging issue. However, ever since mulberry plantations were found infested by the leaf-roller, *Diaphania pulverulentalis* (Hampson) (Lepidoptera: Pyralidae) in 1995 for the first time in Karnataka (Geetha Bai *e al.* (1997) importance of mulberry pests was felt. The leaf-roller is causing extensive damage to sericulture industry in this state and its neighbouring states, viz., Tamil Nadu and Andhra Pradesh. Of late, sericulturists in these areas have realized the importance of mulberry pests in reducing cocoon production.

Defoliators cause heavy loss to sericulture industry. They not only adversely affect the quality and quantity of mulberry, but also can play an important role in transmission of silkworm diseases, since they are susceptible to them. The most important defoliators in irrigated mulberry plantations are the leaf-roller, *D. pulverulentalis*, the Bihar hairy caterpillar, *Spilarctia obliqua* Walker (Lepidoptera: Arctiidae) and cutworm *Spodoptera litura* (F.) (Lepidoptera: Noctuidae). The wingless grasshopper, *Neorthacris acuticeps nilgirensis* Uvarov (Orthoptera: Acrididae) is an important defoliator in rain-fed mulberry plantations. Mealy bugs, jassids, scale insects, thrips and white flies suck sap from mulberry leaves. The most important sap sucker of mulberry leaves is *Maconellicoccus hir-*

* Karnataka State Sericulture Research & Development Institute, Bangalore-560 062

sutus (Green) (Hemiptera: Pseudococcidae), which causes "Tukra". Stem borers bore through the stem portion of mulberry plants causing damage to them. The bark borer, *Indarbela quadrinotata* Walker (Lepidoptera: Indarbelidae) is a severe pest of tree mulberry and its damage to the tree bark causes considerable reduction in growth and yield of the crop. Stem girdler beetle, *Sthenias grisator* F. (Coleoptera: Cerambycidae) has a habit of ringing the stem portion of mulberry plants. The bark and wood are cut around the main stem or branch leaving a clear girdle. The portion above the girdle wilts and dies since grubs tunnel into the branches and feed. Termites attack bottom portion of older mulberry plants.

Silkworm pests

The most important pest of *B. mori* in India is the uzifly, *Exorista bombycis* (Louis) (Diptera: Tachinidae). Dermestid beetles, the most important being, *Dermestes ater* De Geer (Coleoptera: Dermestidae) and an earwig predator, *Labia arachidis* Yersin (Dermaptera: Labiidae) attack *B. mori* in grainages.

Biological control using natural enemies of pests affecting sericulture industry has tremendous scope since this is an eco-friendly method. A large number of natural enemies have been recorded on these pests and detailed studies on their bio-ecology and potential to contain the pests have been conducted. However, only some of them are being used for management of these pests in Integrated Pest Management Programmes. Scope for further work and possibilities of using natural enemies and botanicals in management of pests affecting sericulture industry are discussed. Only commonly occurring and important pests are dealt with in detail.

Mulberry pests

Defoliators (Lepidoptera)

The leaf-roller, *Diaphania pulverulentalis* (Hampson) (Lepidoptera: Pyralidae)

The leaf-roller is one of the most important pests of mulberry in Karnataka, Tamil Nadu and Andhra Pradesh, since it causes heavy loss to sericulture industry, ranging from 10 to 15%.

Occurrence: The leaf-roller was first recorded from Ngas, Tenasserim (Nagaland) and described (Hampson, 1896) This pest was reported to infest mulberry plantations in Malaysia (Sengupta *et al.*, 1990). *D. pulverulentalis* was recorded as a pest of mulberry for the first time in Karnataka during 1995 (Geetha Bai *et al.,* 1997). Rajadurai *et al.*, (1999) reported *D. pulverulentalis* to be a severe pest of mulberry in Karnataka, Andhra Pradesh and Tamil Nadu. Siddegowda *et al.,* (1995) reported *Diaphania* sp. as a pest of mulberry in Karnataka, Andhra Pradesh and Tamil Nadu in South India.

The leaf-roller infestation begins after the onset of South-West monsoons. Even though it occurs from, June and continues till next April, infestation is very high from July to December. Infestation by this pest coincides the best silkworm rearing season.

Nature of damage: The leaf-roller feeds on tender mulberry leaves, used for rearing chawki silkworms and also damages the terminal bud and hence the plant ceases to grow and the leaves at the lower portion mature fast. This pest is susceptible to muscardine causative pathogen in the field, viz., *Beauveria bassiana* and *Aspergillus tamarii*. It is highly susceptible to the dreaded pebrine disease of silkworms caused by *Nosema bombycis* Nageli (Microsporida: Nosematidae) in laboratory studies (Geetha Bai and Ramegowda, 1999). The pest caterpillars infected with pebrine spores in the field can contaminate mulberry leaves. When silkworms are fed on such pebrine spore contaminated mulberry leaves, it can lead to pebrine infection in silkworms.

Symptoms: Young caterpillars found in the apical portion of mulberry plants secrete white, delicate, silky filament, which binds the leaf blades of tender leaves together and feed on soft, green tissues of the leaf (Fig. 1). The tender leaf is rolled-up and the caterpillar is found within the rolled leaf formed by silky white filaments, and hence the name, leaf-roller. Grown up caterpillars feed voraciously on tender leaves and their faecal matter can be seen on the leaves below the affected portion.

Life-cycle: Moths are greyish brown in colour (Fig. 2) and measure about 1 cm in length. Wings have brownish, wavy stripes. Gravid moths lay one or two eggs on the lower surface of tender mulberry leaves, along the veins, near the apical portion of shoots. Each female lays 310 to 460 eggs. Incubation period of leaf-roller egg is 3-4 days. There are five larval instars. Grown up caterpillars are light green in colour and have black markings on the lateral and dorsal regions of abdominal segments (Fig. 3). The head is black in colour. Grown up caterpillars measure about 1.5 cms in length. Larval duration is 10-15 days. Pupae are slender, elongated and brown in colour (Fig.4). Pupal duration is 8-10 days. The pest completes its life-cycle in 21-29 days.

Management

Cultural and mechanical methods:

1. Clip affected apical portions of infested plants, collect them and dip them in 0.5% soap solution to destroy the caterpillars.
2. Clip the terminal buds when silkworms are in fourth moult to avoid leaf-roller mothd ovipositiong.
3. Collect and burn dry leaves and weeds in infested mulberry plantations to destroy the pest pupae.
4. Dress mulberry shoots before transporting them to other areas for new plantations.
5. Promote activity of bird predators by providing them with water in and around mulberry plantations.

Biological control

Releases of Trichogramma chilonis and Phanerotoma noyesi: Release three "Tricho" cards and 50 adult *P. noyesi,* 15 to 20 days after pruning or leaf-harvesting. Cut "Tricho" cards into 20 pieces and staple them on lower surface of mulberry leaves. After 10 days, use two more "Tricho" cards. Each "Tricho" card has approximately 20,000 developing *T. chilonis.* Do not spray chemical pesticides after releases of parasitoids.

Application of neem: When infestation is moderate (about 20%) spray neem seed kernel extract (NSKE) (10 ml of NSKE {1,500 ppm}in 1 litre of water and five gms detergent soap powder), 20 days after pruning, shoot harvesting or leaf plucking, and subsequently twice at weekly intervals. When infestation is high, spray NSKE (four ml of NSKE {10,000 ppm} in 1 lit. of water and 5 gm detergent soap powder) and subsequently twice at weekly internals.

Chemical control

Silkworm, being a highly domesticated insect, is sensitive to chemical pesticides. Use them sparingly, with utmost care, when absolutely necessary. Spray DDVP or Dichlorovos 0.076% 1ml of DDVP (76% EC) in one liter of water with 5 gm of detergent soap powder. Ten days after spraying mulberry leaves can be fed to silkworms. Do not chemical pesticides when bio-control agents are used for management of the pest.

Recommendation of TNAU

Sl. No.	Component	Days after pruning
1.	Flooding the mulberry field to expose the leaf webber pupae	0
2.	Release of pupal parasitoid, *Tetrastichus howardi @ 20,000/acre*	1

3.	Release of egg parasitoid, *Trichogramma chilonis* @ 2 cc/acre	10
4.	Spraying of dichlorvos @ 1 ml/lit (200ml/acre)	30
5.	Mechanical clipping and burning of affected shoots	40

Recommendations of CSR&TI, Mysore

Irrigated

Step	Days after pruning	Activity to be taken up	Quantity required/acre
1.	3	Release of pupal parasitoid *Tetrastichus howardi*	50,000
2.	10	Release of egg parasitoid *Trichogramma chilonis* @ 18,000 eggs /card	5 cards
3.	15	Spraying DDVP @ 0.076% in 200 lit	200 ml DDVP
4.	20	Release of *T. howardi* + *T. chilonis*	50,000 + 3 cards
5.	25	Spraying neem pesticide (0.03% in 200 lt.	1 liter neem pesticide
6.	30	Release of *Trichogramma chilonis*	3 cards
7.	Top clipping (bud and next leaf) when the silkworms are in 4th moult		

For Kolar area (Specific)

Step	Days after pruning	Activity to be taken up	Quantity required/acre
1.	3	Release of pupal parasitoid *Tetrastichus howardi*	50,000
2.	10	Release of egg parasitoid *Trichogramma chilonis* @ 18,000 eggs /card	5 cards
3.	15	Spraying DDVP @ 0.076% in 200 lit	200 ml DDVP
4.	20	Release of *T. howardi* + *T. chilonis*	50,000 + 3 cards
5.	Top clipping (bud and next leaf) when the silkworms are in 4th moult		

In Kashmir the mulberry pyralid, *Diaphania pyloalis* Walker is an important pest of mulberry.

The Bihar hairy caterpillar, *Spilarctia obliqua* (Walker) (Lepidoptera: Arctiidae)

The Bihar hairy caterpillar is a sporadic and severe pest of mulberry, causing considerable damage. It is a polyphagous pest, infesting a wide variety of economic crops, such as agricultural, fiber crops and medicinal plants.

Occurrence: S. obliqua has been reported in India, Bangladesh, China, Myanmar, Philippines and Sri Lanka. It occurs during monsoons and winter seasons and aestivates during summer. Infestation is severe from August to January.

Nature of damage and symptoms: Young caterpillars feed on green, soft tissues of leaves, leaving behind only the veins. Grown up caterpillars are solitary and feed voraciously on leaves, leaving behind the petiole and mid-rib of the leaf reducing the leaf yield considerably.

Life-cycle: Moths are stout and have light brown coloured wings, brick red coloured abdomen, peppered with a row of black spots on the dorsal and lateral sides. Gravid females lay 1000 to 2000 eggs in a group on the lower surface of mulberry leaves. Eggs are pale green in colour and hatch in 5-7 days. There are six to seven instars. Young caterpillars are gregarious up to third instar, whereas later age larvae are solitary. Grown up caterpillars are completely covered with bunches of bristles. Anterior and posterior ends are black in colour, while, the rest of the body is reddish brown. Larval duration is 27-32 days. Grown up caterpillars weave loose silken brownish cocoons and pupate inside them. Pupation takes place in crevices between stones on bunds and in mulberry plantations, tree trunks and also corners of nearby buildings. Pupal duration is 12-14 days. Pupae are brown in colour. Life-cycle is completed in 44-53 days.

Management: An Integrated Pest Management strategy, consisting of cultural, mechanical, biological and chemical methods have to be adopted for effective management of *S. obliqua.*

Cultural and Mechanical methods

1. Collect and destroy egg masses and young caterpillars.
2. Use light traps to attract moths and kill them by placing trays with 0.5% soap solution or kerosene water below the light source.

Biological control: Release *Trichogramma chilonis* twenty days after pruning. Release five "Tricho" cards each at an interval of 3 days. Cut the "Tricho" card into 12-16 pieces and staple to the under surface of mulberry leaf uniformly in the mulberry plantation.

Chemical control: Spray 0.15% DDVP (2 ml of DDVP {76% EC} in 1 litre of water with 5 gm of detergent soap powder) 20 days after pruning or leaf harvesting. Mulberry leaves can be fed to silkworms 15 days after spraying. Use the insecticide carefully only when it is absolutely necessary.

Cutworm, *Spodoptera litura* (Lepidoptera: Noctuidae)

The cutworm, S. *litura* is a sporadic and severe pest of mulberry plantations, where vegetable crops are commonly grown. It is a polyphagous pest infesting several corps of agricultural and horticultural importance.

Occurrence: *S. litura* is widely distributed throughout tropical and temperate Asia, Australasia and the Pacific islands. It attacks mulberry plantations from August to February.

Nature of damage and symptoms: Caterpillars attack young shoots and cut them and hence the name cutworm. The shoots dry up subsequently and fall off. They feed voraciously on mulberry leaves.

Life-cycle: Moths are stout, have brown coloured patches, with wavy white markings on forewings and hind wings are white. Gravid moths lay about 200-300 eggs in clusters on the lower surface of mulberry leaves and cover them with bristles and secretions. Incubation period is 4-5 days. Grown up caterpillars are pale greenish brown in colour with longitudinal yellow stripes and dark markings. They are nocturnal. During day time they hide in crevices of loose soil at the base of the plant and come out in the night to feed. Larval duration is 2-3 weeks. They pupate in soil and are brown in colour. Pupal duration is about two weeks. Life-cycle is completed in 36-40 days.

Management: An Integrated Pest Management strategy, consisting of cultural, mechanical, chemical methods and use of pheromone trap needs to be adopted for management of *S. litura.*

Cultural and Mechanical methods:

1. Collect and destroy egg masses and young caterpillars.
2. Plough the pest infested mulberry plantations and dig at the base of mulberry plants to expose pest caterpillars and pupae to predators and sunlight.
3. Use light traps to attract moths and kill them by placing trays with 0.5% soap solution or kerosene water below the light source.
4. Flood irrigation wherever possible will kill caterpillars hiding in soil and pupae.

Pheromone trap: Use "Spodolure" a pheromone trap @ 2 lures /acre twice at an internal of 15 days starting from 25 days after pruning or leaf harvest, to attract and kill male moths.

Chemical control: Spray 0.15% DDVP (2 ml of DDVP (76% EC) in 1 litre of water with 5 gm detergent soap powder) 20 days after pruning or leaf harvesting. Mulberry leaves can be fed to silkworms 15 days after spraying. Use the insecticide carefully only when it is absolutely necessary.

Other lepidopterans of lesser importance infesting mulberry are the Leaf-weber or Bell moth, *Archips micaeana,* the Moringa hairy caterpillar, *Eupterote mollifera,* Tussock caterpillar, *Euproctis fraterna,* Wasp moth, *Amata passalis,* Red headed caterpillar, *Amsacta moorei* etc.

Orthoptera: Grasshoppers are known to infest mulberry plantations. Eighteen species of grasshoppers were recorded on mulberry plantations in Karnataka. Of these the wingless grasshopper is the important pest of mulberry.

The wingless grasshopper, *Neorthacris acuticeps nilgirensis* Uvarov (Orthoptera: Acrididae)

It is a severe pest of mulberry and is polyphagous, feeding on jawar, sunflower horse gram pigeon pea etc.

Occurrence: N. a. nilgirensis is a severe pest of rainfed mulberry plantations in South India, particularly Chamarajnagar and Mysore districts of Karnataka state in South India. Several major outbreaks of this pest have been recorded in these areas. This pest has also been recorded in Coimbatore in Tamil Nadu and also in Malaysia. It occurs throughout the year and the peak of prevalence is during October and November.

Nature of damage and symptoms: N. a. nilgirensis feeds extensively on sprouting buds and mulberry leaves. When there is an outbreak of this pest, it denudes the mulberry plants and also feeds on the green portion on the bark of affected plants. Such severely affected plants sometimes die.

Life-cycle: Gravid females deposit 6 to 18 egg pods, 2-3 cm deep in soil. Egg pods are conical and have an earthern covering. Each egg pod contains 10 to 18 eggs. They oviposit from August to September and January to February. Eggs hatch in about a month after commencement of South-West monsoons in May-June. There are six nymphal instars. The first four nymphal instars are straw coloured. Fifth instar nymphs develop two red angular patches on the lateral margin of pronotum. The sixth instar nymph is light brown in colour. Dark green band bordered by yellowish white stripes develop on side of the body, extending across side of the head to the base of the antennae, the lower stripes and at the base of coax of hind leg. Ventral margin of pronotal lobes are yellowish white, with two large irregular patches. The nymphal duration is 13,14, 13, 18, 18 and 18 days during the first to sixth instars, respectively. Adult grasshoppers resemble the sixth instar nymph except for larger size. Longevity of females varies from 45-60 days, whereas males live for shorter duration. The life-cycle is completed in 150-175 days. Eggs undergo aestivation in summer. Two generations occur in one year in the field.

Management

Cultural and Mechanical methods:

1. Undertake ploughing of mulberry plantations soon after commencement of monsoons. The egg pods exposed to natural enemies and sunlight are killed.
2. Remove weeds in and around mulberry plantations to destroy alternate shelters .
3. Encourage bird predators by providing water in containers in and around mulberry plantations.

Application of neem: Spray infested mulberry plantations and bunds with 2% neem formulation (Azadirachtin=0.03% EC), 20 days after pruning or leaf harvesting. If the infestation is high, undertake a second spraying, 10 days after the first spray. One hundred litre of emulsion is required per acre of mulberry plantation. Feed mulberry leaves 10 days after the last spray.

Chemical control: Spray 0.0.076% DDVP (1 ml of DDVP (76% EC) in 1 litre of water with 5 gm detergent soap powder) 20 days after pruning or leaf harvesting. If the infestation is high, undertake a second spraying 10 days after the first spray. Mulberry leaves can be fed to silkworms 10 days after spraying. Use the insecticide carefully only when it is absolutely necessary.

Sap suckers

Mealy bugs, jassids, scale insects, thrips, bugs and white flies suck sap from mulberry leaves. The most important sap sucker pest of mulberry is *Maconellicoccus hirsutus* (Green) (Hemiptera: Pseudococcidae), which causes "Tukra".

The pink mealy bug, *Maconellicoccus hirsutus* **(Green) (Hemiptera: Pseudococcidae)**

M. hirsutus, is one of the most destructive sucking pests of mulberry, due to the extent of qualitative and quantitative damage it inflicts on mulberry plantations. It is a polyphagous pest infesting many crops of agricultural and horticultural importance. Mealy bugs are called "hard to kill pests of fruit trees". Eggs of the mealy bugs, protected by waxy secretions are almost impossible to reach with insecticides. Late instar nymphs and adult mealy bugs are also not affected by foliar application of insecticides due to waxy coating.

Though chemical insecticides have failed to check the mealy bug damage, being sessile, it is more amenable to biological control.

Occurrence: This pest is widely distributed in Karnataka, Tamil Nadu, Andhra Pradesh, and Maharashtra states. It is also recorded in Indonesia. Though "Tukra" is observed throughout the year, it is severe in summer.

Symptoms and nature of damage: Malformation of the apical shoot due to flattening and thickening of the affected portion of shoot, reduction in internodal distance and wrinkling and curling up of apical leaves is observed. Infested leaves are dark green in colour and subsequently turn pale yellow. These leaves become brittle. White mealy substance is seen on infested portion of plants. *M. hirsutus* suck the sap, resulting in stunted growth of the plant and reduction in leaf yield

Life-cycle: Gravid females deposit about 250 eggs. Incubation period is 5-8 days. Crawlers are orange in colour and the duration of this stage is one day. They settle on the plant and secrete a white waxy substance on their body. Females have three nymphal instars, whereas males have four. Life-cycle is completed in 24-26 days. Adult males are slender and possess a pair of wings and caudal filaments, each. The pink mealy bug predominantly reproduces parthenogenetically.

Management

Cultural and Mechanical methods:

1. Clip the affected portion of infested plants, collect them in polythene containers and dip in 0.5% soap solution for 3 to 4 minutes to destroy the pest.
2. Do not grow alternate host plants such grapes guava, hibiscus, cotton near mulberry plantations.

Biological control: Release predatory ladybird beetle, *Cryptolaemus montrouzieri* (Muls.) (Coleoptera: Coccinellidae) @ 125 adults /acre twice each during August-September and October-November.

Chemical control: Spray 0.15% DDVP (2 ml of DDVP (76% EC) in 1 litre of water with 5 gm detergent soap powder) on pruned or leaf harvested plants, twice at an interval of 10 days. These leaves can be fed to silkworm 15 days after spraying the pesticide.

Silkworm pests

Silkworm pests attack developing or adult *B. mori* in rearing houses or grainages.

The uzifly, *Exorista bombycis* (Louis) (Diptera: Tachinidae)

The most important pest of *B. mori* in India is the uzifly, *Exorista bombycis.*

Occurrence: Presence of the pest has been reported in Bangladesh, China, Japan, India, South

Korea, Thailand and Viet Nam. The pest occurs throughout the year, but its incidence is high from July to November.

Symptoms and nature of damage: Presence of egg and /or black scar on the body of parasitised silkworms and maggot emergence hole in cocoons indicate uzifly infestation. Silkworms parasitised in early instars are killed before attaining spinning stage, while those parasitised in the late V instar spin cocoons of weak built and from such cocoons uzifly maggots emerge by piercing, thus rendering cocoons unfit for commercial purpose.

Life-cycle: Gravid flies lay about 300 eggs. Normally only one or two eggs are laid on each silkworm. Incubation period is 2-3 days and young maggots pierce into the host body and devour body tissues. After 5-8 days of parasitic life, they emerge from the host and they lead 12-20 hrs of post-feeding life and pupate in loose soil, crevices on the rearing house floor. Pupal duration is 10-12 days. Life-cycle is completed in 18-24 days.

Management

Cultural and Mechanical methods:

1. Collect and destroy uzifly infested silkworms, uzifly maggots and pupae.
2. Keep the rearing house floor free from crevices.
3. Uzifly infested silkworms spin cocoons a day or two earlier than other silkworms. These cocooms should be harvested separately and stifled.

Exclusion method: Fix nylon net to doors, windows and ventilators of rearing houses, besides ante-room at the entrance of the rearing houses. This method is commonly used.

Biological control: A large number of parasitoids have been recorded from uzifly. Most of them have earlier been recorded from housefly, *Musca domestica* and other muscid flies.

1. *Nesolynx thymus,* an indigenous hymenopteran ecto-pupal parasitoid, to kill the uzifly pupae, is being mass produced in cocoon markets and released for management of uzifly.
2. Releases of one lakh adult *N. thymus* adult females needs to be carried out three times corresponding to IV and V instrs and within one or two days after cocoon harvest at 8,000, 16,000 and 76,000 adults, respectively. Parasitoids need to be released after sunset in the rearing house, places of mountage storage, near mountages with spinning worms and also near manure pits.
3. A recent study at KSSRDI has culminated in recommedation of releases of 30,000 adult *N. thymus* or *Exoristobia philippinensis* or *Trichopria* sp./100 disease free layings (dfls).

However, popularization of these bio-control agents at the level of sericulturists is at initial stages.

Chemical control:

Uzi powder: This product developed by KSSRDI is an ovicidal dust formulation. It is dusted on the body of silkworms on the 2nd day during III instar, 2nd and 4th days during IV instar and 2nd, 4th and 6th days in V instar. Four to five kg uzi powder is required for rearing silkworms from 100 layings.

Uzi trap: It is a chemo-trap effective in trapping and killing both the sexes of uzi flies. The uzi trap in the form of a tablet is dissolved in one litre of water and the resultant yellow solution is kept in light coloured flat containers and is placed near windows at the window height, inside and outside, in the silkworm rearing house starting from III instar till spinning stage of silkworm for trapping the uzifly.

Pests of *B. mori* in grainages

A large number of insect pests are known to attack *B. mori* in grainages. Dermestid beetles, the most important being, *Dermestes ater* De Geer (Coleoptera: Dermestidae) and an earwig predator, *Labia arachidis* Yersin (Dermaptera: Labiidae) infest *B. mori* in grainages in Karnataka.

***Dermestes ater*:**

Occurrence: This pest is cosmopolitan in distribution and it occurs throughout the year.

Nature of damage and symptoms: D. ater is omnivorous, feeding on eggs and silk moths, besides boring into cocoons to feed on pupae. It also feeds on dead organic matter available where pierced cocoons are stored. Remnants of eggs, pupae and silk moths left behind after feeding and cocoons pierced by the beetle are seen.

Life-cycle: Adult beetles are black in colour and measure 6-7 mm in body length. They are capable of sustained flight over long distances. They breed prolifically in pierced cocoons stored in grainages, after grainage operations. Gravid females start ovipositing about five days after eclosion. They oviposit inside pierced cocoons (PC), in crevices of trays, walls and furniture. Eggs are milky white, elongated and incubation period is 3-6 days. Grubs from newly hatched eggs are white, which gradually turn to brown in first instar and are covered with bristles. They turn black from second instar onwards. They moult 4-6 times and larval duration is 27-28 days. Grubs are spindle shaped and morphologically they are similar except in size. Duration of pupal stage is 7-8 days. Life-cycle is completed in 37-42 days.

Natural enemies: A mite predator, *Paracarophenax dermestidarum* Rack (Acarina: Acarophenacidae) was found preying on eggs of *D. ater* in Karnataka (Geetha Bai, 1991). The predatory mites cling on the ventral surface of the beetle, especially the head and thorax. The number of mites on the body of a beetle ranged from 1 to 64 and infestation was 48.65%. When gravid female beetles oviposit, the mites disengage themselves from the beetle and prey on the eggs. The mites puncture the egg chorion and suck the egg contents and destroy them. Fully fed mites appear as milky white, shining globules. *Aspergillus flavus* Link, an entomogenous fungus was isolated from *D. ater*. This is pathogenic to *B. mori*. This pest can play a role in spreading this pathogen.

Labia arachidis

Occurrence: This predator has been recorded in grainages in Karnataka and could occur in other areas also. This pest also occurs throughout the year.

Nature of damage and symptoms: L. arachidis adults prey on silk moths laid out for mating and oviposition. They rip open the abdominal region of gravid moths and feed on the injured moths. The injured moths die before completing oviposition. Haemolymph of these moths turns black and stains the egg cards.

Life-cycle: Adult earwigs are brown, slender and have a pair of caudal cerci at the posterior end. Males are narrow and smaller than females. They measure about 6-8 mm in length and are unable to fly. They breed in crevices of rearing trays, stands, doors, windows and furniture. Mated females oviposit when they are 4-7 days old and lay 1 to 33 eggs. There are 1 to 4 ovigenic cycles. Eggs remain in clusters and are creamy white, irregular in shape and have a leathery, transluscent chorion. It is an ovo-viviparous species. Whitish nymphs with two prominent black eyes are visible in just laid eggs. Eggs hatch 4-7 min. after laying. It is a cannibalistic species. Other earwigs prey on young nymphs. The mother earwigs exhibit maternal care. She rests on the eggs and nymphs and prevents other earwigs from preying on her offspring. Maternal care lasts for 3-5 days. Ovo-viviparity and maternal care are adaptations to protect offspring from cannibalism. Young nymphs from just hatched eggs are white in colour and gradually turn light brown and then brown. There are four instars. Duration of development ranges from 29-53 days. Adult longevity varies from 29-119 days.

Natural enemy: A microsporidian pathogen, *Nosema* sp. has been recorded from *L. arachidis.*

Management

Cultural and Mechanical methods:

1. Clean the grainage premises and place of storage of cocoons thoroughly, so that continuity of life-cycle of the pest is disrupted.
2. Collect the pest insects using vacuum cleaner and destroy them.
3. Sun-dry pierced cocoons to kill eggs and young grubs and nymphs of pests before storage.
4. Remove pierced cocoons from the grainage premises within four days from the date of completion of grainage operations of a lot and store away from the grainage building.
5. Dispose off pierced cocoons at least once a month. Provide wire mesh for windows to avoid movement of pest insects from storage place of pierced cocoons.
6. Use plastic trays (HDP), instead of wooden and bamboo trays in grainages, to avoid breeding of insects in crevices.

Chemical control:

1. Dust commercial grade beaching powder (30% chlorine) at the rate of 200 g/sq. ft. on the floor near the wall to kill dermestid grubs.
2. Treat gunny/cotton bags used for storage of PC by dipping them in 0.028% Deltamethrin (1 gm of Deltamethrin {2.8% EC} in 1 liter of water), followed by drying in shade.
3. Spray 0.028% Deltamethrin on the floor and of PC storage room.
4. Dust 5% Malathion on pierced cocoons, before storage. Five kg dust is sufficient to cover 50 kg of pierced cocoons.
5. Dip wooden trays harbouring beetles and earwigs in 0.076% DDVP (1 ml of DDVP {EC=76%} in 1 liter of water) for 2-3 mins. After 10 days, wash them thoroughly in water, sun-dry and re-use.
6. Spray stands and room with 0.076% DDVP.

Conclusion

Even though there is a lot of scope for utilizing biocontrol agents for management of pests of both mulberry and mulberry silkworm, a very few sericulturists use them. Efforts on a massive scale are necessary to popularize biocontrol agents to contain these pests and step up silk production. Biological control has to be integrated suitably to fit into Integrated Pest Management of pests affecting sericulture industry.

Biointensive insect pest management strategy for rice under System of Rice Intensification

R. Nalini,* D.S. Rajavel* and M. Shanthi*

Field trials were conducted during *rabi* 2006 – 2007 and 2007 – 2008 to evolve a management strategy for the rice insect pests raised under System of Rice Intensification (SRI). The experiments were laid out with CO 43 paddy variety in Exploded block design with two treatments (IPM and non IPM plots) each having a plot size of 36 x 5 m with five microplots in each. In the IPM plot, various pest management methods were judiciously combined which include seed treatment with Super Pseudomonas @ 10 g/kg; application of Mahua / Neem cake 250 kg/ha in two splits, basal application of *Azospirillum* 2 kg/ha, Phosphobacteria 2 kg/ha, Silica solubilizing bacteria 2 kg/ha and *Pseudomonas* 2.5 kg/ha; release of *Trichogramma japoni*cum @ 5 cc/release/week on 30 and 37 DAT; *Trichogramma chilonis* @ 5 cc/release/week on 30, 44 and 51 DAT and one round of insecticide spray. Non IPM plot received inorganic fertilizer application and three rounds of insecticide spray. The brown planthopper population was 0.34 no./tiller in IPM plot compared to 0.37 no./tiller in non IPM plot. The population of green leafhopper was 0.95 and 1.13 no./hill in IPM and non IPM plots respectively. The intensity of damage by stem borer was 5.10% in IPM plot, while it was 7.39%, in non IPM plot. Similar trend was observed in leaf folder and gall midge damage. The cost incurred towards plant protection in IPM plot was Rs. 4275/ha and in non IPM plot Rs.4651/ha. The yield recorded in IPM plot was 5270 kg/ha and in non IPM plot 4940 kg/ha. The predatory mirid bug *Cyrtorhinus lividipennis* and spider population was maximum of 8.52 and 2.12 no./hill in IPM plot compared to 4.84 and 1.08 no./hill in non IPM plot. The stem borer egg mass parasitization by Trichogrammatid in IPM plot was 32.25 per cent whereas in non IPM plot it was 10.09 per cent. The activity of other predators *viz*., ground beetle, coccinellid beetles, damselflies, rove beetles, cricket, water strider, long horned grasshopper and parasitoids *viz.*, Ichneumonid, Braconid, Chalcid, Eulophid, Scelionid, Dryinid and Pipunculid were high in the IPM plot. Addition of organic matter, microbials, *Trichogramma* sp. in SRI with IPM resulted in reduction of pest load, enhancement of natural enemies activity and increased yield; in addition the food web in the rice ecosystem was undisturbed parting no hazardous impact on the environment.

Introduction

System of Rice Intensification (SRI) is gaining acceptance and beginning to spread around the rice growing world. In Madagascar, it was developed over 20 years ago (Stoop *et al.,* 2002). Today, its benefits have been demonstrated in at least 24 countries in Asia, Africa and Latin America. In India, the first on farm comparison trials were made in 2003 at Andhra Pradesh state, within three years, the area under SRI in the state has reached more than 40,000 ha (ANGRAU, 2006). In Tamil Nadu also this system of rice cultivation is intensifying owing to its increased water and crop productivity. The other interesting facet include higher profitability due to lower costs of production accompanied with higher yield and increased factor productivity of land, labour, water and capital. With the cultivation of larger plants and the production of more grain, the challenge of crop protection increases. SRI

* Department of Agricultural Entomology, Agricultural College and Research Institute, Madurai 625 104

farmers report that losses through pests and diseases are minimal and do not need chemical protection (Chabousson, 2004). Considering this, the present study aimed to evolve a biointensive insect pest management strategy for rice raised under SRI.

Materials and methods

Field experimentation

Two field experiments were conducted during *rabi* 2006 – 2007 and 2007 – 2008 at the Research Block of the Agricultural College and Research Institute, Madurai. The experiments were laid out with CO 43 paddy variety in Exploded block design with two treatments *viz*., IPM and non IPM each having a plot size of 36X5 m. Each plot had five microplots of one square metre size and all the observations were taken within the microplots. In IPM plot, the various pest management methods were judiciously combined which include seed treatment with Super Pseudomonas @ 10 g/kg; application of Mahua / Neem cake 250 kg/ha in two splits basally and 30 days after transplanting; basal application of *Azospirillum* 2 kg/ha, Phosphobacteria 2 kg/ha, Silica solubilizing bacteria 2 kg/ha and Pseudomonas 2.5 kg/ha; release of *Trichogramma japoni*cum @ 5 cc/ release/week on 30 and 37 DAT; *Trichogramma chilonis* @ 5 cc/release/week on 30, 44 and 51 DAT and one spray of monocrotophos 36 SL 1000 ml/ha. Non IPM comprised of inorganic fertilizer application (150: 50: 50 NPK kg/ha) and three rounds of insecticide spray *viz*., Monocrotophos 36 SL 1000 ml/ha, Carbaryl 50 WP 1 kg/ha and Chlorpyriphos 20 EC 1250 ml/ha.

Observations on insect pest population / intensity of damage

The population of brown planthopper, *Nilaparvata lugens* Stal.; green leafhopper *Nephotettix* spp. and the intensity of damage by the stem borer, *Scirpophaga incertulas* (Walker), leaf folder *Cnaphalocrocis medinalis* (Guen.) and *Marasmia* spp., gall midge, *Orseolia oryzae* (Wood-Mason) were recorded on 30, 60 and 90 days after transplanting (DAT).

Observations on natural enemies

The observation on natural enemies was done on 30, 60 and 90 DAT and the mean worked out. The activity of the insect predators and spider complexes were recorded either by direct counting or using sweep net. Similarly, the activity of parasitoids was recorded from the field collected insect pest population and per cent parasitization was computed.

Economics and cost benefit ratio

The grain yield was recorded and gross return was calculated based on the market price. The cost incurred towards the plant protection and fertilization was computed and compared for IPM and non IPM plots. The cost benefit ratio was worked out using the cost of plant protection and fertilization alone and the output at the market rate to arrive at the benefit for each rupee spent on insect pest management.

Results

Insect pest population / intensity of damage

Brown planthopper (BPH): The incidence of BPH remained much below the Economic Threshold Level (ETL) in both the IPM and non IPM plots. The population ranged from 0.18 to 0.50 no./tiller in IPM plot and from 0.18 to 0.49 no./tiller in non IPM plot. The overall mean population was 0.34 no./tiller in IPM plot compared to 0.37 no./tiller in non IPM plot and showed only slight variation.

Green leafhopper (GLH): Its population also remained below the ETL in both the IPM and non IPM plots. The population ranged from 0.26 to 1.56 no./hill in IPM plot while 0.34 to 1.72 no./hill in non IPM plot. The overall mean population of green leafhopper was 0.95 and 1.13 no./hill in IPM and non IPM plots respectively.

Yellow stem borer: The intensity of damage by yellow stem borer was high in non IPM plot and exceeded the ETL in both the years of study. On 60 DAT in first year and 90 DAT in second year the damage in non IPM plot exceeded the ETL. It implies that, IPM strategies were able to suppress stem borer incidence significantly in comparison with non IPM practices. The overall mean intensity of damage by stem borer was 5.10 per cent in IPM plot, while it was 7.39 per cent, in non IPM plot.

Leaf folder: The per cent leaf damage by the leaf folder was lower in IPM plot compared to non IPM plot. The damage by leaf folder did not exceed ETL in IPM and non IPM plots of first year field trial 2006 - 2007 while it exceeded in both the IPM and non IPM plots during the second year field trial 2007 - 2008. The per cent leaf damage ranged from 0.14 to 16.67 in IPM plot and from 0.54 to 20.31 in non IPM plot. The overall mean percentage of leaf damage was 4.13 in IPM plot and 5.85 in non IPM plot.

Gall midge: The gall midge incidence in both the field trials was high and exceeded the ETL. The mean per cent of onion shoot was 11.99 and 9.60 in IPM plot and 18.68 and 12.49 in non IPM plot during 2006 - 2007 and 2007 - 2008, respectively. The overall mean per cent damage was 10.80 in IPM plot and 15.59 in non IPM plot.

Natural enemies

In IPM plot the activity of natural enemies was significantly high when compared to the non IPM plot. In IPM plot the predators *viz*., mirid bug *Cyrtorhinus lividipennis*, ground beetle *Ophionea* sp., coccinellid beetles *Micraspis*, *Harmonia*, *Cheilomenes*, Damselflies *Agriocnemis* sp., rove beetle, *Paederus fuscipes*, cricket *Metiocha* sp., water strider *Limnogonus* sp., long horned grasshopper *Conocephalus* sp. and spider complexes were found predating on different stages of insect pest in high numbers compared to non IPM plot. Among the parasitoids, Trichogrammatid Ichneumonid and Braconid were dominant in both IPM and non IPM plot. The stem borer egg mass parasitization by Trichogrammatid in IPM plot was 32.25 per cent whereas in non IPM plot it was 10.09 per cent. Ichneumonid and Braconid were the next dominant in IPM plot (11%) compared to the non IPM plot (2.4%). The other parasitoids *viz*., Chalcid, Eulophid, Scelionid, Dryinid and Pipunculid were recorded to the tune of 2 - 4 per cent in IPM plot compared to less than 0.5 per cent in non IPM plot.

Economics and cost benefit ratio

The grain yield recorded in IPM plot was 5270 kg/ha and in non IPM plot 4940 kg/ha. It directly reflected in the high total returns in IPM plot compared to the non IPM plot. The cost benefit ratio was high in IPM plot (1:7.63) compared to non IPM plot (1: 6.43). Combination of organic amendments, biofertilizers and bioagents of lower input cost made the cost benefit ratio to be higher in IPM plot. In non IPM plot, the number of insecticide sprays influenced the cost benefit ratio *i.e.,* increased input cost reduced the cost benefit ratio.

Discussion

In IPM plot, the scientifically proven plant protection components *viz*., organic nutrients, organic amendments, biofertilizers and bioagents were integrated and the crop suffered less damage due to insect pest's *viz*., brown planthopper, green leafhopper, yellow stem borer, leaf folder and gall midge. In addition, the natural enemy activities were numerically high. This observation corroborate with the findings of NCIPM Annual report, 2006 – 2007.

Table 1. Incidence of major pests in IPM and non IPM plots.

Treatment	First trial 2006 – 2007				Second trial 2007 – 2008				Grand mean
	Period of observation (Days after transplanting)			Mean	Period of observation (Days after transplanting)			Mean	
	30	60	90		30	60	90		
BPH (No./tiller)									
With IPM	0.39	0.43	0.50	0.44	0.34	0.18	0.18	0.23	0.34
Without IPM	0.44	0.46	0.49	0.46	0.42	0.20	0.18	0.27	0.37
GLH (No./hill)									
With IPM	0.74	1.06	0.26	0.69	1.00	1.08	1.56	1.21	0.95
Without IPM	0.76	1.36	0.34	0.82	1.16	1.40	1.72	1.43	1.13
Stem borer (%)									
With IPM	3.97	8.18	6.84	6.33	0.07	2.85	8.69	3.87	5.10
Without IPM	7.86	10.42	8.22	8.83	0.20	3.74	13.94	5.96	7.39
Leaf folder (%)									
With IPM	1.96	3.44	1.78	2.39	0.14	0.79	16.67	5.87	4.13
Without IPM	2.73	4.56	5.66	4.32	0.54	1.27	20.31	7.37	5.85
Gall midge (%)									
With IPM	6.12	16.61	13.24	11.99	0.00	5.40	23.39	9.60	10.80
Without IPM	12.35	19.71	23.99	18.68	0.27	7.25	29.94	12.49	15.59

Table 2. Natural enemy population in IPM and non IPM plots

Natural enemy	IPM	Non IPM
Predators		
Mirid bug, *Cyrtorhinus lividipennis* (No./hill)	8.52	4.84
Spiders (No./hill)	2.12	1.08
Ground beetle, *Ophionea* sp. (No./hill)	0.96	0.33
Coccinellid beetles, *(Micraspis, Harmonia, Cheilomenes)* (No./hill)	1.56	0.33
Damselflies, *Agriocnemis* sp., (No./5 sweeps)	3.80	1.20
Rove beetle, *Paederus fuscipes* (No./hill)	1.27	0.42
Cricket, *Metiocha* sp. (No./5 sweeps)	2.40	0.40
Water strider, *Limnogonus* sp. (No./sq.m.)	3.00	0.64
Long horned grasshopper *Conocephalus* sp. (No./hill)	1.20	0.40
Parasitoids		
Trichogrammatid (%)	32.25	10.09
Ichneumonid (%)	9.24	2.13
Braconid (%)	2.99	0.43
Other parasitoids (%) (Chalcid, Eulophid, Scelionid, Dryinid, and Pipunculid)	2 - 4	< 0.05

Table 3. Economics of rice production in IPM and non IPM plots

Variables	IPM	Non IPM
Mean yield (kg/ha)	5270	4940
Cost incurred towards plant protection and fertilization (Rs. /ha)	4275	4651
Gross returns (Rs.) @ Rs. 7/kg	36,890	34,580
Net returns (Rs.)	32,615	29,929
Cost Benefit ratio	1:7.63	1: 6.43

Influence of organic nutrients, organic amendments and biofertilizers

The key factor for superior SRI performance compared to conventional method comes from the soil which is rich in organic matter and microbial activity. Organic matter is mainly supplemented through Farm Yard Manure (FYM) in rice fields that increased levels of leucoanthocyanins, catechins, flavanol glycosides and phenol carboxylic acids in plants (Lyashenko *et al*., 1982) and this would be responsible for lesser incidence of many rice insect pests.

Earlier reports also revealed that incorporation of *Azospirillum* significantly decreased the incidence of BPH, GLH and leaf folder in rice (Anuradha, 1989). Similarly, the combined application of *Azospirillum* with organic manure decreased the feeding rate of BPH and adversely affected its growth and development (Athisamy and Venugopal, 1995). The role of neem cake in reducing the incidence of planthoppers and leaf hoppers, gall midge, stem borer and leaf folder has been reported by Athisamy (1994) and Ambethgar (1996). Mohankumar *et al.* (1995) reported that organic manure application lowered gall midge incidence than inorganic fertilizer application.

The application of FYM, neem cake and biofertilizers would have increased the plant phenol content which in turn had negative effect on the insect multiplication and development. This is confirmed with the findings of Athisamy (1994) where in plots which received *Azospirillum* and FYM the amount of total phenol and silicon was more. Moreover, Mohan *et al.* (1988) reported that *Azospirillum* would have favourably activated Phenyl Ammonia Lyase enzyme concerned in the biosynthesis of phenolics resulting in increased plant phenolics in plants which are responsible for lower damage.

Influence of agronomic practices

The novel practices of SRI include wider spacing, single seedling planting and unflooded condition which have negative effect on insect build up. This is in line with the findings of Saroja and Raju, 1982 and TNAU (2006). The agronomic practices such as closer planting and nitrogen application increased the leaf folder incidence whereas wider planting and potash application decreased the leaf folder attack (Saroja and Raju, 1982). Nalini *et al.* (2006) observed that in conventional method of rice cultivation, incidence of stem borer, leaf folder and gall midge were high compared to SRI method. In SRI, intermittent irrigation, cono weeding and wider spacing were identified to have negative impact on arthropod diversity (TNAU, 2006).

Influence of bioagents

In IPM plot the activity of natural enemies was numerically high when compared to the non IPM plot. Earlier researches also confirm this result. According to TNAU (2007) combined use of PGPR consortia, release of *Trichogramma japonicum* and *T. chilonis*, Azadirachtin 1% (2ml/l) and leaf colour chart based nitrogen application registered lowest stem borer and leaf folder compared to farmers practice. The *T. japonicum* and *T. chilonis* were reported on leaf folder eggs to the tune of 62 and 38 per cent. Moreover, the Super Pseudomonas [Improved Flourescent Pseudomonad bioformulation (Pfl+TDK1+Py15)] with chitinase and ACC deaminase activities was found more effective against leaf folder. Ragini *et al.* (1995), Chandramani (2003) and Usha Rani (2005) also agree with the findings of organic farming supporting more *Ophionea*, *Paederus*, cocicinellid beetles, mirids and spiders.

Influence of insecticides

In non IPM plot, irrespective of insecticide spray the stem borer, leaf folder and gall midge damage exceeded the ETL. It is because the insect pests have started developing resistance to insecticides of repeated usage at sub lethal dose. In addition, the natural enemies' population was drastically less. It agrees with the findings of Chalapathi Rao *et al.* (2006) where use of flufenoxuron and lambda-cyhalothrin effectively controlled leaf folder but recorded low predator population *viz.*, rove and ground beetle. The concept of IPM is basically to make need based use of insecticides and the crop protection practices ought to be such that there is pest residue to maintain the population of natural enemies of the pest in a balanced state.

Economics and socio-economic attribute

The grain yield, total returns and cost benefit ratio recorded in IPM plot was high compared to non IPM plot. Combination of organic amendments, biofertilizers and bioagents of lower input cost made the CB ratio to be higher in IPM plot. In non IPM plot, the number of insecticide sprays directly influenced the CB ratio; the increased input cost reduced the CB ratio. These results concur with Subramanian and Rangarajan (1990) and Athisamy (1994). *Azospirillum* promotes growth of plant mainly by the production of phytohormornes *viz.*, auxins, gibberlins and cytokinins in addition to biological nitrogen fixation (Cicciari *et al.*, 1989).

Conclusion

In the long run, SRI with organic and various soil enrichment practices will help to improve soil health and plant vigour. Integrating the ecofriendly approach of biocontrol with SRI will bring down the insect pest population without affecting natural enemy fauna in the rice ecosystem. Reduced insecticide usage will result in huge monetary savings without any environmental degradation and serious health risk of farming families and village community.

References

Ambethgar, V., 1996. Effectiveness of neem (Azadiracta indica A. Juss) products and insecticides against the leaf folder *Cnaphalocrocis medinalis* (Guenee). *J. Ent. Res.*, 20 (1), 83-85.

ANGRAU, 2006. Report of a meeting on the System of Rice Intensification (SRI) at Acharya N.G.Ranga Rau, 29 June 2006 (available at *ciifad .cornell. edu/sri/ countries/India/inapfingerman06.pdf*).

Anuradha, S., 1989. Effect of fertilizers and varieties on pest incidence inplant and ratoon rice. M.Sc.(Ag.) thesis, Tamil Nadu Agricultural University, Agricultural College and Research Institute, Madurai, India 102 p.

Athisamy, M., 1994. Effect of biofertilizer (*Azospirillum*) and organic amendments on major pests of rice. M.Sc.(Ag.) thesis, Tamil Nadu Agricultural University, Agricultural College and Research Institute, Madurai, India 102 p.

Athisamy, M. and Venugopal, M.S., 1995. Effect of *Azospirillum* and organic amendments on the incidence of major pests of rice. In: National symposium on Organic Farming held on 27 – 28 October, Madurai, India. pp. 169 -172.

Chabousson, F., 2004. Healthy crops: A new agricultural revolution. Charnley, UK, Jon Anderson.

Chalapathi Rao, N.B.V., Vidya Sagar Singh and Subash Chander, 2006. Effect of various combinations on rice leaf folder and to coleopteran predators. *J. ent. Res.*, 30, 219 – 223.

Chandramani, P., 2003. Studies on induced resistance through organic sources of nutrition to major pests of rice. Ph.D. thesis, Tamil Nadu Agricultural University, Agricultural College and Research Institute, Madurai. 144p.

Cicciari, I., Lippi, O., Pietrosanti, T. and Pietrosanti, W., 1989. Phytohormone like substance produced by single and mixed diazotrophic cultures of *Azospirillum* and *Azatobacter. Plant and Soil*, 15, 151 – 153.

Lyashenko, N.I., Solody, U.K.G.D., Godovany, A.A., Verzhbitskii, V.I. and Moskal Chuk, N.I., 1982. The effect of fertilizers on phenol metabolism and on plant resistance to aphids. *Fizlologiya Blokhimiya Turnyh Rastenil*, 14, 337 – 377.

Mohan, S., Purushothaman, D., Jayaraj, S. and Rangarajan, A.V., 1988. PAL ase activity in the roots of Sorghum bicolor (L.) inoculated with *Azospirillum. Curr. Sci.*, 57, 492 – 493.

Mohankumar, S., Sundara Babu, P.C. and Venugopal, M.S., 1995. Effect of organic and inorganic forms of nutrients on the occurrence of rice gall midge and its parasitoid. In: National symposium on Organic Farming held on 27 – 28 October, Madurai, India. p. 72.

Nalini, R., Suganya Kanna, S., Rajavel, D.S. and Durai Singh, R., 2006. Insect pest scenario in System of Rice Intensification (SRI) technique in the Periyar Vaigai Project area of Tamil Nadu. In: National symposium on "System of Rice Intensification (SRI) – Present status and Future Prospects" held on 17 - 18 November, 2006 at Acharya NG Ranga Agricultural University, Rajendranagar, Hyderabad, Andra Pradesh 111p.

NCIPM (National Centre for Integrated Pest management), 2006-2007. Annual Report. Indian Agricultural Research Institute, New Delhi. p. 7- 13.

Ragini, J.C., Thangaraju, D. and David, P.M.M., 1995. Effect of organic manures on some predators in the rice ecosystem. *IRRN.*, 24(3), 15.

Saroja, R. and Raju, N., 1982. Effect of nitrogen fertilizer levels and spacing on rice gall midge and leaf folder damage. *IRRN,* 7(4), 10.

Stoop, Willem, Norman Upfhoff and Amir Kassam, 2002. A review of agricultural research issues raised by the System of Rice Intensification (SRI) from Madagascar: Opportunities for improving farming systems for resource-poor farmers. Agricultural systems, 71, 249 – 274.

Subramanian, R. and Rangarajan, M., 1990. Response of rice to Azospirillum brasilense and chemical fertilized farms in India. *IRRN*, 15(3), 27.

TNAU, 2007. Annual Rice Research Meet Report, meet held during 25 – 27 April 2007 at Tamil Nadu Agricultural University, Coimbatore. p. 4-5.

TNAU, 2006. .Annual Rice Research Meet Report, meet held during 25 – 27 April 2006 at Tamil Nadu Agricultural University, Coimbatore. p. 3

Usha Rani, B., 2005. Integration and evaluation of resistant sources with organic nutrients for the management of major insect pests of rice. Ph.D. thesis, Tamil Nadu Agricultural University, Agricultural College and Research Institute, Madurai. 187p.

Selection of *Trichogramma* for improved host searching ability

J.S. KENNEDY,* M. PRAKASM* AND N. SATHIAH*

Attempt has been made to improve the host searching ability of *Trichogramma* by acclimation and conditioning for ten generations. The fittest wasps were selected in each generation and advanced. Thus, initially *Trichogramma* having an ability to forage 50 cm were trained to forage for 180 cm at the end of tenth generation. The experiments were conducted in three conditions including control, where the wasps had no interaction with wind flow, *Trichogramma* allowed to forage along the wind direction and *Trichogramma* allowed to forage across the wind direction. Of the wasps released, 81.6 per cent were able to parasitize the eggs kept at 50 cm in control. The same 75.5 per cent and 83.2 per cent, when *Trichogramma* were released against and along with wind flow respectively. There was a negative relationship between the distance trained and the wind flow on the effect of parasitization of *Trichogramma.* The *Trichogramma* were able to parasitize 54.4 per cent only of *Corcyra* eggs after having selected to forage for 180 cm when they were allowed to move along with the wind. On the other hand, only 38.7 per cent and 42.74 per cent of parsitization was observed when *Trichogramma* were made to move against the wind and control respectively.

INTRODUCTION

Despite wide spread use of *Trichogramma,* there are relatively few cases, where successful control of pests can be unequivocally attributed to release of these parasitoids. There are many documented failures of *Trichogramma* release despite a few notable successes (Smith *et al.*, 1986; Li 1994; Singh and Jalali, 1994). The variable success of *Trichogramma* field release can be attributed to a number of factors such as pesticide application, weather conditions at the time of release, qualities of release materials and lack of hosts-parasitoid synchrony. This encompasses diverse issues such as technical aspects related to rearing, storage, and shipping of release materials, characteristic of *Trichodermma* chosen for release programme and quality parameters such as enhanced host searching ability, insecticide tolerance, heat resistance, etc. One of the major limiting factors of *Trichogramma* is its poor searching ability. *Trichogramma* can forage a maximum of a meter radius and it has to find its host, otherwise it will become futile. This problem is addressed here and an attempt has been made to develop a viable technology to train *Trichogramma* to improve its host searching ability.

MATERIALS AND METHODS

Host culturing techniques: Corcyra cephalonica Stainton is an alternative host for *Trichogramma* spp. It was cultured following the procedure of Singh and Jalali (1994). The adult moths were allowed inside an oviposition cage of 21 x 25 cm size, covered with wire mesh at the bottom, and two windows for ventilation. Adults, were fed with 10% honey solution. The eggs of *C. cephalonica* were sprinkled over half ground cumbu grains fortified with 5 g of yeast and 100 g of groundnut kernel powder in a plastic basin of 11 x 37.5 cm and covered with khada cloth. Care was taken to maintain the culture free of storage mites and diseases by mixing 5 g wettable sulphur and streptomycin sulphate at the rate

* Department of Entomology, Tamil Nadu Agricultural University, Coimbatore 641 003

of 0.05 per cent. The emerging adults were collected and used for culturing the host and parasitoid. The culture was maintained at room temperature (18.28^0 C).

Techniques of Parasitoid culturing: Trichogramma sp. was cultured in the laboratory as per Singh and Jalali (1994) using 46 x 30 cm polythene bags. Freshly laid eggs of *C. cephalonica* were sterilized under ultra violet lamp (UV) for 20 minutes. They were evenly glued to strips of card board of 30 x 18 cm, marked into 30 strips of 2.0 x 6.0 cm size at the rate of 0.22 cc per strip. This was done by sprinkling through a wire mesh strainer from above. The cards were air-dried and excess eggs were brushed off with a fine hair brush and exposed to newly emerged parasitoids in a polythene bag. The parasitoids were fed with 10 per cent honey swabbed in perforated Mylar sheet. The culture was maintained at 26$\pm$ 1^0 C and 60 per cent relative humidity.

Preparation of loose eggs for experiments: A clean glass jar of 19 x 8.5 cm size was swabbed inside with cotton moistened with water. Three cc of sterilized fresh eggs of *C. cephalonica* were uniformly sprinkled over moistened inner surface of the glass jar and air dried. Parasitoids that emerged from 0.5 cc of eggs were allowed inside the jar and the mouth of the jar was covered with muslin cloth. The adults were fed with 50 per cent honey swabbed in perforated Mylar sheet on the fourth day after exposure. Parasitoid eggs that turned black in colour were removed from the glass jar with a fine camel hair brush and used for the experiments. The loose parasitized eggs were transferred individually to Durham's glass rubes of 2.8 x 5.0 cm size and plugged with cotton for sexing and observing for adult longevity without anesthetisation.

Selecting Trichogramma for improved host searchability: To enhance the host searching ability of parasitoids, acclimation and conditioning was carried out for ten generations and the fittest parasitoids were selected and advanced. This experiment was carried out in a wind tunnel (Cantech, India). *Trichogramma* infested egg cards were kept at the feeding entrance (Figures). They were allowed to parasitize the sterilized eggs of *C. cephalonica* at the other end at 50 cm. Successful parasitoids were selected and advanced to the subsequent generations. The searching distance was increased to 75 cm. The same process of selections was done up to 10 generations. Each time the fittest wasps were selected and advanced to the next generations and the search distance was increased as furnished in Table 1. The experiments were done in the wind tunnel, where the parasitioids were released along and across the directions of the wind. Control was maintained where there was no wind movement.

Table 1. Details of selection of *Trichogramma* for improved host searching ability

Generation	*Trichogramma* selected at
F_0	50 cm
F_1	75 cm
F_2	100 cm
F_3	110 cm
F_4	120 cm
F_5	130 cm
F_6	140 cm
F_7	150 cm
F_8	160 cm
F_9	170 cm
F_{10}	180 cm

RESULTS

Attempt has been made to improve the host searching ability of *Trichogramma* by acclimation and conditioning for ten generations. The fittest wasps were selected in each generation and advanced. Thus, initially *Trichogramma* having an ability to forage 50 cm were trained to forage for 180 cm at the end of tenth generations.

The experiments were conducted in three conditions including control, where the wasps had no interaction with wind flow, *Trichogramma* allowed to forage along with wind direction and *Trichogramma* allowed to forage across the wind direction. Of the wasps released, 81.6 per cent were able to parasitize the eggs kept at 50 cm in control. The same was 75.5 per cent and 83.2 per cent, when *Trichogramma* were released against and along the wind flow respectively.

The effects of wind flow against the distance trained on *Trichogramma* in terms of parasitization are presented in Table 2. There was a negative relationship between the distance trained and the wind flow on the effect of parasitization of *Trichogramma*. *Trichogramma* were able to parasitize 54.4 per cent only of *Corcyra* eggs after having selected to forage for 180 cm, when they were allowed to move along with the wind. On the other hand, only 38.7 per cent and 42.74 per cent of parasitization was observed when *Trichogramma* were made to move against the wind and control respectively. There was a statistically significant effect of wind flow on the parasitization of *Trichogramma* ($F=13453.32$, $P< 0.0001$). The effect of distance trained on the parasitization of *Trichogramma* was also statistically significant ($F=7352.24$, $P< 0.0001$). Likewise, there was also a statistically significantly interaction effect of wind and distance trained on the parasitization of *Trichogramma* ($F=253.64$, $P<0.0001$). Post hoc Tukey's HSD test exceeded statistically significant differences between the wind manipulations. Along the wind improved parasitization, while, against the wind significantly reduced it. Similarly, a statistically significant ($P<0.0001$) effect of distance trained was also observed. Every attempt to increase distance resulted in a significant reduction in percentage parasitization.

Table 2. Parasitization of *Trichogramma* at different conditions after selection

Distance (cm)	Control	Against the wind	Along the wind
50	81.6	77.5	83.2
75	78.4	65.4	77.4
100	65.4	61.3	75.8
110	59.4	58.8	75.2
120	57.7	58.0	72.0
130	57.3	56.1	69.5
140	56.5	56.4	67.8
150	48.6	43.5	65.5
160	43.3	40.5	63.7
170	42.8	40.3	58.2
180	42.7	38.7	54.4

There was however a slight reduction observed in terms of adult emergence of the selected *Trichogramma* viz., 78.3 per cent, 74.2 per cent, and 79.3 per cent parasitized eggs in control, against the wind and along the wind respectively. Progressive reduction in adult emergence of the *Trichogramma* was observed, when the host searching distance increased. The hatchability in *Trichogramma* selected to forage for 100 cm were 62.3 per cent, 57.9 per cent and 71.7 per cent respectively in control, against the wind and along the wind. These values further reduced as the host searching distance increased to 180 cm, where only 40.3 per cent, 37.4 per cent, and 49.8 per cent of emergence respectively were observed (Table 3). There was statistically significant effect of wind flow on the emergence of *Trichogramma* (F=11074.95, P< 0.0001). The effect of distance trained on the emergence of *Trichogramma* was also statistically significant (F=5563.78, P< 0.0001). Likewise, there was also statistically significant interaction effect of wind flow and distance trained on the emergence of *Trichogramma* (F=263.56, P 0.0001). Post hoc Tukey's HSD test exceeded statistically significant difference between the wind manipulations. Along the wind improved, and against the wind significantly produced emergence of *Trichogramma.* Similarly a statistically significant (P< 0.0001), effect of distance trained was also observed. Every attempt to increase distance trained resulted in a significant reduction in adult emergence.

Table 3. *Trihcogramma* adult emergence at different conditions after selection

Distance (cm)	Control	Against the wind	Along with the wind
50	78.3	74.3	79.3
75	72.9	61.4	74.3
100	62.3	57.9	71.7
110	54.2	56.8	69.0
120	54.5	50.5	68.0
130	53.4	55.9	66.7
140	52.0	54.7	65.1
150	43.7	41.0	63.6
160	40.4	37.9	63.5
170	40.7	38.9	56.2
180	40.3	37.4	49.8

Discussion

Despite the widespread use of *Trichogramma,* there are relatively few cases, where successful control of a pest can be unequivocally ascribed to the releases of this parasitoid. There are many documented failures of *Trichogramma* release despite a few notable successes (Smith *et al.,* 1990; Li. 1994). The variable success of *Trichogramma* field releases can be attributed to a number of factors (Smith, 1996). These include pesticide applications, weather conditions at the time of release, lack of host parasitoid synchrony and the quality of release materials. This last area is complex and encompasses diverse issues such as technical aspects related to rearing, storage and shipping of the release materials as well as the characteristics of the *Trichogramma* chosen for program.

The search of *Trichogramma* largely depends on host searching, which is also a major limiting factors in its exploitation in biocontrol. To enhance the foraging ability of the wasps, they were subjected to acclimation and conditioning through ten generations and the host searching ability could be increased significantly from 50 cm to 180 cm, although a significant reduction in parasitzation and adult emergence was observed.

Activity of *Trichogramma* females can be considerd as behavioural responses to the physiological state (Connolly, 1967). A female's searching hehavior probably be a consequence of an innate "appetite to parasitize" or "ovipoistional drive", which apparently is largely dependent on the number of eggs of the females oviducts (Hassell and Stouthwood, 1978). In the present study, searching activity of *Trichogramma* females quickly resulted in encounters with the host eggs due to the relatively small size of experimental unit and subsequent parasitzation activity. Since the experiments were conducted at room temperatures where the temperature, has not been controlled, wind flow was the major factor that influenced the parasitization the parasitization rate of the active females were determined as different aspects of parasitization activity. Wind flow against the directions of *Trichogramma* flight showed down the parasitization activity, adult emergence and hence low proportional activity.

The differential mechanisms for such activity may be the results of two different strategies in *Trichogramma* to overcome the limiting effect of wind flow on reproductive fitness. In one strategy, the high proportional activity was maintained, when wind flow was along the direction of *Trichogramma* flight. As *Trichogramma* has to spend more of its energy in flight against the wind, it had lesser activity and hence its host searching ability and adult emergence were affected. However, selecting a strain under stress condition for high host searching ability will be of greater advantages. The latter strategy may compensate for a low proportional activity (Pak and Van Heiningen, 1989).

One of the main aims of this laboratory work was to develop suitable methods to select *Trichogramma* with emphasis on desirable quality such as high host searching ability. Screening variables of *Trichogramma* strains using such methods would reduce the strain to be tested in the field. If suitable method can be internationally standardized, information can be exchanged between laboratories that would help to improve the selection of candidate strain for different purposes (Hassan, 1995).

References

Cannolly, K., 1967. Locomotors activity in Drosophila, 111. A distinction between activity and reactivity, *Anim, Behav.,* 15, 149-152.

Hassan, S.A., 1995, Introduction to the effectiveness and assessment session les collq *INRA*, 73, 101-11.

Hassell. M.P. and Southwood, T.R., 1978. Foraging strategies of insects, *Ann. Rev. Ecol. Syst*., 9, 75-98.

Li, Y.L., 1994. World wide use of *Trichogramma* for Biocontrol control on different crops. A survey, pp 37-53. In: Biological control with egg parasitoids CAB international, Oxon, U.K.

Pak, G.K. and Heiningen, T.G., 1989. Behavioral variations among strain of *Trichogramma* sp. Adaptability to field temperatures conditions. *Entomol. Exp. Appl.,* 38, 3-13.

Singh, S.P. and Jalali, S.K, 1994. Trichogrammitidy, PDBC Annual Report.

Smith, S.M., Carrow, J.R. and Laing, J.E., 1990. Inundative release of the parasitoid *T. minutum* (Hym; Trichogrammatidae) against forest insects pest such as the spruce budworm. *Choristonerua fumiferna* (Lep: Torticidae) Tu Ontario project 1982-1986. *Mem. Entomol. Soc. Canada*, 153, 1-82.

Smith, S.M., Hubner, M. and Carrow, J.R., 1986. Factor affecting inundative releases of *T. minutum* against the spruce budworm; proc. Xvii inter. Cong. Entomol. Humurg, FRG, against 20-26, 1986. *Z. Ang. Entomol*., 101, 29-39.

Smith, S.M., 1996. Biological control with *Trichogramma* advances, success, and potential of their use. *Ann. Rev. Entomol*, .41, 375-406.

Predatory Potential of *Micromus igorotus* Banks (Neuroptera: Hemerobilidae), of Sugarcane Woolly Aphid, *Ceratovacuna lanigera* Zehntner

Sathiah, N.,* J.S. Kennedy,* and V. Padmini*

Laboratory experiments on the predatory potential of *Micromus igorotus* at constant and fluctuating temperature revealed that predation was higher at an incubation temperature of 25^0 C than at room temperature conditions. The grubs consumed 672.35 nymphs and 748.89 grown up woolly aphids in the first set of treatments while, it decreased to 360.72 to 386.82 in the second set of treatments. In the present study after eclosion and up to 7 days the daily predation rates ranged from 54.80 to 89.06 aphids/days. The mean predation per day irrespective of the treatments ranged from 40.08 to 83.21. A high degree of voracity obtained in the present studies at two temperatures is a fair determination of approximate temperature prevailing in sugarcane fields. This limited study has indicated maximum attack rates for the combination of predator and prey stages. In the present investigation *M. igorotus* fed similarly on nymphs and grown up stages irrespective of the conditions of incubation temperature. The high degree of voracity reported in this investigation offers scope for *M. igorotus* as a biocontrol agent in augmentation. Experiments on the predatory potential of different population of *M. igorotus* under laboratory conditions have revealed that the differences in predatory potential were not significant. So far, there are no reports available on the predatory potential of *M. igorotus* geographical population. The concept behind taking up this evaluation was to identify the most voracious population that can be deployed in biological control of woolly aphid.

Introduction

Currently, the sugarcane woolly aphid, *Ceratovacuna lanigera* Zehnter has emerged as a serious pest in India and particularly in Tamil Nadu. The aphid attacked crop at any age. A loss of 20% in yield and CCS was observed at the 9th month when the crop was attacked from 6th month onwards. There was significant reduction in both quantitative and qualitative parameters (Gupta and Goswami, 1995). In sugarcane ecosystem the prevalence of natural enemies like *Dipha aphidivora* (Meyrick) (Pyralidae: Lepidoptera), *Synonycha grandis* (Thunberg) (Coccinellidae: Coleoptera), *Coccinella septempunctata* (Linnaeus) (coccinellidae: Coleoptera), *Chrysoperla carnea* (Stephens) (Chrysopidae: Neuroptera); *Micromus* sp. (Banks) have been observed on woolly aphids. Among them, the pyrallid predator, *D. aphidivora* was reported to be more voracious on this aphid (Malathi, 2006). The hemerobiid, *Micromus igorotus* Banks was found in large numbers in sugarcane woolly aphid infested field. Work on this predator under Indian condition is scanty. There is an urgent need for evolving suitable management strategy for this pest. Keeping this in view, studies were designed.

Materials and Methods

Stock culture of Micromus igorotus : The laboratory cultures of *M. igorotus* were established by rearing grubs collected from sugarcane fields infested with woolly aphids. Five pairs of freshly emerged

* Department of Entomology, Tamil Nadu Agricultural University, Coimbatore - 641003

adults (*ca.* 1 day) were transferred to plastic containers (18 x 9 cm) for mating and oviposition. The bottom of the container was lined with filter paper. As females preferred to lay eggs on loose cotton, a piece of cotton wool was attached to the ceiling as an ovipositional substrate. Adults were fed with live aphids and water was provided in a glass vial with a cotton wool. The containers had sufficient provision for aeration.

Eggs laid on the cotton in the stock culture were collected and placed in polypots for development within 24 h of laying. After hatching, grubs were fed with live aphids and maintained till pupation. The culture process was repeated thereafter from egg to adult stages continuously. From this stock, insects were used for further studies.

Studies on biology of M. igorotus: The studies were initiated with fresh stock of adults obtained from the culture. Five pairs of adults in five replicates were transferred to plastic containers (18 x 9 cm) and maintained at room temperature, 27°C ± 2.08 as described earlier. The oviposition containers were checked daily for oviposition. Eggs were removed from the oviposition substrate daily and counted. The eggs obtained on a single day were used in the experiment. The observations were initiated with 250 eggs. Grubs reared from eggs were checked daily for molt to the next instar, pupation and emergence. After the adult emergence, adult longevity, fecundity, fertility, hatchability, preoviposition, oviposition and egg periods were noted. To determine the number of instars and the duration of each instar the head width of grubs from another stock maintained separately as described above from eclosion were measured daily after eclosion using image analyzer (Carl Zeiss®). The second experiment was conducted similarly at 25°C in an incubator. In this experiment, duration of each metamorphic stage in addition to reproductive parameters was observed.

Predator potential of M. igorotus: To find out the predator potential of different larval instars of *M. igorotus*, on nymphs and adults of prey insects, a study was conducted with the following treatments.

a. Predator fed with nymphal stages of woolly aphid and maintained at 25°C.
b. Predator fed with grown up stages of woolly aphid and maintained at 25°C.
c. Predator fed with nympal stages of aphid and maintained at room temperature, 27°C ± 2.08
d. Predator fed with grown up stages of woolly aphid and maintained at room temperature, 27°C ± 2.08.

Freshly hatched *Micromus* grubs were confined individually in polypots (1 oz) lined with filter paper for mobility. Sufficient number of preys was provided to the grubs daily. The number of prey insects consumed was recorded regularly at 24 h interval. The daily predation of each instar of the *M. igorotus* was found with different doses of prey insect *viz.*, 10, 20, 30 and 40 for the first, second and third instar of predatory larvae, respectively. The study had five replications with 10 grubs/ replication.

Predatory potential of M. igorotus population collected from different locations: Surveys by earlier workers have indicated the prevalence of *Micromus* population in different cane growing districts of Tamil Nadu. In order to find out the variation in the predatory potential of these populations, a laboratory study was conducted. *Micromus* population was collected from Vadipatti of Madurai district, Udumalapet of Coimbatore district and compared with Coimbatore population. The experiment was conducted at room temperature. The treatments were replicated seven times with ten individuals per replication. The wild populations were reared in the laboratory and the experiment was conducted in F1 generation. The observations were carried out on the number of preys consumed during the larval period.

Results

Biology of M. igorotus: Laboratory investigation on the biology of *M.* igorotus was carried out at two temperatures in two experiments. The results are presented in tables 1 and 2.

Biological parameters at room temperature (2°C ± 2.08): The observations on the different metamorphic stage (Plate 6) showed that the incubation period ranged from 3 to 4 days with a mean of 3.60 days. The larval duration ranged from 5 to 8 days with a mean of 6.66 days (Table 1). The grubs during this period passed through three instars before induction of population (Plate 7). The mean duration of each stadia were 1.73, 2.80 and 2.13 days, respectively (Table 1). Measurements taken on the head capsule width showed a mean range of 0.41 to 1.27 mm for the three stages. The head capsule width of one stage did not overlap with another stage (Table 3). This species of *Micromus* had a short prepupal period of 1.20 days only before induction to population. The pupal period lasted 5.47 days. Among other parameters, at 27^0 C ± 2.08 the pupation and adult emergence were 91.54 and 86.17 per cent, respectively. Considering the total life cycle, females lived longer (32.80 days) than males (29.20 days). The adult female had a preoviposition period of 3.60 days, while the oviposition period was extended upto 30 days. During the life span, the fecundity ranged from 420 to 928 eggs with a mean of 662 eggs/ female. The fertility ranged from 82.35 to 96.30 per cent with a mean of 91.52 per cent. The hatchability followed similar trend with a mean of 91.69 per cent.

Biological parameters at constant temperature of rearing: Observations on the population of *M. igorotus* maintained at 25^0 C showed that the incubation period was 4.47 days, while the total larval and pupal periods were 7.67 and 7.93 days, respectively. These per cent pupation and adult emergence were 87.33 and levels of 93.02 and 92.56 per cent respectively.

Table 1. Biological parameters of *M. igorotus* reared at room temperature (27^0 C ± 2.08)

Particulars/ stage	* Mean ± SE
Incubation period (days)	3.80 ± 0.11 (3-4)
Larval period	
First instar (days)	1.73 ± 0.12 (1-2)
Second instar (days)	2.80 ± 0.09 (2-3)
Third instar (days)	2.13 ± 0.12 (2-3)
Total larval period (days)	6.66 ± 0.11 (5-8)
Prepupal period (days)	1.2 ± 0.11 (1-2)
Pupal period (days)	5.47 ± 0.13 (5-6)
Total life cycle	
Male	29.20 ± 0.62 (27-33)
Female	32.90 ± (0.73) 25-35)
Pupation (%)	91.54 ± 0.93 (90.00-95.83)
Adult emergence (%)	86.17 ± 1.95 (80.00-92.31)
Preoviposition period (days)	3.60 ± 0.13 (3-4)
Oviposition period (days)	23.13 ± 0.98 (20-30)
Fecundity	662 ± 60.12 (420-928)
Fertility (%)	91.52 ± 1.28 (82.35-96.30)
Hatchability (%)	91.69 ± 0.55 (88.99-93.75

* Mean of 25 values; Figures in parentheses represent range

Table 2. Biological parameters of *M. igorotus* reared at 25°C

Particulars/ stage	* Mean ± SE
Incubation period (days)	4.47 ± 0.13 (4-5)
Total larval period (days)	7.67 ± 0.13 (7-8)
Pupal period (days)	7.93 ± 0.15 (7-9)
Pupation (%)	88.66 ± 1.37 (80.00-95.24)
Adult emergence (%)	88.25 ± 0.78 (85.00-92.86)
Fertility (%)	93.02 ± 1.31 (76.47-95.24)
Hatchability (%)	92.56 ± 1.41 (77.42-94.74

* Mean of 25 values; Figures in parentheses represent range

Predatory potential of M. igorotus at different temperatures: The experiment on predation during entries larval cycle showed that predation was significantly higher in population maintained at 25⁰ C than at room temperature (27⁰ C ± 2.08). The number of aphids consumed when the grubs were reared at 25⁰C was ranging from 072.35 to 748.89 (X = 40.08-42.98). It was found that *Micromus* grubs consumed significantly more number of grownup stages of woolly aphids (748.89) and nymphal stage (672.35) when held at 25⁰ C. However, significantly lower numbers were eaten at room temperature (Table 3). The study also revealed that predation of aphids observed on daily basis was significantly higher at 25⁰ C with no differences among the stages. The mean predation ranged from 25.95 during the eighth day to 89.06 on fifth day.

Predatory potential of different population of M. igorotus: Observations of the predatory potential of populations of *M. igorotus* collected from different location and evaluated in F_1 generation at room temperature condition revealed insignificant differences. The predation ranged from 364.47 to 369.11 aphids in six days and the values were on par. Considering the days, it was observed that predation increased as the grubs grew and declined when the grubs prepared themselves for pupation. The peak predation was observed when the grubs were 4-5 days old (83.13-85.63 grubs/individual). The predatory potential was similar to earlier study (Table 4).

Discussion

Predatory potential of M. igorotus: Laboratory experiments on the predatory potential of *M. igorotus* at constant and fluctuating temperature revealed that predation was higher at an incubation temperature of 25⁰ C than at room temperature conditions. The grubs consumed 672.35 nymphs and 748.89 grown up woolly aphids in the first set of treatments while, it decreased to 360.72 to 386.82 in the second set of treatments. Lingappa et al. (2004) mentioned that the feeding potential was 20 to 25 aphids/day. In the present study after eclosion and up to 7 days the daily predation rates ranged from 54.70 to 89.06 aphids/days. The mean predation per day irrespective of the treatments ranged from 40.08 to 83.21 (Table 3). A high degree of voracity obtained in the present studies at two temperatures is a fair determination of approximate temperature prevailing in sugarcane fields. This limited study has indicated the maximum attack rates for the combination of predator and prey stages. In laboratory tests, Samson and Blood (1980) reported that predation increased exponentially with successive immature stages of predator like *Chrysopa signata* Scheider. Similar trend was also observed here. In the present investigation of *M. igorotus* fed similarly on nymphs and grown up stags irrespective of the conditions of incubation temperature. This aphid species is reported to possess high innate capacity to reproduce parthenogenetically. Grown up stages have been noted to lay large number of off springs (Sunil Joshi and Virakthmath, 2004). Predation of nymphs can control the population

Table 3. Predatory potential of *M. igorotus* grubs at different temperatures

Treatments	Aphids consumed on day after hatching*										
	1	2	3	4	5	6	7	8	9	Mean	Total
Predator fed with nymphal stages of woolly aphid and maintained at 25°C	82.67^{cA}	94.00^{bA}	94.44^{BA}	9.77^{aA}	99.67^{aA}	99.67^{aA}	92.16^{bA}	11.97^{dB}	0.00^{cB}	71^{B}	672.35^{A}
Predator fed with grown up stages of woolly aphid and maintained at 25°C	79.33^{cA}	87.33^{dA}	92.00^{cA}	94.67^{bB}	95.99^{abA}	98.00^{aA}	98.67^{aA}	91.82^{cA}	10.07^{fA}	83.21^{A}	748.89^{A}
Predator fed with grown up stages of woolly aphid and maintained at room temperature 27°C±2.08	42^{cB}	48.00^{cB}	61.33^{bB}	83.02^{aB}	85.24^{aB}	41.11^{cC}	0.00^{dD}	0.00^{dC}	0.00^{dB}	40.08^{c}	360.72^{B}
Predator fed with grown up stages of woolly aphid and maintained at room temperature, 27°C±2.08	37.33^{dB}	47.27^{cB}	51.00^{cC}	67.00^{bC}	75.33^{aC}	80.89^{aB}	27.97^{eB}	0.00^{fC}	0.00^{fB}	42.98^{C}	386.82^{B}
Mean	60.33	68.15	74.69	85.62	89.06	79.92	50	25.95	2.77		

* Mean of five replications; In a column means followed by similar alphabets in upper case are statistically not significant by LSD (P=0.05); In a row means followed by similar alphabets in lower case are statistically not significant by LSD (P=0.05)

Table 4. Predatory potential of *M. igorotus* at room temperatures, 27°C ± 2.08 collected from different locations

Population	Aphids consumed on day after hatching*							
	1	2	3	4	5	6	Mean	Total
Coimbatore	41.82^{d}	54.03^{c}	62.76^{b}	83.70^{a}	85.92^{a}	40.88^{d}	61.64	369.11
Udumalapet	41.55^{d}	51.98^{c}	61.72^{b}	82.91^{a}	85.59^{a}	40.72^{d}	60.75	364.47
Vadipatti	41.28^{d}	52.30	62.32^{b}	82.79^{a}	85.38^{a}	40.59^{d}	60.78	364.66
Mean	41.55	52.77	62.27	83.13	85.63	40.73		

* Mean of seven replications; In a row means followed by similar alphabets are statistically not significant ; In a column means are not significantly different

during several investigations. The high degree of voracity reported in this investigation offers scope for *M. igorotus* as a biocontrol agent in augmentation.

Experiments on the predatory potential of different population of *M. igorotus* under laboratory conditions have revealed that the differences in predatory potential were not significant (Table 4). So far, there are no reports available on the predatory potential of *M. igorotus* geographical population. The concept behind taking up this evaluation was to identify the most voracious population that can be deployed in biological control of woolly aphid. The studies are inconclusive and preliminary in nature and needs further investigation as screening of population could yield a most fit population.

References

Gupta, M.K. and Goswami, P.K., 1995. Incidence of sugarcane woolly aphid and its effect on yield attributes and juice quality. *Indian Sugar,* 44, 883-885.

Lingappa, S., Patil, R.K. Tipanavar, P.S., Balikai, R.A. and Prasadkumar, 2004. Development of strategies for the management of wolly aphid, *Ceratovacuna lanigera* Zehnter (Homoptera: Aphididae) on sugarcane. Project Report submitted to department of Agriculture, Government of Karnataka, April 2004

Malathi, A., 2006. Studies on the sugarcane woolly aphid (SWA) *Cerovacuna lanigera* Zehntner (Homoptera: Aphididae) and its predator, *Dipha aphidivora* Meyrick (Pyralliade: Lepidoptera). *M.Sc, thesis,* TNAU, Coimbatore-641 003

Sunil Joshi and Viraktamath, C.A., 2004. The sugarcane woolly aphid, *Ceratovacuna Lanigera* Zehnter (Homoptera: Aphididae) its biology, Pest status and control. *Curr, Sci*., 87, 307-316.

Seasonal Dynamics of *Cotesia flavipes* Cameron (Braconidae: Hymenoptera): A promising Larval Endoparasitoid of *Sylepta aurantiacalis* Fisch in the Cashew Ecosystem

V. Ambethgar*

The cashew leaf folder, *Sylepta aurantiacalis* Fisch (Pyraustidae: Lepidoptera), is among the important defoliating pest in most of the cashew growing areas of Tamil Nadu. The gregarious larval endoparasitoid, *Cotesia flavipes* Cameron (Braconidae: Hymenoptera) is an important natural mortality factor for *S. aurantiacalis* larvae in cashew orchards. Studies were conducted during 2003-2007 at the Regional Research Station, TNAU, Vridhachalam, Tamil Nadu to investigate the seasonal dynamics of *C. flavipes* on the larval population of *S. aurantiacalis* by collecting and rearing the bio-stages of the pest and parasitoid. The five years study reveals that the mean level of parasitization ranged from 9.2 to 52.2 per cent in different months. The larval parasitoid showed a positive correlation with the host larval population throughout the study periods. Minimum temperature and relative humidity also have positive influence on *C. flavipes* parasitism where as maximum temperature and rainfall showed negative influence. The occurrence of *C. flavipes* in fairly high proposition synchronizing with the pest epidemics highlights its contribution to the natural control of cashew leaf folder. This study opens up the scope for utilizing this parasitoid as a component of integrated management of leaf folders in cashew ecosystem.

Introduction

Insect pests have been recognized as a major biotic stress responsible for significant reduction in yield of cashewnut in all cashew growing areas of the country. The crop is infested by more than 60 species of insect pests in different stages of growth and development. Nearly six species of leaf folders infest cashew across cashew growing areas of India. Of these the pyralid leaf folder *Sylepta aurantiacalis* Fisch (Pyralidae: Lepidoptera) has become a major defoliating pest of cashew orchards in Tamil Nadu (Ambethgar *et al.,* 1999). The pest is active throughout the year in young cashew trees, which produce flush leaves continuously round the year. Highest incidence of 45% shoot damage has been recorded in fields (Ambethgar *et al.,* 2001). In grown up orchards, however, the pest activity is seasonal, and confined to flushing through fruiting periods only. The infestation begins after commencement of South-West monsoon rains during August-September synchronizing with the emergence of new flushes, and the infestation reaches the peak after North-East monsoon rains during February-March of the succeeding year, when cashew trees are in mixed phases of flushing and blooming. There after, the gravity of infestation declines gradually towards the nut maturity phase, and disappear after harvesting of nuts in summer months due to lignification of foliage, which contains higher levels of phenols.

The adult of *S. aurantiacalis* is a medium size, yellowish moth with brownish wavy linings on both the pair of wings, and its larva is glistering green with brownish head (Fig.1). Female moth

* Regional Research Station, Tamil Nadu Agricultural University, Vridhachalam-606 001

lays several eggs singly or in small clusters of 4-6 numbers preferentially on tender leaves. The eggs hatch in 3-4 days. The eclosing neonate larvae are ivory white in colour and full grown larvae are shiny green colour. The larvae soon after emerging from eggs roll up the tender leaves from tip to downward or longitudinally towards dorsal side throughout the length or breadth to form a tubular niche. The leaves are tightly fastened with silken thread made of salivary secretion of caterpillars. In the first two instars, the larvae display gregarious behaviour with 3-6 larvae found per rolled leaf. From third instar onwards, it forms solitary niche, and each rolled leaf normally contains single caterpillar; rarely two or three caterpillars may be present in single leaf. An individual rolled leaf resembles the shape of a candle. The caterpillar remains inside the leaf roll and feeds on the green mesophyllar tissues by scrapping. The fed green matter of the leaf is visible through the transparent integument of the caterpillar, which appears green in colour. As a result of the damage, the terminal leaves dries up causing loss in active photosynthetic area. This affects the production of new flushes in clonal nurseries and transplanted orchards. Severely affected plants present a scorched appearance (Fig.2), which can be spotted even from a distance. On regular seasonal flushes, the emergence of inflorescence is hampered resulting in poor bearing and reduced nut yield.

Fig. 1. Adult moth and larva of Sylepta aurantiacalis

Fig. 2. Infestation of cashew leaf folder, *Sylepta aurantiacalis*

The spreading of this pest into new areas has evoked an increasing interest in its control strategies. Several pest management practices are currently in practice in cashew ecosystem with special emphasis on chemical control. The practice of using such toxic insecticides on massive scale tends to pose exposure risks and health hazards. Hence, the possibilities of its biological control were explored. Routine surveys for natural enemies in cashew growing tracts of Tamil Nadu have identified prevalence of the gregarious larval endoparasitoid, *Cotesia flavipes* Cameron (Braconidae: Hymenoptera) on leaffolder population with a mean parasitism of 48.0% (Ambethgar *et al.,* 1999). This stimulated the research interest to study the seasonal dynamics of this important natural parasitoid in cashew ecosystem. Estimation of extent of parasitization is the foremost step to quantify the natural field mortality for any biocontrol programme. The present investigation on the seasonal dynamics of *C. flavipes* will provide vital information for biological suppression of leaf folder complex in cashew.

Materials and Methods

Study location

The study was conducted at the Regional Research Station, Vridhachalam (11°30'N and 79°26'E) in the experimental orchard of the All India Co-ordinated Research Project on Cashew. The total area of the orchard is approximately 25 ha established with four clonally propagated cashew varieties *viz.,* VRI 1, VRI 2, VRI 3 and VRI 4, and more than 268 germplasm accessions. The age of the trees ranged from 3-18 years with a plant density of 200-525 per ha. The orchard is located at an elevation of 42.67 meters above MSL. The climate of this area is characterized by a warm, dry summer and wet winter with mean maximum temperature of 34.2°C and mean minimum temperature of 23.7°C. The average annual temperature is 29.3°C. The average annual rainfall of the region is about 950mm.

Field survey and rearing of parasitoids

Field surveys were conducted on the seasonal dynamics of the larval endoparasitoid, *C. flavipes* in totally unsprayed cashew trees for five years from January 2003 to December 2007. The experimental orchard was divided into four quadrants and the larvae of *S. aurantiacalis* were collected regularly from ten trees selected randomly from each quadrant. One hundred and twenty larvae were collected at weekly intervals commencing from first week of January onwards. The field collected larvae were held individually in specimen tubes (10 x 2cm) covered with cotton plugs and brought to the laboratory for estimation of extent of parasitism. The field collected larvae were reared continuously at temperature of 27°C- 30°C and with RH 70% until emergence of parasitoid and pupation of host larvae. The data on the number of healthy and parasitized larvae were recorded out of a total number of larvae collected per month to determine the levels of parasitization during different time intervals from January-December every year. Month wise mean per cent parasitism was estimated using the following equation.

$$\text{Parasitism (\%)} = \frac{\text{Number of parasitized larvae}}{\text{No. of healthy larvae + No. of parasitized larvae}} \times 100$$

To determine the influence of weather factors, correlations were worked out using weather data recorded at the research station.

Weather parameters and data analysis

The corresponding monthly weather parameters *viz.,* maximum and minimum temperature, morning and evening relative humidity, rainfall, and number of rainy days recorded from the Meteorological Observatory of the Research Station were utilized in this study. Based on the meteorological data, correlation analysis was done between extent of parasitism and abiotic factors. Meteorological parameters having significant correlation at 5% level and above were subjected for further graphical analysis to examine the information generated by statistical techniques.

Results and Discussion

The braconid parasitoid, *Cotesia flavipes* Cameron also known by the name *Apanteles flavipes* (Cameron), is a common larval endoparasitoid of cereal stem borers in Asian, the Neotropics and African countries (Overholt *et al.,* 1994; Kimani and Overholt, 1995 and Omwega *et al.,* 1995). In India, this parasitoid has earlier been recorded on a range of lepidopterous pests including: *Scirpopahga nivella*, *Chilo auricilius* (Gupta, 1953), *Chilo zonellus* (Swinhoe) and *Chilo infuscatellus* (Sharma *et al.,* 1966), *Chilo partellus* (Mohan *et al.,* 1992), *Diacrisia obliqua* (Muthukrishnan and Senthamizhselvan, 1987), *Porthesia scintillans* Walker (Senthamizhselvan and Muthukrishnan. 1989). In the present study, the ecology and seasonal dynamics of *C. flavipes* on the field populations of *S. aurantiacalis* in cashew ecosystem is discussed.

Morphology of *C. flavipes*

The *Cotesia* adults are tiny (about 6-7mm), dark wasps and resemble small flying ants. The two pairs of wings were thin, papery and translucent; the hind wings being smaller than the forewings (Fig.3). The adult parasitoids have chewing-lapping mouth parts. The antennae are about 1.5mm long and curved upward. Female parasitoids were slightly larger than males; and the abdomen of female narrows to a downward curving extension called the ovipositor with which the female parasitic wasp lays eggs.

Bioecology of *C. flavipes*

The adult *C. flavipes* wasps are free living. A female wasp lays as many as 12-19 eggs inside the body of host, preferably in first or second instar larvae. The parasitized larvae become sluggish within six hours after parasitization. Embryogenesis is completed 3-4 days after oviposition. The eggs hatch and the grubs develop gregariously for two instars within the host larvae, exiting en masse when they moult to the third instar.

Larvae (grubs): Upon hatching, tiny neonate grubs float freely within the hemolymph. The developing grubs draw substance primarily from the host hemolymph, and remain until the host larvae host has reached fourth or fifth instar. At this time, the parasitoid grub moults to the second instar and undergo rapid development, feeding upon fat-body tissues as well as hemolymph. Fully developed grubs of parasitoid after completion of their feeding stage, emerged from the host caterpillars by penetrating the body wall with the help of their mandibles. When the grubs have almost passed out of the host, they adhere to the host integument by the last abdominal segment, which is not completely withdrawn from the exit hole. They immediately begin to secrete off-white silky-substance which was used to spin the cocoon besides the host larvae (Fig.3). The grubs first spin a loose outer web about their bodies, beginning at the last abdominal segment, working their way to the head. The first few threads are looped onto hairs of the caterpillar's body for temporary support. The grubs then emerge fully from the host and continue to spin within the outer framework until the characteristic off-white cocoon is complete, when they then pupate. The pupae are in an irregular mass attached to the host larva inside the rolled leaves.

Pupae: Twenty-four hours after the grubs have emerged from the host, the brown eyes of the pupae can be observed through the cuticle of the third instar grub. Pupal ecdysis takes place two days after leaving the host. The colour of the pupal body at this stage is ivory- white. During pupal development, melanization proceeds from the eyes to the head, then to thorax and dorsal side of the abdomen, and finally to the antennae. Each developmental process takes approximately 10-12 hours. The length of the pupal period varied varies with the ambient temperature. Under field conditions at Vridhachalam, the pupal period lasted 8-10 days. In the laboratory, the pupae of *C. flavipes* reared in *S. aurantiacalis* larvae took 9.7 (±2.3) days to develop at 20°C and 6.8 (± 1.5) days to develop at 25°C.

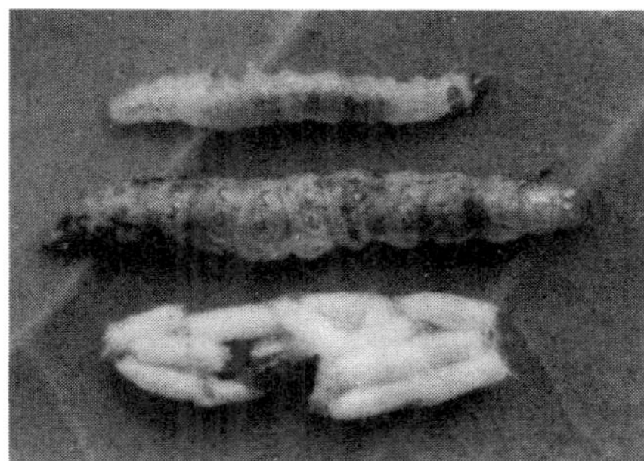

Fig. 3. Adult of *Cotesia flavipes* (left) and parasitized larvae of *Sylepta* with cocoons of *C flavipes* (right)

Adults: The adult parasitoid wasp emerge by first cutting a circular lid at the end of the cocoon with their mandibles and then pushing it off as they force their way out. Males tend to emerge before females. After emergence males often remain on the cocoon cluster, or join other males on their cocoon cluster, and engage in the characteristic precopulatory behavior of 'wing vibrating'. Females are ready to mate upon emergence. They tend to walk away from their cocoon clusters, often falling or flying off the surface of the substratum. Fertilized females immediately begin to search for hosts. Once host is encountered, the parasitoid drums the surface of the caterpillar for a few seconds. The abdomen is bend forward between the hind legs so that the ovipositor is brought into contact with the host integument. At the moment, the ovipositor is thrust into the host, and the wings of the parasitoid are splayed vertically over its thorax. In the field, adults feed on nectar from flowers. The longevity in the field is unknown, but in the laboratory females live 22-28 days if fed a solution of sugar or honey water. If deprived of food and water they die within a few days.

Seasonal occurrence of *C. flavipes*

The results of the five year field study conducted on the seasonal activity of the *C. flavipes* on the field population of *S. aurantiacalis* are summarized in Table 1. The parasitic wasp was active throughout the year in cashew orchards. The average parasitization accounted for 27.0, 24.0, 22.8, 20.8 and 18.5% during 2003, 2004, 2005, 2006 and 2007 respectively. The observations recorded from these five years indicated that peak activity of the parasitoid was occurred from February to March. Maximum parasitization of 52.2% was recorded during March alone. The extent of parasitism during April-July and October-November is very low due to prevalence of high temperature and rainfall respectively. Field observations showed that larval parasitism by *C. flavipes* was low during initial flushing periods (+) irrespective of the age of the trees. With the production of post-monsoon flushes, the extent of parasitism gradually increased towards the productive flushing phase (++) through peak flushing of cashew (+++), which supported increased number of host larvae.

In grown-up trees, however, both host larval population and extent of parasitization declined towards the progression in lignifications of foliage. In young cashew trees of below three years old, which produce flush leaves continuously round the year, the extent of larval parasitization was observed to be uniform through out the year, indicating that natural larval parasitization in cashew ecosystem is mainly depend on the higher production of flush leaves, which constantly support the host larval population. Thus, the pattern of parasitism suggests a close relationship between production of flush leaves and population build up of host larvae are of prime importance. Intensive surveys throughout the year are now being conducted at several locations across cashew growing tracts of Tamil Nadu for precise quantification of leaf roller mortality caused by *C. flavipes*.

Table 1. Natural parasitization of *S. aurantiacalis* larvae by *C. flavipes* during different years

Month	Ecology		Per cent larval parasitization					
	Flushes	Larvae	2003	2004	2005	2006	2007	Mean
January	+++	+++	42.5	38.3	36.6	35.0	33.3	37.1
February	+++	+++	52.5	44.1	43.3	42.5	39.2	44.3
March	+++	+++	62.5	53.3	51.7	48.3	45.0	52.2
April	+	+	15.0	13.3	12.5	10.0	8.3	11.8
May	+	+	11.7	10.8	10.0	7.5	5.8	9.2
June	+	+	15.0	14.1	12.5	10.8	8.3	12.1
July	+	+	20.0	19.2	18.3	15.8	13.3	17.3
August	+	+	26.6	25.0	24.1	21.7	19.2	23.3
September	++	++	36.6	33.3	30.0	26.6	25.0	30.3
October	++	++	11.7	10.0	9.1	7.5	6.6	9.6
November	++	++	13.3	10.8	10.0	8.3	5.0	10.6
December	++	++	16.6	15.8	15.0	15.8	13.3	15.8
Mean	-	-	27.0	24.0	22.8	20.8	18.5	-

+: low, ++: medium, +++: high

Influence of larval population on parasitization

The larval parasitization had significant positive correlation with leaf roller larvae (Table 2). The data shows that the extent of parasitization was found to have direct bearing on larval population load. With the rapid decline in larval population during hot summer months (May-June) and rainy period (October-November), the extent of parasitization also declined in the absence of the host larvae.

Table 2. Correlation coefficient of natural parasitization by *C. flavipes* in relation to larval population and weather factors

Weather elements	Per cent parasitization					
	2003	2004	2005	2006	2007	Pooled
Larval population (nos.)	0.682*	0.723*	0.685*	0.716*	0.663*	0.693*
Maximum temp. (°C)	-0.324	0.647	-0.309*	-0.348	-0.375	-0.400
Minimum temp. (°C)	0.402	0.342	0.413	0.324	0.295	0.355
Forenoon RH (%)	0.530^{S}	0.524	0.316^{NS}	0.598	0.403	0.474*
Afternoon RH (%)	0.612	0.233 ^{NS}	0.596	0.372 ^{NS}	0.385 NS	0.439
Rainfall (mm)	-0.236	-0.298	-0.266	-0.355	-0.389*	-0.308

* Significant at 5% level; NS Non-significant

With the revival of the larval population of *S. aurantiacalis* after cessation of rains, the extent of parasitization also gradually increased, thus exhibiting close association between the larval population and that of its parasitoid. The leaf roller population was reduced considerably with the increased activity of the parasitoid, *C. flavipes* and confirm to the previous report by Ambethgar *et al.* (1999).

Influence of weather factors on *C. flavipes*

With respect to weather factors, maximum temperature had a negative influence on the activity of *C. flavipes* in terms of parasitism (Table 2). The analysis indicated that for every one degree increase in temperature, the extent of parasitism decreased significantly by 1.43%. Similarly, there was also significant negative influence of precipitation on the build up of parasites. However, the influence of relative humidity on parasitism by *C. flavipes* was found to be positive. The study indicated that temperature range of 23.6 to 28°C and relative humidity of 68.5 to 87.3% appeared to be highly favourable for the activity of *C. flavipes*. Ambethgar *et al.* (1999) reported that temperature ranging from 26.6 to 30.8°C was favourable for high activity of *C. flavipes* in cashew ecosystem. In the laboratory rearing, it was observed that higher constant temperatures had adverse effect on the survival and activity of the parasite. But such adverse effects were not noticed under field conditions. The influence of weather upon the behavior of *C. flavipes* in cashew ecosystem has not been fully investigated, but the present study showed that the parasitoid is generally less active in rainy, overcastting cloudy weather than warm sunny weather. Published works clearly show that *C. flavipes* is amendable for mass rearing and field release programme for biological control of various lepidopterous insects in diverse agro-ecosystem (Gifford and Man, 1967; Inayatullah, 1987; Srikanth *et al.,* 2000; Jalali and Singh, 2001).

Conclusion and Future Thrust

The present study indicates that in the presence of leaf folder population, *C. flavipes* acts as a constant mortality factor for the pest throughout the year at varying levels. Thus, the occurrence of *C. flavipes* in fairly high proposition highlights its contribution to the natural control of leaf folder in cashew ecosystem. The investigation also suggests that *Cotesia* has a better prospect to the biological control programme as the parasitoid is quite promising to check the leaf folder multiplication right from early in the season. Further investigations on the mass rearing, field release techniques, conservation and augmentation of the parasitoid would contribute much for the integrated management of leaf folders in cashew ecosystem.

Acknowledgements

The author is indebted to the Director, National Research Centre for Cashew (NRCC), Puttur-574 202 (D.K), Karnataka and the Professor and Head, Regional Research Station, Vridhachalam-606 001, Tamil Nadu for the facilities provided.

References

Ambethgar, V., Lakshmanan, V. and Naina Mohammed, S.E., 1999. Parasitoid to check cashew leaf folder, *Science and Technology: Agriculture, The Hindu,* July 22, 1999: P24.

Ambethgar, V., Swamiappan, M. and Rabindra, R.J., 2001. Evaluation of certain synthetic insecticides and neem products against cashew leaf folder, *Sylepta aurantiacalis* Fisch. (Pyralidae: Lepidoptera). *Indian Journal of Plant Protection*, 29 (1&2), 106-109.

Gifford, J.R. and Man, G.A., 1967. Biology, rearing and a trial release of *Apanteles flavipes* in the Florida Everglades to control the sugarcane borer. *Journal of Economic Entomology,* 60, 40-47.

Gupta, B.D., 1953. Resume of work done under the insect pest scheme during 1946-47 to 1950-51. Indian Centre Sugarcane Commercialization, New Delhi, 111pp.

Inayatullah, C., 1987. Development and survival of *Apanteles flavipes* (Cameron) (Hymenoptera: Braconidae) on *Chilo partellus* (Swinh.) (Lepidoptera: Pyralidae) larvae reared on grain flour diet. *Insect Science and Application*, 8, 95-97.

Jalali, S.K. and Singh, S.P., 2001. Life table studies on the natural enemies of *Chilo partellus* (Swinhoe) (Lepidoptera: Pyralidae). *Journal of Biological Control,* 15(2), 113-117.

Kimani, S.W. and Overholt, W.A., 1995. Biosystematics of the *Cotesia flavipes* complex (Hymenoptera: Braconidae): interspecific hybridization, sex pheromone and mating behaviour studies. *Bulletin of Entomological Research,* 85, 379-386.

Mohan, B.R., Verma, A.N. and Singh, S.P., 1992. Biology of *Apanteles flavipes* (Cameron)- A potential parasitoid of *Chilo partellus* (Swin.) infesting forage sorghum. *Journal of Insect Science*, 5, 144-146.

Muthukrishnan, J and Senthamizhselvan, M., 1987. Effect of parasitization by *Apanteles flavipes* on the biochemical composition of *Diacrisia obliqua*. *Insect Science and Application*, 8, 235-238.

Omwega, C.O., Kimani, S.W., Overholt ,W.A. and Ogol, C.K.P.O., 1995. Evidence of the establishment of *Cotesia flavipes* (Hymenoptera: Braconidiae) in continental Africa. *Bulletin of Entomological Research,* 85, 525-530.

Overholt, W.A., Ochieng, J.O., Lammers, P. and Ogedah, K., 1994. Rearing and field release methods for *Cotesia flavipes* Cameron (Hymenoptera: Braconidae), a parasitoid of tropical graminaceous stemborers. *Insect Science and Application*, 15, 253-259.

Senthamizhselvan, M. and Muthukrishnan, J., 1989. Bioenergetics of *Apanteles flavipes* Cameron (Hymenoptera: Braconidae), a parasitoid of *Porthesia scintillans* Walker (Lepidoptera: Lymantriidae). *Insect Science and Application*, 10(3), 295-299.

Sharma, A.K. Saxena, J.D. and Subba Rao, B.R., 1966. A catalogue of the hymenopterous and dipterous parasites of *Chilo zonellus* (Swinhoe) (Crambidae: Lepidoptera). *Indian Journal of Entomology,* 28, 510-542.

Srikanth, J., Easwaramoorthy, S., Kumar, R. and Shanmugasundaram, M., 2000. Pattern of laboratory parasitization rates of *Cotesia flavipes* Cameron (Hymenoptera: Braconidae) in borers of sugarcane and sorghum. *Insect Science and Application,* 20(3), 195-202.

Seasonal observations on aonla aphid, *Schoutedonia emblica* (Patel and Kulkarny) and their natural enemies

M. Shanthi*, D.S. Rajavel,* R.K. Murali Baskaran* and R. Nalini*

Aonla aphid, *Schoutedonia emblica* (Patel and Kulkarny), is one of the major constraints in the cultivation of aonla. A Fixed Plot survey was conducted in three year old aonla (BSR 1) trees at the Agricultural College and Research Institute, Madurai. Occurrence of aonla aphids and their natural enemies were recorded at fortnightly interval for two years (2005-07). *S. emblica* activity was recorded from March to November. Maximum infestation (50% trees) was noted during June to October; peak population of aphids (100 nos. / 5 cm terminal shoot) was observed in June. Occurrence of *Cheilomenes sexmaculata* (Fab.), *Scymnus (Pullus) nr. castaneus* Sicard (Coccinellidae: Coleoptera), hover fly, *Paragus serratus* (Fab.) (Syrphidae: Diptera) were recorded as major predators; spider, *Oxyopes* sp., *Argiope* sp. (Oxyopidae: Araneae) and preying mantis (*Euantissa* sp.) as minor predators on nymphs and adults of *S. emblica*. Population of *C. sexmaculata* and *S. nr. castaneus* ranged from 1.0 to 4.0/ colony and *P. serratus* maggots @ 1.0 to 2.5 nos. / aphid colony. The predator population *viz*., *C. sexmaculata*, *S. nr. coccivora* and *P. serratus* (r= 0.68; 0.63 and 0.89, respectively) was positively correlated with aonla aphid incidence.

Introduction

Aonla or Indian Gooseberry, *Emblica officinalis* Gaertn. is one of the important minor commercial fruit, recognized as king of Arid fruits. In Tamil Nadu, the area under aonla cultivation is increasing. The tree is being infested by 41 species of insect pests belonging to five different orders (Coleoptera, Hemiptera, Lepidoptera, Diptera and Thysanoptera). Aonla aphid, *Schoutedonia emblica* (Patel and Kulkarny) is one of the major constraints in the cultivation of aonla. *Cerciaphis emblica* (Patel and Kulkarny) and *Schoutedonia (Setaphis) bougainvilliae* (Theobald) are the synonyms of *S. emblica* (George, 1927). It attacks terminal leaves, young shoots and sucks the cell sap, causes death of growing shoots resulting in poor fruiting during subsequent years (Patel and Kulkarny, 1953). The activity of natural enemies of aonla aphid, *S. emblica* was reported by several workers from Pusa, Uttar Pradesh and Central Gujarat (Fletcher, 1916; Bhatia and Shaffi, 1933; Masareet Haseeb, 2005 and Bharpoda *et al*., 2006). To evolve an effective IPM strategy with the benefit of bio-control agents, it is very essential to have information about their natural enemies in different agro-climatic zones. In Tamil Nadu, studies on aonla aphid and its natural enemies are lacking. Keeping in view of the importance of the crop and damage caused by the target pest, observations on aonla aphid, *S. emblica* and its natural enemies was made.

Materials and Methods

A fixed plot survey was carried out in the aonla orchard at the Agricultural College and Research Institute, Madurai, Tamil Nadu. Three year old locally predominant aonla variety (BSR 1) was used for this study. Fortnightly observations were recorded under unsprayed conditions for two years, during December 2005 to November 2007.

* Department of Agricultural Entomology, Agricultural College and Research Institute, Madurai - 625 104

Out of the randomly selected twenty trees, the number of trees infested by aonla aphid was recorded and per cent infestation was estimated. In each infested tree, the population of aphids was registered from 5 cm long terminal shoot of selected four branches facing four directions (East, West, North and South) and mean population of aphid per 5 cm terminal shoot was worked out.

The diversity of natural enemies feeding on aonla aphid, *S. emblica,* its seasonal occurrence and population were noted during the period of study. The coccinellid grubs and syrphid maggots were collected in glass tubes individually and reared on aonla aphids under laboratory condition and the emerged adults were identified. Prey natural enemy relationship was studied by ascertaining correlation of the natural enemy population with aphid population. The impact of abiotic factors like minimum temperature, maximum temperature, relative humidity and rainfall on population fluctuation of aonla aphid, *S. emblica* and its natural enemies were studied using simple correlation. The meteorological data recorded at Meteorological observatory, Agricultural College and Research Institute, Madurai during the period of study were utilized for the study.

Results

Incidence of aonla aphid, *S. emblica* and its relationship with abiotic factors

Aonla aphid, S. emblica, infestation on trees: Fortnightly observations on aonla aphid, *S. emblica,* infestation and their population on BSR 1 are presented in Table 1. Both nymphs and adults infested the new flush of shoots. Incidence of aonla aphid coincided with formation of new flush. Aonla aphid, *S. emblica* infested 5.0 to 50.0 per centage of trees in 2005-2007. The infestation was observed for nine months, which started in the first fortnight of March (10.0%) and reached to its peak (50.0%) during second fortnight of June to first fortnight of October. Minimum temperature of 24.77±1.5°C and maximum temperature of 34.69±1.5°C prevailed during the period. The aonla aphid infestation disappeared after November. Tree infestation was positively correlated with minimum and maximum temperature (r = 0.78 and 0.64, respectively). Relative humidity varied from 69.57±7.2 per cent and it was negatively correlated with aonla aphid tree infestation (r = -0.69). Correlation between tree infestation and rainfall was not significant (r = 0.20).

Population build-up of S. emblica: Throughout the period of infestation, the population of aonla aphids ranged from 10.0 to 100.00 nos. per 5 cm long terminal shoot. Initially, the population of aphids was 35.75 nos./ 5 cm terminal shoot and the population reached to its highest (50.0 to 100.0 nos./ 5 cm terminal shoot) from April to August. Average minimum and maximum temperature during the period was 25.17±1.24 °C and 35.73 ± 1.84°C and average relative humidity was 69.39 ± 7.34 per cent. Peak population (100.0 nos./ 5 cm terminal shoot) of aphids was registered during June, which coincided with minimum and maximum temperature of 25.0°C and 35.2°C, respectively and 67.2 per cent relative humidity. With the build-up of aonla aphid, *S. emblica* population, the extent of tree infestation was also increased (r = 0.79). Relationship between abiotic factors and population of aonla aphid, *S. emblica,* was highly significant and positively correlated with minimum temperature (r = 0.82), maximum temperature (r = 0.76) and negatively correlated (r = -0.70) with relative humidity. The effect of rainfall on population build-up of aonla aphid, *S. emblica* was non- significant but they were positively correlated (r = 0.15) (Table 2). By the end of November the mean population of aphids decreased to 10.0 nos./ 5 cm terminal shoot. Different species of ants were found harvesting honeydew from aonla aphid colonies.

Occurrence of natural enemies and its relationship with abiotic factors

A list of natural enemies on aonla aphid, *S. emblica* recorded during the survey is presented in Table 3. Among them, lady bird beetles *Cheilomenes sexmaculata* (Fab.), *Scymnus (Pullus) nr. castaneus* Sicard, hover fly, *Paragus serratus* (Fab.) were the major predators on nymphs and adults of *S. emblica.*

Lady bird beetle, *S.(P.) nr. castaneus* is reported for the first time on aonla aphid, *S. emblica*.

Table 1. Incidence of aonla aphid and their natural enemies during 2005 – 2007

Month	Infested tree (%)	Mean population (No./5 cm terminal leaflet)	Population (No./colony)		
			C. sexmaculata grub	*S.(P.) nr. castaneus* grub	*P. serratus* maggot
Jan. I F.N.	0	0	0	0	0
Jan. II F.N.	0	0	0	0	0
Feb. I F.N.	0	0	0	0	0
Feb. II F.N.	0	0	0	0	0
Mar. I F.N.	10	35.75	2.5	0	1.25
Mar. II F.N.	20	46.25	3.5	0	1
Apr I F.N.	25	51.5	4	4	2.25
Apr II F.N.	25	49.75	1.5	3.25	2.25
May I F.N.	25	61.25	1.5	2.25	2.5
May II F.N.	25	75.25	1.75	2.5	2.5
Jun. I F.N.	35	100	2.5	2.25	2.5
Jun. II F.N.	50	100	2.25	1.5	2.5
Jul. I F.N.	50	85.25	2	1.5	2.5
Jul. II F.N.	50	81.25	2	1	2.25
Aug. I F.N.	50	71.5	1.5	1	1.25
Aug. II F.N.	50	61.25	1.25	1	1
Sep. I F.N.	50	45.25	1.25	1	1
Sep. II F.N.	50	34.25	1.25	1	1
Oct. I F.N.	50	35.5	1	1	1
Oct. II F.N.	20	55.25	1	1	1
Nov. I F.N.	15	10.25	1	0	1
Nov. II F.N.	5	10	1	0	0
Dec. I F.N.	0	0	0	0	0
Dec. II F.N.	0	0	0	0	0

F.N. – Fortnight

Table 2. Correlation coefficients between aonla aphid, their natural enemies and abiotic factors

Factors	Aonla aphid infested trees (%)	*S. emblica* population (No./ 5 cm terminal shoot)	Population (No./ colony)		
			C. sexmaculata grub	*S.(P.) nr . castaneus* grub	*P. serratus* maggots
Aonla aphid infested trees (%)	1				
S. emblica population	0.78**				
C.sexmaculata grub	0.47*	0.68**	1		
S. (P.) nr. castaneus grub	0.44*	0.63**	0.58**	1	
P. serratus maggot	0.63**	0.89**	0.72**	0.81**	1
Minimum temperature	0.78**	0.82**	0.54**	0.75**	0.87**
Maximum temperature	0.64**	0.76**	0.69**	0.68**	0.80**
Relative Humidity %	-0.69**	-0.70**	-0.40*	-0.53**	-0.65**
Rainfall (mm)	0.20	0.15	-0.01	0.05	0.04

* Significant at P = 0.05; ** Highly significant at P = 0.01

Table 3. Natural enemies of aonla aphid, *S. emblica*

S. No.	Common name of predator	Scientific name	Family and Order	Status
1.	Lady bird beetles	*Cheilomenes sexmaculata (=Menochilus sexmaculatus)* (Fab.) *Scymnus (Pullus) nr. castaneus* Sicard - New record	Coccinellidae, Coleoptera	Major predator
2.	Hover fly	*Paragus serratus* (Fab.)	Syrphidae, Diptera	Major predator
3.	Preying mantis	*Euantissa* sp.	Mantidae, Dictyoptera	Minor predator
4.	Spider	*Oxyopes* sp. *Argiope* sp.	Oxyopidae, Araneae	Minor predator

Lady bird beetle, C. sexmaculata: The lady bird beetle, *C. sexmaculata* is a common aphidophagous predator. Its grub was found actively feeding on the aonla aphid, *S. emblica*, throughout the period of infestation. *C. sexmaculata* grub population varied from 1.0 to 4.0 nos./ colony (Table 1). The peak population (4.0 nos./ colony) was recorded during first fortnight of April. The population of *C. sexmaculata* grub was positively correlated with the aonla aphid population (r = 0. 68) (Table 2). The minimum and maximum temperature played a highly significant positive role (r=0.54 and 0.69, respectively) and relative humidity a significant negative role (r=-0.40) on the population of *C. sexmaculata*, but, rainfall had no significant role.

Lady bird beetle, S. (P.) nr. Castaneus: In this study, the lady bird beetle, *S. (P.) nr. castaneus* was reported for the first time on aonla aphid. The population of grubs varied from 1.0 to 4.0 nos./ colony. The peak occurrence of grubs (4.0 nos./ colony) was during April. Association between *S. (P.) nr. castaneus* grub and aonla aphid population was positive and highly significant (r=0.63). The abiotic factors *viz*., minimum temperature and maximum temperature was highly significant and positively correlated with population of *S. (P.) nr. castaneus* (r =0.75 and 0.68, respectively). Relative humidity was negatively correlated with *S. (P.) nr. castaneus* population (r = -0.53) and rainfall had no influence on occurrence of *S. (P.) nr. castaneus*.

Hover fly, P. serratus: The hover fly maggots were the very important aphidophagous insect found feeding on aonla aphid, *S. emblica*. The activity of *P. serratus* was recorded throughout the period of aonla aphid infestation, from March to November. The population of *P. serratus* maggots in aphid colony varied from 1.0 to 2.5 nos.. The highest population of *P. serratus* maggots (2.5 nos.) was registered from May to July, which coincided with the highest population of aonla aphids, *S emblica* (61.25 to 100.00 nos./ 5 cm terminal shoot) and there existed highly significant positive relationship between them (r = 0.89). Their occurrence was significantly influenced by minimum temperature, maximum temperature (r = 0.87 and 0.80, respectively) and relative humidity (r = - 0.65), but rainfall did not play any role on their prevalence.

Discussion

The present survey of aonla trees var. BSR 1 has generated valuable information on the status of aonla aphid and their natural enemies occurring in the aonla field in Madurai, Tamil Nadu.

Incidence of aonla aphid, S. emblica: The fortnightly survey on aonla aphid, *S. emblica* at aonla orchard, Agricultural College and Research Institute, Madurai during 2005–07 revealed that though the aonla aphid was active for nine months from March to November, maximum tree infestation (50%) was registered from second fortnight of June to first fortnight of October. High population of aphids (50.0 to 100.0 nos./ 5 cm terminal shoot) was recorded during April to August. It was supported by the findings of Patel and Kulkarny (1953), that *S. emblica* attacks terminal leaves and young shoots from April to September in North India. Lal *et al.* (1996) also accounted that peak incidence of *S. emblica* was observed during first fortnight of September, where the population of aphid was maximum (6.62/ twig) in Krishna. Lal *et al.* (1997) reported that incidence of this pest was seen from July to October with the peak period in September. In the present study, peak population of 100.0 nos./5cm terminal shoot was registered during June. Masarrat Haseeb *et al.* (2000) recorded, 125 to 160 nymphs and adults per 5 cm of apical portion of a twig.

Diversity and occurrence of predators on aonla aphid, S. emblica: During the fixed plot survey, six predators were recorded on aonla aphid, *S. emblica*. Among them, three were the major predators *viz.*, lady bird beetles, *C. sexmaculata, S.(P.) nr. castaneus* and hover fly, *P. serratus* and others were minor. Lady bird beetle grubs, adults and hover fly maggots were the common predators offering some natural control of aonla aphid, *S. emblica*. Earlier, the activity of natural enemies *viz.*, hover fly, *P. serratus*, *Syrphus* sp., lady bird beetles, *B.rumoides suturalis* (Fab.), *C. sexmaculata* and *Coccinella septempunctata* Linn. were reported on aonla aphid, *S. emblica* (Fletcher, 1916; Bhatia and Shaffi, 1933; Masareet Haseeb, 2005 and Bharpoda *et. al..*, 2006). There was no detailed study on their seasonal occurrence and population.

The lady bird beetle, *S.(P.) nr. castaneus* was reported for the first time on aonla aphid, *S. emblica*, in this study. *C. sexmaculata* on *S. emblica* was earlier reported by Masareet Haseeb (2005) and Bharpoda *et al.* (2006) from Uttar Pradesh and Central Gujarat. Occurrence of *P. serratus* on aonla aphid was earlier reported only from Pusa by Fletcher (1916) and Bhatia and Shaffi (1933).

Population of lady bird beetles, *C. sexmaculata* and *S.(P.) nr. castaneus* ranged from 1.0 to 4.0 nos. / colony and hover fly, *P. serratus* varied from 1.0 to 2.5 nos./ colony. The results showed that aonla aphid, *S. emblica* – predator association were positive and more evident during March to November. Highly significant positive correlation were found between fortnightly aphid and predator population *viz.*, *C. sexmaculata, S.(P.) nr. castaneus* and *P. serratus* (r = 0.68, 0.63 and 0.89, respectively). The occurrence of predators synchronized with the aonla aphid, *S. emblica* incidence. Thus, there is an indication that the incidence of both prey and predator overlap and the predator density increased with prey population.

Competition between the lady bird beetles *C. sexmaculata* and *S. (P.) nr. castaneus* was less, both of them occurred at the same time, they were positively correlated (r=0.58). Correlation between *P. serratus* and *C. sexmaculata*, between *P. serratus* and *S. (P.) nr. castaneus* were also positive and highly significant (r=0.72 and 0.81, respectively). The results revealed that there was no competition between the predators *viz.*, *C. sexmaculata, S. (P.) nr. castaneus* and *P. serratus*.

Influence of abiotic factors: The effect of abiotic factors on the population build up of aonla aphid, *S. emblica* was highly significant. The minimum and maximum temperature played positive role in the population build up of *S. emblica* (r = 0.82 and 0.76, respectively), relative humidity played negative role (r =-0.70) and rainfall had no significant role on it. Mean minimum, maximum temperature of 25.17 ± 1.24°C, 35.73 ± 1.84°C and 69.39 ± 7.34 per cent relative humidity, during April to August, were congenial for aphid multiplication. The present findings concurred with the findings of Singh *et al.* (2005), who reported that maximum fecundity of aonla aphid was recorded when the average minimum, maximum temperature and relative humidity were 27.0, 35.7° C and 71.3%. The results showed that predator population *viz.*, lady bird beetles *C. sexmaculata, S. (P.) nr. castaneus* and hover fly maggot, *P. serratus* were more active when minimum, maximum temperature were high with low relative humidity.

Acknowledgements

The authors gratefully acknowledge Dr. J. Poorani, Scientist, Project Directorate of Biological Control, Bangalore for identification of natural enemies on aonla aphids.

References

Bharpoda, T.M., Koshiya, D.J. and Korat. D.M., 2006. Survey of insect pests and their natural enemies on aonla (*Emblica officinalis* Gaertn.) *Insect Environ.,* 12 (2), 93-94.

Bhatia, H.L. and Shaffi, M., 1933. Life histories of some Indian Syrphidae. *Ind. J. Agric. Sci.*, 2, 543-570.

Fletcher, T.B. 1916. One hundred notes on Indian insects. *Bull. Agrl. Res. Instt. Pusa No.* 59, pp. v+39.

George, C.J. 1927. South Indian Aphididae. *J. Pro. Asiat. Soc. Bengal N.Ser.*, 23, 1-12.

Lal, M.N., Joshi, G.C. and Singh, S.S., 1996. Studies on incidence of *Indarbela* spp. *Cerciaphis emblica* and *Betousa stylophora* on aonla in Eastern Uttar Pradesh. *Prog. Hort.,* 28 (1-2), 66-71.

Lal, M.N., Joshi, G.C. and Singh, S.S., 1997. Studies on the occurrence of insect pests of aonla in Eastern Uttar Pradesh. *Prog. Hort.,* 29(1-2), 85-87.

Masarrat Haseeb, 2005. Insect pests of amla and their management. In: *Amla in India*: (Eds. S.S.Mehta and H.P. Singh), Aonla Growers Association, Salem, pp. 128-138.

Masarrat Haseeb, Abbas, S.R., Srivastava, R.P. and Sharma, S., 2000. Studies on insect pests of amla. *Ann. Pl. Prot. Sci.,* 8, 85-88.

Patel, G.A. and Kulkarni, H.L., 1953. *Cerciaphis emblica* sp. nov. a new aphid pest of *Emblica officinalis. J. Bom. Nat. His. Soc.*, 51, 435-438.

Singh, D.B., Singh, H.M. and Singh, V.K., 2005. Studies on bionomics of aonla aphid, *Schoudetonia bougainvillea* (Theobald). *Ind. J. Entomol.,* 67(3), 189-192.

Reaction of *Bt* and non *Bt* cotton to American Bollworm, *Helicoverpa armigera*

R. Pandi,* D.S. Rajavel,* S. Mahendran** and P.S. Subramaniam**

Observations were made on *Helicoverpa armigera* Hubner (Noctuidae :Lepidoptera) damage in *Bt* and non *Bt* cotton crop raised during Winter 2006-07 and Summer 2007-08 at the Agricultural College and Research Institute, Madurai (weather condition of 31°C and 65% RH). Bunny *Bt* cotton (NCS 929 *Bt*) and Ankur *Bt* cotton (RCH 371 *Bt*) were compared with respective non *Bt* hybrid. *H. armigera* damage (square damage and boll damage) was recorded at weekly intervals on all the cotton entries. From the pooled data it was concluded that both during first and second seasons the bollworm damage was significantly low on *Bt* cotton than on non *Bt* cotton. The square damage in *Bt* cotton (I season) was 1.75% where as it was 14.35% in non *Bt* cotton. The same trend in square damage was observed during second season also (0.89% in *Bt* and 5.31% in non *Bt*). The percent boll damage was very low in *Bt* cotton when compared to square damage (0.52% in I season and 0.12% in II season). The non *Bt* cotton recorded 7.73% boll damage during I season and 2.77% damage during II season. From the present study it is evident that *Bt* cotton is least damaged by *H. armigera* and there by use of insecticides in cotton ecosystem is minimized.

Introduction

Of the various causes of low production and productivity of cotton in India, the depredations by bollworms, particularly *H. armigera* are most threatening pest in the cultivation of crop. Out of the total pesticides used in the country as much as 54% are used for control of pests in cotton alone, out of which nearly 60% are used for the control of bollworms (Jagmail and Kaushik, 2007). This leads to high economic and ecological damages in terms of backlashes of these pesticides such as insecticide resistance, pest resurgence, effect on non-target species, environmental pollution and health hazards. Inevitable but excessive use of pesticides and above their failure to eliminate American bollworm has attracted the cotton growers towards this transgenic crop as a breather. Genetically modified organisms in case of *Bt* cotton is practically known to reduce the pesticide load from 50 to 70 per cent in different parts of world, as there is less need of insecticides sprays for controlling bollworms (Wadhwa and Gill, 2007). Studies on *H. armigera* abundance on *Bt* cotton were initiated to understand the role of these host plants in influencing the population changes of this pest.

Materials and Methods

Field experiments were conducted at the Agricultural College and Research Institute, Madurai during winter 206-07 and summer 207-08 (Weather condition of 31ºC and 65% RH). *Bt* and non - *Bt* hybrids of American cotton, Bunny *Bt* cotton (NCS 929 *Bt*) and Ankur *Bt* cotton (RCH 371 *Bt*) were compared with respective non – *Bt* cotton hybrids. The incidence of *H. armigera* damage was recorded on twenty randomly selected plants from each trial plot at weekly intervals. The total number of healthy and damaged squares and bolls were recorded. The collected data were pooled and expressed as per cent damage.

---

* Dept. of Agricultural Entomology, ** Dept. of Agronomy, Agricultural College and Research Institute, Madurai - 625 104

Results and Discussion

The American bollworm, *H. armigera* inflicted a low damage on both the *Bt* cotton hybrids tested *viz*., Bunny *Bt* cotton and Ankur *Bt* cotton, irrespective of seasons effect, than non – *Bt* cotton (Tables 1 and 2).

Table 1. American Bollworm damage on Bunny *Bt* and non *Bt* cotton hybrid – Season I

Week	*Bt* cotton		Non *Bt* cotton	
	Square damage (%)	Boll damage (%)	Square damage (%)	Boll damage (%)
1	3.43	1.09	9.81	5.56
2	1.14	0.72	15.07	6.54
3	3.21	0	16.67	26.81
4	1.22	1.08	22.84	18.61
5	2.86	0.86	9.68	6.67
6	3.85	0.89	39.29	4.15
7	0	0	15.79	1.23
8	0	0	0	0
9	0	0	0	0
Mean	1.75	0.52	14.35	7.73

Table 2. American Bollworm damage on Ankur *Bt* and non *Bt* cotton hybrid –Season II

Week	*Bt* cotton		Non *Bt* cotton	
	Square damage (%)	Boll damage (%)	Square damage (%)	Boll damage (%)
1	0	-	0	-
2	0	-	0	-
3	0	-	0	-
4	4.52	0	11.67	0
5	2.35	0	6.17	0
6	2.38	0	4.08	0
7	0	0	3.74	0
8	1.33	0	3.49	0
9	0	0	4.35	0
10	1.02	0	6.19	3.33
11	1.43	0.81	4.00	4.27
12	2.13	0	5.98	6.29
13	0	0	6.02	4.27
14	0	0	4.35	3.64
15	0	0.97	11.90	3.33
16	0	0	3.45	8.43
17	0	0	14.81	5.19
Mean	0.89	0.12	5.31	2.77

The square damage by the bollworm were 1.75% and 0.89% in the Bunny *Bt* cotton and Ankur *Bt* cotton respectively. Whereas the damage on non-*Bt* cotton accounted for 14.35 and 5.31% in season I and season II where non *B t* Bunny and Ankur was raised.

On the other hand, the boll damage by *H. armigera* in both the season on *Bt* cotton *viz*., Bunny *Bt* cotton (0.52%) and Ankur *Bt* cotton (0.12%) was low. Conversely, non-*Bt* cotton of respective hybrids showed 7.73 and 2.77% boll damage.

Sonawane *et al*. (2007) reported that higher boll damage was observed in non *Bt* cotton while *Bt* cotton was less. Similar results were obtained in the present investigation. The reduction in *H.*

armigera damage might be due to the larval mortality that occurred in neonates in transgenic cotton (Anonymous, 2001). The present study confirmed that *H. armigera* damage in *Bt* cotton was negligible as compared to the non *Bt* cotton crop.

References

Anonymous, 2001, Annual Report of CICR ; Nagpur, 9 p.

Jagmail, S. and Kaushik, S.K.,* 2007. *B.t.* cotton in India ; *Indian Farming,* 2, 26-28.

Sonawane, J.R.D., Khande, M. Bisane, K.D. and Tayade, P.P., 2007. Population abundance of *Helicoverpa armigera* (Hubner) on *Bt* and non *Bt* cotton; *Insect Environment,* 13(2), 87-89.

Wadhwa, S. and Gill, R.S., 2007, Effect of *Bt* cotton on biodiversity of natural enemies; *J. Biol. Control,* 21(1), 9-16.

Tetratrophic Association of *Aphis gossypii* Glover with its Food Plants and Natural Enemies in Northeast Bihar

Md. Equbal Ahmad,* Krishna Murari Kumar,* Nishat Parween* and Sanjeev Kumar*

The fauna of aphids and their natural enemies have been surveyed during 2003-2008 in most of the districts of northeastern Bihar. During the survey, 40 species of aphids were recorded on more than 100 plants in the target area. Among these, *Aphis gossypii* Glover was most common aphid pest in the target area and recorded on 49 food plants. 12 species of food plants viz., *Acalypha torta, Amaranthus tricolor, Basella alba, Callistephus chinensis, Catharanthus roseus, Citrus limettioides, Cucurbita hispida, Dahlia pinnata, Hordeum vulgare, Malvastrum coromandelianum, Phaseolus vulgaris, Tinospora cordifolia* were recorded as new host plants from India. Most of the host plants are recorded first time from Bihar. The most suffered food plants were *Abelmoschus esculentus, Capsicum frutescence, Clerodendrum infortunatum, Coccinia grandis, Hibiscus rosasinensis, Lagenaria siceraria, Lawsonia inermis, Momordica charantia, Ocimum tenuiflorum, Punica granatum, Solanum melongena* and *Tagetes* sp. Four species of parasitoids viz., *Aphelinus gossypii* (Timberlake), *Aphelinus albipodus*, Hayat and Fatima, *Lipolexis oregmae(=scutellaris)* (Gahan) and *Binodoxys indicus* (Subba Rao and Sharma) were recorded on *A. gossypii*. Only *A. gossypii, L. oregmae* and *B. indicus* were found abundantly with high rate of parasitisation. Three species viz., *Alloxysta pleuralis* (Cameron), *Phaenoglyphis* sp. and *Syrpophagus* sp. were also recorded as hyperparasitoids of above mentioned parasitoids.

Introduction

The basic approach in biological control is the survey of pest and their natural enemies that provide basic information about the correct identification of pest species, their seasonal distribution, interaction among food plants, parasitoids and hyperparasitoids in varying ecological conditions. These complexes are the source of information through which biological studies of the parasitoids can be made for proper utilization in biological control programme.

Northeast Bihar has rich agricultural land where farmers face a lot of problems due to heavy infestation of aphids. But in this area, only little work has been done earlier by Ahmad and Singh (1996a, 1997) and Ahmad *et al.* (2005). Keeping in view, different localities of districts of northeast Bihar viz., Araria, Bhagalpur, Begusarai, Katihar, Khagaria, Kishanganj, Madhepura, Munger, Naugachia, Purnea and Saharsa were extensively surveyed for the records of aphids, their food plants and natural enemies to know the true tetratrophic relationship between them and their occurrence. In this present paper the tetratrophic interaction of *Aphis gossypii* is discussed in detail.

Materials and Methods

For the study of tetratrophic interaction of aphids, different agricultural, horticultural as well as wild plants were surveyed in the target area. The parts of plants infested with aphids were cut and transported to the laboratory. Few aphids were preserved in 70% ethyl alcohol for taxonomic study

* University Department of Zoology, T.M. Bhagalpur University, Bhagalpur – 812 007

and rest of alive aphids were kept for culture. The parasitized aphids were sorted and kept separately in BOD incubator at 22°C until the emergence of adult. The aphids were identified with the help of software aphid taxa key developed by CAB, London, key provided by Raychaudhari, (1980) and the parasitoids were identified following Stary and Ghosh (1983).

Results and Discussion

During the extensive survey in different localities of districts of northeast Bihar (2003-2008), 40 species of aphids were recorded on more than 100 plants in the target area. Among these, *Aphis gossypii* Glover was the most common aphid pest in the target area and recorded on 49 food plants of 23 families. However earlier it was recorded on only 6 families viz., Apiacea, Asteraceae, Cucurbitaceae, Fabaceae, Malvaceae and Solanaceae (Ahmad and Singh, 1997). In the present survey, the most suffered families were Cucurbitaceae (7 plants), Asteraceae (6 plants), Malvaceae (4 plants), Solanacea (4 plants) and Fabaceae (3 plants). Earlier, *A. gossypii* was recorded on only 18 food plants in the target area (Ahmad and Singh, 1997). Thus most of the food plants are recorded first time from the target area. 12 species of food plants viz., *Acalypha torta, Amaranthus tricolor, Basella alba, Callistephus chinensis, Catharanthus roseus, Citrus limettioides, Cucurbita hispida, Dahlia pinnata, Hordeum vulgare, Malvastrum coromandelianum, Phaseolus vulgaris, Tinospora cordifolia* were recorded as new host from India (Table-I). The intensity of infestation varied too much on different food plants. Very high intensity of infestation was observed on *Abelmoschus esculentus, Capsicum frutescence, Clerodendrum infortunatum, Coccinia grandis, Hibiscus rosasinensis, Lagenaria siceraria, Lawsonia inermis, Momordica charantia, Ocimum tenuiflorum, Punica granatum, Solanum melongena* and *Tegetes* sp. The sites of infestation also varied with the type of food plants (Table-I). Mostly leaves, nascent stems are vulnerable part but sometimes on few host plants, flower and inflorescence were also attacked. The seasonal study revealed that *A. gossypii* are found throughout the year except in June due to high temperature. It started to appear during July and increased the population very fast. Its peak population was observed during November to February. The number of food plants attacked during the peak population was also high as compared to other months (Graph I).

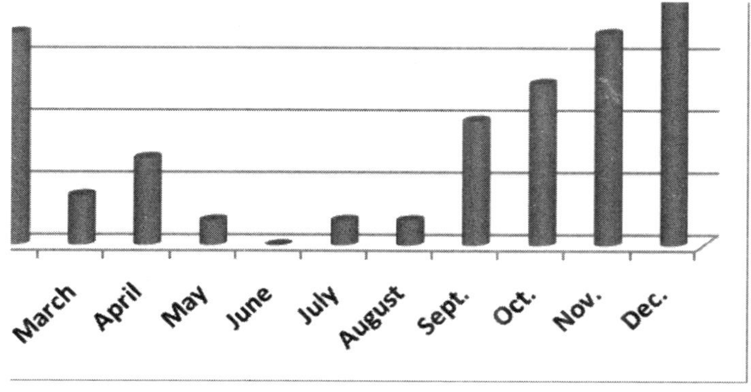

Graph 1

Parasitoids: Four species of parasitoids viz., *Aphelinus gossypii* (Timberlake), *Aphelinus albipodus*, Hayat and Fatima, *Lipolexis oregmae (=scutellaris)* (Gahan) and *Binodoxys indicus* (Subba Rao and Sharma) were recorded on *A. gossypii*. *A. albipodus* is recorded first time as parasitoid of *A. gossypii* in the target area. Other parasitoids were already reported by Ahmad and Singh (1996a). Several new host association of these parasitoids were observed from India (Table-II). *L. oregmae* and *B. indicus* were found abundantly in most localities with moderate to very high rate of parasitisa-

Table I. Records of *Aphis gossypii* with their food plants and degree of infestation from northeast Bihar

Food plants/family	Sites of infestation	Degree of infestation	Place and date of collection
Abelmoschus esculentus (Malvaceae)	Leaves, flowers, inflorescences and fruits	++, + + +, ++++	Most of the places , (July to Nov., 2005-2007)
Acalypha indica * (Euphorbaceae)	Upper surface of leaves, inflorescence	+ +	Bhagalpur,(Jan.2008)
Ageratum conyzoides (Asteracea)	Lower surface of leaves, Pedicel	++, +++	P.G. Zool., Sahebganj (Bhagalpur), Kishanganj (Oct., to Jan, 2007-2008)
Amaranthus tricolor * (Amaranthaceae)	leaves	+	Nashratkhani (Bhagalpur), (Sept.,2007)
Altermanthera sp. (Amarantheceae)	Old and new leaves	+ +	Araria, (April, 2006)
*Basella alba** (Basellaceae)	Lower surface of leaves	+	Kursella (Katihar), (Jan., 2008)
Brassica oleracea var. *botrytis (*Brassicaceae*)*	Leaves	+ +	Maur, (Lakhisarai), (Dec.,,2005)
Cajanus cajan (Fabaceae*)*	Leaves	+ + ++++	Adabari (Khagaria), Bakhari (Begusarai), (Aug., to Dec., 2006-2007)
Calendula sp. (Asteraceae)	Pedicel, petals of flowers	++	Univ. Campus (Bhagalpur), (Feb., 2008)
*Callistephus chinensis** (Asteraceae)	Leaves and flower	+ +	Qajiwalichak,(Bhagalpur), (Feb., 2005)
Calotropis gigantea (Asclepiadacae)	leaves and flowers	++	Mukundpur, (Bhagalpur) , (Oct., 2003)
Catharanthus roseus * (Apocynaceae)	Leaves and flowers	+, + +	Qajiwalichak,,Zool.Dept., Khanzarpur (Bhagalpur), (Sep., to March, 2004-07)
Capsicum frutescens (Solanaceae)	Stems, leaves and flowers.	+ +, + + +, + + + +	Most of the places (Aug., to Feb., 2003-08)
Chrysanthemum indicum (Asteraceae)	Leaves and flowers	+ + +	Qajiwalichak, Zool.Deptt., (Bhagalpur), (Dec., to Feb., 2004-05)
Citrus limon (Rutaceae)	New leaves	++	Murarpur, (Bhagalpur), (Jan.,2006)
*Citrus limettoides** (Rutaceae)	Nascent leaves	+ +	Krishnapuri,(Monger), (Jan.,2006)
Clerodendrum infortunatum (Verbenaceae)	New and old leaves	+ +, + + +, ++++	Policeline, Dasrathpur (Monger), Sultanganj, Univ. campus, (Bhagalpur), (Sept. to Feb., 2003-08)
Coccinia grandis (Cucurbitaceae)	Stem, leaves and pedicel.	+ , + + + + +	Krishnapuri, Ratanpur (Monger), Naugachia, Railway line (Bhagalpur), Belaganj (Khagaria), Baisi (Purnea), (Sept. to Jan. 2004-08)

Colocasia sp. (Araceae)	Lower surface of old and new leaves	+ ++	Debtha (Khagaria).Firangola (Kishanganj) (Jan.2006 and Apri, 2008)
Cucumis sativus (Cucurbitaceae)	Old and New leaves	+ + + +	Qajiwalichak, Sahebganj (Bhagalpur) (Nov., 2005 and July,2007)
Cucurbita maxima (Cucurbitaceae)	Lower surface of leaves, flowers and tendril	+	Farka (Sabour), (Feb.,2008)
*Cucurbita hispida** (Cucurbitaceae)	Lower surface of old and new leaves	+, + +	Shastrinagar, Baryarpur, Dasrathpur (Monger), Nauranga (Khagaria), (Oct.,2005 - 06)
*Dahlia pinnata** (Asteraceae)	Leaves	+	Belaganj (Khagaria), (April, 2006)
Helianthus annus (Asteraceae)	Flower and lower surface of new leaves	+ +	Qaziwalichak (Bhagalpur), (March, 2006)
Hibiscus rosasinensis (Malvaceae)	Leaves, pedicel, buds, flower,	+ +, + + +	Lahritola, Vikramshila, Nargah (Bhagalpur), Shastrinagar (Monger), Begusarai, (Nov., to March, 2004-08)
Hibiscus sabdariffa (Molvaceae)	Lower surface of old and new leaves and flowers	+ + + + + + , + +	Dasrathpur, Kajara (Monger), Gangania (Bhagalpur),(Nov., to Jan., 2005-06)
*Hordeum vulgare** (Poaceae)	leaves	+	Railway line (Bhagalpur), (March, 2005)
Impatiens balsamina (Balsaminacea)	Stem, lower and upper surface of leaves and flower	+ + +	Qaziwalichak (Bhagalpur), (Nov., 2007)
Lablab purpureus purpureus (Fabaceae)	Old and new leaves	+, ++	Jalakaura (Khagaria), Baisi (Purnea) (Jan., 2008)
Lagenaria siceraria (Cucurbitaceae)	Both surface of leaves	+ +, + + + , + + + +	Most of the places (Oct., to April, 2004 -08)
Lawsonia inermis (Lathyraceae)	Lower and upper surface of nascent leaves	+ + , + + +, + + + +	Dasrathpur, Kajara (Monger), Sahebganj, Belaganj(Khagaria), (Sept., to Jan., 2003-07)
Luffa aegyptiaca (Cucurbitaceae)	Lower surface of leaves	+ +	University campus, Amarpur (Bhagalpur) (Dec., 2007 and Feb., 2008)
Lycopersicon esculantum (Solanaceae)	Lower surface new leaves and flowers	+ + , + + +	Akbarnagar, Gangania, Railway line (Bhagalpur), Ratanpur (Monger) , (Dec., to Jan., 2005-06)
Malvastrum corom- andelianum *(Malvaceae)	Stem and leaves	+ + +	Zool. Deptt. (Bhagalpur) (Feb.,2005)
Momordica charantia (Cucurbitaceae)	Lower surface of leaves	+, + +, + + +	Most of the places (Nov. to Feb., 2005-08)
Ocimum tenuiflorum (Labiatae)	Lower and upper surface of leaves, stem and flower	+ + , + + +	Most of the places (Nov. to Feb., 2003-08)

37. *Phaseolus vulgaris** (Fabaceae)	Lower surface of leaves	+ +	Vinodpur (Katihar), (Jan., 2008)
38. *Psidium guajva* (Myrtaceae)	Lower surface of nascent leaves and pedicel	+ + , + + +	Baryarpur, Mangarha (Monger), Belaganj, (Khagaria) (Oct., to April, 2005-06)
Punica granatum (Punicaceae)	Old and new leaves, and flower	+ +, + + +	Krishnspuri, Mangarha, Ratanpur, (Monger), Naugachia (Bhagalpur), (Oct., to Dec.,2003-05)
Raphanus sativum (Brassicaceae)	Leaves	+	Ratanpur (Monger) , (Dec.,2005)
Rosa sp. (Rosaceae)	Lower surface of old and new leaves	+ +	Ratanpur (Monger), (Dec.,2005)
Rumax sp. (Poligonaceae)	Lower surface of leaves	+	Ratanpur (Monger), (Dec.,2005)
Solanum melongena (Solanaceae)	Lower surface of leaves, fruits and pedicel	+ +, + + + + + + +	Most of the places (Oct. to Feb., 2005-08)
Solanum tuberosum (Solanaceae)	Old and new leaves and stems	+ + + + +	Most of the places (Dec. to Feb., 2005-08)
Tagetes sp. (Asteraceae)	Leaves, and flowers	+ + + , + + + +	Qaziwalichak (Bhagalpur), Shastrinagar, (Monger) (Nov. to Feb., 2004-05)
Tagetes erectus (Asteraceae)	Flower	+	Fringola (Kishanganj), (Jan., 2008)
*Tinospora cordifolia** (Menispemaceae)	Leaves, stem, and inflorescence.	+ + +	Babupur (Bhagalpur), (Sept., 2005)
Trichosanthes dioica (Cucurbitaceae)	Leaves, nascent bud and fruits	+ + , + + +	Naugachia, Railway line (Bhagalpur). (Sept., 2005)
Wild plants (Unidentified)	Leaves and flowers	+	Gangania (Munger), (January, 2006)

* New host record from India; + = Little infestation; ++ = Moderate infestation; +++ = high infestation; ++++ = Very high infestation

tion. *A. gossypii* was also recorded from most of the localities with only low to moderate intensity of parasitism. The intensity of parasitism of these parasitoids vary in different localities on different food plants (Table-I). But sometimes variation in the parasitism was observed even on the same aphid infesting similar plants. It is due to different kinds of agricultural practices including variable amount of fertilizer, insecticides etc.(West and Mordue, 1992).

B. indicus is widely distributed parasitoid in India and it is polyphagous parasioitds of aphids (Stary and Ghosh, 1983). In the target area it was recorded on several aphid species. Moderate to very high rate of parasitism of *B. indicus* was observed against *A. gossypii* on *A. conyzoides, C. frutescence, C. infortunatum, C. hispida, Lablab purpureus, L. siceraria, Luffa aegyptica, O. tenuiflorum, Rosa sp., S. melongena, Solanum tuberosum* and *Tegetes* sp. On other host plants little infestation was observed (Table-II). It was recorded mainly during the last week of October to May in the target area. Different aspects of its biology, ecology and control efficiency have been reviewed by Singh and Agarwala (1992). Singh and Rao (1995) observed successful control of *A. gossypii* by *B. indicus* on cucurbits plants.

L. oregmae is also widely distributed in India. Its moderate to high rate of parasitism was observed against *A. gossypii* on *A. esculentus, C. frutescence, C. grandis, Colocasia* sp., *Momordica charantia, Rosa* sp., and *S. melongena.* It was recorded mainly during the September to April in the target area. In the target area, its population was observed high on *Aphis craccivora* (Ahmad and Kumar, 2007). Earlier, few aspects of *L. oregmae* have been studied by Ramaseshiah and Bhat (1970), Shuja-Uddin (1977), Singh *et al.* (1999, 2000) and Pandey and Singh (2005).

A. gossypii is a native of Hawai and was used in biological control of *A. gossypii* in Pakistan. In the target area, its high to very high parasitisation was observed on *A. gossypii* against only few food plants viz., *Luffa acutangula, L. aegyptica* and *M. charantia.* On other host plants little infestation was observed (Table-II). Its occurrence was recorded during October to April. The details of host associations of the parasitoid have been reported from India by Agarwala (1983), Bhagat (1985) and Ahmad and Singh (1992a, 1993).

A. albipodus is recorded on *A. gosypii* against only one host *S. melongena* with very little parasitisation rate.

Hyperparasitoids: Hyperparasitoids constitute fourth trophic level in plant-aphid-parasitoid-hyperparasitoid complex but their presence considered harmful to the parasitoid and so biological programme have carefully and routinely excluded hyperparasitoids during introduction of exotic parasitoid. Three species viz., *Alloxysta pleuralis* (Cameron), *Phaenoglyphis* sp. and *Syrpophagus* sp. were recorded as hyperparasitoids of above mentioned parasitoids. The study of tetratrophic interaction revealed that most of the association of these hyperparasitoids is recorded first the time from India (Table-II).

Table 2. Tetratrophic interaction of *Aphis Gossypii*

Food plants	Parasitoids/ Degree of parasitism	Hyperparasitoids/ Degree of Hyperparasitism
1. *Abelmoschus esculentus*	*Aphelinus gossypii* (+) *Lipolexis oregmae* (+++)	–
2. *Ageratum conyzoides*	*Binodoxys indicus* (+++)	*Syrphophagus* sp. (++)
3. *Cajanus cajan*	*A. gossypii* (++) *B. indicus* (+++)	*Alloxysta pleuralis*(++) *Phaenoglyphis* sp. (++)
3. *Catharanthus roseus**	*A. gossypii* (++)*	–
4. *Capsicum frutescens*	*Aphelinus* sp. (+) *B. indicus* (+++) *L.oregmae* (+++)*	– *A. pleuralis* (+++) *Phaenoglyphis* sp. (+)*
5. *Citrus limon**	*B. indicus* (++)*	*Phaenoglyphis* sp. (+)*
6. *Clerodendrum infortunatum**	*B. indicus* (+++)* *L. oregmae* (++)*	– *A. pleuralis* (++)*
7. *Coccinia grandis**	*B. indicus* (+)* *L. oregmae* (++++)*	*A. pleuralis* (+)*
8. *Colocasia* sp*.	*Aphelinus* sp. (+)* *L. oregmae* (+++)*	– –

9. *Cucurbita hispida**	*B. indicus* (+++)*	*A. pleuralis* (+)*
10. *Hibiscus rosasinensis**	*B. indicus* (+)* *L. oregmae* (+)* *Prochiloneurus testaceus**	– – –
11. *Hibiscus sabdariffa**	*B. indicus* (+–)*	–
12. *Hordeum vulgare**	*B. indicus* (+)*	*A. pleuralis* (+++)*
13. *Impatiens balsamina**	*Aphelinus* sp. (++)*	*A. pleuralis* (+)*
14. *Lablab purpureus purpureus*	*A. gossypii* (++) *B. indicus* (+++)	*A. pleuralis* (++)* *Phaenoglyphis* sp. (+)*
15. *Lagenaria siceraria*	*A. gossypii* (++) *B. indicus* (++++) *L. oregmae* (++)*	*A. pleuralis* (+++) *Phaenoglyphis* sp. (++)* –
16. *Lawsonia inermis*	*B. indicus* (+) *L. oregmae* (++)*	*A. pleuralis* (+)* –
17. *Luffa acutangula**	*Aphelinus* sp. (+++)*	–
18. *Luffa aegyptiaca*	*A. gossypii* (+++)* *B. indicus* (++++) *L. oregmae* (++)	*A. pleuralis* (++++)* *Phaenoglyphis* sp. (+++)* –
19. *Lycopersicon esculantum*	*B. indicus* (++)	–
20. *Momordica charantia*	*A. gossypii* (++++)* *B. indicus* (++) *L. oregmae* (+++)*	*A. pleuralis* (++++) * *Phaenoglyphis* sp. (++)* –
21. *Ocimum tenuiflorum**	*A. gossypii* (+)* *B. indicus* (++++)* *L. oregmae* (+)*	*A. pleuralis* (+)* *Phaenoglyphis* sp.(+)* –
22. *Psidium guajva*	*L. oregmae* (+)	–
23. *Rosa sp.**	*B. indicus* (+++)* *L. oregmae* (+)*	*A. pleuralis* (+)* –
24. *Solanum melongana*	*Aphelinus albipodus* (+) *B. indicus* (++++) *L. oregmae* (++++)	*A. pleuralis* (+)* – –
25. *Solanum tuberosum*	*B. indicus* (+++)	*A. pleuralis* (++)* *Phaenoglyphis* sp. (+)*
26. *Tagetes* sp.	*A. gossypii* (+) *B. indicus* (+++)*	*A. pleuralis* (+)* *Phaenoglyphis* sp. (+)*
27. Wild plants (Unidentified)	*Aphelinus* sp. (+++)	–

* New association from India; + low parasitism/hyperparasitism; ++ Moerate parasitism/hyperparasitism; +++ High parasitism/hyperparasitism; ++++ Very high parasitism/hyperparasitism

A. pleuralis was reported first the time from India by Singh and Sinha (1979). It was also reported in the target area on *B. indicus* only by Ahmad and Singh (1996a) through several aphid host. But it was recorded on *B. indicus* through *A. gossypii* only on two host plants viz., *Cajanus cajan* and *L. aegyptiaca.* In the present survey, it is recorded on *A. gossypii, B. indicus* and *L. oregmae* through *A. gossypii* on several food plants with little to high intensity of hyperparasitism (Table-II). It was generally found in the month of Novemebr to April in the target area. Its peak population was observed during January to March. The details of association, biology, host selection and ecology were studied by Ahmad and Singh (1992b, 1996b); Singh and Srivastava (1987).

Phaenoglyphis sp. is widely distributed in India (Singh and Tripathi, 1991). Earlier it was reported on *B. indicus* against *A. gossypii* through only one food plant, *C. cajan* in the target area by Ahmad and Singh (1996a). But in recent survey, it was recorded on *B. indicus* as well as on *L. orgemae* through *A. gossypii* on several food plants with very little to moderate intensity of infestation (Table-II). Maximum hyperparasitism was recorded through *L. aegyptiaca.* It was mostly recorded during November to April.

Syrpophagous sp. is very rare in the target area and recorded on *B. indicus* through *A. gossypii* on only one food plants, *A. conyzoides* with little intensity of infestation.

Conclusion

Thus it is concluded that in the target area, *A. gossypii* is a serious pest and causes direct injury to plants by retarding their growth, failure of fruit development and transmission of viral diseases. These aphids should be controlled to increase production. Three species of parasitoid viz. *A. gossypii, B. indicus* and *L. oregmae* are very common and potent; they must be studied in detail for possible use in biological control of *A. gossypii.*

Acknowledgements

The authors are thankful to Head, University Department of Zoology, T.M. Bhagalpur University for providing facilities, Prof. Md. Hayat, Department of Zoology, A.M.U., Aligarh for identification of Aphelinids and Department of Science and Technology (DST), New Delhi for financial assistance.

References

Ahmad, M.E. and Singh, R., 1992a. New host association of a parasitoid *Aphelinus gossypii* Timb. (Aphelinidae: Hymenoptera) from India, *Newsletter, The Aphidological Society, India,* 10(1), 4-5.

Ahmad, M.E. and Singh, R., 1992b. New host association of Chalcidoid hyperparasitoids. *J. Adv. Zool.,* 13, 66-67.

Ahmad, M.E. and Singh, R., 1993. *Aphis gossypii* Glover on different food plants and its association with the parasitoids and hyperparasitoids in north-eastern Uttar Pradesh. *Entomon.* 12, 65-69.

Ahmad, M.E. and Singh R., 1996a. Records of aphids, parasitoids from the north Bihar and associtations with their hosts and food plants. *J. Adv. Zool.* 17, 26-33

Ahmad, M.E. and Singh, R., 1996b. Trophic relations of aphid hyperparasitoids in north-eastern Uttar Pradesh. *Entomon.* 21(1), 37-42.

Ahmad, M.E. and Singh R., 1997. Records of aphids and their food plants, parasitoids and hyperparasitoids from the north Bihar . *J. Adv. Zool.,* 18(1), 54-61

Ahmad, M.E., Kumar, K.M and Kumar, S., 2005. Records of aphids of Northeast Bihar -I. *J. Aphidol.* 19, 121.

Ahmad, M.E. and Kumar, K.M., 2007. Food plants and natural enemies of *Aphis craccivora* Koch (Hem.: Aphididae) in northeast Bihar. *J. Aphidol.* 21(1-2), 89-95.

Agarwala, B.K., 1983. Host range of *Trioxys indicus* Subba Rao and Sharma (Hymenoptera : Aphidiidae) *Akitu, New series,* 49,1-6.

Bhagat, R.C., 1985. Host aphid complex of Aphelinid parasite in Kashmir valley. *Geobios,* 4, 178-179.

Pandey, S. and Singh, R., 2005. The reproductive behavior of *Lipolexis scutelaris* Mack. (Hym : Braconididae), a parasitod of *Aphis gossypii* Glover (Homoptera : Aphidiidae) adjustment of sex ratio in response to host size. *J. Adv. Zool.,* 26(1)7-19.

Ramaseshiah, G. and Bhat, T.V., 1970. Supply of natural enemies of *Sipha flava* (for Trinidad, West Indies). *Report Commonw. Inst. Biol. Cont.* 1969, 55.

Raychaudhari, D.N., 1980. *Aphids of North-east India and Bhutan*. pp. 459, Zoological Society Publication, Calcutta

Shuja- Uddin, 1977. Observation on normal and diapausing cocoons of the genus *Lipolexis* Foerster (Hymenoptera : Aphelinidae) from India. *Bull. Lab. Ent. Agr. Portici*, 34, 51-54.

Singh, R. and Agarwala, B.K., 1992. Biology, ecology and control efficiency of the aphid parasitoid, *Trioxys indicus* : a review and bibliography. *Biol. Agric. Hort.* 8, 271-298.

Singh, R. and Rao, S.N., 1995. Biological control of *Aphis gossypii* Glover (Homoptera : Aphididae) on cucurbits by *Trioxys indicus* Subba Rao and Sharma (Hymenoptera : Aphidiidae) *Biol. Agric. Hort.*, 12 : 227-236

Singh, R. and Singh, T.B., 1979. First record of *Alloxysta* sp. a hyperparasitoids of *Trioxys indicus* Subba Rao and Sharma and (Aphidiidae: Hymenoptera). *Curr. Sci.,* 48, 1008-1009.

Singh, R. and Srivastava, P.N., 1987. Potential host-habitat location by *Alloxysta pleuralis* (Cameroon) (Alloxystidae :Hymenoptera). *Z. Ang. Zool.,* 74, 337-341

Singh, R. and Tripathi, R.N., 1991. Records of aphids hyperparasitoids in India. *Bioved., 1*, 141-150.

Singh, R., Upadhyay, B.S., Devendra Singh and Choudhary, S.C., 1999. Aphid (Homoptera: Aphidiidae) and their parasitoids in north-eastern Uttar Pradesh. *J. Aphidol.,* 13(1-2), 49-62.

Singh, R., Singh, K and Upadhyay, B.S., 2000. Honey dew as a food source for an aphid parasitoid *Lipolexis scutelaris* Mackauer (Hymenoptera : Brachonidae) *J. Adv. Zool.,* 21, 76-82

Stary, P. and Gosh, A.K., 1983. Aphid parasitoid of India and adjacent countries (Hymenoptera : Aphediidae) *Tech. Monograph, 7*. Zoological survey of India , Calcutta, pp.96

West and Mordue, A.J., 1992. The influence of Azadirachtin on the feeding behavior of cereal aphids and slugs. *Ent. Exp. Appl.,* 62, 75-79.

Screening of certain plant products for toxic effect to the cotton bug, Dysdercus cingulatus F.

S. Govindan* and K. Shanthi*

A screening test was carried out to assess the toxic effect of acetone extract of plant leaves such as *Catharanthus roseus*, *Annona squamosa*, *Tridax procumbens* and *Calotropis gigantea* against the cotton pest *Dysdercus cingulatus* Fab. using topical application method. Sublethal (72h LD_{16} value) effect of these extracts on reproductive potential and haemolymph proteins of female insect was studied. The 72 h LD_{50} values were 1950, 1420, 704 and 27 µg/insect for *Catharanthus*, *Annona*, *Tridax* and *Calotropis* extract respectively. Hence, *Calotropis* extract was found to be highly toxic to *D. cingulatus*. As the post treatment time increased from 24 to 96 h, the median lethal concentration decreased. The plant extracts treated insects were observed to exhibit poor fecundity (12 - 24%) and hatchability (16 - 35%) but high sterility (60 - 84%). Fecundity, hatchability and sterility effect wise these extracts were graded as *Calotropis* > *Tridax* > *Annona* > *Catharanthus*. The haemolymph protein content of treated female bug was (12.5 - 25.4 µg/µl) which was less than control (49.6µg/µl). Among the treatments, *Calotropis* treated insect showed maximum protein reduction over control.

Introduction

The red cotton bug, *Dysdercus cingulatus* Fabricius (Heteroptera: Pyrrhocoridae), is a serious pest of cotton in Southeast Asian countries (Kohno and Bui Thi, 2004). If not managed properly, both nymph and adult insects feed gregariously on cotton bolls and leaves. As a result, the bolls open irregularly and affect the quality of ginning and oil content of seeds. Moreover, the red cotton bug introduces a bacterium, *Nematospora gossypii* into bolls causing red staining of the lint, besides depositing excreta, which make the seeds unfit for sowing (Vasantharaj David and Kumaraswami, 1996). Several plant derivatives and crude extracts are reported to be active as sterilants against many insects (Satyanarayana and Kumuda Sukumar, 1988; Magdum *et al.,* 2001). Plant extracts induce various ranges of ovipositional inhibitory effects on several insects tested (Prabhu and John, 1975; Rao and Gujar, 1995; Verma, 2004; Kumar *et al.,* 2007). In the present report, insecticidal properties of leaves extracts of *C. roseus*, *A. squamosa*, *T. procumbens* and *C. gigantea* were tested on *D. cingulatus* using topical application method.

Materials and methods

Stock culture: The nymphs and adults of *D. cingulatus* were collected from the cotton field in the vicinity of Madurai. They were reared in plastic container of 7 cm diameter and 8 cm height with perforated lid in the laboratory at room temperature (28° ± 1° C), fed with soaked cotton seeds and kept as stock culture.

Preparation of plant extracts: Dried powdered leaves of *Annona squamosa*, *Calotropis gigantea, Catharanthus roseus* and *Tridax procumbens* was subjected to acetone extraction at 55°C for 8 h by using Soxhlet apparatus. The extract was again redistilled and concentrated on a water bath and weighted accurately in a monopan balance and than it was stored in a refrigerator for future use. The

* Advanced Research Centre, Department of Zoology, Yadava College, Madurai – 625014

extracted leaf powder was dissolved in the required volume of acetone separately to make the desired concentration of extract in 1 µl of acetone for the bioassay study.

Bioassay: One microlitre of acetone containing the selected concentration of material was topically applied on the abdominal sternum of newly emerged (6 h old) adult *D. cingulatus.* A group of 100 insects were tested for each concentration. Five different concentrations were used in each plant extract to obtain the mortality response. Separate control using acetone (1 µl/insect) alone was run simultaneously for comparison. Mortality was counted after 24, 48, 72 and 96 h in each concentration. Probit analysis was carried out using computer program Co-Stat given in NCPC Technical bulletin No.1 (1986).

Treatment of plant extracts for reproductive potential: The virgin female and young male of 6 h old were topically treated with LD_{16} values of the chosen plant extract dissolved in acetone (1 µl/ insect) using a micropipette. The treated pair was allowed to mate in a plastic container for egg-laying. These insects were maintained under the laboratory conditions as the stock culture. The counts were recorded and tabulated for the treated insects which failed to mate, eggs laid by treated and control insect, incubated and hatched eggs. The fecundity, hatchability and sterility were expressed in percentage by following the formulae of Kumar and Jaipal (1991).

$$\text{Sterility (\%)} = \frac{\text{Insects actually failed to mate}}{\text{Total pairs enclosed for mating}} \times 100$$

$$\text{Fecundity (\%)} = \frac{\text{No. of eggs laid by treated females}}{\text{No. of eggs laid by control female}} \times 100$$

$$\text{Hatchability (\%)} = \frac{\text{No. of eggs successfully hatched}}{\text{No. of eggs kept for incubation}} \times 100$$

Haemolymph collection: The haemolymph was collected from both male female and insects, 3 days after treatment, by cutting the thoracic leg and draining the haemolymph into an Eppendorf tube separately with a sufficient quantity of phenylthiourea and stored at -20°C for future use.

Protein Estimation: Total haemolymph protein content was quantitatively estimated following the method of Lowry *et al.* (1951) using phenol - reagent of Folin–Ciocalteau. The optical density was measured at 620nm. Bovine serum albumin was used as standard.

Results

Bioassay: The LD_{50} value of *C. gigantea* leaf extract to *D. cingulatus* was 21µg/insect at 24 h; and it decreased to 18, 15, and 0.7 µg/insect at 48, 72 and 96 h treatment period respectively. The percentage of mortality increased when the dose level was increased. For e.g., *C. gigantea caused* 5% mortality at a dose of 21 µg/insect for 24 h, when the dose was increased to 25, 38 and 93 µg/insect the mortality rate increased to 10, 16 and 50%, respectively (Table 1). Out of four plant extracts, the highest toxicity was observed in *C. gigantea* based on LD_{50} followed by *T. procumbens, A. squamosa* and *C. roseus* against *D. cingulatus*. The *C. gigantea* leave extract, at a dose of 30 µg/insect (= 3%), caused 100% insect mortality in four days.

Effects on reproductive potential: Among the four plant extracts, *C. gigantea* significantly (P <0.05) affected the fecundity, hatchability and sterility at maximum level; which is followed by *T. procumbens, A. squamosa and C. roseus* (Table 2).

Effects on haemolymph proteins: The data obtained from the sub lethal effect of plant leaf extracts at LD_{16} value on the haemolymph protein content of *D. cingulatus* are presented in Table 3. The data on percentage decrease of haemolymph protein over control three days after topical application in male and female are presented in Table 4. The results indicate that the haemolymph protein was found significantly low in bugs treated with leaf extract of *C. gigantea*. Similar trend was observed in decrease of haemolymph protein over control in the treatment with leaf extract of *C. gigantea.*

Table 1. Lethal Doses of plant leaf extracts (μg/insect) against *D. cingulatus*

Plant leaf extracts used	LD_5	LD_{10}	LD_{16}	LD_{50}
A. squamosa at 24 h	550	625	859	1854
A. squamosa at 48 h	405	563	729	1693
A. squamosa at 72 h	342	419	634	1420
A. squamosa at 96 h	270	372	479	495
C. gigantea at 24 h	21	25	38	93
C. gigantea at 48 h	18	22	26	78
C. gigantea at 72 h	15	3.0	5	27
C. gigantea at 96 h	0.7	1.2	2.0	11
C. roseus at 72 h	168	354	789	1950
C. roseus at 96 h	36	86	170	1798
T. procumbens at 24 h	405	517	628	1231
T. procumbens at 48 h	189	282	386	1052
T. procumbens at 72 h	55	78	114	704
T. procumbens at 96 h	41	66	96	354

(1 μg substance in 1 μl solvent = 0.1% i.e. 10 μg/ μl = 1. 0%)

Table 2. Effect of plant leaf extracts at 72h LD_{16} on the reproductive potential of *D. cingulatus*

Plant leaf extracts	Eggs laid Mean Number	Fecundity (%)	Hatchability (%)	Sterility (%)
Control	123 ± 5	100	97	3
A. squamosa	27 ± 3*	22*	32*	64*
C. gigantea	15 ± 2*	12*	16*	84*
C. roseus	29 ± 2*	24*	35*	60*
T. procumbens	22 ± 1*	18*	28*	72*

*Significantly different when compared to respective control ('t' test at 5%level)

Table 3. Effect of plant leaf extracts at 72h LD_{16} on haemolymph protein content (mg/ml ± SD) of *D. cingulatus,* three days after topical application

Sex	Control	*A. squamosa*	*C. gigantea*	*C. roseus*	*T. procumbens*
Male	20.5±0.52	13.5*± 0.36	9.4*± 0.64	14.5*± 0.2	10.6*± 0.27
Female	49.6± 0.88	17.34*± 0.75	12.5*± 0.83	25.4*± 1.8	18.5*± 0.74

*'t' (5% level) is significantly different between control and treated

Table 4. Percentage decrease of haemolymph protein of *D. cingulatus* over control, three days after topical application of plant leaf extracts.

Sex	*A. squamosa*	*C. gigantea*	*C. roseus*	*T. procumbens*
Male	34.15	54.14	29.27	48.29
Female	65.04	74.79	48.75	62.7

Discussion

The insecticidal activity varied with plant species against *D. cingulatus* and each plant extract was positively correlated with its dose. The above extracts might have affected the ovarian development of *D. cingulatus* as evidenced from the drop in the fecundity of the treated insects. Deformities in the ovaries of *D. cingulatus* and *D. koenigii* due to application of extracts of plants possessing juvenomimetic activity have been reported by Prabhu and John (1975); Saxena and Mathur (1976), respectively. After topical application of plant extract viz., *Cassia fistula* and *Eucalyptus globulus,* there was reduction in the number of ovarioles and mature eggs, loss of ovarian weight, small follicular epithelium and resorbed oocyte in the female of *D. cingulatus* and there was agglutination, malformation and immobility of sperm in the male, resulted in reduced fecundity and no viability of eggs (Verma 2004). In the present study, the hatchability of the eggs laid by the treated *D. cingulatus* with plant extracts tested was significantly reduced. The fecundity and fertility of the *D. cingulatus* might be disturbed in multiple ways, that is either by a direct effect on the germ cells or by indirect hormonal disruption, whether of neuro -secretion or of the juvenile or the moulting hormone.

The percentage of haemolymph protein reduction over control varied with plant extracts. Decrease in haemolymph protein could be attributed to reduction of protein synthesis in treated insect by deranging protein synthetic machinery (Bhagwan *et al.,* 1995; Boreddy *et al.,* 2000). This could be attributed to an adaptation of insect to overcome the phytochemical stress, the insect utilizes energy derived from the breakdown of protein circulating in haemolymph and is in conformity with the finding of Bhola and Srivastava (1986) and Anitha *et al.* (1999). This decrease may be due to increased neuromuscular activity of treated insect resulted in higher demands for energy (Jadhav 2005). It can be concluded that the decline was due to higher metabolic activity and imbalance between the rates of anabolism and catabolism in the plant extract treated *D. cingulatus*.

In conclusion, it can be stated that among the selected plant leaf extracts *C. gigantean* significantly inhibited the reproductive potential at sublethal concentration followed by the extracts of *T. procumbens, A. squamosa* and *C. roseus.*

Acknowledgements

This work was supported by the UGC-SERO, Hyderabad to SG (MRP/F. No 706/05/ L.No.1706/Feb. 05). The authors are grateful to the HOD, the Principal and the Management of Yadava College, Madurai for providing the necessary facilities.

References

Anitha, B., Arivazhagan, M., Nalini Sundari, M.S. and Durairaj, G., 1999. Effect of Alkaloid abrine, isolated from *Abrus precatorius* Linn. seeds on mealy bug *Maconellicoccus hirsutus* Green. *India J. Exp. Biol.*, 37, 415-17.

Bhagawan, C.N., Grover, P., Hameed, A. and Sukumar, K., 1995. Developmental defects protein reduction caused by *Ailanthus* extracts in red cotton bug *Dysdercus koenigii*. *Indian J. Exp. Biol.*, 33, 287- 290.

Bhola, R. K. and Srivastava, K.P., 1986. Electrophoretic analysis of the proteins of haemolymph, fat body and ovaries in the red cotton bug, *Dysdercus koenigii* (Heteroptera: Pyrrhocoridae) during the first egg cycle. In *Recent Advances in Insect Physiology, Morphology and Ecology* (Eds. S. C. Pathak and Y. N. Sahai), Today and tomorrow's Printers and publishers, pp. 71-80.

Boreddy, Y., Chitra, K.C. and Eswar Reddy, N.P., 2000. Studies on sublethal concentration (LC_{30}) of *Annona* seed extract on total proteins of *Spodoptera litura* Fab. *Entomon*, 25, 351-355.

Jadhav, S., 2005. Effect of azadirachtin on total free amino acids in the haemolymph of larva of *Corcyra cephalonica* St. *Entomon*, 30, 231-234.

Kumar, Y. and Jaipal, S., 1991. Evaluation of some JH analogues in deterrance of morphogenesis of *Dysdercus koenigii* F. *J. Insect Sci.*, 4(1), 23-26.

Kumar, R., Srivastava, M. and Dubey, N.K., 2007. Evaluation of *Cymbopogon martinii* oil extract for control of post harvest insect deterioration in cereals and legumes. *J. Food Prot.*, 70(1), 172-178.

Kohno, K. and Bui Thi, N., 2004. Effects of host plant species on the development of *Dysdercus cingulatus* (Heteroptera: Pyrrhocoridae). *App. Entomol. Zool.*, 39, 183-187.

Lowry, O.H., Rosenbrough, N.J., Farr, A.L. and Randall, R.J., 1951. Protein measurement with the folin phenol reagent. *Journal of Biological chemistry*, 193, 265-275.

Magdum, S., Banerjee, S., Kalena, G.P. and Banerjee, A., 2001. Chemosterilant activity of naturally occurring quinines and their analogues in the red cotton bug *Dysdercus koenigii*. *J. App. Entomol.*, 125, 589.

Prabhu, V.K.K. and John, M., 1975. Ovarian development in juvenilised adult *Dysdercus cingulatus* affected by some plant extracts. *Ent. Exp and appl.*, 18, 87-95.

Rao, R.V.S. and Gujar, G.T., 1995. Toxicity of Plumbagin and Juglone to the cotton stainer *Dysdercus koenigii*. *Ent. Expt. et. Appl.*, 77, 189-192.

Satyanarayana, K. and Kumuda Sukumar., 1988. Phytosterilants to red cotton bug *Dysdercus cingulatus* F. *Current Sci.*, 57, 918-919.

Saxena, B.P. and Mathur A.C., 1976. Loss of fecundity in *Dysdercus cingulatus* F. due to vapours of *Acorus calamus* L. *Experientia*, 32, 315.

Singh, R., 2004. Pests of cotton. In *Elements of Entomology* (Ed. G. C. Sachan), Rastogi publication, Meerut, India, pp. 331-332.

Vasantharaj David, B. and Kumaraswami, T., 1996. *Elements of economic entomology*. Popular book depot, Madras, pp. 103.

Verma, A., 2004. Effect of certain plant extracts on gonodal maturation of *Dysdercus koenigii*. *J. Exp. Zool. India*, 7, 13-20.

Effect of *Pseudomonas fluorescens* and Plant Products as Seed Treatment against the Leafhopper, *Amrasca devastans* (Distant) in Cotton

N. Murugesan,* N. Shunmugavalli* and A. Kavitha*

The present investigation was conducted to evaluate *Pseudomonas fluorescens* and neem oil along with eight synthetic insecticides as seed treatments against *Amrasca devastans* in cotton. Acephate 75 SP, *Pseudomonas fluorescens,* carbosulfan 25 DS, carbosulfan 25 EC, dimethoate 30 EC, ethofenprox 10 EC, imidacloprid 17.8 SL, monocrotophos 36 SL, neem oil and phosalone 35 EC were the treatments. An untreated check was also maintained. The dose was 10 ml or g per kg of seeds.

Of the 10 seed treatments evaluated, imidacloprid, monocrotophos and *P. fluorescens* were found to be effective in reducing the leafhopper population by more than 50 per cent. Imidacloprid was found to be the most effective treatment recording the least population of 0.8/3 leaves and was followed by monocrotophos (1.23/3 leaves) that was on a par with *P. fluorescens* (1.42/3 leaves). All other treatments were unable to reduce the leafhopper population by less than 50 per cent. In the On Farm Trial conducted at Thirupanikarisalkulam, all the seed treatments were able to reduce the leafhopper population. Imidacloprid was found to be the most effective one recording the least mean population of leafhoppers (0.53 /3 leaves). Imidacloprid and monocrotophos were able to reduce the leafhopper population by 72.54 and 59.59 per cent respectively. Other treatments *viz.,* acephate, *P. fluorescens*, phosalone, ethofenprox, dimethoate, neem oil, carbosulfan EC and carbosulfan DS resulted in less than 50 per cent reduction in leafhopper population compared to untreated check. Laboratory studies have shown that imidacloprid, monocrotophos and *P. fluorescens* improved germination and increased the shoot length. Neem oil had adverse effect on shoot length.

Introduction

In India, 45 percent of the pesticides are applied (David, 2008) in cotton alone. Pesticide load in crop ecosystem has culminated in many undesirable effects such as resistance, resurgence, residues *etc.* , disturbing the agro-ecosystem. Sprays and soil application of pesticides are costly and cumbersome to adopt. So it is imperative to find out an ecofriendly and need based use of chemical pesticides as a component of integrated pest management. Seed treatment is an easy, economic and feasible method for pest control (Murugesan. and Annakkodi. 2007). It protects against insect pests and is ecofriendly to bio control agents like coccinellids and chrysopids under field condition (Murugesan. and Annakkodi. 2007). Hence, the aim of the present study was to evaluate seed treatment with antagonistic organisms, botanicals and different insecticides, against the leafhopper, *Amrasca devastans* Distant in cotton.

Methodology

The efficacy of seed treatment with *Pseudomonas fluorescens,* neem oil and synthetic insecticides (Table 1) was evaluated in field experiments laid out in Randomised Block Design with eleven

* Cotton Research Station, TNAU, Srivilliputtur - 626 125

treatments – at Agricultural College and Research Institute (TNAU) Killikulam- as detailed below and were replicated thrice. The treatments were,

Table 1 Treatments

Sl.No	Common Name	Insecticide / Botanical name	Dose/Conc. (kg^{-1} of seed)
1.	Acephate (Asataf 75 SP)	O, S- dimethylacetyl phosphoroamidothioate	10 g
2.	*Pseudomonas fluorescens* (*Pf* 1)	*Pseudomonas fluorescens* -	10 g
3.	Carbosulfan (Marshal 25DS)	2,3- dihydro-2,2 – dimethyl-7- benzofuran-7-yl (dibutylaminothio) methylcarbamate	10 g
4.	Carbosulfan (Marshal 25EC)	2,3-dihydro-2,2-dimethyl-7-benzofuran –7-yl (dibutylaminothio) methylcarbamate	10 ml
5.	Dimethoate (Rogor 30 EC)	0,0-dimethyl 1 S (N methyl carbomoyl methyl) phosphorodithioate	10 ml
6.	Ethofenprox (Nukil (10EC)	2-(4-ethoxyphenyl)-2-methylpropyl 3-phenoxybenzyl ether	10 ml
7.	Imidacloprid (Confidor 17.8 SL)	1-(6-chloronicotinyl)-2 nitroiminoimidazolidine	10 ml
8.	Monocrotophos (Nuvacron 36SL)	Dimethyl (E)-1-methyl-2-(methyl carbamoyl) vinyl phosphate	10 ml
9.	Neem oil	*Azadirachta indica* A. Juss.	10 ml
10.	Phosalone (Zolone 35 EC)	S-6-Choloro-2,3 dihydro-2-oxobenzoxazolin-3 yl methyl 0,0-diethyl phosphorodithioate	10 ml
11.	Untreated check	-	-

The acid delinted (concentrated sulphuric acid @ 100 ml kg^{-1} of seed) seeds were used for the experiments. To treat one kg of seed 0.5 g of *Acacia* gum powder and 20 ml of water were used. Gum was dissolved in water and mixed with the stipulated quantity of insecticides / plant products / antagonistic organisms. The seeds were thoroughly mixed with gum + insecticide mixture, dried under shade and kept for 24 hours before sowing. Untreated acid delinted seeds served as untreated check (UTC). Population of nymphs of *A. devastans* was recorded 10 DAS at ten days interval on ten randomly selected plants in each plot. In each plant three leaves - one each from top, middle and bottom strata- were observed and mean population per three leaves was worked out. The effect of seed treatments on seedling growth parameters was studied with Paper Towel method.

Statistical Analysis

The data gathered were transformed into angular or square-root values for statistical scrutiny, wherever necessary (Gomez and Gomez, 1984). The experiments were subjected to statistical scrutiny following the method of Panse and Sukhatme (1989) and Gomez and Gomez (1984) and the means were compared with Least Significant Difference (L.S.D.).

Results

The results of the field and laboratory experiments conducted with seed treatment with *Pseudomonas fluorescens,* neem oil and synthetic insecticides are presented below (Tables 2 and 3).

Trial in Research Farm

Incidence of *A. devastans:* Of the 10 seed treatments evaluated, imidacloprid, monocrotophos and *P. fluorescens* were found to be effective in reducing the leafhopper population by more than 50 per cent. Imidacloprid was found to be the most effective treatment recording the least population of 0.8/3 leaves and was followed by monocrotophos (1.23/3 leaves), which was on a par with *P. fluorescens* (1.42/3 leaves) (Table 2). All other treatments were able to reduce the leafhopper population by less than 50 per cent only.

Table 2. Influence of seed treatment with different insecticides on the incidence of *devastans*

Treatment	Conc. (%)	Leafhopper population (No./ 3leaves)	Per cent reduction over untreated check
Acephate 75 SP	10gm	1.59 (1.39) cd	46.64
Pseudomonas fluorescens	10gm	1.42 (1.34) bc	53.34
Carbosulfan 25DS	10gm	1.88 (1.49) de	36.91
Carbosulfan 25EC	10ml	1.88 (1.48) de	36.91
Dimethoate 30 EC	10ml	1.82 (1.46) cde	38.93
Ethofenprox 10EC	10ml	2.02 (1.51) de	32.21
Imidacloprid 17.8 SL	10ml	0.80 (1.10) a	73.15
Monocrotophos 36SL	10ml	1.23 (1.26) b	57.72
Neem oil	10ml	2.10 (1.56) e	29.53
Phosalone 35 EC	10ml	1.95 (1.48) de	34.56
Untreated check	-	2.98 (1.79) f	-
Mean	-	1.79 (1.44)	-
Significance		0.01	
CD (p=0.05)		0.12	

Figures in parenthesis are square root ($\sqrt{x} + 0.5$) transformed values. In a column, means followed by a common letter are not significantly different at 5% level (LSD).

On Farm Trial (Farmer's holding)

Incidence of A. devastans: In the on farm trial all the seed treatments were able to reduce the leafhopper population (Table 3). Imidacloprid was found to be the most effective recording the least mean population 0f 0.53 per three leaves. Imidacloprid and monocrotophos were able to reduce the leafhopper population by 72.54 and 59.59 per cent respectively. Other treatments *viz.,* acephate (1.01/3 leaves), *P. fluorescens* (1.05/3 leaves), phosalone (1.19/3 leaves), ethofenprox (1.24/3 leaves), dimethoate (1.28/3 leaves, neem oil (1.34/3 leaves), carbosulfan EC (1.32/3 leaves) and carbosulfan DS (1.57/3 leaves) resulted in less than 50 per cent reduction in leafhopper population compared to untreated check (1.93/3 leaves).

Germination and seedling growth parameters (Laboratory study): Influence of seed treatments on germination, shoot length and root length was evident (Table 4).

Germination: Imidacloprid recorded the highest germination (82.42%) and it was followed by monocrotophos (74.36%), which was on a par with *P. fluorescens* (69.11%). Acephate (63.40%) was on a par with *P. fluorescens.* Phosalone (56.74%), carbosulfan DS (51.11%), dimethoate (48.88%), carbosulfan EC (42.22%), ethofenprox (42.22%) and neem oil (42.97%) neither affected nor improved germination.

Table 3. Influence of seed treatment with different insecticides on the incidence of *devastans* – (On Farm Trial)

Treatment	Conc. (%)	Leafhopper population (No./ 3leaves)	Per cent reduction over untreated check
Acephate 75 SP	10gm	1.01 (1.19) c	47.67
Pseudomonas fluorescens	10gm	1.05(1.20) cd	45.60
Carbosulfan 25DS	10gm	1.57 (1.38) f	18.65
Carbosulfan 25EC	10ml	1.32 (1.31) e	31.61
Dimethoate 30 EC	10ml	1.28 (1.28) e	33.68
Ethofenprox 10EC	10ml	1.24 (1.27) e	35.75
Imidacloprid 17.8 SL	10ml	0.53 (0.99) a	72.54
Monocrotophos 36SL	10ml	0.78 (1.10) b	59.59
Neem oil	10ml	1.34 (1.29) e	30.57
Phosalone 35 EC	10ml	1.19 (1.26) de	38.34
Untreated check	-	1.93 (1.50) g	-
Mean	-	1.20 (1.25)	-
Significance		0.01	
CD (p=0.05 %)		(0.06)	

Figures in parenthesis are square root ($\sqrt{x} + 0.5$) transformed values. In a column, means followed by a common letter are not significantly different at 5% level (LSD).

Shoot length: Imidacloprid (17.40 cm) was the most effective treatment resulting in the longest shoots; monocrotophos (16.44cm) was the next best treatment; however, it was on a par with acephate (15.66cm). *P. fluorescens* (15.15cm) equaled acephate and dimethoate (14.35cm). Dimethoate, phosalone (13.86cm), carbosulfan DS (13.30cm), carbosulfan EC (13.37cm) and ethofenprox (13.17cm) neither affected nor improved shoot length compared to untreated check (13.62cm). Neem oil reduced the shoot length by 21.59 per cent.

Root length: Roots were longer with imidacloprid (17.12cm) and monocrotophos (16.53cm) seed treatments. Acephate (14.53 cm) and *P. fluorescens* (14.38cm) were the next best ones which were better than dimethoate (13.11cm) and carbosulfan DS (12.68cm) that were equal among themselves. Seed treatment with carbosulfan DS (12.68cm), phosalone (11.70cm), carbosulfan EC (11.39cm) and ethofenprox (11.58cm) neither improved nor affected germination.

Discussion

Seed treatment with insecticides and plant products to manage crop pests is an alternative approach to minimize pesticide hazards. It has advantages such as, easy application, low cost, less pollution, selectivity and least interference in the natural equilibrium over soil or foliar application. Several earlier workers also reported better growth of the plants of imidacloprid 17.8 SL treated seeds in cotton (Dandale *et al.*, 2001; Gupta and Roshan Lal, 1998). Neem oil reduced the shoot length by 21.59 per cent. Such adverse effect on the interference of seed treatment on germination and seedling growth (Mitra *et al.*, 1970; Das and Chandrika , 1972; Murugesan. and Annakkodi. 2007) also available. The present study, on the effectiveness of seed treatment with imidacloprid, gains support from earlier studies (Dandale *et al.*, 2001; Karabhantanal et al., 2001) in cotton. Soil *Pseudomonads* are excellent biological control agents against diseases caused by soil borne plant pathogenic fungi and phyllosphere organisms. The mechanism of plant disease control by *P. fluorescens* includes production of antibiotics, siderophores, volatile compounds like HCN and ammonia, induction of systemic resistance and competition for nutrients (Muthusamy, 1999). Talc based *P. fluorescens*

Table 4 Influence of seed treatment on germination and seedling growth parameters (laboratory study)

Treatment	Dose (kg^{-1})	Germination (%)	+ /- over untreated check (%)	Shoot length (cm)	+ /- over untreated check (%)	Root length (cm)	+ /- over untreated check (%)
Acephate 75 SP	10gm	63.40(52.79)cd	30.97	15.66bc	14.98	14.53b	22.72
P. fluorescens	10gm	69.11(56.26)bc	42.78	15.15cd	11.23	14.38b	21.45
Carbosulfan 25DS	10gm	51.11(45.64)efg	5.58	13.30f	-2.35	12.68cde	7.10
Carbosulfan 25EC	10ml	42.22(40.88)g	-12.79	13.17f	-3.30	11.39e	-3.80
Dimethoate 30 EC	10ml	48.88(44.36)efg	0.97	14.35de	5.36	13.11c	10.73
Ethofenprox 10EC	10ml	42.22(40.88)g	12.79	13.17f	-3.80	11.58e	-2.20
Imidacloprid 17.8 SL	10ml	82.42(65.31)a	70.25	17.40a	27.75	17.12a	44.60
Monocrotophos 36 WSC	10ml	74.36(59.59)b	53.61	16.44b	20.71	16.53a	39.61
Neem oil	10ml	42.97(40.95)fg	-11.24	10.68g	-21.59	11.46e	-3.21
Phosalone 35 EC	10ml	56.74(48.93)de	17.21	13.86e	1.76	11.70e	-1.18
Untreated check		48.41(44.09)efg	-	13.62ef	-	11.84e	-
Mean		56.53 (48.99)	-	14.23	-	13.30	-
Significance		0.01	-	0.01	-	0.01	-
CD (P=0.05)		5.13	-	0.84	-	1.05	-

Figures in parenthesis are arcsin-transformed values. In a column, means followed by a common letter are not significantly different at 5% level (LSD).

formulation is recommended in seed treatment (@10g kg ha^{-1}) for the control of wilt of chickpea, red gram and blast of rice in Tamil Nadu (Swamiappan, 1999). Foliar spray of talc-based product at 0.2 per cent (2 g lit^{-1}) is recommended for the control of blast of rice (Muthaiyan, 2000). The present study brought out the effectiveness of *P. flourescens* as seed treatment, probably for the first time, against *A. devastans* on cotton.

References

David, B.V., 2008. Biotechnological approaches in IPM and their impact on environment. *J. Biopestici.,* 1(1),1-5.

Dandale, H.G., Thakare, A.Y., Tikar, S.N., Rao, N.G.V. and Nimbalkar S.A., 2001. Effect of seed treatment on sucking pests of cotton and yield of seed cotton. *Pestology*, 25(3), 20-23.

Das, N.M., and Chandrika, S., 1972. Effect of seed treatment with systemic insecticides on the germination of paddy and the growth of seedlings. *Agric. Res. J. Kerala*, 10, 152-156.

Gomez, K.A. and Gomez A.A., 1984. Statistical Procedures for Agricultural Research. Wiley-Interscience Publication,. John Wiley and Sons, New York. 680 p.

Gupta, G.P. and Roshan Lal, 1998. Utilization of newer insecticides and neem in cotton pest management system. *Ann. Pl. Prot. Sci.*, 6(2), 155-160.

Karabhantanal, S.S., Bheemanna, M. and Somashekhar, 2001. Combating insect pests of rainfed cotton through integrated pest management practices. *Pestology*, 25(8), 7-9.

Mitra , D.K., Raychaudhuri, S.P., Everett, T.R.,Ghosh A. and Niazi, F.R., 1970. Control of the rice green leafhopper with insecticidal seed treatment and pre- transplant seedling soak. *J. Econ. Ent.,* 63, 1958-1961.

Murugesan, N. and Annakodi, P., 2007. Seed Treatment with Insecticides, botanicals and antagonistic organisms against the leafhopper, Amrasca *devastans* (Distant). pp. 183-188 In: (Ed: S.Baskaran) Proc. National Seminar on Applied Zoology, Ayya Nadar Janaki Ammal College, Sivakasi.

Muthaiyan, M.C., 2000. Biocontrol of rice blast disease with *Pseudomonas fluorescens*. *Pestology*, 24(3), 20-23.

Muthusamy, M., 1999. Biofungicides- effective tools for the management of plant diseases. *Pestology* Spl. Issue, 185-193.

Panse, V.G and Sukhatme, P.V., 1989. Statistical Methods for Agricultural Workers. Indian Council for Agricultural Research, New Delhi. 359 p.

Swamiappan, R., 1999. Application technology for *Pseudomonas fluorescens.* In: Mass Multiplication of Bio-control agents. Dhandapani, N., P. Devasenapathy, M.V.Ranghaswami and J.Oliver (Eds.), Director of Extension Education, Tamil Nadu Agri. Univ. Coimbatore. p59.

Evaluation of Bovine Organics and Botanicals against pests and Non-target Organisms in Bhendi Crop

N. MURUGESAN* AND M. DHANAVENDHAN*

The efficacy of bovine organics *viz.*, cow dung extract (CDE) 5%, cow urine (CU) 5% and buttermilk (BM) 2% against pests of bhendi and their impact on the non-target organisms such as coccinellids and honey bees on bhendi crop were studied. They could reduce incidence of leafhoppers by sole treatment. Spraying with CDE + CU + BM + imidacloprid was effective against fruit borer. Cow dung extract spray against fruit borer was economically viable with higher cost benefit ratio (1: 3.6) than other treatments. Cost benefit ratios of cow dung extract, cow urine, imidacloprid, CDE + CU + BM and CDE + CU + BM + imidacloprid were 1:3.6, 1:2.5, 1:0.7, 1:0.7 and 1:0.3 respectively. Bovine organics had no influence on phyllosphere bacteria. BM spray enhanced yeast colonization. CU spray reduced fungi and actinomycetes. Neem oil affected the fungi and actinomycetes. Spray mix of BM and CDE enhanced the colonies of actinomycetes. Bovine organics were safer to the coccinellids natural enemies and honey bees.

INTRODUCTION

Cow plays a vital role in the life of farmers at the time of auspicious functions. Cow urine solution is sprayed in the premises while performing poojas, especially during house warming ceremony *etc*. Cow dung slurry is sprinkled on the soil in front of the houses to-ward off harmful organisms particularly germs. The cow excretion is well known for its antiseptic properties. Few reports on utility of *Vrkshayurveda* (Traditional Plant Science) recipes like *Panchagavya*, *Dashagavya* are available (Vijayalakshmi and. Shayam Sundar, 1994). Just like plant products, the cow excretion also possesses antifungal properties (Sundarraj *et al.*, 1996) and antiviral principles (Kurucheve, 1989). It's efficacy against insect pests is little known. Hence, this present study was undertaken to investigate the influence of bovine organics and botanicals against the pests of bhendi. The key pests of bhendi are the fruit borer (*Earias* spp.), the leafhopper (*Amrasca devastans*), the aphid (*Aphis gossypii*) and red spider mite (*Tetranychus cinnabarinus*).

METHODOLOGY

The field experiments were conducted at Agricultural College and Research Institute, Killikulam. Bhendi variety, *Arka Anamica* was raised at 45 × 30 cm spacing. The experiments were laid out in Randomized Block Design. The preparations were sprayed at drenching level with hand operated Knapsack sprayer. The spray fluid used ranged from 500-750 lit. ha^{-1} depending on the crop growth period. The treatments evaluated against pests, coccinellid beetles and honeybees were: **1.** CDE (5%) **2.** CU (5%) **3.** Imidacloprid 17.8 SL (0.004%) **4.** CDE(5%) + CU(5%) + BM (1.0%) **5.** CDE + CU + BM + imidacloprid and **6.** Untreated check.

Bovine organics Cow dung (5%) or cow urine (5%) extract was prepared by mixing stipulated quantity cow dung (50 gm) or cow urine (50 ml) with a litre of water. Butter milk 5 per cent was

* Cotton Research Station, Tamil Nadu Agricultural University, Srivilliputtur- 626 125.

prepared by mixing 10 ml of butter milk with a litre of water.

The details of observations recorded are presented hereunder.

A. devastans , *A. gossypii* and *T. cinnabarinus:* The population was recorded in five randomly selected plants, five days after spray.

***Earias* spp.:** At each harvest, the healthy and infested fruits were sorted out. The number and weight were recorded. The percentage of infested fruits in terms of number and weight was calculated.

The treatments evaluated against phyllosphere microflora were CDE, CU, Buttermilk, CDE + CU, CDE + BM, CU + BM, CDE + CU + BM, neem oil, imidacloprid ,water spray and untreated check.

Coccinellids: Observations on number of coccinellid beetles (adults) were recorded on five randomly selected plants in each plot at five days after spray.

Honeybee visit: Observations on number of bees visited per minute was recorded at 9.00 AM on five randomly selected plants in each plot.

Phyllosphere Microflora: The leaf blotting (or) Imprint technique was adopted to perform analysis of microbial population on leaf surface. Leaf bits of 1 × 1 cm were prepared using cork borer. One day after spray the upper and lower surface of leaf bits were pressed separately on appropriated medium in Petridish as detailed hereunder. The appropriate mediums were employed (Anonymous, 1957) for bacteria, fungi, actinomycetes respectively. The observation was recorded 24 hrs after the spray.

Yield: Yield in terms of kg/plot was recorded. Yield per hectare also worked out.

Cost Benefit Ratio: Cost benefit ratio was worked out (Heinrichs *et.al.*,1981) as detailed hereunder.

$$\text{Cost Benefit Ratio} = \frac{\text{Value of treated crop - Value of untreated check}}{\text{Cost of plant protection}}$$

Results

Leafhoppers

The influence of treatments as well as periods of observation on leafhopper population was significant. The interaction effect was not evident (Table 1). The leafhopper population varied from 1.73 (CDE + CU + BM + imidacloprid at I spray) to 8.23/leaf (UTC at I spray). Reckoning the overall mean, the CDE + CU + BM + imidacloprid (1.79/leaf) and imidacloprid (1.90 /leaf) reduced the population by 76.66 and 75.23 per cent respectively over untreated check (7.67/leaf). The next best were, CDE (3.74) and CDE + CU + BM (4.01/leaf), which was on a par with former one. The sole effect of bovine organics could reduce incidence by 51.34 (CDE) and 34.42 (CU) per cent only.

Aphids

The effect of treatments as well as observation at different spray was evident. The interaction effect was not obvious (Table 2). The aphid population ranged from 1.97 (imidacloprid at I spray) to 6.94/ cm^2 (CDE at II spray). Considering the sprays together, the overall mean showed that imidacloprid alone (2.04/cm^2) or CDE + CU + BM + imidacloprid (2.27/cm^2) reduced the aphid population by 64.89 and 60.93 per cent respectively. The other treatments neither increased nor decreased the aphid population.

T. cinnabarinus

The variability in mite population due to treatments was evident. The effect of different sprays was significant; interaction effect was not significant (Table 3). The mite population varied from 8.97 (imidacloprid at I spray) to 40.53/cm2 (UTC at II spray). Reckoning the sprays together, overall

Table 1. Effect of bovine organics on the incidence of *A. devastans*

Treatment	Dose (%)	Population of leafhopper (No/ leaf) 5 days after spray I	II	Mean	Per cent + / - over UTC
CDE	5.0	4.28 (2.07)	3.20 (1.79)	3.74 (1.93) b	-51.24
CU	5.0	5.32 (2.31)	4.73 (2.17)	5.03 (2.24) c	-34.42
Imidacloprid 17.8 SL	0.004	1.81 (1.35)	1.98 (1.41)	1.90 (1.38) a	-75.23
CDE + CU + BM	5.0 + 5.0 + 1.0	4.10 (2.02)	3.92 (1.98)	4.01 (2.00) b	-47.72
CDE + CU + BM + imidacloprid 17.8 SL	5.0 + 5.0 + 5.0 + 0.004	1.84 (1.35)	1.73 (1.32)	1.79 (1.33) a	-76.66
Untreated Check (UTC)	--	8.23 (2.87)	7.10 (2.66)	7.67 (2.77) d	--
Mean	--	4.26 A (1.99)	3.78 B (1.89)	4.02 (1.94)	
	Significance	CD (P=0.05)			
Treatment (T)	0.01	0.18			
Period (P)	0.05	0.10			
T× P	NS	--			

Figures in parenthesis are square root transformed values. In a column/row means followed by a common letter are not significantly different at 5% level (LSD).

mean showed that imidacloprid alone (10.05/cm^2) followed by imidacloprid with combined bovine organics (10.80/cm^2) reduced the mite population by 73.52 and 71.54 per cent respectively. The other treatments were also better than untreated check but exhibiting less than fifty per cent reduction in mite population.

In a column/row means followed by a common letter are not significantly different at 5% level (LSD).

Fruit borer infestation

Number basis: The effect of treatments and periods of observation on fruit borer infestation on number basis was significant. The interaction effect was not obvious (Table 4). The infestation varied from 4.17 (imidacloprid at I spray) to 27.65 per cent (untreated check at II spray). Reckoning the overall mean, imidacloprid alone (4.23%) and CDE + CU + BM + imidacloprid (5.20 %) were the effective ones that reduced the infestation by 84.62 and 81.09 per cent respectively over untreated check. They were followed by CU (12.02 %), CDE (13.60 %) and CDE + CU + BM (14.43 %). which were also better than untreated check. The former two were equal among themselves.

Table 2. Effect of bovine organics on the incidence of *A. gossypii*

Treatment	Dose (%)	Aphid population/cm² 5 days after spray I	II	Mean	Per cent + / - over UTC
CDE	5.0	6.20 (2.49)	6.94 (2.63)	6.57 (2.56) b	13.08
CU	5.0	6.23 (2.50)	6.88 (2.62)	6.56 (2.56) b	12.91
Imidacloprid 17.8 SL	0.004	1.97 (1.40)	2.10 (1.45)	2.04 (1.42) a	-64.89
CDE + CU + BM	5.0 + 5.0 + 1.0	5.98 (2.45)	6.37 (2.52)	6.18 (2.48) bc	6.37
CDE + CU + BM + imidacloprid 17.8 SL	5.0 + 5.0 + 5.0 + 0.004	2.01 (1.42)	2.53 (1.59)	2.27 (1.50) a	-60.93
Untreated Check (UTC)	--	5.28 (2.30)	6.33 (2.51)	5.81 (2.41) b	--
Mean	--	4.61 A (2.09)	5.19 B (2.22)	4.90 (2.16)	
	Significance	CD (P=0.05)			
Treatment (T)	0.01	0.15			
Period (P)	0.01	0.07			
T× P	NS	--			

Figures in parenthesis are square root transformed values.

Weight basis: The influence of treatments as well as observations at different spray significantly differed on weight basis. The interaction effect was not evident (Table 4). The per cent infestation varied from 3.80 (CDE + CU + BM + imidacloprid at I spray) to 28.06 per cent (untreated check at II spray). Considering the observation at different spray together, CDE + CU + BM + imidacloprid (4.01 %) and imidacloprid (4.53 %) exhibited higher performance than other treatments. They were followed by CU (12.00 %) and CDE + CU + BM (14.93 %). CDE (18.30 %) was better than untreated check but inferior to former treatments.

Coccinellid beetle: The effect of treatments as well as periods of observation on the population of coccinellid beetles was evident (Table 5). The beetle population varied from 1.60 (CU at II spray) to 2.42/plant (CDE + CU + BM + imidacloprid at II spray). Reckoning the overall mean, imidacloprid alone (2.28/plant) and CDE + CU + BM + imidacloprid (2.26/plant) registered higher beetle number than other treatments, CDE (1.85/plant), CDE + CU + BM (1.81/plant) and CU (1.64/plant). The later ones neither decreased nor increased the coccinellid population.

Honey bee : The effect of treatment on honey bee visits was significant (Table 6). The activity varied from 20.00 (imidacloprid) to 24.00/5 plants/minute (CU). Frequency of bee visit was normal in plots treated with cow urine as well as CDE + CU + BM (25.00/5 plants/minute). Imidacloprid (20.00/5

Table 3. Effect of bovine organics on the incidence of *T. cinnabarinus*

Treatment	Dose (%)	Mite population (No./cm^2) 5 days after spray I	II	Mean	Per cent +/- over UTC
CDE	5.0	30.52 (5.52)	35.00 (5.92)	32.76 (5.72) c	- 13.68
CU	5.0	34.00 (5.83)	38.77 (6.23)	36.39 (6.03) d	- 4.11
Imidacloprid 17.8 SL	0.004	8.97 (2.99)	11.12 (3.34)	10.05 (3.17) a	- 73.52
CDE + CU + BM	5.0 + 5.0 + 1.0	19.53 (4.42)	26.50 (5.15)	23.02 (4.78) b	- 39.34
CDE + CU + BM + imidacloprid 17.8 SL	5.0 + 5.0 + 5.0 + 0.004	9.10 (3.02)	12.49 (3.53)	10.80 (3.27) a	- 71.54
Untreated Check (UTC)	--	35.37 (5.95)	40.53 (6.37)	37.95 (6.16) d	--
Mean	--	22.92 A (4.62)	27.40 B (5.09)	25.16 (4.85)	
	Significance	CD (P=0.05)			
Treatment (T)	0.01	0.13			
Period (P)	0.01	0.08			
T× P	NS	--			

Figures in parenthesis are square root transformed values. In a column/row means followed by a common letter are not significantly different at 5% level (LSD).

plants/min) or imidacloprid with combined bovine organics (21.00/5 plants/minute) reduced the visit by 20.00 and 16.00 per cent respectively over untreated check.

Effect of Bovine Organics on Phyllosphere Microflora

Bacteria : The influence of treatments as well as surfaces was evident; interaction effect on phyllosphere bacteria was also obvious (Table 7). The number of colonies per cm^2 varied from 14.50 (CDE + CU + BM on lower surface) to 83.00 colonies /cm^2 (CDE + CU on upper surface). The upper surface (44.32 colonies/cm^2) harboured more colonies than lower surface (32.62 colonies/cm^2). Reckoning the overall mean, the addition of neem oil to CDE + CU was found to exert influence (89.50/cm^2); *i.e.*, the colonies increased by 141.89 per cent over untreated check (37.00/cm^2). BM and CDE + BM enhanced the number of colonies by 39.86 and 16.89 per cent respectively over untreated check. The other treatments could not exert any influence on phyllosphere bacteria. Similar was the trend at each surface.

Fungi: The variability in number of fungal colonies due to treatments was evident. The effect of surface was not observed. However, the combined influence of treatments and surfaces was apparent (Table 8). The colony density ranged from 3.50 (CU at lower surface) to 58.50 colonies/cm^2 (CDE at upper surface). Both upper surface (18.53 colonies /cm^2) and lower surface (15.03 colonies /cm^2)

Table 4. Effect of bovine organics on fruit borer infestation

Treatment	Dose (%)	Fruit infestation (%) Number basis Spray I	Spray II	Mean	Per cent +/- over UTC	Weight basis Spray I	Spray II	Mean	Per cent +/- over UTC
CDE	5.0	12.12 (3.48)	15.07 (3.86)	13.60 (3.67) bc	- 50.55	17.94 (4.24)	18.65 (4.32)	18.30 (4.28) d	- 33.52
CU	5.0	11.61 (3.41)	12.42 (3.52)	12.02 (3.47) b	- 56.29	11.95 (3.46)	12.05 (3.47)	12.00 (3.46) b	- 56.41
Imidacloprid (17.8 SL)	0.004	4.17 (2.04)	4.29 (2.07)	4.23 (2.06) a	- 84.62	3.90 (1.97)	5.16 (2.27)	4.53 (2.12) a	- 83.54
CDE + CU + BM	5.0 + 5.0 + 1.0	14.15 (3.76)	14.71 (3.83)	14.43 (3.80) c	- 47.53	14.23 (3.77)	15.63 (3.95)	14.93 (3.86) c	- 45.77
CDE + CU + BM + Imidacloprid (17.8 SL)	5.0 + 5.0 + 1.0 + 0.004	5.15 (2.27)	5.25 (2.29)	5.20 (2.28) a	- 81.09	3.80 (1.95)	4.22 (2.05)	4.01 (2.00) a	- 85.43
Untreated check	--	27.35 (5.23)	27.65 (5.26)	27.50 (5.24) d	--	27.00 (5.20)	28.06 (5.30)	27.53 (5.25) e	--
Mean	--	12.43 A (3.36)	13.23 A 93.47)	12.82 (3.42)		13.14 A (3.43)	13.96 B (3.56)	13.55 (3.50)	

Figures in parenthesis are square root transformed values. In a column/row means followed by a common letter are not significantly different at 5% level (LSD).

	Number basis Significance	CD (P=0.05)		Weight basis Significance	CD (P=0.05)
Treatment (T)	0.01	0.25	Treatment (T)	0.01	0.13
Period (P)	NS	--	Period (P)	0.01	0.07
T x P	NS	--	T x P	NS	--

In a column/row means followed by a common letter are not significantly different at 5% level (LSD).

harboured equal number of colonies. Considering upper and lower surfaces together, the overall mean showed that cow dung extract (44.00/cm^2) alone enhanced the fungal colonies by 120 per cent, while CU (4.75/cm^2) and neem oil (10.25/cm^2) reduced the fungal colonies by 76.25 and 48.75 per cent respectively over untreated check. The other treatments could not influence the fungal colonies.

Actinomycetes: The influence of treatments as well as surfaces on the density of actinomycetes colonies, interaction effect on phyllosphere actinomycetes was also obvious (Table 9). The number of colonies varied from 13.50 (CU on lower surface) to 105.00 colonies/cm^2 (CDE + BM on upper surface). The upper surface (47.56 colonies /cm^2) harboured more colonies than lower surface (23.00 colonies/cm^2). Reckoning the overall mean, CDE + BM (73.25 colonies/cm^2) favoured the actinomycetes by enhancing the density by 81.99 per cent over untreated check; on the other hand, the treatments, neem oil (23.00/cm^2) and CU (17.75/cm^2) suppressed the colonies by 42.86 and 55.90 per

Table 5. Effect of bovine organics on coccinellid beetle

Treatment	Dose (%)	Population of Beetle (No/ plant) 5 days after spray I	II	Mean	Per cent + / - over UTC
CDE	5.0	1.74 (1.32)	1.95 (1.40)	1.85 (1.36)c	7.56
CU	5.0	1.68 (1.30)	1.60 (1.26)	1.64 (1.28)e	- 4.65
Imidacloprid 17.8 SL	0.004	2.20 (1.48)	2.35 (1.53)	2.28 (1.51)a	32.56
CDE + CU + BM	5.0 + 5.0 + 1.0	1.78 (1.33)	1.83 (1.35)	1.81 (1.34)cd	5.23
CDE + CU + BM + imidacloprid 17.8 SL	5.0 + 5.0 + 5.0 + 0.004	2.10 (1.45)	2.42 (1.55)	2.26 (1.50)b	31.40
Untreated Check (UTC)	-	1.76 (1.33)	1.68 (1.30)	1.72 (1.31)de	--
Mean		1.88 A (1.38)	1.97 B (1.40)	1.92 (1.38)	
	Significance	CD (P=0.05)			
Treatment (T)	0.01	0.12			
Period (P)	0.01	0.07			
T× P	NS	--			

Figures in parenthesis are square root transformed value. In a column / row, means followed by a common letter are not significantly different at 5% level (LSD).

Table 6. Effect of bovine organics on Honey bee

Treatment	Dose (%)	Visits / 5 plants /min (5 DAS)	Per cent + / - over UTC
CDE	5.0	24.00 (4.90)c	- 4.00
CU	5.0	24.00 (4.90)c	4.00
Imidacloprid 17.8 SL	0.004	20.00 (4.47)e	- 20.00
CDE + CU + BM	5.0 + 5.0 + 1.0	25.00 (5.00)b	0.00
CDE + CU + BM + imidacloprid 17.8 SL	5.0 + 5.0 + 5.0 + 0.004	21.00 (4.58)d	- 16.00
Untreated Check (UTC)	-	25.00 (5.00)b	-
Mean		23.5 (4.84)	
Significance		0.05	
CD (P=0.05)		0.40	

Figures in parenthesis are square root transformed value. In a column / row, means followed by a common letter are not significantly different at 5% level (LSD). DAS- Days after spraying

Table 7. Effect of bovine organics on phyllosphere bacteria

Treatment	Dose (%)	Bacterial colonies / cm²		Mean	% + / - over UTC
		Upper	Lower		
Cow dung extrct (CDE)	5.0	34.50	53.50	44.00	18.91
		(5.87)	(7.29)	(6.58)	
Cow urine (CU)	5.0	21.50	34.50	28.00	-24.32
		(4.61)	5.54)	(5.08)	
Buttermilk (BM)	1.0	38.00	65.50	51.75	39.86
		(6.12)	(8.08)	(7.10)	
CDE + CU	5.0 + 5.0	83.00	21.00	52.20	41.08
		(9.05)	(4.57)	(6.81)	
CDE + BM	5.0 + 1.0	56.50	30.00	43.25	16.89
		(7.52)	(5.36)	(6.44)	
CU + BM	5.0 + 1.0	63.50	17.50	40.50	9.46
		(7.97)	(4.17)	(6.07)	
CDE + CU + BM	5.0 + 5.0 + 1.0	27.50	14.50	21.00	-43.24
		(5.22)	(3.81)	(4.52)	
Neem oil	3.0	37.00	19.00	25.00	-32.43
		(5.91)	(4.32)	(5.12)	
Imidacloprid	0.004	24.50	23.00	23.75	-35.81
		(4.91)	(4.70)	(4.80)	
Water	--	24.00	23.50	23.75	-35.81
		(4.79)	(4.36)	(4.57)	
Untreated check (UTC)	--	41.50	32.50	37.00	--
		(6.44)	(5.59)	(6.02)	
Mean	--	44.32A	32.62B	38.47	
		(6.46)	(5.45)	(5.95)	
	Significance	CD (P=0.05)			
Treatment (T)	0.01	1.61			
Period (P)	0.01	0.55			
T× P	0.05	2.28			

Figures in parenthesis are square root transformed value. In a column / row, means followed by a common letter are not significantly different at % level LSD).

cent respectively when compared to untreated check.. Addition of buttermilk to CDE had a profound influence *i.e.,* the colonies increased by 81.99 per cent. The other treatments were unable to exert a profound influence; the differences are too small to be statistically significant.

Yield and Economics: The performance of all the treatments was better than untreated check. Imidacloprid (6375 kg/plot) recorded higher yield than other treatments (Table10). The other treatments could not exert any profound influence on yield. Spraying with cow dung extract (5.0 %) reaped the higher cost benefit ratio (1:3.6); it was followed by spraying with CU (5.0 %) (1:2.5). Imidacloprid alone (1:0.7) as well as imidacloprid with bovine organics were not economically feasible than others bovine organics; the cost benefit ratios were 1:0.3 to 1:0.7 (Tables 10)

Table 8. Effect of bovine organics on phyllosphere fungi

Treatment	Dose (%)	Fungal colonies / cm² Leaf surface		Mean	Per cent + /- over UTC
		Upper	Lower		
Cow dung extrct (CDE)	5.0	58.50 (7.64)	29.5 (5.43)	44.00 (6.53)	120.00
Cow urine (CU)	5.0	6.00 (2.44)	3.50 (1.83)	4.75 (2.13)	-76.25
Buttermilk (BM)	1.0	22.00 (4.68)	17.00 (4.09)	19.50 (4.39)	-2.50
CDE + CU	5.0 + 5.0	21.50 (4.60)	12.50 (3.52)	17.00 (4.06)	-15.00
CDE + BM	5.0 + 1.0	9.50 (3.08)	17.50 (4.16)	13.50 (3.62)	-32.50
CU + BM	5.0 + 1.0	19.00 (4.35)	14.00 (3.74)	16.50 (4.04)	-17.50
CDE + CU + BM	5.0 + 5.0 + 1.0	38.50 (6.19)	17.00 (4.12)	27.75 (5.15)	38.75
Neem oil	3.0	4.50 (2.12)	16.00 (3.90)	10.25 (3.01)	-48.75
Imidacloprid	0.004	24.50 (4.89)	11.00 (3.22)	17.75 (4.06)	-11.25
Water	--	17.50 (4.18)	11.50 (3.38)	14.50 (3.78)	-27.5
Untreated check (UTC)	--	20.00 (4.47)	20.00 (4.46)	20.00 (4.46)	
Mean		18.53A (4.05)	15.03A (3.75)	16.78 (3.90)	
	Significance	CD (P=0.05)			
Treatment (T)	0.01	1.11			
Surface (S)	NS	—			
T× S	0.01	1.57			

Figures in parenthesis are square root transformed value. In a column / row, means followed by a common letter are not significantly different at 5% level (LSD).

Discussion

Effect on pests and beneficial organisms

CDE, CU, CDE + CU and CDE + CU + BM performed better than untreated check against leafhoppers, mites and fruit borer. Efficacy of imidacloprid against fruit borer was not deterred by the addition of bovine organics. Efficacy of imidacloprid 17.8 SL against *Earias* spp in cotton has earlier been reported (Kavitha, 2000). Considering the number as well as weight of fruit basis, spraying with either cow urine or cow dung extract alone or CDE + CU reduced the fruit infestation. Buttermilk with bovine excretion also controlled fruit borer infestation. Bovine excretions might improved the efficacy of buttermilk through synergistic effect.

Table 9. Effect of bovine organics on phyllosphere actinomycetes

Treatment	Dose (%)	Actinomycetes colonies / cm² Leaf surface Upper	Lower	Mean	Per cent + /- over UTC
Cow dung extrct (CDE)	5.0	48.50 (6.95)	18.00 (4.24)	33.25 (5.59)	-17.39
Cow urine (CU)	5.0	22.00 (4.69)	13.50 (3.66)	17.75 (4.17)	-55.90
Buttermilk (BM)	1.0	44.50 (6.67)	16.00 (3.99)	30.25 (5.33)	-24.84
CDE + CU	5.0 + 5.0	58.00 (7.53)	17.00 (4.12)	37.50 (5.82)	- 6.83
CDE + BM	5.0 + 1.0		41.50 (6.44)	73.25 (8.34)	81.99
CU + BM	5.0 + 1.0	43.00 (6.52)	20.50 (4.52)	31.75 (5.52)	-21.12
CDE + CU + BM	5.0 + 5.0 + 1.0	25.00 (4.97)	35.50 (5.96)	30.25 (5.47)	-24.84
Neem oil	3.0	31.50 (5.53)	14.50 (3.78)	23.00 (4.66)	-42.86
Imidacloprid	0.004	48.00 (6.92)	20.50 (4.52)	34.25 (5.72)	-14.91
Water	--	51.00 (7.14)	28.00 (5.18)	39.50	- 1.86
Untreated check (UTC)	--	51.00 (7.14)	29.50 (5.34)	40.25 (6.29)	
Mean		47.56A (6.76)	23.00B (4.71)	35.28 (5.74)	

	Significance	CD (P=0.05)
Treatment (T)	0.01	0.15
Surface (S)	0.01	0.39
T× S	0.05	1.62

Figures in parentheses are square root transformed value. In a column / row, means followed by a common letter are not significantly different at 5% level (LSD).

Table 10. Influence of bovine organics on bhendi yield and economics

Treatment	Dose (%)	Yield (Kg/ha)	+ /- over UTC	Materials Required /ha	Cost of treatment (Rs/ha)	Increased yield (Kg/ha)	Value of extra yield (Rs/ha)	CB ratio
CDE	5.0	6360 b	0.13	25 Kg CD	25.00	10.00	90.00	1: 3.6
CU	5.0	6364 b	0.26	25 lit CU	50.00	14.00	126.00	1: 2.5
Imidacloprid 17.8 SL	0.004	6375 a	0.39	89 ml	341.00	25.00	225.00	1:0.66
CDE + CU + BM	5.0 + 5.0 + 1.0	6360 b	0.13	25 Kg + 25 lit + 5 lit	125.00	10.00	90.00	1:0.72
CDE + CU + BM + imidacloprid 17.8 SL	5.0 + 5.0 + 1.0 + 0.004	6365 b	0.26	25 Kg + 25 lit + 5 lit + 89 ml	466.00	15.00	135.00	1: 0.3
Untreated Check (UTC)	--	6350 b	--	--	--	--	--	--
CD (P=0.05)		15						

Price / kg = cow dung- Rs. 1; cow urine- Rs. 2; buttermilk- Rs. 10; bhendi- Rs. 9.

Fig.1 Effect on Coccinellids

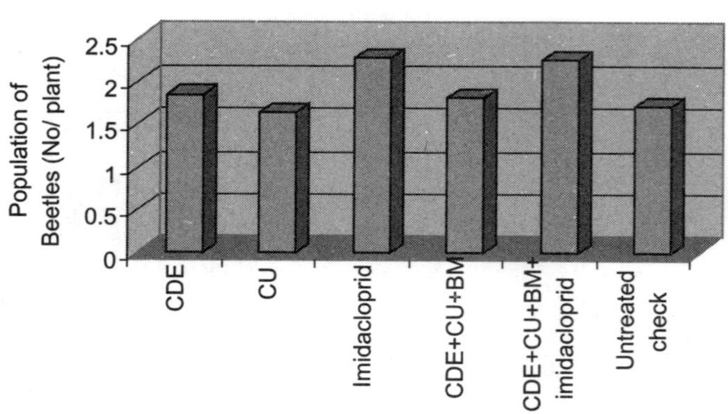

The bovine organics ***viz.,*** cow dung extract, cow urine, buttermilk and milk were the pivotal sources in pest and disease management in the traditional plant science, *Vrkshayurveda* (Vijaylakshmi, *et. al.*,1997)

Lower incidence of *Diabrotica speciosa* has been observed in maize plants applied with cow manure (Vardasa *et.al.* 1989). Reports on the efficacy of different recipes of bovine organics with botanicals were effective against wide range of pests are also available (Baskaran and Narayanasamy, 1995)). Bovine organics with neem oil, garlic extract were earlier reported as effective against cotton and vegetable pests

Population of coccinellid was the highest in imidacloprid treated plants; it may probably prove the safety nature of imidacloprid to coccinellid beetles. Bovine organics were also found to be safer to coccinellid beetles. This is in conformity with earlier findings. from Katole and Patil (2000) who reported that imidacloprid is safer to coccinellids. Bovine organics did not reduce the frequency of bee visits (Fig.2). Imidacloprid either alone or with bovine organics recorded lesser intensity of visits than other treatments. It may be due to the toxicity of imidacloprid to honey bees. This report is in accordance with Mayer *et al.* (1994).

Fig.2 Effect on honey bee visits

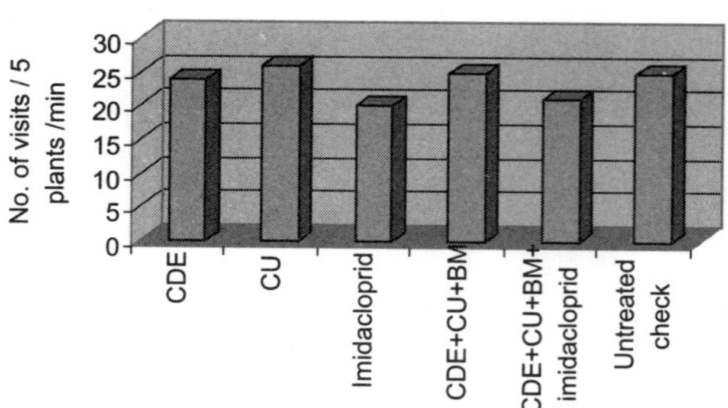

Earlier, adverse effect of imidacloprid has been observed on honeybees, leaf cutting bees and bumble bees in cotton.

Effect of Bovine Organics on Phyllosphere Microflora

Cow dung extracts + cow urine + neem oil recorded more bacterial colonies per cm^2 of leaf surface than all other treatments. Cow urine inhibited bacterial growth but when it mixed with cow dung extract its effect was *vice versa*. This may probably due to the ease of decomposition process for the bacteria; the low C / N ratio of urine and the simple nature of the nitrogenous compounds helps the bacteria to multiply and utilize the cow urine for its nourishment (Gaur *et al.*, 1990). Imidacloprid alone and its combinations with bovine organics had no influence. Earlier Sridhar (2000) that documented that imidacloprid is found to suppress rhizosphere bacteria, *Pseudomonas flourescens*.

Cow urine inhibited the fungal colonies highly. On the otherhand, cow dung extract favoured the fungal colonies by 120.00 per cent. The higher C / N ratio in cow dung might have favoured the fungi which can utilize the cow dung.

The higher number of actinomycetes colonies were found after cow dung extract + buttermilk spray. Cow urine as well as neem oil treatments inhibited the colonies. Imidacloprid recorded higher yield than other treatments. The influence of bovine organics on yield was little; but found to be economically viable. Earlier higher effectiveness of imidacloprid against *Earias* spp. with less cost benefit ratio of 1:1.7 has been recorded in in cotton (Kavitha , 2000) In the present study, cow dung extract spray against fruit borer was found to be economically viable with higher cost benefit ratio (1:3.6) than other treatments. Cost benefit ratios of cow urine, imidacloprid, CDE + CU + BM and CDE + CU + BM + Imidacloprid were, 1:2.5, 1:0.7, 1:0.7 and 1:0.3 respectively.

Bovine organics exhibited better performance than the untreated check in the management of major pests of bhendi. The yield also got enhanced. Eventhough inferior to the synthetic insecticide, imidacloprid-17.8 SL, spraying cow dung extract against fruit borer was economically viable with higher cost benefit ratio (1:3.6) than other treatments.

The present study highlights the influence of bovine organics and botanicals on the non-target organisms prevailed in bhendi eco system. Results from the present investigation can provide a timely lead towards the development and perfection in IPM programme with special emphasis on organic farming for bhendi. This being the backdrop, the investigation and the results are of great interest

This investigation brought out the promising nature of bovine organics, especially cow dung extract (5.0%), cow urine (5.0 %) and CDE(5.0 %) + CU (5.0 %) + BM (1.0 %) in pest management on bhendi. The penultimate bovine organics recipe developed, in this study *viz*., CDE (5.0 %) + CU (5.0 %) + BM (1.0 %), is nomenclated as *ACK-Dheengavya* (ACK-Agricultural College, Killikulam; *Dheengavya*- Three from cow).

References

Anonymous, 1957. Manual of Microbial Methods. Mc.Graw Hill Book Co., New York, U.S.A. 315 p.

Baskaran, V.P. and Narayanasamy, P., 1995. Traditional Pest Control. Caterpillar Publ. Annamalai Nagar. p90.

Gaur, A.C., Neelakandan, S. and Dargan, K.S., 1990. Organic Manures. Publ. and Information Divn., ICAR, New Delhi. 159 p.

Heinrichs, E.A., Chelliah, S., Valencia, S.L., Arcer, M.B., Fabellar, L.T., Aquino, G.B. and Pickin, S., 1981, Manual for Testing Insecticides in Rice. Intl. Rice Res. Instt., , Los Banos, Manila, Philippines, 113p.

Katole, S.R. and Patil, P.J., 2000. Bio-efficacy of imidacloprid and thiamethoxan as seed treatment and foliar spray to some predators. Pestology, 24 (11), 11.13.

Kavitha, 2000. Management of the leafhopper, *Amrasca devastans* (Distant) in cotton. M.Sc.(Ag.), Thesis, Tamil Nadu Agri. Univ., Killikulam

Kurucheve, V., 1989. Studies on antiviral principles for the control of tomato spotted wilt virus. Ph.D.Thesis, Tamil Nadu Agricultural University, Coimbatore.

Mayer, D.F., Pattern, K.D., Macfarlane, R.P. and Shants, C.H., 1994. Difference between susceptibility of four pollinator species (Hymenoptera: Apoidea) to field weathered insecticide residues. Melanderia, , 50, 24-27.

Rosaiah, B., 2001. Performance of botanicals against the pest complex of bhendi (*okra*). Pestology, 25(4), 17-19.

Sridhar, S., 2000. Impact of pesticides on soil microorganisms. In: *Environmental Impact of Pesticides in Agro-Ecosystem.* G. Santharam, P. Jayakumar, S. Kuttalam, S. Chandrasekaran, T. Manoharan (Eds.), Centre for Pl. Prot. Studies, Tamil Nadu Agri. Univ., Coimbatore, pp.78-83.

Sundarraj, T., Kurucheve,V.and Jayaraj, J., 1996. Screening of higher plants and animals faeces for the fungitoxicity against *Rhizoctonia solani* Kuhn. Indian Phytopath., 49, 398-403.

Surekha, J. and Rao, P.A., 2000, Management of fruit borer of bhendi with organic sources of NPK and certain insecticides and its effect on bhendi yield, Pestology, 24(8), 35-39.

Vardasa, L.A., Schiaretto D.C. and. Calafiori, M.H., 1989. Influence of *Diabrotica speciosa* Germar on maize with inorganic and chemical fertilizers. Ecossistema., 14, 158-162.

Vijayalakshmi, K..and Shayam Sundar, K.M., 1994. Pest control and disease management- in *Vrkshayurveda.* CAPRAT, *LSPSS Monograph*, 12, 86.

Vijaylakshmi, K., Subhashini, B. and Shivani Koul, 1997. How garlic helps sunflower? Spice India, 10(6), 15-16.

Bio efficacy of plant oils and leaf extracts for the management of coconut eriophyid mite, *Aceria guerreronis* Keifer

K. Ramaraju,* M.K. Varadarajan,* and E. V. Bhaskaran*

Two field experiments were conducted at Sundapalayam and Thondamuthur near Coimbatore during 2001–02 to evaluate the bioefficacy of plant oils and leaf extracts against coconut eriophyid mite. Among the plant oils tested, spot application of neem oil 3% was found to be effective causing 50 and 56 per cent reduction in mite population over control after three rounds of spraying at Sundapalayam and Thondamuthur, respectively. The overall mean mite population was reduced significantly in the trees sprayed with three rounds of neem oil 3 per cent (50%) which was on par with neem oil 2 per cent + garlic extract 2 per cent (48%) at Sundapalayam. This was followed by NSKE 44 per cent. The per cent reduction in mite population in the trees sprayed with other plant oils (pungam, illuppai and pinnai) ranged between 30 and 39 per cent with the corresponding populations ranging between 13 and 14 mites/4 sq. mm as against 21 mites in control at Sundapalayam. The per cent damaged green nuts in bunch 4, 5 and 6 was least in neem oil 3 per cent (36%) followed by neem oil 2 per cent + garlic extract 2 per cent (43%) and illuppai oil (46%) treated trees as against control (100%) at Sundapalayam. Further, the mean grade index recorded at the time of harvest was also low in neem oil 3% (2.7) was on par with neem oil 2%+ garlic extract 2% (2.8), which statistically differed from other treatments. A similar trend was also recorded in the second experiment. The standard check triazophos (0.2%) was also found effective. The other plant oils *viz*., pinnai and pungam were less effective and the leaf extracts (*Ipomea, Jatropha* and *Acorus*) were not effective.

Introduction

In India, Coconut is grown in an area of 1.78-m ha producing 12,252 million nuts with an average productivity of 6982 nuts per hectare (DES, 2001). Among the coconut growing states in India, Tamil Nadu ranks third in area (3.2 lakh hectares) and production (3,816 million nuts. Nine species of eriophyid mites have been recorded to infest coconut fronds and nuts from different parts of the world (Amrine Jr. and Stasny, 1994). Among them, coconut eriophyid mite (CEM) *Aceria guerreronis* Keifer is considered to be one of the very serious pests and found to cause heavy damage resulting in the loss of copra yield. Regular topical spraying of the highly toxic acaricides and insecticides *viz*., monocrotophos, triazophos, carbosulfan, cyhexatin, and chinomethionate etc., at monthly intervals reduced the mite population and nut damage significantly (Ramaraju *et al.,* 2000; Julia and Mariau, 1979). Repeated application of chemicals, at short intervals, is uneconomical and impracticable besides environmentally unsafe. Therefore, it is imperative to search for the alternative methods that are economical and ecofriendly in nature. Hence keeping these views in mind, the present study was undertaken.

---

* Department of Agricultural Entomology, Tamil Nadu Agricultural University, Coimbatore – 641 003

Material and methods

Two field experiments were conducted to evaluate the bioefficacy of plant oils and leaf extracts against coconut mite in two different locations *viz.*, Sundapalayam and Thondamuthur near Coimbatore. The experiment at Sundapalayam includes ten treatments *viz.*, T1-neem oil 3%; T2- pinnai oil 3%; T3- pungam oil 3% ; T4- Illuppai oil 3%; T5- neem seed kernel extract (NSKE) 5%; T6- *Ipomea* leaf extract 5%; T7- *Jatropha* leaf extract 5%; T8- neem oil 2% + garlic extract 2%; T9- pungam oil 3% + achorus leaf extract 1% and T10- untreated control. In the second experiment at Thondamuthur, T8 and T9 were replaced and triazophos 0.2% was included. Each treatment consisted of eight replications and was laid out in a Randomized Block Design (RBD). Three rounds of sprays were given at 45 days interval. Observations on the mite population were recorded on 15, 20 and 45 days after each spraying. Damage assessment was done on tagged bunches (i.e. 4, 5 and 6) 30 days after the last application.

Preparation of plant oils and leaf extracts

Neem oil (*Azadirachta indica* A. Juss.) and neem kernels for the preparation of kernel extract (NSKE), pinnai oil (Alexandrian laurel- *Calophyllum inophyllum* L.), pungam or karanj oil (*Pongamia pinnata* L.), illuppai oil (Mahua) (*Madhuca longifolia* Macbr) var. *latifolia* Boxb. and Chev.) were purchased from the local market. These oils were emulsified with water, 0.3 % 'Teepol' and tested at 3 per cent concentration. For preparation of NSKE, the method reported by Abdul Kareem *et al.* (1974) was followed.

Preparation of leaf & garlic extract

Leaf samples were collected from the *Pungam* and *Jatropha* plants and cleaned well with running cold water. About 400 g of leaf sample was soaked over night in 8 litre of water for 12 h and filtered through muslin cloth for spraying. Similarly, 160 g of garlic was crushed and soaked with eight litre of water for overnight and filtered with muslin cloth.

Assessment of mite population

In Llive mite populations (both nymphs and adults) were recorded in 4 sq. mm area on the inner most bracts (4, 5, and 6th bracts) and meristamatic nut surface (at three places) where maximum populations were observed in each sample.

Damage assessment

Before the initiation of spraying, the youngest bunch was marked with yellow coloured paint for taking the bunch damage assessment. Post treatment observations on the mite damage on the 4, 5, and 6th bunches were made by counting the total number of nuts it bears and the infested nuts showing mite damage symptoms. The per cent damaged buttons or per cent infestation in each bunch was worked out by using the following formula:

$$\text{\% infestation in green nuts} = \frac{\text{Number of infested buttons}}{\text{Total number of buttons}} \times 100$$

The nut damage on the matured nuts of bunch 9,10 and 11 were made at harvest by following the damage scale 1-5 adopted by Julia and Mariau (1979) and mean grade index was worked out for each of the treatments.

Grade	Damage / symptom category
1	Nuts with no mite damage
2	Nuts with 1 – 10% of mite damage
3	Nuts with 11 – 25% of mite damage
4	Nuts with 26 – 50% of mite damage
5	Nuts with >50% of damage with reduced size and greatly distorted

The mean grade index was worked out by using the formula,

$$\text{Mean grade index (MGI)} = \frac{G1P1+G2P2+G3P3+G4P4+G5P5}{P1+P2+P3+P4+P5}$$

Where, G- grade, P- total number of nuts in corresponding grade

Results and discussions

The results of the experiments conducted at Sundapalayam and Thondamuthur near Coimbatore revealed the continuous presence of mites through out the period of study in both the locations. The overall mean mite population reduced significantly in the trees sprayed with three rounds of neem oil 2 per cent + garlic extract 2 per cent 50.2% which was on par with neem oil 3 per cent (48.3%) at Sundapalayam. This was followed by NSKE 43.9 per cent. The per cent reduction in mite population in the trees sprayed with other plant oils such as *pinnai* oil, *pungam* oil and *illuppai* oil ranged between 21.0 and 37.1 per cent with the corresponding population ranged between 12.9 and 16.2 mites / 4 sq.mm. as against 20.5 mites in control at Sundapalayam (Table 1). The per cent damaged green nuts in bunch 4, 5 and 6 was least in pinnai oil 3 per cent (35.5 %) followed by illuppai oil (42.5%) and neem oil (44.5%) treated trees as against control (100%) at Sundapalayam (Table 2). Further, the mean grade index recorded at the time of harvest was also low in neem oil 2% + garlic extract 2% (3.2) which was on par with illuppai oil 3% (3.2) followed by neem oil 3% (3.5). A similar trend was also recorded in the second experiment. The overall mite population reduced by 56.0 per cent over control in neem oil treated trees which was on par with triazophos 0.2 per cent (55.6%) and illuppai oil (50%) after three rounds of spraying (Table 3). The per cent damage in the green nuts was also low in triazophos 0.2 per cent (28.0%) and was on par with neem oil 3 per cent (31.3%). Similarly, among oils the mean grade index was also lowest in neem oil 3% (3.3) as against untreated check (4.1) (Table 4).

Investigations on the mortality of coconut eriophyid mites due to different eco-friendly agents *viz*., neem products, non-edible oils of plant-origin and fish oil rosin soap (FORS) clearly showed their effectiveness in the control of mites. Ramaraju *et al*. (2001) reported that neem oil 3%, NSKE 5%, and FORS 4% recorded a cumulative reduction of 52.6 to 54.7 per cent over control after four sprays. Earlier workers have also reported that the spot application of neem oil (3%), neem azal T/S 1% (0.005%), neem oil + garlic extract (2%) and FORS (4%) satisfactorily controlled the eriophyid mite population and reduced the nut damage (Ramaraju *et al*., 1999, 2000; Nair 2000; Saradamma *et al*., 2001; Palaniswamy *et al*., 2004). Fernando *et al*. (2000) reported that spraying of neem oil in combination with garlic 2 per cent mixture and neem azal 1 % recorded 60 per cent reduction of eriophyid mite population and recommended for continued application. It was evident from the results that a maximum per cent reduction in mite population (56%) was achieved in neem oil treated palms. Illuppai, pinnai and pungam oil were less effective. This could be attributed due to the antifeeding, growth distrupting and ovipositional deterrence of NO and NSKE (Schmutterer *et al*. 1981). Mansour

Table 1. Bioefficacy of plant oils and leaf extracts on the population of *A. guerreronis* (Location 1. Sundapalayam)

Treatments	Population No./ 4 sq.mm.- Days after treatment																	
	I spray						II spray					III spray						% reduct. over control
	Pre	15	30	45	Mean	% reduct. over control	15	30	45	Mean	% reduct. over control	15	30	45	Mean	% reduct. over control	Cuml. mean	
Neem oil 3%	17.1	10.3	7.3	5.0	7.5^{a}	60.9^{a}	7.6	9.3	11.7	9.5^{a}	51. 5^{a}	11.8	12.1	16.9	13.6^{ab}	40.1^{b}	10.2^{a}	50. 2^{a}
Pinnai oil 3%	23.1	8.5	17.4	6.6	10.8^{c}	43.8^{c}	10.9	17.0	18.7	15.5^{c}	20..9^{d}	14.2	15.0	17.6	15.7^{ab}	30.8^{c}	14.3^{c}	30. 2^{d}
Pungam oil %	16.5	16.0	15.4	9.0	10.5^{c}	45.3^{c}	10.5	14.6	9.8	11.7^{b}	40. 3^{b}	18.8	21.7	10.2	16.7^{ab}	26.4^{cd}	12.9^{b}	37.1^{c}
Illuppai oil 3%	22.7	12.1	13.3	4.9	10.1^{c}	47.4^{bc}	10.3	13.3	21.1	14.9^{c}	23. 9^{d}	10.6	14.7	12.4	12.6^{a}	44.5^{ab}	12.5^{ab}	39. 0^{c}
NSKE 5%	16.7	23.8	2.6	6.4	10.9^{c}	43.2^{c}	10.8	12.4	14.2	12.5^{b}	36. 2^{c}	4.3	15.2	14.0	11.2^{a}	50.7^{a}	11.5^{a}	43. 9^{b}
Ipomea leaf extract 5%	15.3	30.1	4.9	9.1	14.7^{d}	23.4^{d}	14.3	11.2	21.6	15.7^{c}	19. 9^{d}	17.3	19.2	12.6	16.4^{ab}	27.8^{c}	15.6^{c}	23. 9^{e}
Jatropha leaf extract 5%	19.3	29.0	16.4	5.8	17.0^{d}	11.4^{e}	17.0	16.2	13.0	15.4^{c}	21.4^{d}	19.3	10.7	23.6	17.9b	21.1^{d}	16.8^{cd}	18. 0^{f}
Neem oil 2%+ garlic extract 2%	17.6	10.9	12.8	3.3	8.9^{b}	53.6^{b}	9.0	12.6	13.3	11.7^{b}	40.3^{b}	10.9	13.3	9.9	11.4^{a}	49.8^{a}	10.6^{a}	48. 3^{a}
Pungam oil 2%+ Acorus leaf extract 1%	17.7	16.4	16.8	11.0	14.7^{d}	23.4^{d}	14.6	15.4	17.9	15.9^{c}	18. 9^{d}	22.5	19.3	11.6	17.8^{b}	21.6^{d}	16.2^{c}	21. 0^{e}
control	24.0	20.8	17.5	19.3	19.2^{e}	-	17.1	19.3	22.3	19.6^{d}	-	21.8	26.1	20.1	22.7^{c}	-	20.5^{d}	-

In a column means followed by the same letter (s) are not significantly different ($p<0.05$) by DMRT

Table 2. Assessment of damage on green and matured nuts

Treatments	% Bunch damage in green nuts (One month after third spray)				MGI at harvest in marked bunches			
	Bunch 4	5	6	Mean	Bunch 9	10	11	Mean
Neem oil 3%	24.7	34.6	47.2	35.5[a]	2.2	2.9	3.1	2.7[a]
Pinnai oil 3%	43.9	63.5	84.5	63.9[cd]	2.8	4.1	4.1	3.7[c]
Pungam oil 3%	16.4	86.7	51.2	51.4[c]	3.9	4.0	2.8	3.6[c]
Illuppai oil 3%	23.1	38.5	71.8	44.5[b]	2.8	2.9	3.6	3.1[b]
NSKE 5%	36.9	36.5	67.4	50.2[c]	3.4	2.8	4.3	3.5[c]
Ipomea leaf extract 5%	63.3	100.0	100.0	81.6[e]	4.0	3.2	3.8	3.6[c]
Jatropha leaf extract 5%	78.9	100.0	100.0	89.4[e]	3.8	4.4	3.5	3.9[c]
Neem oil 2%+ garlic extract 2%	34.8	66.8	25.9	42.5[b]	2.5	3.1	2.9	2.8[a]
Pungam oil 2%+ Acorus leaf extract 1%	100.0	90.0	100.0	95.0[e]	2.8	3.2	3.7	3.9[c]
Control	100.0	100.0	100.0	100.0[e]	3.6	3.8	4.6	4.0[c]

In a column means followed by the same letter (s) are not significantly different (p<0.05) by DMRT

Table 4. Assessment of damage on green and matured nuts (Experiment 2)

Treatment	% Bunch damage in green nuts (One month after third spray)				MGI at harvest in marked bunches			
	Bunch 4	5	6	Mean	Bunch 9	10	11	Mean
Neem oil 3%	21.7	31.7	42.3	31.3[a]	2.7	3.4	3.9	3.3[ab]
Pinnai oil 3%	50.0	49.7	61.3	53.6[bc]	3.8	4.2	3.8	3.9[b]
Pungam oil 3%	41.3	51.5	55.6	49.4[b]	4.1	4.0	4.0	4.0[b]
Illuppai oil 3%	31.7	41.8	54.5	42.6[b]	4.2	3.9	3.7	3.9[b]
NSKE 5%	42.7	61.9	78.4	61.0[c]	3.1	3.7	3.4	3.4[ab]
Triazophos 0.2 %	20.3	25.7	38.1	28.0[a]	2.5	3.4	3.5	3.1[a]
Jatropa leaf extract 5%	40.4	49.7	55.4	48.4[b]	4.5	3.2	4.0	3.9[b]
Ipomea leaf extract 5%	47.2	61.5	71.7	60.1[c]	3.1	3.9	4.2	3.7[b]
Control	51.7	76.9	78.4	69.0[d]	4.3	4.5	3.7	4.1[b]

In a column means followed by the same letter (s) are not significantly different (p<0.05) by DMRT

and Ascher (1983) reported that pentane and acetone extracts of kernel of neem caused mortality of carmine spider mite *Tetranychus cinnabarinus* Boisdual even seven days after spraying. Moreover, the population reduction was not uniform and highly fluctuated in all treatments. Only a maximum of 50 per cent reduction in mite population was achieved in the effective treatments. The reason for this is due to the multiplication of unaffected mite population inside the bracts. However, application of eco friendly agents is completely safer to environment and natural enemies. The variation recorded in the per cent bunch damage and mean grade index at harvest in Sundapalayam trial may be due to the multiplication of mite population found inside the bracts at a later stage. Hence, in order to achieve maximum control continued application of eco-friendly agents is recommended (five to

Table 3. Bioefficacy of plant oils and leaf extracts on the population of *A. guerreronis*. (Location 2. Thondamuthur)

Treatment	(Population No. / 4 sq. mm.) – Days after treatment																	
	I spray						II spray					III spray					Overall mean	% reduct. over control
	Pre	15	30	45	Mean	% reduct. over control	15	30	45	Mean	% reduct. over control	15	30	45	Mean	% reduct. over control		
Neem oil 3%	10.9	9.5	9.3	10.6	9.8^{a}	53.9^{a}	13.3	12.1	14.7	13.4^{b}	45.3^{c}	14.5	16.9	15.2	15.5^{c}	35.1^{d}	12.9^{b}	55.6^{a}
Pinnai oil 3%	12.4	11.9	12.1	11.0	11.7^{a}	45.1^{ab}	13.0	14.1	13.9	13.7^{b}	44.1^{c}	9.4	11.8	18.9	10.3^{a}	56.9^{b}	11.9^{ab}	48.7^{ab}
Pungam, oil 3%	9.2	10.4	12.3	9.5	10.7^{a}	49.8^{a}	9.7	10.1	14.3	11.4^{ab}	53.5^{b}	18.3	11.7	14.6	14.9^{b}	37.7^{d}	12.3^{b}	47.0^{ab}
Illuppai oil 3%	11.0	8.1	11.1	12.9	10.7^{a}	49.8^{a}	12.7	13.1	13.7	13.2^{b}	46.1^{c}	6.4	14.5	11.8	10.9^{a}	54.4^{b}	11.6^{ab}	50.0^{a}
NSKE 5%	11.2	8.8	12.4	12.8	11.3^{a}	46.9^{ab}	11.8	12.1	10.7	11.5^{ab}	53.1^{b}	10.1	12.9	15.7	12.9^{b}	46.0^{c}	11.9^{ab}	48.7^{ab}
Jatropha leaf extract 5%	9.6	7.2	12.9	8.8	9.6^{a}	54.9^{a}	11.4	8.7	9.1	9.7^{a}	60.4^{a}	11.5	15.3	13.2	11.7^{ab}	51.0^{b}	10.3^{a}	44.4^{c}
Ipomea leaf extract 5%	10.8	10.1	12.5	10.9	11.2^{a}	47.4^{ab}	12.4	15.9	16.4	14.9^{b}	39.2^{cd}	14.9	17.5	12.6	15.0^{b}	37.2^{d}	13.7^{bc}	40.9^{c}
Triazophos 0.2 %	14.2	12.3	10.9	12.3	11.6^{a}	45.5^{ab}	10.2	9.8	10.3	10.1^{a}	58.8^{a}	7.5	10.0	9.5	9.0^{a}	62.3^{a}	10.2^{a}	56.0^{a}
Control	11.8	18.9	21.2	23.8	21.3^{b}	-	23.8	24.1	25.7	24.5^{c}	-	23.1	21.7	27.0	23.9^{d}	-	23.2^{c}	-

In a column means followed by the same letter (s) are not significantly different ($p<0.05$) by DMRT six rounds/year) which was in accordance with the report made by earlier workers (Ramaraju *et al.,* 2002; Nair, 2000; Fernando *et al.,* 2000).

References

Abdul kareem, A., Sadakkuthalla, S., Venugopal, M.S. and Subramanium, T.R., 1974. Efficacy of two organotin compounds and neem extract against the sorghum shoot fly. *Phytoparasitica,* 2, 127-129.

Amrine Jr., J.W. and Stasny, T.S., 1994. Catalog of the Eriophyoidea (Acarina : Prostigmata) of the world. Indira Publishing House, West bloom field, Michigan, USA, 804 p.

Director of Economics and Statistics (DES), 2001. Ministry of Agriculture, Government of India, India. p. 425

Fernando, L.C.P., Wickramananda I.R. and Aratchige N.S., 2000. Status of coconut mite, *Aceria guerreronis* in Sri Lanka. In: *Proceedings of the International workshop on coconut eriophyid mite (Aceria guerreronis).* Coconut Research Institute, Sri Lanka, 1-8 pp.

Julia, J.F. and Mariau, D., 1979. Nouvelles recherches en cote d Ivoire sur *Eriophyes guerreronis* K., acarien ravageur des noix du cocotier. *Oleagineux* 34: 181-189.

Mansour, F.A. and Ascher K.R.S., 1983. Effects of neem seed kernel extracts from different solvents on the caramine spider mite *Tetranychus cinnabarinus*. *Phytoparasitica,* 11, 177-185.

Nair, C.P.R., 2000. Studies on coconut eriophyid mite *Aceria guerreronis* K. in India. In: Proceedings of the international workshop on coconut eriophyid mite (*Aceria guerreronis*). Coconut Research Institute, Sri Lanka, 9-12 pp.

Palaniswamy. S., Ramaraju. K. and Bhaskaran, V., 2004. Evaluation of ecofriendly agents against eriophyid mite *Aceria guerreronis* K. on coconut. In: *National conference on Role of Bio-Pesticides, Bio-agents, and Bio-fertilizers for sustainable agriculture and horticulture*. February 13 – 16, 2004 at Lucknow. p. 52.

Ramaraju, K., Natarajan K., Sundara Babu P.C. and Murali Ragini G.T., 1999. Management of coconut eriophyid mite, *Aceria guerreronis* in Tamil Nadu. *J. Acarol.*, 14, 82-83.

Ramaraju, K., Natarajan, K., Sundarababu, P.C. and Palaniswamy, S., 2000. Studies on coconut eriophyid mite *A.guerreronis* K. in Tamil Nadu, India. In: *Proceedings the international workshop on coconut eriophyid mite,* Coconut Research Institute, Sri Lanka. 13-31pp.

Ramaraju, K., Rabindra, R.J., Karuppuchamy, P., Palaniswamy, S., Kempraj, T., Balasubramani, V. and Bhaskaran, E.V., 2001. Evaluation of ecofriendly agents against coconut eriophyid mite *Aceria guerreronis*. Paper presented at "*National Seminar on Emerging Trends in Pest and Diseases and their Management*" Oct 11-13, 2001, TNAU, and Coimbatore. p. 81.

Ramaraju, K., Palaniswamy, S. and Bhaskaran, E.V., 2002. Evaluation of neem azal T/S 1% and TNAU neem oil 60E.C.(C) against eriophyid mite, *Aceria guerreronis* K. on coconut. Paper presented at "*Neem 2002 World Neem Conference Souvenir Volume 1*" organized by Neem foundation in collaboration with government agencies, Industry and research Institute held at Mumbai, India. pp.113-118

Saradamma, K., Naseema Beevi S., Hebsy Bai., Mathew, T.B. and Sudharma, K., 2001. Natural products and bioagents for the management of coconut eriophyid mite, *Aceria guerreronis* Keifer. In: *Proc. Sym. Biocontrol based pest management for quality crop protection in the current millennium*. PAU, Ludhiana. 18-19, July 2001. 190-191 pp.

Schmutterer, H, K. Ascher R. S. and Rembold H. (Eds)., 1981. Natural pesticides from the neem tree (*Azadirachta indica* A. Juss). In: *Proc. Ist. Neem Conf., Rottach- Egern*, FRG, 1980, 297p.

Field Evaluation of Botanicals and Fish oil rosin soap for the Management of Major Pests of Senna

R.K. Murali Baskaran,* D.S. Rajavel,* M. Shanthi,*
S. Kumar,* K. Suresh* and S. Senthil Kumaran*

A field experiment was conducted at Orchard, Agricultural College and Research Institute, Madurai during April – October 2007 to evaluate the field efficacy of botanicals and fish oil rosin soap for the management of major pests of senna (*Cassia angustifolia* Vahl.). Eight rounds of application of seven botanicals and one animal origin insecticide at fortnightly interval were made and the incidence of thrips (*Kurtomathrips morrilli* Moulton), aphid (*Aphis craccivora* Koch.), defoliators, (*Catopsilia pyranthe* L., *Eurema hecabe* Moore) and pod borer (*Etiella zinckenella* (Treitschke)) on senna were recorded. The results indicated that Azadirachtin 0.15% EC (Neem Gold) (1.5 ml/lit.) was effective in reducing the incidence of sucking pests of senna, recording mean leaflet damage of 4.3 and 5.7 for thrips and aphid, respectively in main senna crop and 3.4 and 3.1 per cent for thrips and aphid in ratoon senna crop. Similarly, the incidence of *C. pyranthe* and *E. hecabe* was low in the plots applied with Azadirachtin, resulting the mean leaflet damage of 2.7 and 1.7 per cent in main senna crops and 2.1 and 1.3 per cent in ratoon senna crop, respectively. The mean per cent pod borer damage was also low in the same treatment. Azadirachtin recorded the highest dry leaf and pod yield of 1635 Kg/ha and 640 Kg/ha, respectively with cost benefit ratio of 1:1.42.

Introduction

Senna (*Cassia angustifolia* Vahl.) is a small perennial, branched under-shrub that belongs to the family Leguminaceae. The leaves and young pods possess sennosides with laxative principles. It is reported to be very safe and effective drug for habitual constipation and has world demand for use as a household drug in pharmaceutical industries (Bamini *et al.*, 2006). It is cultivated traditionally over 10,000 ha in semi-arid lands in coastal districts of Tirunelveli, Ramanathapuram and Madurai in Tamil Nadu. In Madurai and Virudhunagar districts, it is being grown from November to May as one of the mixed crops in coriander and bengal gram under rainfed condition and the same is maintained as perennial crop from October to September in Tirunelveli and Tuticorin districts under irrigated condition, comprising 6-7 harvests of leaves and pods (Mohaideen Pitchai, 2007). Senna is reported to be attacked by five different species of insect pests, thrips (*Kurtomathrips morrilli* Moulton), aphid (*Aphis craccivora* Koch.), defoliators (*Catopsilia pyranthe* L.; *Eurema hecabe* Moore) and pod borer (*Etiella zickenella* (Treitschke)) at various stages of growth (Usha Rani and Kalyanasundaram, 2005; Jhansi Rani and Sridhar, 2005; Bamini *et al.*, 2006; Murali Baskaran *et al.*, 2007). Since leaves and pods are used in Medical field, management of insect pests attacking senna through chemical insecticides is not adviceable.

Materials and Methods

A field experiment was conducted at the Orchard of the Agricultural College and Research Institute, Madurai during April – October 2007 to evaluate the field efficacy of neem products and animal origin

* Department of Agricultural Entomology, Agricultural College and Research Institute, Madurai 625 104

insecticide (Fish oil rosin soap) for the management of major pests of senna (local variety). Seven botanicals, neem oil 3%, mahuva oil 3%, karanj oil 3%, Azadirachtin 0.15% EC (Neem Gold) @ 1.5 ml/lit., NSKE 5% and Neem cake extract 5% and Fish oil rosin soap @ 25 g/lit. (FORS) were tried in a randomized block design with plot size of 4 x 3 m and spacing of 45 x 30 cm. Each treatment was replicated thrice and compared with untreated check. Four rounds of application of above said treatments were given on main senna crop, at fortnightly interval starting from 30 days after sowing. On 90th day after sowing, leaves (thick and bluish leaves) and pods (golden yellow colour) were harvested by leaving 10-15 cm of stem from ground level. Recommended cultural practices were undertaken to grow senna as ratoon crop and harvested on 90th day after first harvest during which another four rounds of application of eight treatments were given. Damage and population of thrips, aphids, defoliators and pod borer were recorded prior and 7 and 14 days after each spray. Population of coccinellids, syrphids and spiders were also recorded. Leaves and pods collected in main and ratoon senna crop were shade dried and yield was recorded. Based on the yield and expenditure, cost benefit ratio was worked out. Arcsine and square root transformations were applied to data on per cent damage and population of insect pests, respectively during statistical analysis and means were separated by DMRT.

Results and Discussion

Efficacy of botanicals and fish oil rosin soap against sucking pests of senna: The results are presented in Table 1. The results indicate that azadirachtin (Neem Gold) 0.15% EC @ 1.5 ml/lit is the best in minimizing the population of both thrips and aphid and damage on main and ratoon senna followed by Neem oil 3% and NSKE 5%.

b. *Efficacy of botanicals and fish oil rosin soap against defoliators of senna:* The results are presented in Table 2. The results indicate that azadirachtin (Neem Gold) 0.15% EC @ 1.5 ml/lit is the best in minimizing the population of the two defoliators and damage to leaf on main and ratoon senna followed by Neem oil 3% and NSKE 5%.

c. *Efficacy of botanicals and fish oil rosin soap against pod borer of senna:* The results are presented in Table 3. The population of pod borer was the lowest in Azadirachtin (Neem Gold) in both main and ratoon crops followed by Neem oil at 3%.

Table 1. Efficacy of botanicals and fish oil rosin soap against sucking pests of senna

Treatment	Main Senna				Ratoon Senna			
	*Kurtomathrips morrilli**		*Aphis craccivora***		*Kurtomathrips morrilli**		*Aphis craccivora***	
	Thrips/ Plant	% leaflet damage	Aphids/ shoot	% shoot damage	Thrips/ Plant	% leaflet damage	Aphids/ shoot	% shoot damage
Neem oil 3% (15 days interval)	2.2 (1.48)[a]	4.5 (12.24)[a]	8.8 (2.96)[b]	6.0 (14.17)[ab]	1.9 (1.37)[b]	3.6 (10.93)[a]	4.8 (2.18)[a]	3.4 (10.62)[b]
Karanj oil 3% (15days interval)	3.3 (1.82)[b]	6.7 (15.00)[b]	12.2 (3.49)[c]	8.3 (16.74)[c]	2.8 (1.67)[c]	5.6 (13.68)[c]	6.6 (2.56)[c]	4.5 (12.28)[d]
Mahua oil 3% (15 days interval)	3.6 (1.89)[c]	8.0 (16.42)[c]	15.5 (3.93)[d]	10.5 (18.90)[d]	3.1 (1.76)[d]	6.1 (14.29)[d]	10.3 (3.20)[e]	6.9 (15.22)[f]
Azadirachtin 0.15% EC 1.5 ml/lit. (15 days interval)	2.0 (1.41)[a]	4.3 (11.96)[a]	8.5 (2.91)[a]	5.7 (13.81)[a]	1.7 (1.30)[a]	3.4 (10.62)[a]	4.6 (2.14)[a]	3.1 (10.14)[a]

NSKE 5% (15 days interval)	2.3 (1.41)^{a}	5.0 (12.92)^{a}	9.5 (3.58)^{b}	6.3 (14.53)^{b}	2.1 (1.44)^{b}	4.1 (10.68)^{b}	5.4 (2.32)^{b}	3.6 (10.93)^{c}
Neem cake extract 5% (15 days interval)	3.5 (1.87)^{c}	8.0 (16.42)^{c}	16.0 (4.00)^{d}	10.7 (19.09)^{d}	3.4 (1.84)^{de}	7.0 (15.34)^{e}	9.3 (3.04)^{d}	6.1 (14.29)^{e}
Fish oil rosin soap 25g/lit. (15 days interval)	4.0 (2.14)^{c}	8.8 (17.25)^{c}	17.5 (4.18)^{e}	11.8 (20.09)^{e}	3.6 (1.89)^{e}	7.4 (15.78)^{f}	9.8 (3.13)^{d}	5.0 (12.92)^{f}
Untreated check	6.5 (2.54)^{d}	10.7 (19.09)^{d}	21.4 (4.62)^{f}	12.6 (20.79)^{f}	4.5 (2.12)^{f}	8.5 (16.95)^{g}	11.5 (3.39)^{f}	7.6 (16.00)^{g}
Mean	3.4 (1.84)	7.0 (15.34)	13.7 (3.70)	9.0 (17.45)	2.8 (1.67)	5.7 (13.81)	7.8 (2.77)	5.0 (12.92)
SEd	0.04	0.21	0.04	0.18	0.04	0.18	0.03	0.14
CD (0.05%)	0.08	0.43	0.08	0.37	0.08	0.36	0.07	0.29

*Mean of four observations; **Mean of three observations; Population: Figures in parentheses are square root transformed values; Damage: Figures in parentheses are arcsine transformed values; In a column, means followed by same letter(s) are not significantly different by DMRT (P=0.05); I spray: 30 DAS; II spray: 45 DAS; III spray: 60 DAS; IV spray: 75 DAS; V spray: 30 DAH; VI spray: 45 DAH; VII spray: 60 DAH; VIII spray: 75 DAH; DAS: Days after Sowing; DAH: Days after harvest

Table 2. Efficacy of botanicals and fish oil rosin soap against senna defoliators

Treatment	Main Senna				Ratoon Senna			
	*Catopsilia pyranthe**		*Eurema hecabe**		*Catopsilia pyranthe**		*Eurema hecabe**	
	Larva/ 5 plants	% leaflet damage	Larva/ 5 plants	% leaflet damage	Larva/ 5 plants	% leaflet damage	Larva/ 5 plants	% leaflet damage
Neem oil 3% (15 days interval)	1.3 (1.13)^{a}	2.8 (9.63)^{a}	0.8 (0.89)^{a}	1.8 (7.71)^{a}	0.9 (0.94)^{b}	2.2 (8.52)^{b}	0.8 (0.89)^{ab}	1.3 (6.54)^{a}
Karanj oil 3% (15days interval)	1.7 (1.31)^{b}	3.8 (11.24)^{b}	1.0 (1.00)^{b}	2.5 (9.21)^{b}	1.1 (1.06)^{d}	2.8 (9.63)^{d}	0.9 (0.97)^{bc}	1.7 (7.49)^{c}
Mahua oil 3% (15 days interval)	2.2 (1.64)^{c}	5.1 (13.05)^{c}	1.2 (1.09)^{cd}	3.2 (10.30)^{c}	1.3 (1.13)^{e}	3.0 (9.97)^{e}	1.1 (0.91)^{c}	2.2 (14.88)^{d}
Azadirachtin 0.15% EC 1.5 ml/lit. (15 days interval)	1.3 (1.12)^{a}	2.7 (9.45)^{a}	0.8 (0.89)^{a}	1.7 (7.48)^{a}	0.7 (0.81)^{a}	2.1 (8.33)^{a}	0.8 (0.89)^{ab}	1.3 (6.46)^{a}
NSKE 5% (15 days interval)	1.3 (1.12)^{a}	2.8 (9.63)^{a}	0.8 (0.89)^{a}	1.8 (7.71)^{a}	0.9 (0.94)^{b}	2.4 (8.91)^{c}	1.0 (1.00)^{c}	1.5 (7.03)^{b}
Neem cake extract 5% (15 days interval)	2.2 (1.48)^{c}	5.1 (13.05)^{c}	1.2 (1.09)^{cd}	3.2 (10.30)^{c}	1.6 (1.26)^{f}	4.0 (11.53)^{f}	1.3 (1.13)^{d}	2.4 (8.91)^{e}
Fish oil rosin soap 25g/lit. (15 days interval)	2.3 (1.51)^{d}	5.5 (13.56)^{c}	1.4 (1.18)^{d}	3.5 (10.98)^{d}	1.5 (1.22)^{h}	4.2 (11.82)^{f}	1.4 (1.18)^{d}	2.6 (9.27)^{e}
Untreated check	2.8 (1.67)^{e}	6.8 (15.11)^{d}	2.1 (1.44)^{e}	4.3 (11.96)^{e}	2.0 (1.41)^{g}	4.9 (12.78)^{g}	1.5 (1.22)^{e}	3.1 (10.14)^{f}
Mean	1.9 (1.37)	4.3 (11.84)	1.2 (1.09)	2.7 (9.34)	1.3 (1.10)	3.2 (10.30)	1.1 (1.05)	2.6 (8.84)

SEd	0.02	0.56	0.02	0.18	0.02	0.16	0.03	0.20
CD (0.05)	0.04	1.11	0.05	0.36	0.03	0.32	0.06	0.40

*Mean of three observations; Population: Figures in parentheses are square root transformed values; Damage: Figures in parentheses are arcsine transformed values; In a column, means followed by same letter(s) are not significantly different by DMRT (P=0.05); I spray: 30 DAS; II spray: 45 DAS; III spray: 60 DAS; IV spray: 75 DAS; V spray: 30 DAH; VI spray: 45 DAH; VII spray: 60 DAH; VIII spray: 75 DAH; DAS: Days after Sowing; DAH: Days after harvest

Table 3. Efficacy of botanicals and fish oil rosin soap against senna pod borer

Treatment	Main Senna *Etiella zinckenella**		Ratoon Senna *Etiella zinckenella**	
	Larva/ 5 plants	% pod damage	Larva/ 5 plants	% pod damage
Neem oil 3% (15 days interval)	1.5 (1.23)^{a}	9.5 (17.95)^{b}	0.9 (0.94)^{a}	6.8 (15.11)^{b}
Karanj oil 3% (15days interval)	1.7 (1.29)^{b}	10.8 (19.18)^{c}	1.1 (1.04)^{abc}	8.3 (16.81)^{d}
Mahua oil 3% (15 days interval)	2.6 (1.61)^{d}	16.4 (23.88)^{e}	1.7 (1.25)^{cd}	11.4 (19.73)^{e}
Azadirachtin 0.15% EC 1.5 ml/lit. (15 days interval)	1.4 (1.19)^{a}	9.3 (17.75)^{a}	0.9 (0.94)^{a}	6.5 (14.77)^{a}
NSKE 5% (15 days interval)	1.5 (1.23)^{a}	9.7 (18.14)^{b}	0.9 (0.94)^{ab}	7.6 (16.00)^{c}
Neem cake extract 5% (15 days interval)	2.3 (1.51)^{c}	14.8 (22.62)^{d}	1.7 (1.23)^{bcd}	13.0 (21.13)^{f}
Fish oil rosin soap 25g/lit. (15 days interval)	2.6 (1.61)^{d}	17.0 (24.35)^{f}	1.9 (1.37)^{d}	13.8 (21.80)^{g}
Untreated check	3.3 (1.81)^{e}	21.0 (27.27)^{g}	2.2 (1.40)^{abcd}	14.6 (22.46)^{h}
Mean	2.1 (1.44)	13.6 (21.56)	1.4 (1.16)	10.3 (18.63)
SEd	0.03	0.11	0.19	0.15
CD (0.05)	0.06	0.23	0.38	0.31

*Mean of two observations; Population: Figures in parentheses are square root transformed values; Damage: Figures in parentheses are arcsine transformed values; In a column, means followed by same letter(s) are not significantly different by DMRT (P=0.05); I spray: 30 DAS; II spray: 45 DAS; III spray: 60 DAS; IV spray: 75 DAS; V spray: 30 DAH; VI spray: 45 DAH; VII spray: 60 DAH; VIII spray: 75 DAH; DAS: Days after Sowing; DAH: Days after harvest

d. Population of natural enemies, as influenced by botanicals and fish oil rosin soap:

The population of natural enemies like coccinellids, syrphids and spiders were more or less equal in all treatments and untreated check except Azadirachtin and neem oil in which little reduction in activity of natural enemies was noticed (Table 4).

e. Yield and cost benefit ratio : Azadirachtin recorded the highest yield of dry leaf (1635 kg/ha) and pod (640 kg/ha) with cost benefit ratio of 1:1.42, followed by neem oil, NSKE, karanj oil, mahua oil, neem cake extract and FORS with cost benefit ratio of 1:1.39, 1:1.34, 1:1.27, 1:1.19,1:1.13 and 1:1.02, respectively.

Table 4. Population of natural enemies in main and ratoon senna sprayed with botanicals and fish oil rosin soap

Treatment	Coccinellids/5 plants				Syrphids/5 plants		Spiders/5 plants	
	Main Senna		Ratoon Senna		Main Senna	Ratoon Senna	Main Senna	Ratoon Senna
	30 DAS	60 DAS	30 DAH	60 DAH	60 DAS	60 DAH	60 DAS	60 DAH
Neem oil 3% (15 days interval)	3.1[c]	7.0[c]	2.9[b]	4.6[c]	2.5[c]	1.9[b]	1.5[d]	1.5[b]
Karanj oil 3% (15days interval)	4.3[ab]	7.9[b]	3.5[a]	5.1[b]	3.1[b]	2.5[a]	2.4[a]	2.4[a]
Mahua oil 3% (15 days interval)	4.1[b]	8.0[b]	3.4[a]	5.2[ab]	3.2[ab]	2.4[a]	2.3[ab]	2.3[a]
Azadirachtin 0.15% EC 1.5 ml/lit. (15 days interval)	3.5[c]	7.1[c]	2.4[c]	4.2[c]	2.3[c]	1.6[b]	1.3[d]	1.4[b]
NSKE 5% (15 days interval)	4.8[ab]	8.0[b]	3.4[a]	5.0[b]	3.2[ab]	2.5[a]	2.1[c]	2.1[a]
Neem cake extract 5% (15 days interval)	4.9[a]	7.9[b]	3.6[a]	5.2[ab]	3.0[b]	2.3[a]	1.9[c]	2.1[a]
FORS 25g/lit. (15 days interval)	4.5[ab]	8.0[b]	3.5[a]	5.0[ab]	3.2[ab]	2.2[a]	2.1b[c]	2.0[a]
Untreated check	4.8[ab]	8.5[a]	3.6[a]	5.4[a]	3.4[a]	2.4[a]	2.4[a]	2.0[a]
Mean	4.3	7.8	3.3	4.9	2.9	2.2	2.0	1.9
SEd	0.19	0.06	0.11	0.14	0.09	0.12	0.12	0.12
CD (0.05)	0.46	0.15	0.25	0.33	0.23	0.29	0.29	0.31

In a column, means followed by same letter(s) are not significantly different by DMRT (P=0.05); I spray: 30 DAS; II spray: 45 DAS; III spray: 60 DAS; IV spray: 75 DAS; V spray: 30 DAH; VI spray: 45 DAH; VII spray: 60 DAH; VIII spray: 75 DAH; DAS: Days after Sowing; DAH: Days after harvest

The order of field efficacy of various botanicals and fish oil rosin soap in managing major insect pests of senna was Azadirachtin > neem oil > NSKE > karanj oil > mahua oil > neem cake extract > FORS. Formulated products of neem were reported to be superior than crude products of neem in managing major insect pests of agricultural and horticultural crops (Regupathy and Ayyasamy, 2005). In the present study, Azadirachtin 0.15% EC was found to be superior in reducing the population of thrips, aphids, defoliators and pod borer of senna.

Various botanicals and fish oil rosin soap except Azadirachtin and neem oil used in the present experiment were safe to coccinellids, syrphids and spiders. Neem oil was reported to be moderately toxic to *Chelonus blackburni* Cameron and *Bracon brevicornis* Wesmael (Srivastava, 1997) and *Rhynocoris marginatus* (Fabricius) (Jhansilakshmi *et al.* 1998). However, formulated products of neem, Neemark and Achook were safe to *Syrphis* sp. and *Coccinella septempunctata* Linn. on tea pests (Sharma and Kashyap, 2002).

Conclusion

Azadirachtin 0.15% EC (Neem Gold) applied for times at fortnightly interval commencing 30 days after sowing in main crop and again four rounds in ratoon crop after lowest on 90th day of main crop, has been found to be the best in minimizing the population of thrips, aphid, defoliators and pod borer and their damage and giving the highest yield of pods.

Acknowledgement

Authors are grateful to the National Medicinal Plants Board, New Delhi for the financial assistance in conduct of field experiments.

References

Bamini, T., Usha Rani, B. and Suresh, K., 2006. Senna (*Cassia angustifolia*)- The Medicinal Crop, *Agrobios Newsletter*, 5(4), 28-29.

Jhansi Rani, B. and Sridhar, V., 2005. Record of aphids (Homoptera: Aphididae) and their natural enemies on some medicinal and aromatic plants, *Pest Management in Horticultural Ecosystems*, 11(1), 71-73.

Mohaideen Pitchai, 2007. Selvam kolikka vaikkum senna sagupadi, *Mooligai Payir*, 76-78.

Mukesh Kumar, Ram Singh, Kumar, M. And Singh, R., 2002., Potential of *Pongamia glabra* Vent. as an

Murali Baskaran, R. K., Rajavel, D. S., Shanthi, M., Suresh, K. and Kumar, S., 2007, Insect Diversity and Damage Potential in Medicinal Plants Ecosystem, *Insect Environment*, 13(2), 76-79.

Regupathy, A. and Ayyasamy, R., 2005. Formulation of botanical pesticides pp. 133-149, In: Green Pesticides for Insect Pest Management (eds. S. Ignacimuthu and S. Jayaraj), Narosa Publishing House, New Delhi 324p.

Sharma, D. C. and Kashyap, N.P., 2002. Impact of pesticidal spray on seasonal availability of natural predators and parasitoids in tea ecosystem, *Journal of Biological Control*, 16, 31.

Srivastava, M., Paul, A. V. N., Rengaswamy, S., Kumar, J. And Parmer, B. S., 1997. Effect of neem seed kernel extracts on the larval parasitoid *Bracon brevicornis*, *Journal of Applied Entomology*, 121, 51.

Usha Rani, B. and Kalyanasundaram, M., 2005. Insect Pests of Senna (*Cassia angustifolia*) and their management, *Indian Journal of Arecanut, Spices & Medicinal Plants*, 8(1), 7-9.

Botanicals - An Ecofriendly Approach for Pest Management in Mulberry

V. Sathyaseelan* and V. Baskaran**

Mulberry, *Morus alba*, the food plant of silkworm (*Bombyx mori*) is exposed to various insecticides for pest management which threatens the silkworm industry by residual toxicity. Hence a field experiment was conducted at Tamil Nadu Agricultural University, Coimbatore to compare the efficacy of different botanicals (Neem oil, Pungam oil, Madhuca oil, Pinnai oil, Neem seed kernel extract 5%, Annona seed kernel extract 5%, *Vitex negundo* leaf extract 10%) and three insecticides (Dichlorvos 76 WSC 0.1%, Malathion 50 EC 0.1%, Phosalone 35 EC 0.1%) against mulberry leaf webber, *Glyphodes pulverulentalis*.H. The mean larval population of mulberry leaf webber before treatment varied from 7.2-9.47/plant. The reduction of mean larval population increased with increase in time after exposure over control in all the treatments. Though drastic reduction @ 2.57- 4.17 larvae/ plant was observed with insecticide treatment, botanicals equally effected the control of leaf webber. Among the botanicals *V.negundo* leaf extract was found to be superior with larval reduction of 4.00/ plant and was on par with insecticide treatments. Neem oil and Neem seed kernel extract were less effective when compared to insecticide treatments.

Introduction

Mulberry, *Morus alba* (L.), the food plant of silkworm *Bombyx mori* (L.) is attacked by a large number of species of insects. Among the major pests, mulberry leaf webber, *Glyphodes pulverulentalis* (Hampson), is one of the most important pests which damage the tender leaves of mulberry plant. Generally insecticide is not advisable for mulberry ecosystem because of residual toxicity and also its effect on silkworm rearing. Botanicals can be an efficient alternative for the insecticides in mulberry garden. Hence the efficacy of insecticides and the botanicals were tested in mulberry garden against the mulberry leaf webber.

Materials and Methods

Efficacy of various botanicals and insecticides against Mulberry leaf webber: A field experiment was laid out in field number 68, Eastern block of Agricultural College & Research Institute, Coimbatore to evaluate the efficacy of different botanicals and insecticides against leaf webber, *G. pulverulentalis* on five year old mulberry variety Kanva 2. The treatments were imposed on the mulberry after 40 days of pruning. The trial was conducted in a randomized block design under irrigated condition during December 2000. There were eleven treatments which were replicated thrice. Each treatment was imposed on the mulberry maintained in a plot size of 40 m^2 with the plant population density of 44 plants per plot.

The treatments were as follows: Dichlorvos (DDVP) 76 WSC 0.1 per cent, Malathion 50 EC 0.1 per cent, Phosalone 35 EC 0.1 per cent, Madhuca oil 3 per cent. Neem oil 3 per cent; Pinnai oil 3 per cent, Pungam oil (PO) (*Pongamia glabra*) (L.) 3 per cent, Neem seed kernel extract (NSKE) 5 per

*Department of Entomology, Faculty of Agriculture, Annamalai University, Annamalainagar – 608 002

**Department of Agricultural Entomology, Tamil Nadu Agricultural University, Coimbatore- 641 003

cent, Annona seed kernel extract (ASKE), (*Annona squamosa* (L.) 5 per cent, and karunotchi, *Vitex negundo* (L.), leaf extract 10 per cent, and untreated check as control.

At the time of pruning, 4 per cent endosulfan dust was applied to basal portion of plants to avoid phoretic ants. Predatory spiders were dislodged by jarring the plants. The cultural operations were carried out as per recommended methods (Bongale, 1993).

Preparation of various eco-friendly botanicals

Karunotchi (Vitex negundo) –Cow urine leaf extract: The leaves were collected from the plant bushes and chopped into pieces and ground into a paste. This was soaked overnight in 3.lt.of clean, filtered cow urine for overnight. In the next day morning, stirred the soaked solution and boiled it for 30 minutes at 70-80°C. Added fresh water when the solution was concentrated too much, allowed it to cool overnight and filtered through khada cloth. This filterate was mixed with suitable surfactant and diluted 50 ml of the final solution mixture in 1 lt. of water and sprayed in the field (Baskaran and Narayanasamy, 1995).

Neem Seed Kernel Extract: Neem seed kernels were collected, washed clearly in pure water and shade dried for few days. The outer rind of the seed was broken and kernels gathered and ground well to a paste form. With this 200 lt. of water and 150 ml of soap solution were added and kept undisturbed for overnight, filtered through khada cloth and sprayed it in the morning.

Mixed the paste of neem seed kernel with 5 lt. of water or clean cow urine and left undisturbed overnight. The soaked solution must be boiled for 30 minutes at 70- 80 ° C. The colour of the solution when turned dark allowed it to cool, filtered and stored the extract in a air tight bottles or sprayed it to the field (Baskaran and Narayanasamy, 1995).

Annona seed kernel extract: Annona seeds were collected and the kernels were broken into paste and the remaining procedure is similar to that of neem seed kernel extract (Baskaran and Narayanasamy, 1995).

The treatments were imposed by using high volume knapsack sprayer (Aspee) with a spray fluid of 500 litre / ha. Teepol 0.1% was added uniformly to the tank mixture and sprayed during evening hours. The larval population on 1, 3, 7 and 14 days after treatment was recorded from ten randomly-selected plants in a plot, from all the replications of the above treatments. The per cent shoot damage was worked out.

Results and Discussion

Field studies on the efficacy of certain insecticides and botanicals against leaf webber: The mean larval population of *G. pulverulentalis* under field condition before and 1, 3, 7 and 14 days treatment and the pooled mean data are presented in Table 1. Acritical analysis of the results indicate on third day after first spray, there was a significant reduction in larval population in phosalone, dichlorvos, malathion and *Vitex negundo* leaf extract when compared to all other treatments. A similar trend has been observed in 7 and 14 days after treatment and when the pooled mean value were also considered. The neem products (Neem oil and NSKE), ASKE and pungam oil had little effect on the larval population.

The results of the field experiment with different insecticides and botanicals arein line with the findings of CSR & TI (2004), Pampore. The results showed that monocrotophos at 0.05 and 0.1% concentration caused mortality of larvae of *G. pyloalis* by 93 and 95 per cent while dichlorvos at 0.04 and 0.02 per cent concentration significantly reduced the larval population by 90 and 75 per cent respectively.

The results are in agreement with the findings of Gupta and Vijay (1986) who reported that monocrotophos, quinalphos and phosalone were found toxic to the larvae of *G. pyloalis*. The leaf extract

Table 1. Efficacy of certain botanicals and insecticides against larval population of Mulberry leaf webber

Treatments	Pre- treatment	Mean larval population/ plant DAT				Pooled mean**
		1**	3**	7**	14**	
Dichlorvos 76 EC	9.00 (3.07)	4.30 (2.19)[b]	3.57 (1.92)[b]	3.13 (1.90)[b]	2.87 (1.02)[b]	3.47[b]
Malathion 50 EC	9.27 (3.12)	6.40 (2.62)[cd]	4.27 (2.18)[c]	4.10 (2.14)[c]	4.17 (2.09)[c]	4.74[c]
Phasolone 35 EC	9.10 (3.08)	3.00 (1.86)[a]	2.27 (1.66)[a]	1.43 (1.37)[a]	2.57 (1.96)[a]	2.32[a]
Mahuva oil 3 %	9.20 (3.11)	6.57 (2.66)[c]	5.90 (2.40)[d]	5.23 (2.39)[de]	7.63 (2.87)[e]	6.33[e]f
Neem oil 3 %	8.63 (3.02)	6.27 (2.51)[c]	4.43 (2.21)[d]	5.40 (2.42)[de]	6.30 (2.32)[d]	5.60[de]
Pinnai oil 3 %	9.30 (3.13)	7.30 (2.79)[d]	5.80 (2.49)[cd]	5.37 (2.42)[de]	6.17 (2.31)[d]	6.16[e]
Pungam oil 3 %	9.47 (3.14)	4.87 (2.31)[b]	4.60 (2.21)[c]	4.60 (2.21)[c]	6.13 (2.30)[d]	5.05[cd]
NSKE 5 %	8.40 (2.98)	5.47 (2.32)[bc]	5.47 (2.32)[d]	4.87 (2.31)[c]	6.87 (2.68)[d]	5.67[de]
Anonna seed kernel extract 5 %	9.23 (3.11)	7.43 (2.83)[d]	6.27 (2.39)[e]	6.13 (2.45)[f]	8.83 (2.98)[f]	7.17[f]
Vitex negunndo 10 %	9.17 (3.15)	5.17 (2.37)[bc]	4.33 (2.19)[c]	4.00 (2.11)[c]	4.87 (2.32)[c]	4.59[c]
Control	7.20 (882)	10.27 (3.28)[e]	8.97 (3.07)[f]	8.33 (2.99)[g]	10.73 (3.82)[g]	9.58[g]

Figures in the parentheses are square root transformed values; ** Significant at 1% level in LSD.

of *Vitex negundo* was found to be toxic to the mosquito *Culex quinquefasciatus* (Kannathasan *et al*, 2007), the Diamond back moth, *Plutella xylostella* (Sheng and Bao, 2006) and the Angumois grain moth, *Sitotroga cerealella* (Tandon and Gupta, 2005), The presence of toxic compounds like terpenes, cinol, sabeniene, sesque terpenes in *V. negundo* extract might be the reason for higher susceptibility of the larvae.(Kannathasan *et al*, 2007; Sheng and Bao, 2006; Tandon and Gupta, 2005) Foliar spray of 0.2 per cent dichlorvos was found highly effective against *G. pyloalis.* (Booth *et al*,2007).

Siddegowda *et al.* (1995) reported that foliar spray of dichlorvos (0.2%) mixed with 0.5 per cent soap solution significantly reduced the larval population of *Diaphania* sp. with a safe waiting period of 17 days for feeding the leaves to silkworm, *B. mori*. Foliar spray of dichlorvos 0.076 per cent was found effective against mulberry leaf webber with a safe waiting period of 10 days after the last spray (KSSR & DI, 2006). A second spray was recommended ten days after first spray when the larval infestation exceeded 20 per cent. Considering the limitation of insecticide application in mulberry, the results indicate that *Vitex negundo* 10% leaf extract can be explored in the management of mulberry leaf webber.

References

Baskaran, V and Narayanasamy, P., 1995. *Traditional Pest Control.* Caterpillar Publications, Annamalai Nagar, Tamil Nadu, pp 90.

Bongale, U.D., 1993. Fertility evaluation and fertilizer recommendations for mulberry garden soils. *Indian Silk*, 31(9), 5-8.

Booth, E.D., Jones, E. and .Elliot, B.M., 2007. Review of the *in vitro* and *in vivo* genotoxicity of Dichlorvos. *Toxicol. Pharmocol*., 49(3), 316-326.

CSR & TI., 2004. *Annual Report*. Central Sericultural Research and Training Institute, Pampore, India. pp 67-73.

Gupta, B.K. and Vijay veer, 1986, Laboratory bioassay of some insecticides as contact poison against third instar larva of *Glyphodes pyloalis* Walker (Lepidoptera : Pyralidae). *Indian Forester*, 108(7), 52-58.

Hiware, C.J., 2006. Effect of Fortification of mulberry leaves with homeopathic drug *Nux vomica* on *Bombyx mori* L. *Homeopathy* 95(3), 148-150.

Kannathasan, K., Senthilkumar, A., Chandrasekaran, M. and Venkatesalu, V., 2007. Different larvicidal efficacy of four species of *Vitex negundo* against *Culex quinquefasciatus* larvae. *Parasitol. Res*., 101, 1721-1723.

KSSR & DI, 2006. Leafroller infestation of mulberry plantations and its management. Karnataka State Sericulture Research and Development Institute, Bangalore (Pamplet).

Sheng, Y.Y. and Bao, T.X., 2006. Toxicity and Oviposition- deterrence of *Vitex negundo* extracts to *Plutella xylostella. J.of Applied Ecology*., 17, 695-698.

Siddegowda, D.K., Vinod Kumar Gupta, Sen, K.V., Benchamin, D., Manjunath, K., Sathyaprasad, S.B and. Datta, R.K., 1995. *Diaphania* sp. infests mulberry in South India. *Indian Silk*, 34(8), 6-8.

Tandon, V.R. and Gupta, R.K., 2005. An experimental evaluation of anticonvulsant activity of *Vitex negundo*. *Indian .J .Physiol Pharmacol.* , 49(2), 199-205.

Antifeedant activity of *Annona squamosa* against teak (*Tectona grandis* Linn. f.) skeletonizer, *Eutectona machaeralis* Walker (Lepidoptera : Pyralidae)

NITIN KULKARNI,* SANJAY PAUNIKAR,* K.C. JOSHI,*
ARUN KAKKAR,** VINOD K. MISHRA,** AND V.L. MAHESHWARI***

The paper reports results of laboratory bioassays with *Annona squamosa* seeds extracted in four solvents for its antifeedant property against the larvae of teak skeletonizer, *Eutectona machaeralis* Walker. Results indicated that all the extracts tested at 500 ppm inhibited feeding ranging from 41% to 73% over feeding in control. Detailed investigation was carried out with seed oil and ethyl acetate extract in range of concentrations from 100 to 3000 ppm. Pet. ether extracts at and above 600 ppm and ethyl acetate fraction at and above 800 ppm resulted in 70% feeding inhibition by the larvae of teak skeletonizer. Effective concentrations for 50% antifeedant effect (EC_{50}) for two promising extracts were also calculated. The paper also includes a brief discussion on prospects of using the present results for further isolation of the compounds with allomonal effect against the teak skeletonizer for developing ecofriendly method of its management in forest nurseries.

INTRODUCTION

Besides neem and other botanicals (Devkumar and Parmar, 1993; Schmutterer, 1995; Kulkarni *et al.*, 2004), pesticidal compounds from Annonaceous plants have recently been realized. Some important reports on insecticidal properties of crude fractions and isolated compounds of this important botanical group include; Ruprecht *et al.* (1990; insecticidal property of annonaceous acetogenins), Cortes *et al.* (1993; isolation Bis-tetrahydrofuran acetogenins), Xing-Wang and Hong (1999; on isolation of twenty acetogenins, Alali *et al.* (1999; reviewed annonaceous acetogenins), Parvin *et al.* (2003; pesticidal activity of pure isolated compound from chloroform extract of *A. squamosa* against *Tribolium castaneum*), Isman and Audrey (2004; toxicity and antifeedant activity of seed extract against *Plutella xylostella*;). Morgan and Mandava (1987) and Devakumar and Parmar (1993) have discussed isolated compounds like Annonaine, squamosine, from *Annona reticulate* having promising pesticidal effects against a few agricultural insect pests.

The custard apple, *Annona squamosa* is one such wide-spread member of the annonaceous plant family, available easily in almost all parts of the country, but has not been investigated in detail, except for some fragmentary reports on application of crude solvent extracts against some agricultural pests. These are: Chitra *et al.* (1990) evaluated pet ether extract on *Henosepilachna vigintioctopuctata* on brinjal leaves, Sreedevi *et al.* (1992) and Tripathi and Singh (1994; crude extract on *S. litura* and *Spilarctia obliqua*), Ambadkar and Khan (1994) evaluated for repellent, deterrent action on stored grain insect pests. The most recent excellent report is on isolation of biologically active flavonoids

* Forest Entomology Division, Tropical Forest Research Institute, P.O. – RFRC, Jabalpur (M.P.) 482021
** Govt. Model Autonomous Science College, Jabalpur (M.P.) 482001
*** Department of Biochemistry, North Maharashtra University, Jalgaon (M.S.)

from the leaves of *A. squamosa* and its bioactivity against *Callosobruchus chinensis* (Kotkar *et al.*, 2002; Salunke and Maheshwari, 2005).

In line with research accomplishments on neem (*A. indica* A. Juss) (Schmutterer, 1995), crude extracts and compounds of *A. squamosa* need to be evaluated against insect pests occurring in different ecosystems for its wide spectrum applicability. The forest insect pests, which are tougher-to-control are one such challenge. There are meager evaluations of this important botanical against forest insect pests: Kulkarni *et al.* (2003) studied antifeedant property of solvent fractions of *A. squamosa* leaves with its field stability. Choubey (2004) studied anti-oviposition property of methanolic seed extract against seed weevil, *Sitophilus rugicollis*.

The teak skeletonizer, *Eutectona machaeralis* causes annual growth loss of 13 to 65% in plantations (Champion, 1934; Nair, 1996) and much higher growth loss to teak seedlings in forest nurseries, i.e., up to 54.77% (Joshi *et al.*, 2001). Despite the above, not much work on their management using alternative methods has been carried out, except a few (Kulkarni, 2001, Kulkarni *et* al., 2003, 2004). In view of the above, the paper reports investigations on bioactivity of solvent fractions of *A. squamosa* against the larvae of this pest, under the globally accepted concept of IPM (Integrated Pest Management) programme.

Materials and Methods

Collection and preparation of plant material

Seeds of *A. squamosa* were collected and shade dried. The powdered material (50 g) was subjected to extraction using Soxhlet's apparatus sequentially in petroleum ether, ethyl acetate, ethyl alcohol and water. After evaporation of solvents, residues were used for bioassays. Each residue represented active compounds or fraction soluble in the respective solvent.

Collection of insects

The larvae of teak skeletonizer, *E. machaeralis* were collected from teak plantations within the campus and adjoining forest areas under Mandla Forest Division, Mandla (M.P.). The insect culture was maintained in the laboratory, during the season. The larvae were kept in glass beakers covered with muslin cloth and reared on fresh leaves in the laboratory. On emergence, the adults were released into wooden cages containing teak seedlings for oviposition. Newly hatched instars were fed with the fresh teak leaves in the laboratory and the culture was maintained.

Insect Bioassay

Newly moulted last instar larvae were separated from the culture and starved for at least 6 h prior to bioassay. The bioassays were carried out in two phases; (1) to investigate the comparative efficacy of all the fractions with that of control, and (2) to investigate range of concentrations of promising fraction for determining the optimum concentration for the activity and Effective Concentration for 50% effect (EC_{50})

For comparing different crude extracts, measured leaf area of disc cut from freshly collected teak leaves were treated with 500 ppm of 5 solvent extracts dissolved in distilled water with Triton X-100 as emulsifier. The treated discs were allowed to dry and then exposed for feeding to single larva in no-choice conditions for 24 hr in a Petri-dish. A separate set of experiment was set up with only distilled water + Triton X-100 as control. The complete set of the experiment was replicated 5 times. Feeding in all cases was observed by measuring leaf area with the help of laser leaf area meter (Model CI-203, CID Inc., USA). Leaf area consumption and per cent antifeedancy were calculated following the method used by Kulkarni and Joshi (1998).

In second set of experiment to investigate best effective concentration, different concentrations of petroleum ether extract ranging from 100 ppm to 3000 ppm, were prepared and leaves treated. Treated leaves were then exposed to the larvae in the same manner and leaf area consumption was calculated as in the first experiment. All the laboratory experiments were carried at 28-30 ^{0}C and 80-90 % RH.

Statistical Analysis

Data obtained in per cent values were suitably transformed into angular values (arc √ sin), Log Base (Log_{10}) or Square Root [√(x+0.5)], to get the symmetrical data. Transformed data were subjected to ANOVA and multiple comparisons of means were done by REGW Analysis (Quinn and Keough, 2002) using Gen Stat Analytical Software, Edition 3. Effective concentration for 50% antifeedant effect (EC_{50}) was also calculated for the promising fractions.

Results and Discussion

Bioassay experiment indicated that all the seed extracts, ie., pet. ether, ethyl acetate, ethyl alcohol, water extract precipitate and water extract filtrate significantly inhibited larval feeding. Petroleum ether extract of the seed material showed maximum antifeedant activity (72.99%) followed by ethyl acetate (69.99) and ethyl alcohol (60.07%). Least activity was shown by the water extracts, both by precipitate or filtrate. However, the antifeedant activity by all the treatments except control was statistically at par with each other (*P*> 0.05) (Fig.1).

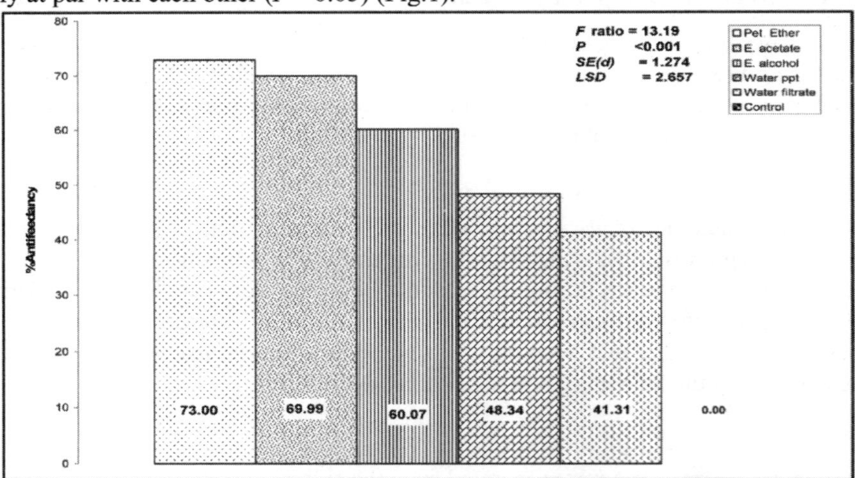

Fig. 1. Antifeedant effect of solvent fractions in no-choice feeding bio-assayexperiment (Bars represent back-transformed Square Root values of means).

Detailed bioassay experiment with range of concentrations of petroleum ether and ethyl acetate extracts showed that there was least inhibition of feeding at the lowest concentration of 100 ppm by petroleum ether and ethyl acetate extracts (respectively, 36.92 and 16.76%). Petroleum ether extract at and above 600 ppm and ethyl acetate extracts at and above 800 ppm inhibited the feeding significantly (respectively, 71.49% and 60.0%) over other treatments and control (*P*<0.05) (Fig. 2&3). Table 1 gives EC_{50} values of the pet ether and ethyl acetate extracts against the last instar larvae of teak skeletonizer.

Table 1. Effective Concentration for 50% antifeedant effect activity of *A. squamosa* seed against *E. machaeralis*

Fractions	EC_{50} (in ppm)	Fiducial limits		X^2	Equation
		Upper	Lower		
Ethyl acetate	524.807	769.804	357.783	1.54677	y = 1.225x+1.6841
Petroleum ether	220.80	425.687	114.528	0.43801	y = 1.063x+2.5579

The inhibition of feeding by other botanicals on teak skeletonizer larvae has also been investigated earlier; *Azadiracta indica* (Kulkarni *et al*., 1996), *Lantana camara* var. *aculeata* and *Aloe vera* (Kulkarni *et al*., 1997; Kulkarni, 2001) with promising results.

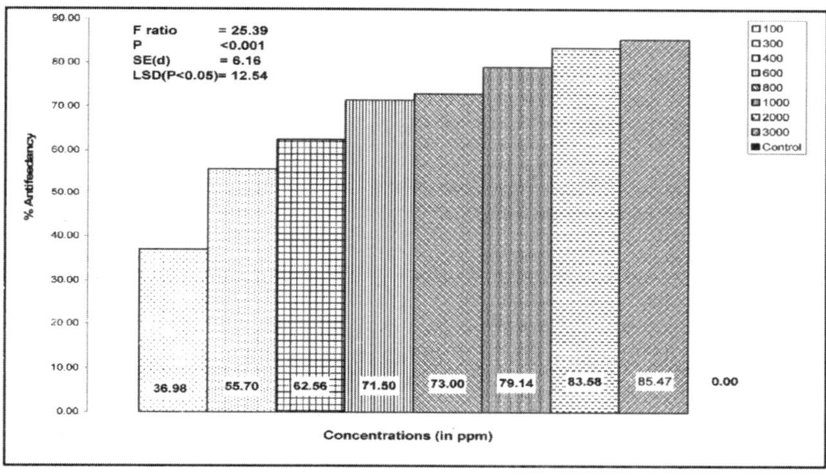

Fig. 2. Antifeedant effect of Petroleum ether fraction in no-choice feeding bio-assay experiment (Bars represent back-transformed angular values of means).

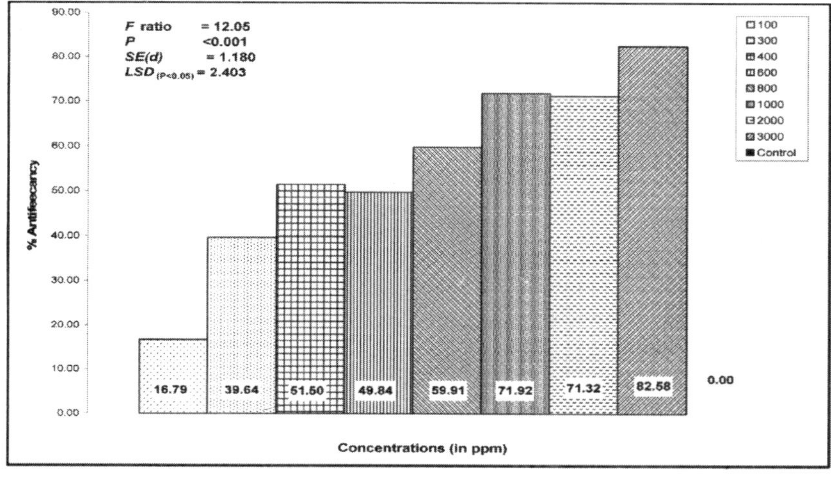

Fig. 3. Antifeedant effect of Ethyl Acetate fraction in no-choice feeding bio-assay experiment (Bars represent back-transformed Square Root values of means).

Teak skeletonizer, as reported by Champion, (1934), Nair (1996) and Joshi *et al.* (2001) cause higher growth loss to teak seedlings in forest nurseries, *i.e.*, up to 54.77%. The results obtained against the teak skeletonizer larvae may be utilized to isolate active compounds present in these crude fractions, which my result in development of biopesticidal product against the pest. The study may prove significant in judicious management of chemical insecticides used widely for controlling this pest in forest nurseries.

Acknowledgements

Authors are thankful to Dr. A.K. Mandal, The Director and Shri S.P. Tripathi, Group Coordinator (Res.), Tropical Forest Research Institute, Jabalpur for the facilities during the work. Thanks are also due to the Principal and Head of the Division (Chemistry), Govt. Autonomous Science College, Jabalpur for constant encouragements. The financial assistance received from the Council of Scientific and Industrial Research (CSIR), is also gratefully acknowledged.

References

Alali, F.Q., Liu, X.X. and McLaughlin, J.L., 1999. Annonaceous acetogenins: Recent progress. *J. Nat. Prod.,*, 62, 504-540.

Ambadkar, P.M. and Khan, D.H. 1994. Screening of responses of adult cigarette beetle, *Lasioderma serricorne* F. (Coleoptera:Anobiidae) to fresh and dried leaves of 51 plant species for possible repellent action. *Ind. J. Entomol.,* 56, 169-175.

Champion, H.G., 1934. The effect of defoliation on the increment of teak saplings. *Forest Bulletin (Silviculture Series)*, 89: Government of India, Delhi. 6p.

Chitra, K.C., Venkataramireddy, P. and Kameshwar, Rao, P., 1990. Effect of certain plant extract in the control of brinjal spotted leaf beetle, *Henosepilachna vigintiocopunctata* Fabr. *J. Appl. Zool. Res.*, I, 39-41.

Choubey, V., 2004. Studies o some control methods against the insect pests of *sal* (*Shorea robusta* Gaertn.f.) in Madhya Pradesh. *Ph.D. Thesis*, Rani Durgavati University, Jabalpur.

Cortes, D., Gigadere, B. and Cave, A., 1993. Bis-tetrahydofuran acetogenins from Anonaceae. *Phytochem.*, 32, 1467-1473.

Devakumar, C. and Parmar, B.S. 1993. In: *Botanical and biopesticides* (B.S. Parmar and Devakumar, C. eds.), SPS India and Westwill Publishing House, New Delhi.

Isman, M.B. and Audrey, L.J., 2004. Toxicity and antifeedant activity of crude seed extracts of *Annona squamosa* (Annonaceae) against lepidopteran pests and natural. *Int. J. Trop. Insect Sci.,* 24, 150-158.

Joshi, K.C., Roychoudhury, N., Kulkarni, N. and Sambath, S., 2001. *Implementation Completion Report*, World Bank, Forestry Research, Education and Extension Project, 48p.

Kotkar, H.M., Mendki, P.S., Sadan, S.V.G.S. and Jha, S.R., Upasani, S.M. and Maheshwari, V.L., 2002. Antimicrobial and pesticidal activity of partially purified flavonoids of *Annona squamosa. Pest Mangt. Scic.*, 58, 33-37.

Kulkarni, N., 2001. Anti-insect bioactivities of some botanicals: their prospects as component of integrated pest control system. In *Recent Trends in Insect Pest Control to Enhance Forest Prod.*, (Eds. P.K Shukla and K.C. Joshi), pp.218-230. *Proc. Work. Entomol. and Biol. Cont.*, Published by Tropical Forest Research Institute, Jabalpur, India. Pp.235.

Kulkarni, N., Joshi, K.C. and Gupta, B.N., 1997. Antifeedant property of *Lantana camara* var. *aculeate* and *Aloe vera* leaves against teak skeletonizer, *Eutectona machaeralis* Walk. (Lepidoptera : Pyralidae). *Entomon*, 22(1), 61-65.

Kulkarni N., Joshi K.C. and Meshram P.B., 1996. Bio-activity of methanolic neem seed extract against the teak skeletonizer, *Eutectona machaeralis* Walk. (Lepidoptera:Pyralidae). *J. Environ. Biol.,* 17, 189-195.

Kulkarni, N., Joshi, K.C. and Shukla, P.K., 2003. Antifeedant activity of *Annona squamosa* Linn. against *Crypsiptya coclesalis* Walk. (Lepidoptera : Pyralidae). *Entomon*, 28, 389-392.

Kulkarni, N., Joshi, K.C. and Shukla, P.K. 2004. Integrated Insect Pest Management of Forest Insects. In *Contemporary Trends in Entomology* (Ed. G.T. Gujar), Campus Books, New Delhi, Pp. 370-410.

Morgan, E.D. and Mandava, N.B. (Eds.), 1987. *Handbook of Natural Pesticides*. Vol. III. *Insect growth regulators* Part B, CRC press, Boca Raton, Florida, 180p.

Nair, K.S.S., Sudheendrakumar, V.V., Varma, R.V., Chako, K.C. and Jayaraman, K., 1996. Effect of defoliation by *Hyblaea puera* and *Eutectona machaeralis* (Lepidoptera) on volume increment of teak. *Proc. IUFRO Symp. On Impact of Diseases and Insect Pests in Tropical Forests*, 1996, Pp. 257-273.

Parvin, M.S., Islam, M.E., Rahman, M.M. ad Haque, M.E., 2003. Pesticidal activity of pure compound annotemoyin-1 isolated from chloroform extract of the plant *Annona squamosa* Linn. against *Tribolium castaneum* (Herbst.) (Coleoptera:Tenebrionidae). *Pak. J. Biol. Sci.*, 6, 1088-1091.

Quinn, G.P. and Keough, M.J., 2002. Experimental design and data analysis for biologists. Cambridge University Press. p.200-201.

Ruprecht, T.K., Hui, Y.H., McLaughlin, J.L. 1990. Annonaceous acetogenins: A review. *J. Nat. Prod.*, 53, 237-278.

Schmutterer, H. (Ed.), 1995. *The neem tree*, VCH Verlagsgesellschaft, Weinheim (FRG), VCH Publishers Inc., New York, 696p.

Solunke, B.K. and Maheshwari, V.L., 2005. Control of a stored grai pest, *Callosobruchus chinensis* (L) using plant flavonoids. In *Conf. on Biopest.:Emerging Trends*, Nov.,11-13th Palampur (India).

Sreedevi, K., Chitra, K.C. and Kameshwar Rao, P., 1992. Efficacy of certain plant extracts in the control of tobacco caterpillar *Spodoptera litura* Fabr. Under laboratory conditions. *J. Appl. Zool. Res.*, 3, 95-96.

Tripathi, A.K. and Singh, D., 1994. Screening of natural products for insect antifeedant activity. Part I. Plant extracts. *Indian J. Ent.*, 56, 129-133.

Xing-Wang, W. and Hong, X., 1999. Annonaceous acetogennins from annonaceae, synthesis and mechanism of action. *Phytochem.*, 48, 1097-1117.

Impact of Neem and Karanj Oil Formulation with *Hyptis suaveolens* Leaf Extract on Feeding, Survival, Histology and Protein Profile of Four Lepidopteran Pests

M. GABRIEL PAULRAJ,* M. PAVUNRAJ* AND S. IGNACIMUTHU*

A laboratory experiment was conducted to find out the effect of different formulations of neem and karanj oils with an active fraction isolated from *Hyptis suaveolens* (L.) Poit (Lamiaceae) ethyl acetate crude extract at 500 ppm concentration on feeding activity, haemolymph protein and midgut histology of fourth instar larvae of four lepidopteran pests viz., *Earias vitella* Stell (Arctiidae), *Helicoverpa armigera* (Hubner) (Noctuiidae), *Leucinodes orbonalis* G. (Pyraustidae) and *Spodoptera litura* Fab. (Noctuidae). Treatment 1 (12.25 mg effective fraction + 44.5% neem oil + 44.5% pongam oil + 10% emulsifier + 1% stabilizer) and Treatment 2 (12.25 mg effective fraction + 89% neem oil + 10% emulsifier + 1% stabilizer) showed significantly high antifeedant activity against *H. armigera* and *E. vitella*. Treatments 1 and 2 recorded 100 per cent antifeedant activity against *E. vitella*. Treatment 2 gave 45.34 per cent antifeedant activity against *H. armigera*. Treatment 1 caused maximum larval mortality in all the four pest species. Among the four insect species, *H. armigera* and *S. litura* were found to be the most susceptible insects to Treatment 1. This treatment killed 66.6 per cent of *H. armigera* and 53.3 per cent of *S. litura* larvae in four days after treatment. Maximum pupal death was recorded in *H. armigera*, *L. orbonalis* and *S. litura* in Treatment 6 (fraction dissolved in acetone). The treatments damaged midgut regions in *H. armigera* larvae after 48 hr of exposure. Complete dystrophication was noticed in the basement membrane and muscular layers of entire midgut and hindgut. From the SDS-PAGE profile of haemolymph protein, it was clear that Treatment 2 increased the intensity of high molecular mass peptides (97 and 43 KDa).

INTRODUCTION

Helicoverpa armigera and *Spodoptera litura* are polyphagous pests occurring throughout India and causing severe yield losses. *Earias vitella* and *Leucinodes orbonalis* are also notorious insects attacking bhindi and brinjal and cause heavy loss to farmers. Chemical pesticides have created a great threat to the health of the present and future generations. They also have created resistance in several pests including the above mentioned four lepidopterans. The plant-derived bioactive compounds are thought to be an important alternative source for chemical pesticides. These secondary metabolites act as antifeedants, insecticides, repellents, attractants, insect growth regulators, ovicides and oviposition deterrents (Rembold and Sieber, 1981; Berenbaum, 1983; Koul *et al.*, 2000; Golawska, 2007). Products obtained from the neem tree *Azadirachta indica* Nees (Meliaceae) have been well studied against many insects throughout the world (Rembold *et al.*, 1982) and the search for new compounds from many other plants for pest control is going on all over the world. *Hyptis suaveolens* (L.) Poit (Lamiaceae) is a medicinal plant used as an appetizing agent, to combat indigestion, stomach pain, nausea, flatulence, colds, and infection of the gall bladder (Fun and Svendsen, 1990; Ahmed *et al.*, 1994; Singh and Handique, 1997). In a preliminary screening experiment Raja *et al.* (2005)

* Entomology Research Institute, Loyola College, Chennai – 600 034.

screened five different organic solvent extracts and water extract of *Hyptis suaveolens* leaves against *Helicoverpa armigera* and *Spodoptera litura.* They found that ethyl acetate extract of *H. suaveolens* showed promising antifeedant activity. In the present study, the active fraction responsible for the antifeedant activity against the four lepidopteran pests was mixed with neem and karanj oils and the antifeedant activity was studied. The effective formulation was tested for its effects on haemolymph protein profile and gut histology of *H. armigera.*

Materials and Methods

Collection of Plant and extraction: Fresh leaves of *Hyptis suaveolens* were collected from different locations in Kancheepuram and Thiruvallur districts, Tamil Nadu. The leaves were dried under shade and ground into powder by using electric blender. Leaf powder (1 Kg) was soaked in three fold (3 L) of ethyl acetate, extracted after 48 hr and filtered by filter paper. The solvent in the filtrate was removed by rotary vacuum evaporator; the crude extract (35 gm) was taken in glass vials and stored in the refrigerator.

Fractionation by column chromatography: A glass column chromatography (4cm diameter; 60cm length) was used to isolate different fractions from the crude extract through silica gel (100-200 mesh, LR). The fractions were separated by 100 per cent hexane followed by combination of hexane: ethyl acetate ranging from 95:5 to 0:100. The fractions were collected in a 200 ml conical flask and concentrated using rotary vacuum evaporator.

Insects: Larval instars of *Earias vitella*, *Helicoverpa armigera*, *Leucinodes orbonalis* and *Spodoptera litura* were collected from their natural host plants from Kancheepuram and Thiruvallur Districts, Tamil Nadu. *S. litura* larvae were collected from groundnut plants and *H. armigera* and *E. vitella* were collected from bhindi fields. *L. orbonalis* larvae were colleceted from brinjal fields. *H. armigera* larvae were reared on a semi-synthetic diet and *S. litura* larvae were reared on castor leaves. *E. vitella* and *L. orbonalis* larvae were reared on bhindi and brinjal fruits respectively. Pupated insects were kept inside oviposition cages and humidity was maintained by keeping moist cotton swabs inside the cages. The emerged adults were fed with 10 per cent honey solution. Cow pea seedlings, castor leaves, bhindi fruits and brinjal fruits were kept inside the oviposition cages for oviposition by *H. armigera*, *S. litura*, *E. vitella* and *L. orbonalis* respectively. Newly hatched larvae were reared by the above mentioned method and a stock culture was established in the laboratory. For bioassay experiments, newly emerged fourth instar larvae from the stock culture were utilized.

Treatments: When the ethyl acetate extract of *H. suaveolens* leaves was subjected to column chromatography, 15 fractions with different Rf values were obtained. From the screening experiments with all these fractions it was found that the 2nd fraction (eluted by 9:1 hexane and ethyl acetate) was very effective against four different pests used in this study. This active fraction was included in different formulations that were prepared by using neem and karanj oils. Emulsifier (DMA-NE) and stabilizer were purchased from UNITOP chemicals. A stock solution of 10 ml of the different treatments was prepared as follows:

Treatment 1: Neem oil (4.45 ml) + Karanj oil (4.45 ml) + active fraction (12.25mg)+ emulsifier (1 ml) + Stabilizer (0.1 ml)

Treatment 2: Neem oil (8.9 ml) + active fraction (12.25 mg) + emulsifier (1 ml) + stabilizer (0.1 ml)

Treatment 3: Karanj oil (8.9 ml) + active fraction (12.25 mg) + emulsifier (1 ml) + stabilizer (0.1 ml)

Treatment 4: active fraction (12.25 mg) + emulsifier (1 ml) + stabilizer (0.1 ml)

Treatment 5: Neem oil (4.45 ml) + Karanj oil (4.45 ml) + emulsifier (1 ml) + stabilizer (0.1 ml)

Treatment 6: active fraction (12.25 mg) dissolved in acetone

Treatment 7: control (emulsifier + stabilizer in water)

Treatment 8: Water control

Antifeedant Bioassay experiments: Antifeedant experiments were conducted by following the method of Isman *et al.* (1990). Only one concentration of formulations (500 ppm of active fraction) was prepared by mixing 7 ml of stock with 17 ml distilled water. Castor leaf disc (3 cm diameter) and cotton leaf disc (3 cm diameter) were used in the case of *S. litura* and *H. armigera* respectively. Antifeedant activity against *E. vitella* and *L. orbonalis* was tested by using bhindi (10 mm thick) (Rao *et al.*, 2003) and brinjal (5 mm thick) fruit discs respectively. Leaf and fruit discs were dipped in 500 ppm concentration solution separately for two minutes, shade dried and supplied to the larvae. In the case of fruit discs, initial weight was taken after dipping in the treatment solutions. A separate set of fruit discs devoid of insects was also maintained for 24 hrs to check the weight loss in the fruit discs at the same experimental condition to correct the consumption. After 24 hr of progressive consumption the unconsumed leaf area was measured by a leaf area meter (Delta-T Devices, Serial No. 15736 F 96, U.K) and final weight of the fruit discs were found out. The consumption by *E. vitella* and *L. orbonalis* on fruit discs was calculated as follows:

Real Consumption	=	Initial weight of fruit disc	–	(final weight + weight loss in fruit disc due to desiccation)
Weight loss due to desiccation	=	Initial weight of fruit disc	–	Final weight after 24 hrs

Feeding activity of treatments was corrected from emulsifier control and the antifeedant activity was measured by the following formula of Isman *et al.* (1990):

$$\text{Per cent antifeedant activity} = \frac{C-T}{C+T} \times 100$$

Where, C= consumption in control and T= consumption in Treatment

Larvicidal bioassay: For larvicidal bioassay the treatments were given as described in the anti-feedant bioassay experiments. After 24 hr of consumption on treated leaves or fruit discs, the larvae were continuously reared on untreated food till pupation. Larval mortality was recorded up to 4days and pupal mortality was also recorded in the treated insects. Per cent larval mortality was adjusted by Abbot's formula (Abbot, 1925):

$$\text{Adjusted morality (\%)} = \left(\frac{\text{N in T after treatment}}{\text{N in Co after treatment}}\right) \times 100$$

here, n = Insect population , T = treated , Co = control

Haemolymph Protein quantification and SDS-PAGE profile: Haemolymph was bled into ice-chilled tubes (1.5 ml) by cutting the prolegs of the larvae after 48 hr of treatment. After removal of haemocytes and debris by centrifugation at 12000 rpm for 10 min at 4°C the haemolymph was immediately stored at –20°C until use. Quantitative protein determination was performed using the method of Lowry *et al.* (1951). SDS-PAGE (Sodium dodecyl sulphate-polyacrylamide gel) electrophoresis was carried out according to Laemmli (1970) using 10% acrylamide for separating and 5% for stacking gel. Gel electrophoresis was performed using a constant current of 75 volts for stacking gel and 120 volts for running gel. Gel was stained overnight in Coomassie Brilliant Blue R-250. A broad range protein molecular weight marker (PMW-B GENEI) was used for sizing molecular weights of unknown proteins present in the haemolymph samples.

Effect of treatments on Midgut tissue: Guts of treated and control larvae were dissected out under physiological saline solution after 48 hr of treatment and fixed immediately in a mixture containing ethanol (70%), formalin (15%) and acetic acid (10%). Midgut and hindgut were separated from the alimentary tract and thin sections (6 μm) were made by following standard methods. The slides were photographed under microscope. Staining was done by Heidenhain's haematoxylin and counter stained with eosin.

Statistical analysis: One way Analysis of Variance (ANOVA) was worked out to find out the significance of the treatments in antifeedant activity. The treatments were separated by Least Significant Difference (LSD) at p=0.05 level.

Results and Discussion

Antifeedant activity of formulations: Table 1 shows the per cent antifeedant activity of different formulations of neem and karanj oils with active fraction of *H. suaveolens* extract. From the results it was clear that the quantity of activity of different treatments varied in a species specific manner. Among the six different treatments, Treatments 1 and 2 showed maximum feeding deterrent effect against all insects tested except *S. litura*. Among the insects tested, *E. vitella* was highly sensitive to all the formulations. The feeding activity of *E. vitella* was reduced by the treatments and the antifeedant activity of all the six treatments was significantly high against *E. vitella* when compared to other three pests. Treatments 1 and 2 recorded significantly high antifeedant activity (100 %) against *E. vitella*. In *H. armigera* the antifeedant activity of T1 and T2 was recorded as 44.35 and 45.24 per cent respectively. In the case of *S. litura*, Treatment 6 presented maximum antifeedant activity (38.44%), but it was on par with T2 (36.77 %), T3 (35.17 %) and T5 (36.16%). *L. orbonalis* was the least affected pest in which the formulations exerted low levels of antifeedant activity ranging from 4.12 per cent (T6) to 27.81 per cent (T1). The utilization of crude plant products and isolated phytochemicals in pest management is increasing day by day due to the increasing public awareness on safe environment and deleterious effects of chemicals on human beings. There is plenty of literature available on insecticidal, antifeedant, oviposition deterrent, ovicidal, repellent and growth inhibiting properties of botanical pesticides (Rembold and Sieber, 1981; Berenbaum, 1983; Koul *et al.*, 2000; Golawska, 2007). Antifeedant activity of botanicals against insects has been studied in many countries. Quantification of antifeedant effect of botanicals is of great importance in the field of insect pest management. From an ecological point of view, antifeedants are very important since they never kill the target insects directly and allow them to be available to their natural enemies and thus help in the maintenance of natural balance. However the monophagous or oligophagous insects may die due to the application of antifeedants on their food plants, which is mainly due to starvation.

Table 1. Per cent antifeedant activity of different oil formulations with active fraction of *Hyptis suaveolens* leaf extract against four lepidopteran larvae (Mean±SE) (n=15)

Treatment	Pest insects			
	Earias vitella	*Helicoverpa Armigera*	*Leucinodes orbonalis*	*Spodoptera litura*
T1	100.00±0^{aA}	44.35±2.66^{aB}	27.81±3.47^{aC}	32.17±2.31^{bcC}
T2	100.00±0^{aA}	45.24±1.80^{aB}	23.74±3.56^{aD}	36.77±2.10^{abC}
T3	87.26±7.20^{bA}	37.03±2.15^{bB}	7.51±3.66^{cC}	35.17±2.43^{abB}
T4	86.03±9.30^{bA}	33.34±3.42^{bB}	6.18±3.11^{cC}	29.49±1.72^{cB}
T5	86.10±8.43^{bA}	20.45±3.66^{cC}	16.02±2.06^{bC}	36.16±1.77^{abB}
T6	86.58±9.54^{bA}	21.34±2.26^{cC}	4.12±3.38^{cD}	38.44±1.72^{aB}
Water Control	0.52±6.31^{c}	2.53±0.84^{d}	3.62±4.61^{d}	2.94±0.74^{d}

Values carrying same capital letters in a row and same small letters in a column are statistically not different by LSD at 5% level

In the present study the active fraction from *Hyptis suaveolens* showed antifeedant property either alone or with neem or karanj oil. However the antifeedant activity was superior when both neem and karanj oils were added to the fraction. This suggested the efficacy of neem and karanj oils. Neem oil is an effective antifeedant because it has some active compounds like azadirachtin. Several reports clearly support the antifeedant activity of neem oil (Isman *et al.*, 1990; Govindachari *et al.*, 1996). Azadirachtin is just one of more than 70 limanoids produced by the neem tree. It is a powerful insect antifeedant and growth regulator. Order Lepidoptera appear most sensitive to azadirachtin's antifeedant effects, with Coleoptera, Hemiptera and Homoptera being less sensitive (Mordue & Blackwell, 1993). Karanjin is the active principle found in karanj oil. Karanj oil is also used as botanical pesticide for many decades. Pavela and Herda (2007) reported that karanj oil exhibited repellent and oviposition deterrent activity against the common greenhouse whitefly *Trialeurodes vaporariorum* in greenhouses.

Insecticidal activity of formulations: The formulations also showed lethal effect on the test insects as evidenced from the larval and pupal mortality recorded during the study. Treatments 1 and 6 recorded maximum larvicidal activity of 66.6 and 40 per cent respectively in *H. armigera* in four days (Figure 1). Next to *H. armigera*, *S. litura* was severely affected by the formulations. About 53.3 and 33.3 per cent larval mortality of *S. litura* was recorded in T1 and T2 respectively. Treatment 2 did not record any larval death in *E. vitella*, *H. armigera* and *L. orbonalis*. Maximum death of *E. vitella* (20 %) and *L. orbonalis* (20%) larvae was recorded in Treatment 1. Maximum mortality was recorded in *H. armigera* and *S. litura* larvae in this formulation (Treatment 1). These studies indicated clearly that the increased activity of the mixed formulation might be due to the synergistic effect of neem and karanj oils along with the active fraction. Kumar *et al.* (2007) reported that the methanolic extract of neem and karanj oil combination showed 70 and 11.36 fold increase in activity over neem oil and karanj oil alone against *Tetranychus* species. George and Vincent (2005) reported that petroleum ether extract of pongamia seeds acted as a powerful synergist with *Annona squamosa* seed extracts against *Culex* mosquitoes. The pongamia extract has been considered as a good synergist and hence has been combined with several pest control agents in the control of various pests (Narasimhan *et al.*, 1998; Pathak and Shukla, 1998).

Pupal mortality due to treatments was recorded only in three treatments viz., T4, T5 and T6. All these three treatments showed lethal effect on *H. armigera* pupa. Treatment 6 recorded a maximum pupal mortality of 40 per cent in *H. armigera*. In *E. vitella*, 25 per cent pupae were killed by Treatment 5 (figure 2).

Protein concentration and SDS-PAGE profile of haemolymph proteins: Heamolymph protein concentration increased after 48 hr of treatment in T1 and T2 compared to control. In control the haemolymph protein concentration was recorded as 42 μg/ml, where as it was 47 and 52 μg/ml in T1 and T2 respectively. Schmidt *et al.* (1998) also reported increased concentration of haemolymph protein in *Melia azedarach* fruit extract treated *Spodoptera littoralis* larvae after six days of treatment. But they found that haemolymph protein concentration drastically decreased after 12 days of treatment. Figure 3 shows the electrophoretic pattern of haemolymph protein of *H. armigera* treated with T1 and T2. In both treatments 1 and 2, the total number of protein bands visible were 10 where as in control, only seven bands were visible. In treatment 2, two protein molecules with nearly 97 and 43 KDa molecular mass were very thick. The appearance of increased number of protein bands due to treatments indicated that the formulations initiated the synthesis of some new proteins in *H. armigera*. Azadirachtin is known to reduce haemolymph protein concentration in insects (Qadri and Narsaiah, 1978; Subrahmanyam and Rao, 1986). Neoliya *et al.* (2007) reported that head protein level in *H. armigera* was reduced when azadirachtin was applied through injection or food. They have proved this result by SDS-PAGE profile. They further reported that 10% azadirachtin treatment by topical application led to the loss of 56, 40, 26 and 22 KDa head polypeptides

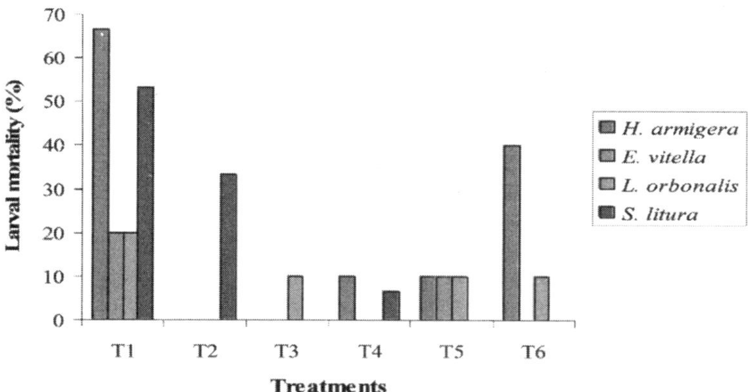

Fig. 1. Per cent larval mortality of four lepidopteran larvae caused by different formulations

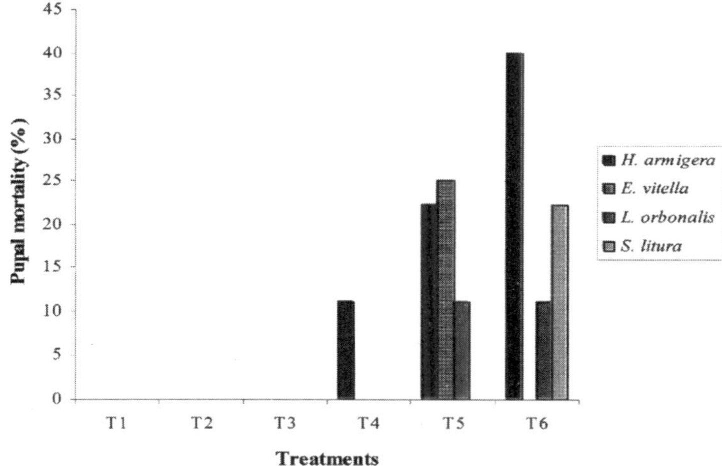

Fig. 2. Per cent pupal mortality of four lepidopteran insects caused by different formulations

Histopathological effect on gut: The histological sections of midgut and hindgut of treated *H. armgiera* clearly indicated that neem oil, karanj oil and active fraction combination affected the gut tissues and many pathological symptoms appeared in the midgut and hindgut regions after 48 hr of treatment. Complete dystrophication was noticed in the basement membrane and muscular layers of entire midgut and hindgut (figure 4). Due to shrunken midgut, with tissues squeezed out from inside the gut, large empty gaps in the midgut and hindgut tissues were distinctly seen. In insects the synthesis and secretion of hydrolytic enzymes and absorption of hydrolyzed products are taking place in the midgut epithelial cells. Most hydrolysis takes place in the midgut lumen. The hindgut is also taking part in absorption and excess digestive enzymes are probably degraded or reabsorbed here. Very few studies have been conducted to test the effect of botanicals and other biopesticides on the gut histology of lepidopteran insects. Prasad and Wadhwani (2006) have studied the histopathological effect of the biopesticide, SlNPV, on mid gut of *S. litura*. They reported that treatments with this biopesticide damaged the inner gut epithelial cells and formed vacuoles inside the gut tissue. These findings coincide with the present findings. Flipsen *et al*. (1995) have also recorded similar effects on midgut tissue in *Spodoptera exigua* due to biopesticides. Murugan *et al* (1998) have reported

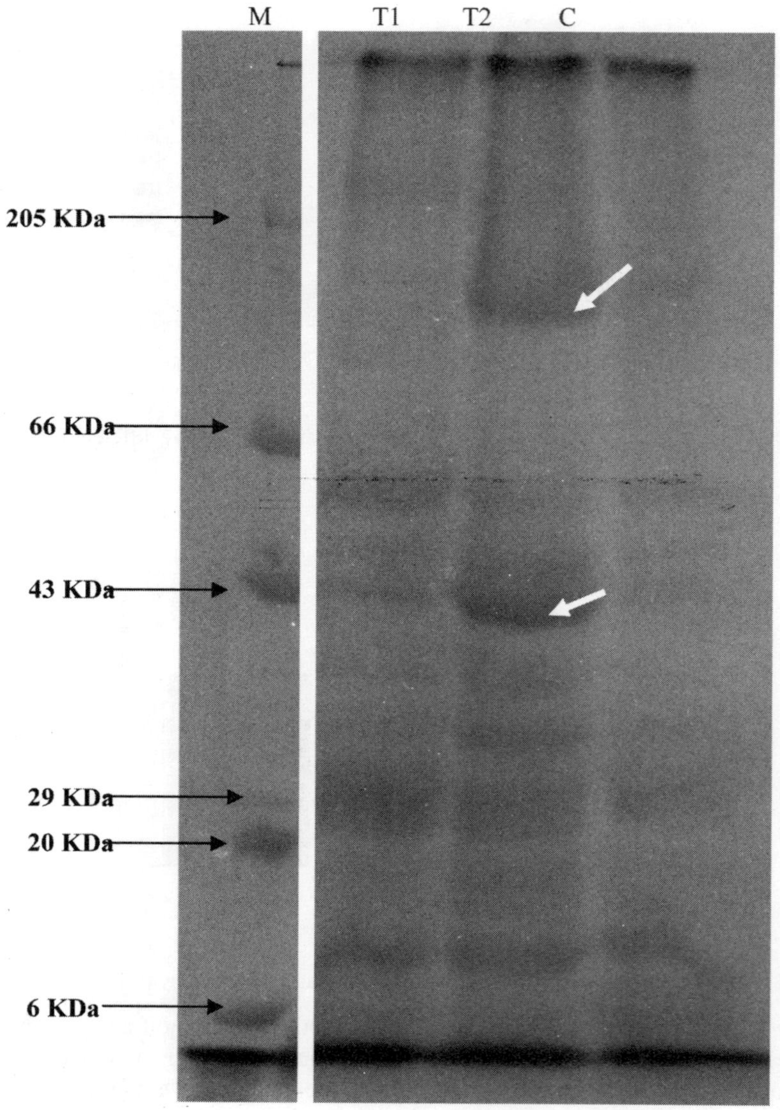

Fig. 3. SDS-PAGE profile of haemolymph protein of *Helicoverpa armigera* larva after 48 hr of treatment with Treatment 1 (T1), Treatment 2 (T2) and control (C). First lane shows the details of protein molecular weight marker (M).

that neem compounds damaged the midgut of *H. armigera* larva. Barbeta *et al.* (2004) tested the effect of plant cyclotides on the gut histology of *H. armigera* and reported similar findings. They reported that plant cyclotides induced disruption of the microvilli, blebbing, swelling, and ultimately ruptured the cells of the gut epithelium. They also stated that histology of this response was similar to the response of *H. armigera* larvae to the *Bacillus thuringiensis* delta-endotoxin.

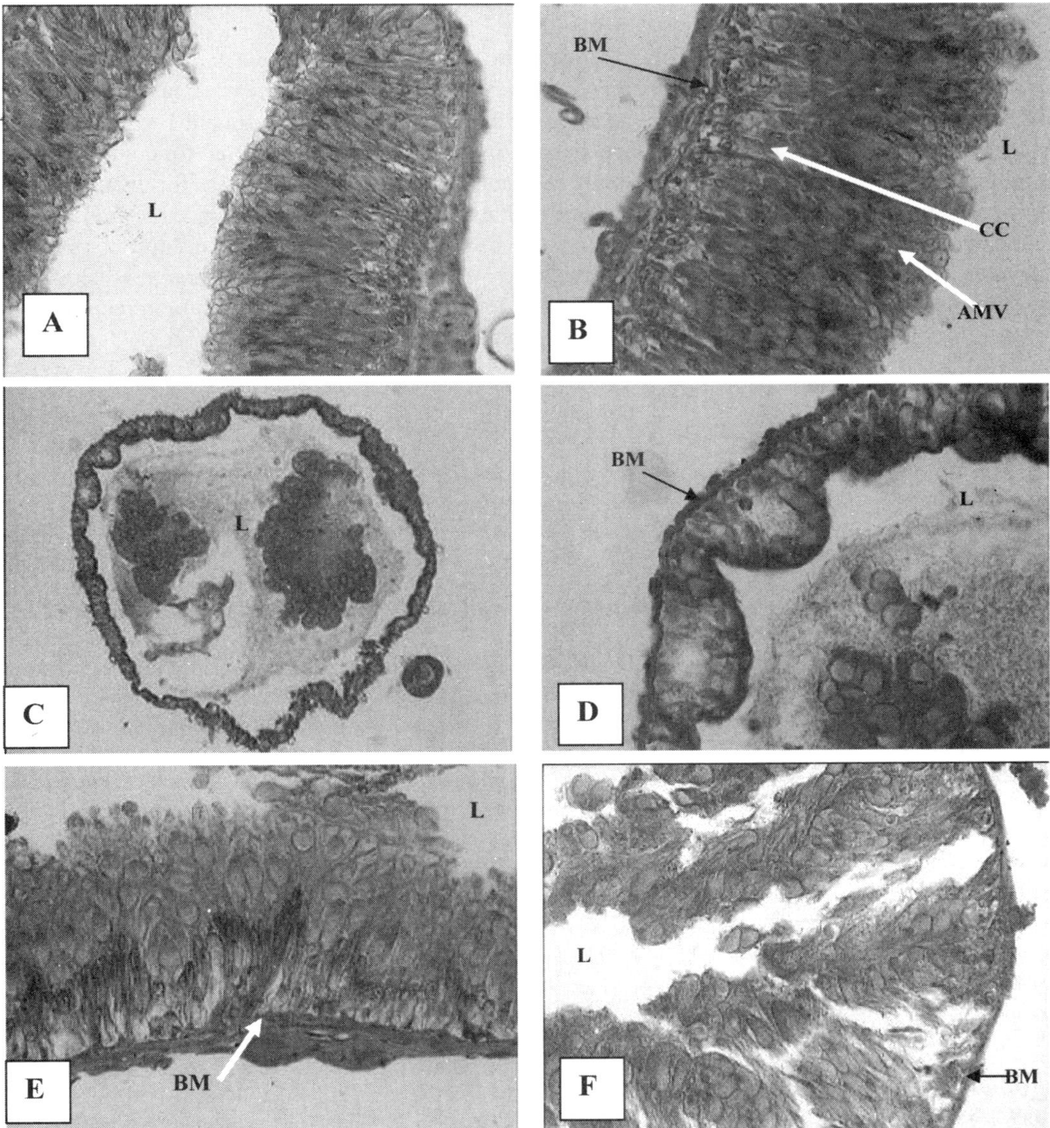

Fig. 4. Histopathological effect of Treatment 1 (neem oil + karanj oil + active fraction from H. suaveolens) on the midgut and hindgut of Helicoverpa armigera fourth instar larva after 48 hr of treatment. (A) & (B) Cross sections of midgut of control larva showing normal arrangement of intact columnar cells (CC) and apical micro villi (AMV) (400X); (C) Cross section of midgut of treated larva (100X); (D) T.S. of midgut of treated larva magnified to 400X. Vesicles shed into the gut lumen (L) and basal membrane (BM) is largely devoid of intact attached cells; (E) hindgut of control larvae (400x); (F) hindgut of treated larvae (400x): columnar cells are destroyed and dystrophication of the epithelial tissue.

Conclusion

The present study clearly gives a picture of the acute lethal effects of the formulation of neem oil, karanj oil with *H. suaveolens* fraction on *H. armigera.* Even though all the six treatments showed antifeedant activity, T1 and T2 were found to be significantly effective. Treatment 1 was also found to be damaging the gut of the larva. Hence these formulations can be considered for controlling the lepidopteran insects. Field trials are necessary to conclude the bioefficacy of these formulations. We have analysed the active fraction by chromatographic and spectroscopic techniques to find out the active compound. Beta Caryophyllene was found to be the major ingredient of the active compound. Caryophyllene is a volatile compound and is quickly evaporated at room temperatures. However the evaporation of this compound could be slowed down by mixing it with neem or karanj oil. In future this compound could be included as a botanical pesticide.

Acknowledgement

The present work has been supported by the Council of Scientific and Industrial Research (CSIR), New Delhi.

References

Abbott, W.S., 1925. A method of computing the effectiveness of an insecticide. *J. Econ. Entomol.*, 18, 265–267.

Ahmed, M., Scora, R.W. and Ting, I.P., 1994. Composition of leaf oil of *Hyptis suaveolens* (L.) Poit. *J. Essent. Oil Res.*, 6, 571-575.

Barbeta, B.L., Marshall, A.T., Gillon, A.D., Craik, D.J. and Anderson, M.A., 2004. Plant cyclotides disrupt epithelial cells in the midgut of lepidopteran larvae. *Pest Management Science*, 60(5), **459–464.**

Berenbaum, M.R., 1983. Effects of tannins on growth and digestion in two species of papilionids. *Entomologia Experimentalis et Applicata*, 34(3), 245-250.

Flipsen, J.T.M., Martens, J.W.M., Van Oers, M.M., Vlak, J.M. and Van lent, W.M., 1995. Passage of *Autographa californica* nuclear polyherosis virus through the midgut epithelium of *Spodoptera exigua* larvae. *Virol*, 208, 328-335.

Fun, C.E. and Svendsen, A.B., 1990. The essential oil of *Hyptis suaveolens* Poit. grown on Aruba. *Flav. Fragr. J.*, 5, 161-163.

George, S and Vincent, S., 2005. Comparative efficacy of *Annona squamosa* Linn. and *Pongamia glabra* Vent. to *Azadirachta indica* A. Juss against mosquitoes. *J. Vect. Borne Dis.,* 42, 159–163.

Goławska, S., 2007. Deterrence and Toxicity of Plant Saponins for the Pea Aphid *Acyrthosiphon Pisum* Harris. *Journal of Chemical Ecology,* 33 (8), 1598-1606.

Govindachari, T.R., Narasimhan, N.S., Suresh, G., Partho, P.D. and Gopalakrishnan, G., 1996. Insect antifeedant and growth-regulating activities of salannin and other C-seco limonoids from neem oil in relation to azadirachtin. *Journal of Chemical Ecology*, 22, 1453-1461.

Isman, M.B., Koul, O., Luczynski, A. and Kaminski, J., 1990. Insecticidal and antifeedant bioactivities of neem oils and their relationship to azadirachtin content. *Journal of Agriculture and Food Chemistry*, 38, 1406-1411

Koul, O., Jain, M.P. and Sharma, V.K., 2000. Growth inhibitory and antifeedant activty of extracts from *Melia dubia* to *Spodoptera litura* and *Helicoverpa armigera* larvae *Indian J. Exp. Biol* 38, 63-68.

Kumar, V., Chandrashekar, K. and Sidhu, O.P., 2007. Synergistic action of neem and karanj to aphids and mites. *Journal of Entomological Research*, 31(2), 121-124

Laemmli, U.K., 1970. Cleavage of structural proteins during the assembly of the head of bacteriophage T4. *Nature,* 227, 680-685.

Lowry, O.H., Rosebrough, J.J., Farr, A.L. and Randall, R.J., 1951. Protein measurement with the folin phenol reagent. *J. Biol.Chem.*, 193, 263-275.

Mordue, A. J. and Blackwell, A., 1993. Azadirachtin: an update. *J. Insect Phys.*, 39, 903-924.

Murugan, K., Sivaramakrishnan, S., Senthil Kumar, N., Jeyabalan, D. and Senthil Nathan, S., 1998. Synergistic interaction of botanicals and biocides Nuclear Polyhedrosis Virus on pest control. *Journal of Scientific and Industrial Research.* 57, 732 – 739.

Narasimhan, V., Rajappan, K., Ushamalini, C. and Abdul Kareem, A., 1998. Efficacy of New EC Formulations of Neem Oil and Pungam Oil for the Management of Sheath Rot Disease of Rice. *Phytoparasitica*, 26(4), 301-306

Neoliya, N.K., Singh, D. and Sangwan, R.S, 2007. Azadirachtin-based insecticides induce alteration in *Helicoverpa armigera* Hub. head polypeptides. *Current Science*, 92(1), 94-99.

Pathak K.M.L. and Shukla, R.C., 1998. Efficacy of AV/EPP/14(herbal ectoparasiticide) against *Canine demodicosis*. *J. Vet. Parasitol,* 12(1), 50–51.

Pavela, R. and Herda, G., 2007. Repellent effects of pongam oil on settlement and oviposition of the common greenhouse whitefly *Trialeurodes vaporariorum* on chrysanthemum. *Insect Science*, 14(3), 219-224.

Prasad, A. and Wadhwani, Y., 2006. Pathogenic virus and insecticides: an effective way of pest control. *Current Science*, 91, 803-807.

Qadri, S.S.H. and Narsaiah, J., 1978. Effects of azadirachtin on the moulting process of last instar nymphs of *Periplaneta Americana* (Linn.). *Indian J. Exp. Biol.*, 16, 1141–1143.

Raja, N., Jeyasankar, A., Venkatesan, S.J. and Ignacimuthu, S., 2005. Efficacy of *Hyptis suaveolens* against lepidopteran pests. *Current Science*, 88(2), 220-222.

Rao, S.N., Raguraman, S. and Rajendran, R., 2003. Laboratory assessment of the potentiation of neem extract with the extracts of sweet-flag and pungam on bhendi shoot and fruit borer, *Earias vitella* (Fab.). *Entomon*, 28(3), 277-281.

Rembold, H. and Sieber, K.P., 1981. Inhibition of oogenesis and ovarian ecdysteroid synthesis by azadirachtin in *Locusta migratoria migratorioides* (R. & F.). *Z. Naturforsch.* 36c, 466-469.

Rembold, H., Sharma, G.K., Czoppelt, Ch. and Schmutterer, H., 1982. Azadirachtin: A potent insect growth regulator of plant origin. *Z. Angew. Entomol.*, 93, 12–17.

Schmidt, G.H., Rembold, H., Ahmed, A.A.I. and Breuer, M., 1998. Effect of *Melia azedarach* fruit extract on juvenile hormone titre and protein content in the haemolymph of two species of noctuid lepidopteran larvae (Insecta: Lepidoptera: Noctuidae). *Phytoparasitica*, 26(4), 283 – 291.

Singh, H.B. and Handique, A.K., 1997. Antifungal activity of the essential oil of *Hyptis suaveolens* and its efficacy in biocontrol measures in combination with *Trichoderma harzianum*. *J. Essent. Oil Res.*, 9, 683-687.

Subrahmanyam, B. and Rao, P.J., 1986. Azadirachtin effects on *Schistocerca gregaria* Forskal during ovarian development. *Curr. Sci.*, 55, 534–538.

Antifeedant and larvicidal effects of *Hydnocarpus alpina* Wt. (Flacourtiaceae) extracts against the larvae of *Helicoverpa armigera* Hub. (Lepidoptera: Noctuidae)

S. Ezhil Vendan,* K. Baskar,* M. Gabriel Paulraj* and S. Ignacimuthu*

Feeding deterrent and larvicidal activities of hexane, chloroform and ethyl acetate crude extracts of *Hydnocarpus alpina* leaf was studied against cotton bollworm *Helicoverpa armigera* (Hubner) using cotton leaf disc no-choice test. Four different concentrations (viz. 0.625, 1.25, 2.5 and 5.0 per cent) were screened. Antifeedant activity was assessed after 24hr and larval mortality was observed after 96hr. The ethyl acetate extract showed maximum antifeedant activity (78.41%) and larvicidal activity (77.78%) at 5 per cent concentration. Bioassay guided fractionation of ethyl acetate extract of *H. alpina* using column chromatography and thin layer chromatography yielded ten different fractions. All ten fractions were further screened at 125, 250, 500 and 1000ppm concentrations for antifeedant and larvicidal activities. Fraction 8 presented maximum antifeedant activity (73.40%) and larvicidal activity (66.67%) at 1000ppm concentration. *H. alpina* seems to be a promising candidate to develop botanical pesticide.

Introduction

The cotton bollworm, *Helicoverpa armigera* (Hubner) (Lepidoptera: Noctuidae) is a well known polyphagous pest. It is one of the biologically successful pests due to its high fecundity, great migration potential, wide host range and diapausing behaviour to overcome unfavourable environmental conditions (Ahmad *et al.*, 2001). The major problem with *H. armigera* in pest management programme is that it has developed resistance to many synthetic insecticides including pyrethroids (Shen *et al.*, 1991; Mc Caffery, 1998). From an ecological point of view, the application of chemical pesticides is not good since it contaminates the environment, kills beneficial arthropods in agroecosystems and creates health hazards to humans (Isman, 2006). In many countries, plant derived products are being used by the farmers from ancient times and it triggered the scientists to search for ecofriendly insecticides from plant kingdom. Several hundred plants have been reported as insect repellents, antifeedants, attractants, insecticides, ovicides and oviposition deterrents (Arnason *et al.*, 1992; Ewete *et al.*, 1996). Many plant extracts and plant compounds have been documented as insecticides, antifeedants and growth regulators against *H. armigera* (Neoliya *et al.*, 2003). Since plant kingdom encloses numerous species, the search for promising botanicals, which will challenge synthetic chemicals in pest management, is necessary. *Hydnocarpus alpina*, commonly known as 'torathi', is traditionally used in the treatment of leprosy (Lima *et al.*, 2005). This plant is typically found in the evergreen forest areas of south Indian hills including Western Ghats and is dispersed mainly by animals (Sundrapandian *et al.*, 2005). In the present study, we have investigated the feeding deterrent effect of *H. alpina* leaf extracts and fractions against *H. armigera* larvae.

* Entomology Research Institute, Loyola College, Chennai – 600 034

Materials and methods

Plant collection and extraction: Fresh leaves of *H. alpina* were collected from Vellangiri hills, Coimbatore district, Tamil Nadu. The collected leaves were washed in water, shade dried at room temperature and ground into powder using an electric blender. About 2½Kg dried powder was sequentially extracted with three fold (6L) amount of hexane, chloroform and ethyl acetate after 72hr of soaking. The extract was filtered through filter paper and the resultant extract was concentrated under reduced pressure using rotary vacuum evaporator.

Insect: H. armigera larvae were obtained from a stock culture in the laboratory. The larvae were reared on semi-synthetic diet (Koul *et al*., 1997) at laboratory conditions (30 ± 2^{0}C; 57-65% RH; 11 ± 1hr photoperiod). Newly moulted third instar larvae were used for the studies.

Column Chromatography: The effective extract, (ethyl acetate extract) was subjected to column chromatography in a silica gel (200 g-acme's 100–200 mesh) glass column. About 24g of the crude extract was loaded on the column and eluted with hexane followed by the combination of hexane: ethyl acetate ranging from 95:5 to 0:100. The fractions were collected in a 200ml conical flask and concentrated using rotary vacuum evaporator. Ten fractions were obtained.

Treatments and concentrations: The crude extract and fractions were dissolved in acetone to prepare different concentrations viz. 0.625, 1.25, 2.5 and 5 per cent (for crude extract) and 125, 250, 500 and 1000ppm (for fractions). A reference control with azadirachtin and a solvent control with acetone were also maintained separately.

Antifeedant bioassay: The antifeedant activity of extracts against *H. armigera* larvae was determined by using leaf disc no-choice method described by Isman *et al.* (1990). Cotton leaf discs (4cm diameter) (var. Surabi) were cut by a cork borer. The leaf discs were dipped separately in different concentrations of crude extracts and fractions for 2min and shade dried for 10min. The treated and control discs were separately put inside a petridish and one newly moulted third instar, 4hr pre-starved *H. armigera* larva, was introduced on one leaf disc taken in petridish. Progressive consumption of leaf area by the larva after 24hr feeding was recorded in control and treated discs using leaf area meter (Delta-T Devices, Serial No. 15736 F 96, UK). The experiments were replicated 10 times and feeding deterrence index for all concentrations of crude, fractions and azadirachtin was calculated using the formula of Isman *et al.* (1990): $(C - T) / (C + T) \times 100$, where C is leaf consumption in control and T is consumption in treated discs.

Larvicidal activity experiment: In a separate set of experiments, third instar *H. armigera* larvae were orally treated with different concentrations of crude extract, fractions and azadirachtin through cotton leaf discs as mentioned above in the antifeedant experiement. A solvent control was also maintained. Larval mortality was recorded in the larvae for 96hr. Ten replications were maintained for each control and treatments. The mortality was adjusted by Abbott's correction factor (Abbott, 1925).

Statistical analysis: The significance of treatments was found out by one way Analysis of Variance (ANOVA) and effective treatment was separated by Tukey's multiple range test. Differences between means were considered significant at $P < 0.05$.

Results and discussion

Antifeedant activity of crude extracts: Table 1 shows the antifeedant activity of hexane, chloroform and ethyl acetate extracts of *H. alpina* against *H. armigera* larvae. Among the three extracts, ethyl acetate extract recorded the maximum antifeedant activity followed by hexane extract. The activity was concentration dependent. The antifeedant activity of ethyl acetate extract was significant ($p<0.05$) at all concentrations and was recorded as 56.2, 62.8, 69.1 and 78.4 per cent at 0.62, 1.25, 2.5 and 5 per cent concentrations. Raja *et al.* (2005), Susuruk *et al.* (2007), Malarvannan *et al.* (2008) and Samarasinghe *et al.* (2008) have reported ethyl acetate extract as an effective treatment that showed maximum feeding deterrent activity against lepidopteran pests. Antifeedants give a first line of crop protection against herbivorous insects. According to Isman (2002) any substance that reduces food consumption by an insect can be considered as an antifeedant or feeding deterrent. In general antifeedants have profound adverse effects on insect feeding behavior (Hummelbrunner, 2001). In the present investigation the food consumption of *H. armigera* in ethyl acetate extract treatment was reduced and it was concentration dependent (Figure 1). In control the consumption was 1252.42mm^2 leaf per day where as in ethyl acetate extract treatment the larva consumed 353.11, 288.49, 229.79 and 161.41mm^2 per day at 0.625, 1.25, 2.5 and 5 per cent concentrations respectively.

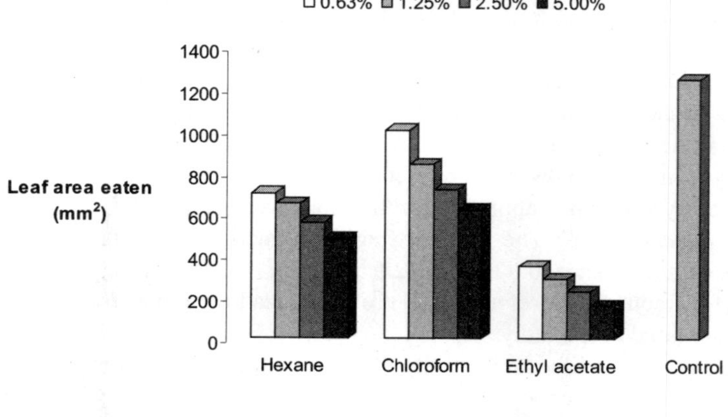

Fig. 1. Mean leaf area consumption by a single third instar larva of *Helicoverpa armigera* after 24 hr of treatment with different concentrations of *Hydnocarpus alpina* crude extracts.

Table 1. Per cent antifeedant effect of crude extracts of *Hydnocarpus alpina* leaves against third instar *Helicoverpa armigera* larvae (Mean ± SD) (n=10)

Extract	Concentration (%)			
	0.625	1.25	2.50	5.0
Hexane	28.43 ± 7.42[b]	31.68 ± 4.99[b]	38.09 ± 3.83[b]	43.54 ± 4.71[b]
Chloroform	11.07 ± 3.74[a]	19.75 ± 5.18[a]	26.99 ± 4.17[a]	32.84 ± 6.10[a]
Ethylacetate	56.21 ± 4.26[c]	62.82 ± 7.13[c]	69.11 ± 4.92[c]	78.41 ± 3.75[c]

Values in each column followed by the same alphabets are not significantly different by Tukey's test at $P \leq 0.05$.

Antifeedant activity of fractions: By column chromatography method, 10 different fractions were isolated from the ethyl acetate extract and all of them were screened against *H. armigera* larva for antifeedant activity. It was found that fraction 8 was the most effective antifeedant (Table 2). At 500ppm concentration, Fraction 8 showed more than 50 per cent feeding deterrent effect and the activity was increased to 73.41 per cent when the concentration was increased to 1000ppm. Azadirachtin (reference control) presented 87.7 per cent antifeedant activity at 1000ppm concentration, which was significantly high ($p<0.05$) compared to fraction 8. Raja *et al.* (2004), Ignacimuthu *et al.* (2006), Pavunraj *et al.* (2006), Malarvannan *et al.,* (2008), Samarasinghe *et al.,* (2008) and Susuruk *et al.* (2007) have reported the antifeedant and insecticidal activities of many plant crude extracts and fractions. Feeding inhibition was increased with increasing concentrations of fractions (Figure 2). The consumption was 1252.42mm^2 leaf per day where as in eighth fraction treatment the larva consumed 707.16, 453.19, 358.33 and 194.782mm^2 per day at 125, 250, 500 and 1000ppm concentrations respectively. Bioassay-guided fractionation of crude plant extracts is an efficient work in phytochemistry (Hostettmann and Wolfender, 1999). Nathan *et al.* (2007) reported that phytocompounds caused deleterious effects to insect in diverse ways, such as through interference with the consumption and/or utilization of food, acute toxicity and enzyme inhibition. The present work clearly indicates that *H. alpina* ethyl acetate extract and the fraction eluted by 30% hexane and 70% ethyl acetate solvent system had good antifeedant activity against *H. armigera* larvae. The antifeedant property of plant extracts is due to the presence of secondary metabolites such as alkaoids, phenolics and terpenoids (Frazier, 1986).

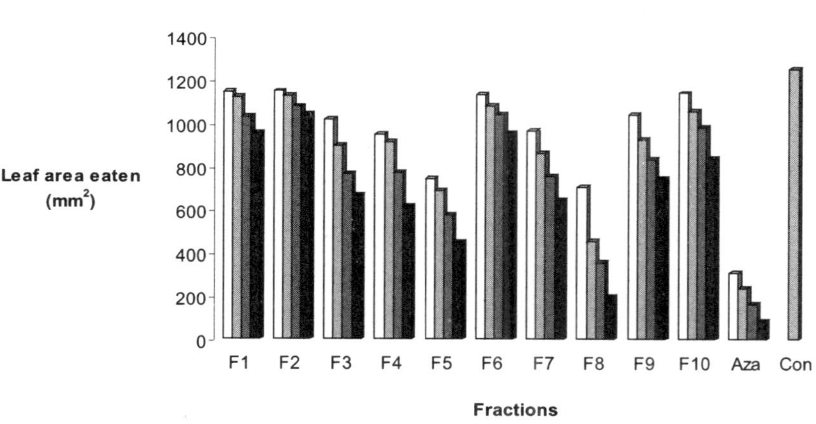

Fig. 2. Average leaf area consumption by third instar larva of *Helicoverpa armigera* after 24hr of treatment with different concentrations of *Hydnocarpus alpina* fractions separated from ethyl acetate extract.

Table 2. Feeding inhibitory activity of ten fractions collected from ethyl acetate extract of *Hydnocarpus alpina* against *Helicoverpa armigera* after 24 hr of treatment. (Mean ± SD) (n=10)

Fractions	Concentration (ppm)			
	125	250	500	1000
F1	4.71 ± 0.89^{ab}	5.76 ± 2.07^{ab}	9.95 ± 2.03^{ab}	14.08 ± 6.47^{abc}
F2	4.65 ± 1.22^{a}	5.59 ± 1.47^{a}	7.77 ± 2.73^{ab}	9.56 ± 1.72^{ab}
F3	10.61 ± 2.57^{ab}	16.73 ± 4.37^{ab}	24.38 ± 4.99^{bc}	31.06 ± 3.56^{bcd}
F4	13.96 ± 2.72^{a}	15.71 ± 2.60^{a}	24.15 ± 6.08^{ab}	34.61 ± 4.29^{abc}
F5	25.63 ± 2.92^{c}	29.32 ± 4.10^{c}	37.52 ± 4.57^{d}	47.53 ± 5.73^{e}
F6	5.32 ± 1.55^{a}	7.55 ± 2.34^{a}	9.55 ± 1.58^{a}	13.61 ± 3.24^{a}
F7	13.11 ± 3.05^{ab}	18.65 ± 4.30^{ab}	25.21 ± 6.35^{bc}	32.55 ± 4.89^{cd}
F8	28.21 ± 3.46^{d}	47.31 ± 5.83^{d}	55.96 ± 6.46^{e}	73.41 ± 6.09^{f}
F9	9.50 ± 1.69^{ab}	15.08 ± 2.31^{ab}	20.33 ± 3.96^{abc}	25.64 ± 4.51^{abc}
F10	4.92 ± 1.19^{bc}	8.74 ± 1.86^{b}	12.2 ± 1.99^{c}	20.09 ± 5.03^{d}
Azadirachtin (40.86%)	60.21 ± 6.43^{e}	68.16 ± 4.83^{e}	76.81 ± 2.37^{f}	87.69 ± 5.94^{g}

Values in columns followed by the same letters are not significant different (Tukey's test; P ≤ 0.05).

Toxic effect of crude extract and fractions: Figures 3 and 4 shows the per cent larval mortality recorded at different concentrations of crude extract and fractions of *H. alpina* leaves. From the data it was clear that the crude extracts and fractions killed larvae and the toxic effect increased with the increased concentration levels. Among the crude extracts, ethyl acetate extract killed maximum number of larvae and the mortality was recorded as 77.78 per cent at 5 per cent concentration. According to Leatemia and Isman (2004), the antifeedants lead to insect mortality due to the effect of combination of starvation and contact toxicity. The screening of different fractions clearly indicated that fraction 8 was the most effective among the 10 different fractions. Fraction 8 recorded 33.33, 50.00, 50.00 and 66.67 per cent larval mortality at 125, 250, 500 and 1000ppm concentrations respectively. However azadirachtin recorded significantly high larval mortality, which was recorded as 55.56, 77.78, 88.89 and 100 per cent larval mortality at 125, 250, 500 and 1000ppm concentrations respectively. Currently the world wide demand for natural insecticides is increasing. The overall results of the present investigation clearly shows that both crude and fractions exhibited antifeedant

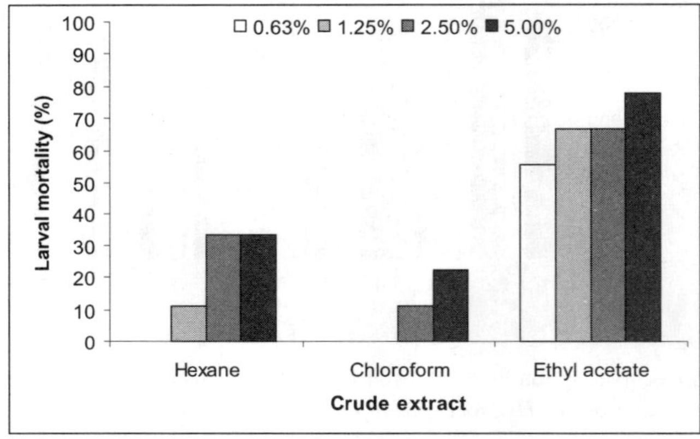

Fig. 3. Larval mortality (%) of *H. armigera* due to *H. alpina* crude extract treatment.

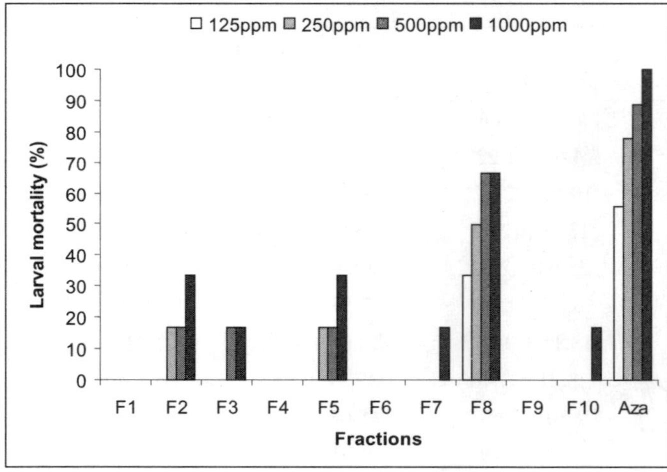

Fig. 4. Larvicidal effect of different fractions collected from ethyl acetate extract of *H. alpina* leaves against *H. armigera* larvae.

and larvicidal activities. Even though the larvae consumed very little quantity of treated leaf, they were affected by the toxin present in the plant extract. Leatemia and Isman (2004) reported that high concentrations of extracts caused high mortality of larvae even though only very small portions of the leaf discs were consumed. The active principle present in the effective fraction, which showed both antifeedant and larvicidal activities, should be analysed, further to which may add new information in the field of botanical pesticides.

ACKNOWLEDGEMENT

The authors are thankful to the Department of Science and Technology, New Delhi for financial support.

REFERENCES

Abbott, W.S., 1925. A method of computing the effectiveness of an insecticide. *Journal of Economic Entomology*, 18, 265-276.

Ahmad, M., Arif, M.I. and Ahmad, Z., 2001. Resistance to carbamate insecticides in *Helicoverpa armigera* (Lepidoptera: Noctuidae) in Pakistan. *Crop Protection,* 20, 427-432.

Arnason, J.T., Mackinnon, S., Isman, M.B. and Durst, S., 1992. Insecticides in tropical plants with non-nerotoxic modes of action. *Recent Adv. Phytochem,* 28, 107-131.

Ewete, F.K., Arnason, J.T., Larson, J. and Philogene, B.J.R., 1996. Biological activities of extracts from traditionally used Nigerian plants against the European corn borer, *Ostrinia nubilalis*. *Entomol. Exp. Appl,* 80, 531-537.

Frazier, J., 1986. The perception of plant allelochemicals that inhibiting feeding, pp. 1-42. In: Molecular Aspects of Insect-Plant Associations, L.B. Brattsten and S. Ahmed (Eds.), Plenum Press, New York.

Hostettmann, K. and Wolfender, J.L., 1999. Application of LC/MS and LC/NMR in the search for new bioactive compounds from plants of the Americas. In K. Hostettmann, M.P. Gupta and A. Marston, (ed.), Chemistry, biological and pharmacological properties of medicinal plants from the Americas. 19-41.

Hummelbrunner, L.A. and Isman, M.B., 2001. Acute, Sublethal, Antifeedant, and Synergestic Effects of Monoterpenoid Essential Oil Compounds on the Tobacco Cutworm, *Spodoptera litura* (Lep., Noctuidae). *J. Agric. Food Chem,* 49, 715-720.

Ignacimuthu, S., Maria Packiam, S., Pavunraj, M. and Selvarani, N., 2006. Antifeedant activity of *Sphaeranthus indicus* L. against *Spodoptera litura* Fab. *Entomon,* 31(1), 41-44.

Isman, M. B., Koul, O., Lucyzynski, A. and Kaminski, J., 1990. Insecticidal and antifeedant bioactivities of neem oils and their relationship to Azadirachtin content. *Journal of Agricultural food Chemistry*. 38, 1407-1411.

Isman, M.B., 2002. Insect antifeedants. *Pesticide Outlook*, 152-156.

Isman, M.B., 2006. Botanical insecticides, deterrents, and repellents in modern agriculture and repellents in modern agriculture and an increasingly regulated world. *Annu. Rev. Entomol,* 51, 45-66.

Koul. O., Shankar, J.S. Mehta, N. Taneja, S.C. Tripathi, A.K. and Dhar, K.L., 1997. Bioefficacy of crude extracts of Aglaia species (Meliaceae) and some active fractions against lepidopteran larvae; *J. Appli. Entomol,* 121, 245-248.

Leatemia, J.A and Isman, M.B., 2004. Toxicity and antifeedant activity of crude seed extracts of *Annona squamosa* (Annonaceae) against lepidopteran pests and natural enemies. *International Journal of Tropical Insect Science*, 24(1), 150-158.

Lima, J.A., Oliveira, A.S., de Miranda, A.L.P., Rezende, C.M. and Pinto, A.C., 2005. Anti-inflammatory and antinociceptive activities of an acid fraction of the seeds of *Carpotroche brasiiensis* (Raddi) (Flacourtiaceae). *Brazilian Journal of Medical Biological Research*, 38, 1095-1103.

Malarvannan, S., Giridharan, R., Sekar, S., Prabavathy, V.R. and Nair, S., 2008. Bioefficacy of crude and fractions of *Argimone mexicana* against Tobacco caterpillar, *Spodoptera litura* Fab. (Lepidoptera: Noctuidae). *Journal of Biopesticides*, 1, 55-62.

Mc Caffery, A.R., 1998. Resistance to insecticides in heliothine Lepidoptera: a global view. *Phil. Trans. R. Soc. Lond. B*. 353, 1735-1750.

Nathan, S.S., Man-Young Choi., Chae-Hoon Paik and Hong-Yul Seo., 2007. Food consumption, utilization, and detoxification enzyme activity of the rice leaffolder larvae after treatment with *Dysoxylum* triterpenes. *Pesticide Biochemistry and Physi*ology, 88, 260-267.

Neoliya, N.K., Shukla, Y.N. and Mishra, M., 2003. New possible insect growth regulators from *Catharanthus roseus*. *Current Science*, 84, 1184-1186.

Nivsakar, M., Cherian, B. and Padh, H., 2001. Alpha-terthienyl: A plant-derived new generation insecticide. *Current Science*. 81, 667-672.

Pavunraj, M., Subramanian, K., Muthu, C., Prabhu Seenivasan, S., Duraipandiyan, V., Mariapackiam, S. and Ignacimuthu, S., 2006, Bioefficacy of *Excoecaria agallocha* (L.) leaf extract against the armyworm *Spodoptera litura* (Fab.) (Lepidoptera: Noctuidae). *Entomon,* 31, 37-40.

Raja, N., Jayakumar, M., Elumalai, K., Jeyasankar, A., Muthu, C. and Ignacimuthu, S., 2004. Oviposition deterrent and ovicidal activity of solvent extracts of 50 plants against the armyworm, *Spodoptera litura* Fab. (Lepidoptera: Noctuidae). *Malas. Appl. Biol*, 33(1), 73-81.

Raja, N., Jayakumar, M., Venkatesan, S.J. and Ignacimuthu, S., 2005. Efficacy of *Hyptis suaveolens* against lepidopteran pests. *Current science*, 88, 220-222.

Samarasinghe, M.K.S.R.D., Chhillar, B.S. and Singh, R., 2008. Effect of methanolic extract and fractions of garlic oviposition and egg hatchig of *Plutella xyostella* (Linnaeus). *J. Insect Sci,* 21(1), 28-33.

Shen, J., Tan, J., Xiao, B., Tan, F. and You, Z., 1991. Monitoring and forecasting of pyrethroid resistance of *Heliothis armigera* (Hubner) in China. *Entomol. Knowledge*, 28, 337-341.

Sundrapandian, S.M., Chandrasekaran, S. and Swamy, P.S., 2005. Phenological behaviour of selected tree species in tropical forests at Kodayar in the Western Ghats, Tamil Nadu, India. *Current science*, 88, 805-810.

Susurluk, H., Caliskan, Z., Gurkan, O., Kirmizigul, S. and Goren, N., 2007. Antifeedant activity of some *Tanacetum* species and bioassay guided isolation of the secondary metabolites of *Tanacetum cadmium* ssp. Cadmium (Compositae). *Industrial Crops and Products*. 26, 220-228.

Efficacy of Leaf Extracts of *Aristolochia bracteata* Retz. on Mortality, Feeding Physiology and Reproductive Ability of Brinjal Spotted Leaf Beetle, *Henosepilachna vigintioctopunctata* (Fabricius)

C. Balasubramanian* and N. Kalimuthu*

The phyto-chemical method for the control of brinjal spotted beetle, *Henosepilachna vigintioctopunctata* was performed by the leaf extracts of *Aristolochia bracteata* with various solvents such as Petroleum ether, Benzene, Ethyl acetate, Chloroform, Carbon-tetrachloride, Acetone and Cumulative treatment. The bioassay on feeding physiology and reproductive efficiency was performed with these extracts against III, IV instar and adult beetles revealed that with regard to all the life forms of the test insect, cumulative treatment was found to have potential effect with respect to all other extracts. The cumulative treatment enhanced 50% mortality with a low concentration i.e. LC_{50}= 65.55 ppm, 113.35 ppm, 84.28 ppm against III, IV instars and adult beetles of *H. vigintioctopunctata* respectively. The electrophorogram revealed third and fourth day incubation appearance of new proteins when treated with benzene extract whereas the cumulative extract treated eggs had less number of bands developed during incubation period. All feeding physiological parameters were diminished when treated with cumulative extracts against the spotted brinjal beetle than other extracts treatment.

Introduction

Pest menace gained momentum with the progression and introduction of innovative tools in agriculture (Pradhan, 1967). Plant products may prove to be useful in formulating sound pest management strategies (Banerji *et al.*, 1985). Because of their biodegradable nature and being harmless to non target organisms in the environment, extensive survey of the flora was undertaken to search for potential plant extracts, which could be used in the management of agricultural, household pests (Ananthakrishnan,1988) and agricultural pests (Sharma *et al* ., 1990). Plant alkaloids have been found to affect physiological systems in higher animals as well as in insects (Saxena and Tikker, 1990). Schneider *et al.* (1982) have mentioned that these compounds in general are useful to control insects.

Information on the effect of plant extracts of *A. bracteata* on grubs and adults of *H. vigintioctopunctata* is not available. In the present investigation, Petroleum ether, Benzene, Carbon tetra-chloride, Ethyl acetate, Chloroform, Acetone and Cumulative effect (altogether mixture) of the same on the feeding physiology and reproductive features like fecundity, hatchability and biochemistry of egg of *H .vigintioctopunctata* are presented.

Materials and Methods

H. vigintioctopunctata beetles were collected from the infested brinjal fields from Paramakudi, Tamil Nadu and they were reared on *Solanum melongena* foliages in the laboratory at 28 ± 1°C and 60 %

* Department of Zoology, Thiagarajar College, Madurai-625 009

R.H. The grubs hatched from the eggs were maintained on *S. melongena* till adults emerged. The stock culture was maintained in the laboratory for several generations.

Leaves of *Aristochia bracteata* Retz., family Aristolochiae were collected from the field and shade dried at room temperature. After complete evaporation, the leaf powder was prepared by using home homogenizer. Fifty grams of the powdered material was extracted in a Soxhlet apparatus for 8 h at 58°C temperature starting in turn with petroleum ether, Benzene, Carbon tetra chloride, Chloroform (to separate lipid & terpenoids) and then using Ethyl acetate, Acetone, Ethanol, methanol and water (for more polar compounds). After 8 hours of extraction the extracts were filtered through Watman filter paper No.1 and concentrated. After complete evaporation of the solvent, the residue of the extract was stored in a refrigerator and used whenever needed, after re-dissolving in acetone. The standard solutions of desired concentration were prepared in acetone to yield a concentration of 100 mg /ml. The known quantity of extract was sprayed on brinjal leaves in different concentration of test samples and the leaves were kept in shade for 15 minutes and the leaves were offered to third, fourth instar grubs and adult beetles.

The mortality counts were taken after 24 h. Control experiments with the application of acetone alone on the food leaves were maintained simultaneously for each experiment. For assessing the percentage of mortality the method described by Abbot (1925) was used. The safe sub-lethal concentration was calculated by multiplying an application factor of 0.25 % with respective LD_{50} values determined from the acute toxicity tests as recommended by the Ontario Ministry of Environment (1974).

The sub-lethal doses of toxicants chosen were placed on a small piece of leaf with the help of a micropipette and acetone was allowed to evaporate (Srivastava and Prasad, 1980). The experimental larvae and adult beetles were starved for 2 - 4 h. and then allowed to consume the extract treated leaf after which untreated leaves were offered continuously to avoid starvation. Thus known quantity of the extract was orally administered to the grubs and adult beetles daily and the controls were fed with leaf pieces similarly but treated only with the concerned solvent.

The scheme of feeding budgets followed in the present work is that of the IBP formula (Petruzwic and Mac Fadeyan, 1970). Newly emerged adult beetles were fed with different sublethal doses of leaf extract treated leaves. Number of eggs laid and percentage hatched over control were calculated for a period of 21 days. Thus the ovicidal activity of the different concentration of extract was assessed. The electrophoretic protein profiles were assessed by the modified method of Laemmli (1970).

Results

The larvicidal activity of *A. bracteata* leaf extracts with different solvents towards the III instar grubs of *H. vigintioctopunctata* showed that, with reference to the LC_{50} values of the various solvents, cumulative treatment brought about 50% mortality at the lowest concentration of 34.29 ppm. With regard to the leaf extracts of *A. bracteata* the cumulative treatment was again found to be more effective which resulted in 50% mortality of the IV instar grubs of *H. vigintioctopunctata* with a minimum concentration of 55.40 ppm. Whereas in the treatment of adult beetles with different leaf extracts, the cumulative treatment was found effective in causing 50% mortality with a lowest concentration of 84.29 ppm. With respect to other solvent treatments, the cumulative treatment was followed by Petroleum ether which requires 65.55 ppm concentration for 50% mortality in the III instar and 113.35ppm in the case IV instar and 142.99 ppm for adult beetles. Petroleum ether treatment was followed by Benzene, Ethyl acetate, Chloroform, Carbon-tetrachloride and Acetone with reference to III and IV instars. But in the case of adult beetle, Petroleum ether was followed by Ethyl acetate (160.56ppm) and then Benzene (170.96ppm), Chloroform, Carbon-tetrachloride and Acetone (Table 1).

Table 1. Concentrations (ppm) required to 50% mortality in three stages of *H. vigintioctopunctata.*

Extracts	III Instars	IV Instars	Adult beetles
Petroleum ether	64.55	113.35	142.99
Benzene	87.19	124.53	160.56
Ethyl acetate	90.62	127.72	170.96
Chloroform	127.70	158.86	203.16
Carbon tetra chloride	130.71	171.88	226.18
Acetone	182.00	266.78	237.65
Cumulative extract (Extracts altogether)	34.29	55.40	84.28

Table 2. Cumulative Effect of leaf extract of *A. bracteata* on fecundity, hatchability and percentage of hatchability of brinjal spotted beetle *H. vigintioctopunctata*

Concentration (ppm)	No. of Eggs laid (N)	No. of eggs hatched (N)	Percentage of hatchability (%)
Control	280	278	99.28%
190 ppm	251	247	98.40%
200 ppm	247	243	98.38%
210 ppm	243	239	98.35%
220 ppm	240	234	97.50%
230 ppm	238	229	96.21%

Egg laying ability of the adult beetles subjected to different sublethal dosage of different solvent extracts of leaf of *A. bracteata* and cumulative effect of the same are assessed. Fecundity and hatchability were found to be reduced to a greater extent in benzene treated beetles. Hatchability was very low in the case of Benzene and petroleum ether at the concentration of 170, 180 and 160 ppm respectively when compared with other solvents. Though, the fecundity was reduced to a greater extent in the case of cumulative treatment hatchability was found to be greater like in other solvents (Table 2). Most significant negative correlation (r=0.9) was obtained between the concentration of the leaf extracts and number of eggs laid.

Electrophoretic analysis of egg protein pattern of the adult beetle of *H. vigintioctopunctata*

Fecundity is an important measure to assess the reproductive capability of pests. The fecundity and hatchability of eggs of *H. vigintioctopunctata* were assessed against different sublethal doses of extracts of *A. bracteata.* Maximum number of egg was laid by the insect in acetone treated insects. The lowest number of eggs was laid by the cumulative extract treated groups; even the fecundity of the benzene treated group was higher than the other groups. The percentage of hatchability was very low than the percentage of hatchability of other extracts on treated groups. It was observed that the period of development vary among the different solvent extracts treated groups. From the above data an attempt has been made on electrophoretic study on benzene and cumulative treated groups, because of the low hatchability and fecundity value. The electrophorogram of protein profile of eggs of *H. vigintioctopunctata* reared on sublethal doses of leaf extracts of *A. bracteata* and control are shown plate 1.

SDS_PAGE profile shows control insect's egg protein of *H. vigintioctopunctata* (Incubation period 0-

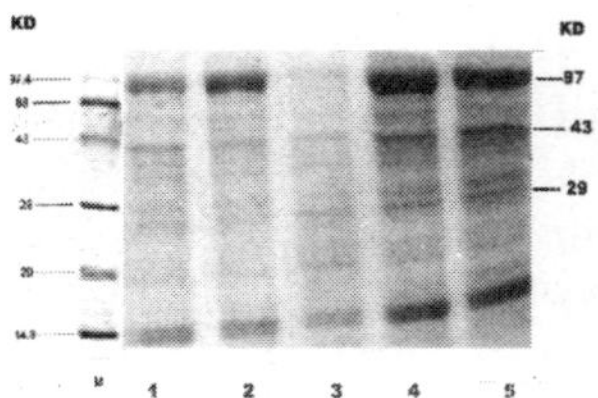

SDS-page profile shows egg protein of *H. vigintioctopunctata* (Benzene treated adult)

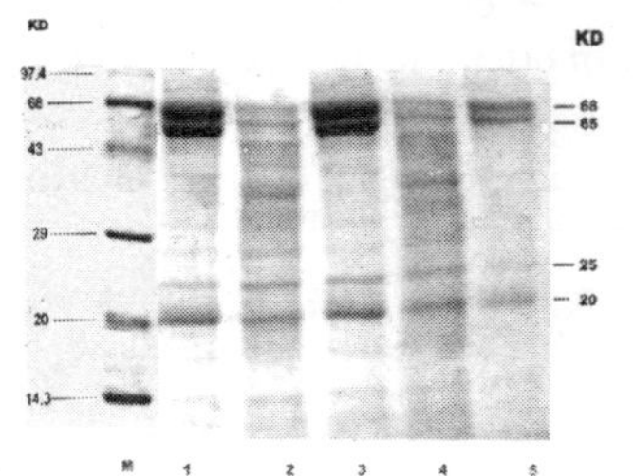

SDS_PAGE profile shows egg protein of *H. vigintioctopunctata* (Cumulative extract treated adult)

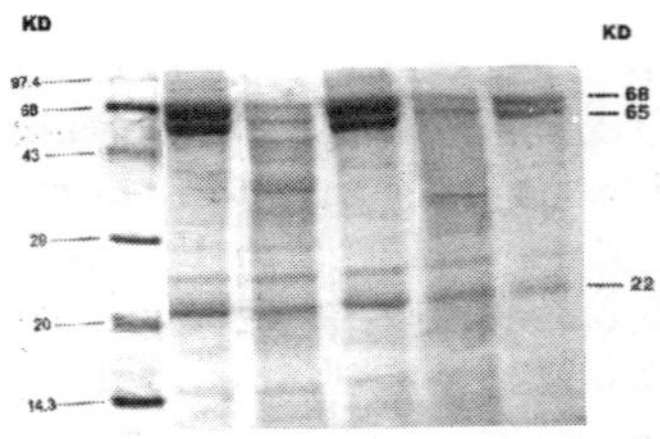

Legend for Plate 3
M- Molecular weight marker proteins
1- Total soluble proteins in '0' day egg
2- Total soluble proteins in '1' day egg
3- Total soluble proteins in '2' day egg
4- Total soluble proteins in '3' day egg
5- Total soluble proteins in '4' day egg

Plate 1. SDS-PAGE profile on *H. vigntiotpounctata* eggs treated with various solvent extracts of *A. bracteta*

In the protein profile of control *H. vigintioctopunctata* the protein bands appeared in the third and fourth day of incubation whereas when treated with benzene, the appearance of new proteins extract was seen from 0-4 day of incubation. In the case of cumulative extract treated insects the polypeptides were accumulated in less number of bands than the control. The storage of 68 kDa was highly utilized when treated with benzene and cumulative extract. The appearances of 22 kDa proteins in large amount in treated than the control. In the case of control 14.3 kDa protein appeared gradually from 0-4 days of incubation. Benzene and cumulative extract treated eggs were utilized wholly during the embryonic development.

INFLUENCE OF SUB- LETHAL DOSES OF DIFFERENT SOLVENTS EXTRACTS OF LEAF OF *A. BRACTEATA* ON FOOD UTILIZATION OF THIRD AND FOURTH INSTAR GRUBS AND ADULT BEETLE OF *H. VIGINTIOCTOPUNCTATA*

Food consumption

Food consumption study was carried out on third and fourth instars and adult beetle of *H. vigintioctopunctata* with various solvents extracts of leaf of *A. bracteata*. They were reared on different concentrations of cumulative extracts (Petroleum ether, Benzene, Carbon tetra chloride, Ethyl acetate, Chloroform Acetone and altogether). The consumption efficiency was considerably less in the all stages over the control (Table 1). The food consumption of control adult and control larvae was 0.417 mg/dry wt/adult, 0.387 and 0.315 mg / dry wt/larvae, respectively. The cumulative extract on III instar larvae consumed less quantity of food when treated at the concentration of 25 ppm (0.172 mg/dry wt/larvae) followed by IV instar larvae reared on cumulative extract treated leaf at high concentration of 45 ppm (0.181 mg/dry wt/larvae) and adult beetle treated at high concentration of 75 ppm (0.197 mg/dry wt/adult). Among the treated concentration the least concentration was required for III instar larvae fed on cumulative extract.

Food assimilation

The food assimilation resembles the results of food consumption for all three stages, It declined when treated with different solvents extracts of leaf of *A. bracteata*. Assimilation for all three stages of treated insects was diminished over the control. The assimilation for all stages varied from the lower concentration to higher concentration. The cumulative extracts reduced the food assimilation with increasing the concentration.

Food conversion

The food converted in to body wt of all stages was considerably declined. The converted food into body wt of III, IV instars and adult beetle was reduced. When compared with other solvents the food converted into body weight was considerably reduced with 0.006, 0.005, 0.004 and 0.001 mg/live wt/larvae in III instar and 0.005, 0.004 and 0.002mg /live wt/larvae in IV instar, with 0.008, 0.005, 0.004, 0.003 and 0.001 mg/live wt/adult beetle at the concentration of 50,55,60,65 and 70 ppm, respectively in cumulative treatment. It is interestingly noted that there was no changes in body weight of fourth instar grubs of *H. vigintioctopunctata* at 45 ppm.

Food Metabolism

Food metabolism of III and IV instars larvae and adult beetle tended to diminish with increasing concentrations for all experimental solvents. The food metabolized by the experimental group was less when compared with control (0.283, 0.334, and 0.386 mg /dry wt). The food metabolism was very much deteriorated in cumulative extract treated groups over the control at high concentration (0.163, 0.172 and 0.158 mg/dry wt) at 25, 45 and 75 ppm, respectively. Though, other solvents diminished the food metabolism in considerable amount, the food metabolism was much reduced in adult beetle treated with Carbon tetra chloride at high concentration of 185 ppm (0.143 mg/dry wt/beetle). Statistical analysis showed that there was a positive correlation between concentrations and metabolism.

Consumption Rate

Greater consumption rate was seen in control group with 623.76, 699.81 and 684.72 mg /dry wt of III, IV instar larvae and adult control beetle of *H. vigintioctopunctata,* when compared with tested solvents. High percentage of reduction in consumption rate was obtained in the fourth instar grubs treated with cumulative extract (333.33 mg/dry wt/ grubs) at high concentration of 45 ppm. At all tested solvents the rate of consumption was diminished from low concentration to high concentration.

Assimilation rate

The rate of assimilation was greater in control, third, fourth and adult beetle of *H. vigintioctopunctata* (598.01, 546.11and 658.45 mg/third, fourth and adult beetle, respectively). The rate of assimilation was higher in control adult beetle when compared with control grub groups. Similar results were obtained in tested solvents. The rate of assimilation was higher in adult tested groups when compared with treated groups of grubs. Assimilation rates of cumulative treated group had significant reduction when compared with other solvents with 322.83, 316.75 and 272.72 mg/live wt. for third, fourth instars and adult beetle of *H. vigintioctopunctata* at high concentration of 25, 45 and 70 ppm, respectively.

Conversion rate

Food conversion in to body weight and the rate of conversion were decreased in third, fourth instars and adult beetle treated with cumulative extract when compared with other solvents (11.36, 9.19 and 14.65 mg/live wt of third, fourth instars and adult beetle, respectively). It was somewhat lesser than that of control in third (21.78mg), fourth (23.50mg) instars and adult beetle (24.63mg) of *H. vigintioctopunctata.*

Metabolic rate

Metabolic rate of solvent extract treated group was lesser than the control groups. Among all solvents, cumulative treated group had lesser metabolic rate (503.66mg at low concentration and 271.01 mg at high concentration and in adult, 380.51 mg at low concentration and 316.75 mg at high concentration) in third instar larvae and 371.21mg at low concentration and 320.86mg at high concentration in fourth instar larvae.

Assimilation efficiency

Assimilation efficiency of control was very meager when compared with various solvent extracts treated *A. bracteata* groups (95.87% in third instars 95.86% fourth instars and 96.16% in adult). Comparatively, cumulative treated groups had greater percent of reduction in assimilation efficiency in fourth instars and adult beetle (95.02%and 80.71% reduction). But in the case of third instars the Acetone extract treated larvae had low assimilation efficiency of 94.96%.

Conversion efficiencies (ECI and ECD)

ECI and ECD of the control third and fourth instar grubs and adult beetle were (3.49% and 3.64%; 3.35% and 3.50%; 3.59% and 3.74%) greater when compare with treatment. The conversion efficiency of all tested solvents significantly varied from low concentration to high concentration. Particularly the cumulative extracts treated groups had conversion efficiency in all three stages (2.84% and 2.97% in third instars; 2.25% and 2.35% in fourth instars and 3.59% and 4.29% in adult, ECI and ECD respectively].

Discussion

Chemical controls of pests using insecticides have been favored so far because of their speedy action and ease of application. Biologically active plant extracts are therefore being used as alternative for studying their potential efficacy to minimize the extent of pollution and for their low cost (ICAR, 2003).

Age of insects is an important consideration in assessing the toxicity of plant extracts (Busvine,

1957). Results revealed that grubs and adult stages showed greater susceptibility than third and final instar grubs. Ratnakaran and Smith, (1976) and Granette and Retnakaran (1977) found that the late instars of spruce bud worm were most susceptible to chitin synthesis inhibitors than the early stage.

Among the three stages, the plant extracts using different solvent extracts and the cumulative treatment of the all solvents had more toxic effect against the grubs as well as the adult forms. The order of toxicity for larval forms was cumulative treatment > Petroleum ether > Benzene > Carbon tetra chloride > Ethyl acetate > Chloroform > Acetone. The order of toxicity for adult forms was cumulative treatment > Ethyl acetate > Petroleum ether > Benzene > Chloroform > Carbon tetra chloride > acetone. Singh *et al.* (1985) reported that petroleum ether extract of root *T. patula* caused 100% mortality in 3rd instar larvae of *Cx. fufican* at a very low concentration of 62.5 ppm. Raghunatha Rao and Dhamu. (1988) reported that Nemidin fraction of *Azadirachta indica* killed all the larvae of *Cx. quinquefasciatus* at the concentration of 62.5 ppm.

Chockalingam *et al.* (1986) reported that the final instar larvae of *S. litura* fed on NOE treated leaves of *R. communis* suffered a heavy mortality. Larvae reared on 100 ppm NOE treated leaves resulted in 90% mortality. Fecundity of adult beetles diminished to a greater extent in solvent extracts treated leaf fed beetles. The sublethal doses of plant extracts caused a decline in the number of eggs laid if orally administered. The number of eggs laid by the insects considerably decreased in the insects treated with Benzene extracts of leaf of *A. bracteata*. The same effect was noticed in the percentage hatchability. Larvae of *Manduca sexta* ingesting neem seed extract showed decrease in fecundity, fertility and longevity (Boles, 1974; Zettler and Le Cato, 1974; Grosch, 1975). Such a reduction the number of eggs may be due to the sterility effect induced by sublethal doses of neem oil treatment (Moriarity, 1968; Flint *et al.* 1978) and impairment of mating.

Reduction in fecundity due to neem oil treatment was observed in *Cnaphalocrocis medinalis* and *Nilaparvata lugens*. Neem seed extracts had also reduced fecundity in insects. Observed reduction in fecundity may be due to the disturbance of the neuro endocrine system as suggested by Slama (1978).

Ethyl acetate extracts of leaves of *A. bracteata* caused considerable reduction in the number of eggs laid by *H. vigintioctopunctata*. But it had no effect on consumption and fecundity. Ascher and Nemny (1974) and Knapp and Herald (1982) reported that the chitin synthesis inhibitors present in the plant ingredients affected the normal hatchability of the eggs.

Impact of toxicants on the food

In general, cumulative leaf extract treated grubs showed significant reduction in food consumption. The effectiveness of cumulative extract was seen in reduced food consumption in third instars rather than fourth instars and adult beetle of *H. vigintioctopunctata*. Reduction in food consumption may be due to the anti-feeding effect of the leaf extracts of the *A. bracteata*. Pradhan *et al.* (1963) reported that the reduction in food consumption may be due to the anti-feeding effect of the neem derivatives.

As the concentration of cumulative as well as other solvents extracts was increased, the adult beetles exhibited a noticeable decline in food consumption (from 0.417 mg in control to 0.197 mg at 70 ppm of cumulative extracts). Lavie *et al.* (1967), Butterworth and Morgan (1968) and Ketkar (1976) reported that the active secondary substances extracted from the neem seed madthe food stuff distasteful and caused the reduction in the food consumption of *H. vigintioctopunctata.*

Reduction in growth of cumulative extracts treated third fourth instars grubs and adult was higher than with the other solvent extracts. It may be due to toxic stress resulting in the expenditure of more energy towards metabolism (0.001 mg at 25 ppm, 0.000 mg at 45 ppm and 0.001 mg at 70 ppm of cumulative extracts) when compared to control (III. 0.011, IV 0.013 and adult 0 .015mg). Adult beetles, III and IV instars of *H. vigintioctopunctata* showed a greater decline in growth when sublethal doses of all tested solvents were tested. The beetles experienced less growth in cumulative treatment treated groups.

The results revealed that cumulative toxicity of leaf extracts was higher than the other tested solvents extracts. Sometimes the treated groups gained more weight when exposed to sublethal dose than the untreated groups. Ramdev and Rao (1980) reported that the larvae of *Achaea janata* treated with sublethal doses of sumithion gained more body weight than untreated ones.

Assimilation efficiency diminished in third and fourth instar larvae and adult of *H. vigintiocto-punctata* reared on all the solvents extracts treated leaves. Greater decline in assimilation efficiency was observed in cumulative extract treated group adult beetles. Similar results obtained by decrease in assimilation efficiency of adult are seem in petroleum ether treated group adults. This attribute is due to increase in calorific content of faeces over that of control. Miller and Kinter (1977) stated that altering protein conformation in cell membrane, and activity of enzymes such as ATP which aid in the absorption of nutrients may also be considered as another explanation for the reduction in assimilation efficiency.

Conversion efficiencies diminished in third and fourth instars and adult beetles when they were reared on solvents extracts treated leaf fed groups. Third instar grubs were susceptible to leaf extracts resulting in the deterioration of conversion efficiencies. Nath *et al.* (1986) reported reduced conversion efficiency in the grasshopper, *Scistocerga gregaria* fed with sublethal doses of smithion. The reduction in conversion efficiency may be due to the herbivorous insects committing more of its energy sources for the tolerance of toxicants than for body growth (Kreiger *et al.,* 1971). Chokalingam *et al.* (1982) reported that the most preferred and heavily consumed food plant was assimilated less whereas the less consumed of the food plants was assimilated more efficiently.

References

Abbot, W. S., 1925. A method of computing the effectiveness of insecticides *J. Econ. Entomol.,* 18, 265.

Ananthakrishnan, T.N., 1988. In dynamics of Insect Plant Interaction (eds. Ananthakrishnan, T.N. and Raman, A.), Oxford and IBH publishing company, New Delhi, pp. 1-11.

Ascher, K.R S and Nemy, N.E., 1974. The ovicidal effect of pH 6.0-4.0 (1-4-Chlorophenyl)-3- (2,6-Diflubenaoyl) Urea in *Spodoptera littoralis* boid. *Phytoparasitica,* 2, 131-133.

Banerji, R., Misra, G and Nigam, S.K., 1985. Role of indigenous plant materials in Pest management. *Pesticides,* March, 32-37.

Boles, H.P., 1974. The effects of doses of pyrethrins on the mating efficiency of the rice weevil *Sitophilus oryzae* (L.) (Curculionidae). *Kan. Entomol. Soc.,* 47, 441-451.

Busvine, J.R., 1971. A critical review of testing techniques. Common Wealth institute of Entomology, London.

Butterworth, J.H. and Morgan, E.D., 1968. Isolation of a substance that suppress feeding in locusts. *Chem. Comm.,* 1, 23-24.

Chockalingam, S., Vasantha, E. and Somasundaram, P., 1982. Efficacy of neem oil exctrative (NOE) against *Spodoptera litura.* (Lepidoptera: Noctuidae) *Oriental Zool.,* 3, 59-64.

Chockalingam, S., M. S. N. Sundari and E. Vasantha., 1986. The use of the exract of *Eucalyptus* to control of *Spodoptera litura.* (Lepidoptera: Noctuidae) *Orient. Zool.,* 3, 59-64.

Fint, H.M., Smith, R.L., Noble, J.M., Shaw, D., de Milo, A.B and Khalil, F., 1978. Laboratory tests of diflubenzuran and four analoques against the pink boll worm and cotton leaf perforator. *J. Econ. Entomol.,* 71, 616-619.

Granette, J and Retnakaran, A., 1979. Stadial susceptibility of eastern spruce budworm, *Choristoneura fumiferena* (Lepidoptera: Tortricidae) to the insect growth regulators. *Dimilin. Can., Entomol.,* 109, 893-894.

Grosch, D.S., 1975. Reproductive performance of *Bracon hebtor* after sublethel doses of carbarly. *J. Econ. Entomol.,* 68, 659- 662.

ICAR Bulletin., 2003. In the control of mosquito vectors prospects of using herbal products.

Ketkar, C. M., 1976. Utilization of neem (*Azadiracta indica*) and its by products. Directorate of non edible oils and soap industry, Khadia village industry Commision Bombay, India. pp. 234.

Knapp, F. W and Herald, F., 1982. Congenetially induced mortality in face flies (Diptera: Muscidae) following adult exposure to diflubenzuron treated surfaces. *J. Med. Entomol.,* 19, 191-194.

Kreiger, R. I., Feeny. P.P and Wilkinsen, C.F., 1971. Detoxification enzymes in the guts of Caterpillars: An evolutionary answers to plant defenses. *Science,* 172, 579-581.

Laemmli, U.K., 1970. Cleavage of structural proteins during the assembly of the head bacteriophage T4. *Nature*., 227, 680-685.

Lavie, D., Jain, M. K and Shapan-gabrielith, S.R., 1967. A locust phagorepellant from two Melia species. *Chem. Comm.,* 18, 910-911.

Miller, D.D and Kinter, W.B., 1977. DDT inhibits nutrients absorption and osmo-regulatory function in *Fundulus heteroclitus*. In: Physiological response of marine biota to pollutants New York, Academic Press, pp. 63-74.

Moriarily, F., 1968. The toxicity and sublethal effects of P, P'-DDT and Dieldrin to *Agalis utricae* L. (Lepidoptera: Nymphalidae) and *Chorthippus brunneus Thunberg* (Saltatoria: Acridicae). *Ann. Appl. Biol.,* 62, 371-393.

Nath, V., Nath, T.N. and Rao, P.J. 1986. Effect of sumithion on consumption and utilization of food by *Schistocerca grigaria* (Forskal). *Ind. J. Entomol.,* 45, 174-179.

Petruswicz, K and Mac Fadyen, A., 1970. Productivity of terrestrial animal. IBP Hand book No.13, Blackwell Scientific Publications, Oxford pp. 190.

Pradhan, S., 1967. Strastegy of integrated pest control. *Indian. J. Entomol.,* 29, 105-122.

Raghunatha Rao, A and K. P. Dhamu., 1988. Statistics workbook for insecticides Toxicology. Suriya Desktop Publishers, Coimbatore. pp.177-179.

Ramdev, Y. P and Rao, P.J., 1980. Effect of sublethal dose of insecticides on consumption and utilization of dry and dietary constituents of Castor *Recinus communis* Linn. By the castor semilooopper, *Achaea janata*. J. *Entomol.,* 42, 567-575.

Retnakaran, A and Smith, L., 1976. Greenhouse evaluation of TH 6040 activity on the forest tent caterpillar. *Bio. Mon. Res.,Can. Fores. Erviro.,* 32:2.

Salma, K., 1978. The principles of anti-hormone action in insects. *Acta. Ent. Bohemoslov.,* 75, 65-82.

Saxena, B.P and Tikker, K., 1990. Impact of natural products on the physiological of phytophogous insects. *Proc. Indian. Acad. Sci.* (Anim. Sci.), 99, 185-198.

Saxena, K.N., 1987. Some of factors governing olfactory and gustatory response of insects. In: olfaction and taste (Eds. T. Hayashi), pp. 799-819, Pergamon, Press Oxford, U.K..

Scheneider, B.P.S., Singh, J.P and Lather, T.S., 1992. A proposed pest management program including neem treatments for combating potato pests in Sudan, pp. 449-459.

Srivastava, U.S and Prasad S.S., 1980. The effects of tropical and oral administration of a juvenoid to the last instar larva of *Spodoptera litura*.(Fab). *Proc. Indian. Acad. Sci.,* 89, 359-370.

Sharma, R.N. Tare, V.S and Despande, S.G., 1990. New chemicals, natural products and their permutations and combinations to combat insect pests. In impacts of environment on animals and aquaculture edited by G. K. Manna and B.B. Jana, 97-100.

Zettler, J.L and Le Cato, G.L., 1974. Sublethal doses of malathion and dichlorphos; Effects on fecundity of the black carpet beetle. *J. Econ. Entomol.,* 67, 19-21.

Reviving Cultural Strategies and Redefining Nutrient Supplies for an effective Insect Pest Management

V. Selvanarayanan*

Cultural control encompasses traditional cultural practices that are manipulated to manage pest populations. Among the various cultural control strategies, habitat management practices like field sanitation, crop rotation, trap cropping, deep ploughing, pruning are widely followed. Besides these, nutrient management is considered vital with regard to insect infestation. The excessive application of nitrogenous fertilizers invites more pest incidence besides their drastic economic and environmental consequences.

Keeping the above in mind, field experiments were conducted at Annamalai University, Tamil Nadu to study the role of nutrient supply and schedule on insect incidence in cowpea and rice. In the study with cowpea cultivar CO 6, various sources of nutrients viz., farm yard manure, biofertilizer (Rhizobium) and NPK fertilizers alone and in combination were evaluated wherein combination of farm yard manure, Rhizobium and synthetic fertilizers recorded the minimum incidence of aphids, thrips and blue butterflies. In another study to optimize nitrogen levels in rice, among the various levels of nitrogen viz., 0, 120, 180 and 240 kg/ha with or without farm yard manure @ 12.5 t/ha evaluated with the rice cultivar ADT 36, the incidence of insect pests namely leaf folder, stem borer and brown plant hopper was less at application of optimum level of nitrogen i.e. 120 kg N with 12.5 t FYM per hectare, whereas, doses above 120 kg N increased the incidence. It is concluded that optimum dose of synthetic fertilizers especially nitrogen supplemented with other organic sources or bio-inoculants is favourable for increased insect tolerance and optimum yield.

Introduction

Cultural methods to control insects involve crop production practices that have the dual purpose of crop production and insect suppression. Cultural control involves traditional cultural practices that make the environment less favourable to the insect pests whereas more favourable to natural enemies. Cultural control may not by itself reduce the pest population to below economic threshold levels, but may aid in reducing losses due to insect pests by other methods.

Cultural control practices (i) make the crop or habitat unacceptable to pests by interfering with their oviposition preferences, host plant discrimination or location by both adults and immatures, (ii) make the crop unavailable to the pest in space and time by utilizing knowledge of the pest's biology and bio-dynamics and (iii) reduce pest survival on the crop by enhancing its natural enemies, or by altering the crop's susceptibility to the pest.

Cultural control methods are preventative rather than curative and involve minimal extra labour and costs. Though various cultural control strategies such as habitat management practices like field sanitation, crop rotation, trap cropping, deep ploughing, pruning are available, few are alone practiced.

Besides the above cultural practices, nutrient management is considered vital with regard to insect infestation. The role of mineral fertilizers in plant nutrition for sustaining and increasing agricultural

* Department of Entomology, Faculty of Agriculture, Annamalai University, Annamalai Nagar – 608 002

production is well recognized. Excessive use of nitrogen leads to higher pest incidence, damage to the environment and low profit from farming. (Jayaraj and Ignacimuthu, 2005). The major disruptive practice now among the farming community is the excessive application of nitrogenous fertilizer. Hence the present study.

Materials and Methods

To study the role of nutrient supply on the incidence of aphids, thrips and pod borer (*Lampides boeticus*) infesting cowpea (cv. CO 6), two field trials were conducted under randomized block design with seven treatments (Table 1) replicated thrice individually in a plot size of 4 x 3 m adopting a spacing of 45 cm between ridges and 30 cm between plants. Incidence of aphids and thrips was recorded by selecting three leaves from the top, middle and bottom portions of each plant and counting the number of insects by using a hand lens. Similarly the number of pod borer larvae per plant and per cent pod damage were also recorded in five plants per replication.

In another study to optimize nitrogen levels in rice, two field trials were laid out during June – September 2006 for *Kuruvai* season and January – April 2007 for *Navarai* season in a randomized block design with seven treatments (Table 3) and three replications individually in a plot size of 4 m × 3 m with the cv. ADT 36. Various levels of nitrogen viz., 0, 120 (100% of recommended N), 180, 240 kg/ha with or without farm yard manure @ 12.5 t/ha and phosphorus and potash fertilizers were applied at recommended level respectively in the individual treatment plots. Nitrogen was applied in the form of urea in two installments, first dose basally and the remaining dose on 25th day after transplanting. Entire phosphorus was given at the time of basal dressing in the form of superphosphate while potash was applied in three split doses: one third was applied basally, one third 25 days after transplanting and the rest at the time of panicle initiation in the form of muriate of potash.

At fortnightly intervals, percentage of leaf damage caused by leaf folder (*Cnaphalocrocis medinalis* Guen.), percentage of dead heart / white ear caused by stemborer (*Scirpophaga incertulas* Walker.) and population of Brown planthopper (*Nilaparvata lugens* Stal.) per hill was recorded (Kavitha, 2007).

Results and Discussion

The rate of influence of nutrient supply on the incidence of sap feeders and pod borers and on yield in cowpea is presented in Tables 1 and 2. The results revealed that the application of synthetic fertilizers alone recorded higher pest incidence and resulted in moderate yield while application of organic manure or bio-fertilizer alone recorded lesser insect incidence as well as lesser yield. The application of N, P and K with *Rhizobium* bio-fertilizer and organic manure recorded the least population of aphids, thrips and pod borer besides recording a lesser pod damage and higher yield as earlier reported by Umaru *et al.* (2002).

Table 1. Influence of nutrient supply on the incidence of sap feeders in cowpea

T. No.	Treatments	Population / plant	
		Aphid	Thrips
1.	Control	10.86^{f}	1.60^{f}
2.	Organic manure (OM) @ 12.5 t/ha	5.06^{ab}	0.46^{b}
3.	Rhizobium (R) @ 20 g/kg of seed	5.53^{bc}	0.66^{c}
4.	NPK @ 10 : 20 : 0 kg /acre	7.86^{e}	1.26^{e}
5.	OM + NPK	5.80^{c}	0.76^{c}
6.	R + NPK	6.53^{d}	0.93^{d}
7.	OM + R + NPK	4.46^{a}	0.26^{a}

Mean values followed by different alphabets differ significantly by DMRT

Table 2. Influence of nutrient supply on the incidence of pod borer and yield of cowpea

T. No.	Treatments	Larval Population / plant	Pod damage (%)	Seed yield (g) / plant
1.	Control	1.80[f]	43.22 (41.10)[f]	7.07[f]
2.	Organic manure (OM) @ 12.5 t/ha	0.60[b]	30.76 (33.68)[d]	7.85[e]
3.	Rhizobium (R) @ 20 g/kg of seed	1.20[d]	26.20 (30.78)[c]	8.54[d]
4.	NPK @ 10 : 20 : 0 kg /acre	1.60[e]	36.83 (37.36)[e]	8.99[d]
5.	OM + NPK	0.60[b]	20.70 (27.06)[b]	9.74[c]
6.	R + NPK	0.80[c]	14.88 (22.69)[a]	11.26[b]
7.	OM + R + NPK	0.30[a]	14.65 (22.50)[a]	13.15[a]

Values in parentheses are arcsine transformed
Mean values followed by different alphabets differ significantly by DMRT

The rate on incidence of leaf folder and stem borer, BPH and yield is presented in Table 3.

Table 3. Influence of N levels on incidence of leaffolder, stem borer, BPH and yield in ADT-36 rice variety

Treatments	Nitrogen levels (kg/ha)	Damage (%)		Population of BPH / tiller	Yield (g/plant)
		Leaf folder	Stem borer		
T_1	0	31.14 (33.91)[b]	28.57 (32.31)[b]	3.09[b]	2.10[d]
T_2	120	34.24 (35.81)[c]	31.36 (34.05)[c]	3.25[c]	2.80[c]
T_3	180	37.05 (37.49)[d]	36.96 (37.44)[d]	3.72[d]	3.00[b]
T_4	240	38.46 (38.33)[d]	42.53 (40.70)[e]	3.92[e]	2.86[bc]
T_5	T_2 + 12.5 t FYM	28.19 (32.07)[a]	25.64 (30.42)[a]	2.97[a]	3.56[a]
T_6	T_3 + 12.5 t FYM	33.72 (35.50)[c]	36.39 (37.10)[d]	3.89[e]	2.93[bc]
T_7	T_4 + 12.5 t FYM	41.27 (39.97)[e]	44.15 (41.64)[f]	4.16[f]	2.90[bc]

Values in parentheses are arcsine transformed
Mean values followed by different alphabets differ significantly by DMRT

Increase in the level of nitrogen was found to escalate the incidence of leaf folder, stem borer and brown planthopper as earlier reported by Singh and Agarwal (1983) and Rekhi and Singh (2004). More deadhearts and white ears were noticed in plants that received N at 240 kg per ha either alone or with FYM. This might be probably due to luxuriant growth of the crop and also due to higher uptake of N by the plant. In line with this, Jiang and Cheng (2003b) reported that survival of rice yellow stem borer was low on plants with low nitrogen content. Zhong – xian *et al.* (2005) reported that the outbreak potential of BPH was induced by the excessive application of N fertilizer in rice fields. If N fertilizer is supplemented with FYM, in general, a comparative reduction in the incidence of insect pests was noticed as witnessed by Abdullah and Hossain (2006) wherein the combined application of organic materials and N-fertilizer produced good yield while minimizing insect prevalence in the rice field.

From the present studies, it may be concluded that the optimum level of nitrogen application schedule i.e. 120 kg N with 12.5 t FYM can very well reduce or avoid the incidence of key insect pests of rice. Excessive nitrogenous fertilizer level i.e. above 120 kg per hectare is not favourable even if supplemented with organic manures like FYM.

References

Abdullah, M.R. and Hossain, K.L., 2006. Effects of urea-N fertilizer dosage supplemented with Ipil-Ipil tree litter on yield of rice and insect prevalence. *J. Forestry Res.,* 17(4), 335-338.

Jayaraj, S. and Ignacimuthu, S., 2005. Progress and perspectives in sustainable pest management. In: *Sustainable Insect Pest Management* (Eds. S. Ignacimuthu and S. Jayaraj), Narosa Publishers, New Delhi, pp.298.

Jiang, M.X. and Cheng, S.A., 2003b. Interactions between the striped stemborer, *Chilo suppressalis* (Walk.) larvae and rice plants in response to nitrogen fertilization. *J. Pest Sci.,* 76(5), 124-128.

Kavitha, S., 2007. Influence of nitrogen levels on incidence of key insect pests in rice. *M. Sc. (Ag.) Thesis*, Annamalai University, India.

Rekhi, R.S. and Singh, R., 2004. Crop yield, disease incidence and pest attack in relation to N dynamics in rice. *IRRN*, 29, 65-67.

Singh, R. and Agarwal, R.A., 1983. Fertilizers and pest incidence in India. *Potash Rev*., 23, 1-4.

Umaru, A.B., Olancyan, G.O., Nwosu, K.I 2002. Effects of nitrogen levels and crop arrangements on the incidence of *Megalurothrips sjostedti* Trybom and *Maruca vitrata* Fab. in maize / cowpea intercrop. *Acta Agronomica Hungarica*, 50(1), 67-74.

Zhong-xian, L., Kong-Luen Heong, Xiao-Ping Yu and Cui Hu.,2005. Effects of nitrogen on the tolerance of brown planthopper, *Nilaparvata lugens* to adverse environmental factors. *Insect Sci*., 12(2), 121-123.

Comparative field efficacy of bait materials for the management of subterranean termites

K. Premalatha,* D.S. Rajavel* and R.K. Murali Baskaran*

A field experiment was conducted at the Agricultural college and Research Institute, Madurai during (June-August 2006 at weather condition 31.1°C and 65.3 % RH) to assess the comparative efficacy of eleven different baits on the attraction of various soil inhabiting species of termites. Four bait materials, gunny bag bits, paddy straw, cow dung and sugarcane baggase were used individually and in different combinations. Tube pots of 16.5 cm height and 11.5 cm width containing bait materials were kept upside down on the places having termite activity. Each treatment was replicated thrice and the number of workers and soldiers caught in the pot were collected daily and counted at fortnightly interval. Various species of termites caught in the pot were also identified. Among the baits evaluated, sugarcane baggase alone and in combination with cow dung, gunny bag bits and paddy straw were effective in attracting the highest number of termites, irrespective of species. Among three species of termites attracted, *Odontotermes obesus* was attracted in large number (925.83), followed by *Coptotermes heimi* (841.93) and *Odontotermes wallonensis* (780.26).

Introduction

Out of 300 species of termites known so far from India, about 35 species had been reported as damaging agricultural crops and timber in buildings. The majority of the pest species were soil inhabiting, either as mound builders or as subterranean nest builders. The major mound building species are *Odontotermes obesus, O. redemanni* and *O. wallonensis*. The major subterranean species were *Heterotermes indicola, Coptotermes ceylonicus, C. heimi, O. horni, Microtermes obesi, M. beesoni and Trinervitermes biformis* (Rajagopal, 2002). Different approaches have been used to control subterranean termites. Among them newest and ecologically safe method is the use of baits. Termite baits effectiveness is dependant upon the willingness of termites to consume the bait when presented with choices of other cellulose food sources.

Badawi *et al.* (1984) made a population study of some species of termites in Saudi Arabia using paper roll bait. The results showed that most abundant species were *Microcerotermes* sp and *Microtermes* sp. Nan – Yao Su *et al.* (1995) developed a commercial prototype monitoring or baiting stations to detect and eliminate foraging populations of *Co. formosanus* and *R. flavipes* near structures. Because of durable plastic housing, the station could be used by pest control professionals for ongoing monitoring and baiting programmes to provide continuous protection of structures from subterranean termites. An effective and inexpensive bait station with cow dung, gunny bag and sugarcane bagasse was evolved to trap large number of termites at the shortest time. Within a week, a colony of *O. obesus*, *M. obesi* and *Macrotermes estherae* could be eliminated with continuous baiting (Rajavel *et al.*, 2007). The present study is aimed to evaluate the bait materials for the management of subterranean termites under field conditions.

* Department of Agricultural Entomology, Agricultural College and Research institute, Madurai - 625 104

Materials and Methods

Tube pots of 16.5 cm height and 11.5 cm width were taken and filled with 400 g of bait materials of old gunny bits, decomposed cow dung, paddy straw and sugarcane baggase. Before filling, all the materials including the pot were made wet. The pot was placed upside down on any place having termite activity at weather condition 31.1°C and 65.3% RH. The mouth of the pot was sealed with soil to avoid ant entry. The set up was left undisturbed from early morning to evening. After 12 h, the pot was removed, and workers and soldiers collected inside the tube pots were counted.

Different combinations of bait materials were prepared as follows.

T1	:	Cowdung	400 g
T2	:	Paddy straw	400 g
T3	:	Gunny bag	400 g
T4	:	Sugarcane baggase	400 g
T5	:	Cowdung + Sugarcane baggase	200 +200 g
T6	:	Cowdung + Gunny bag	200 +200 g
T7	:	Cowdung + Paddy straw	200 +200 g
T8	:	Paddy straw + Gunnybag	200 +200 g
T9	:	Paddy straw + Sugarcane baggase	200 +200 g
T10	:	Gunny bag + Sugarcane baggase	200 +200 g
T11	:	Cowdung + Gunny bag + Paddy straw + Sugarcane baggase	50 + 50 + 50 + 50 g

These baits were kept at three locations *viz.,* Insectary, Central Farm and Staff quarters of Agricultural College and Research Institute, Madurai. The observation on number of workers and soldiers caught in each pot was taken daily and tabulated as fortnightly intervals from June to August, 2006.

Results

The results of observations on attraction of different termite species to bait materials in three different locations are presented in Table 1. The results indicate that maximum number of termites were attracted by the bait cowdung + gunny bag + paddy straw + sugarcane baggase followed by gunny bag + sugarcane baggase. The next best have been cowdung + sugarcane baggase and cowdung + gunny bag. All the three termite species *viz.*, *O. obesus*, *O. wallonensis* and *Co. heimi* were attracted to the baits. Among the termite species, bait catch was high in *O. obesus* (925.83) followed by *Co. heimi* (841.93) and *O. wallonensis* (780.26) in the bait cowdung + gunny bag + paddy + straw + sugarcane baggase.

Discussion

While evaluating different materials to trap termites in a bait station cowdung + paddy straw + gunny bag + sugarcane baggase combination attracted more irrespective of the species. It was observed that bait containing sugarcane baggase attracted more termites than paddy straw, gunny bag and cowdung. It is reasonable to conclude that sugar content in baggase attracted more termites. This result was in accordance with Rajavel *et al.* (2007) who studied an effective and inexpensive bait station and reported that when sugarcane baggase was added to the bait the termite catch was maximum and the species attracted were *O. obesus*, *M. obesi* and *Macrotermes estherae*.

Gunny bag is the next preferred material by termites followed by cowdung. It was contradictory with the study of Khaderkhan (1990) in which he found that wood was more preferred than cowdung. Though the paddy straw bits attracted lowest number of termites it can be well utilized for termite management. It was proved by Nelson *et al.* (1994) and they utilized the bait of paddy straw bits to

Table 1. Evaluation of bait materials to attract different termite species during 2006, at different locations of AC and RI, Madurai (Mean of Four observation)

Treatments	Mean number of termites attracted*		
	O. obesus at Insectary	*O. wallonensis* at Central Farm	*Co. heimi* at Staff Quarters
T_1 Cowdung	427.93 (20.68)	273.98 (16.55)	332.56 (18.23)
T_2 Paddy Straw	319.94 (17.89)	172.33 (13.12)	232.84 (15.25)
T_3 Gunny bag	548.51 (23.42)	398.60 (19.96)	457.93 (21.39)
T_4 Sugarcane baggase	550.57 (23.46)	399.01 (19.97)	459.62 (21.43)
T_5 Cowdung + Sugarcane baggase	823.35 (28.69)	603.26 (24.56)	731.50 (27.04)
T_6 Cowdung + Gunny bag	767.93 (27.71)	634.60 (25.19)	680.79 (26.09)
T_7 Cowdung + Paddy straw	605.68 (24.61)	498.83 (22.33)	557.64 (23.61)
T_8 Paddy straw + Gunny bag	656.68 (25.62)	492.99 (22.20)	551.52 (23.48)
T_9 Paddy straw + Sugarcane Baggase	706.18 (26.57)	542.52 (23.29)	602.03 (24.53)
T_{10} Gunny bag + Sugarcane Baggase	867.60 (29.46)	674.33 (25.96)	771.55 (27.77)
T_{11} Cowdung + Gunny bag + Paddy straw + Sugarcane baggase	925.83 (30.42)	780.26 (27.93)	841.93 (29.01)

*Mean of three replications; Figures in parentheses are $\sqrt{x}$ transformed values

manage harvester termites of upland rice. The variation in bait catch within the species might be due to population density of the colony.

The bait system evaluated in the present study can be utilized for monitoring the termite population, collection of termites for experimental purpose and delivery of Insect Growth Regulators into the colony for the management of termites.

References

Badawi, A., Faragalla, A.A. and Dabbour, A., 1984. Population studies of some species of termites in A1-Kharj oasis, Central Region of Saudi Arabia; *Zeitschrift fur angewandte Entomologie,* 97, 253-261.

Khaderkhan, H., 1990. Studies on ecology of *Odontotermes brunneus* (Hagen) (Termitidae:Isoptera) and infectivity of Mycopathogens against termites in Agro-forestry; *Ph. D. Thesis* Tamil Nadu Agricultural University Coimbatore 207p.

Nan-Yao Su, Thoms, E.M., Ban, P.M. and Scheffrahn, R.H., 1995. Monitoring / baiting station to detect and eliminate foraging populations of subterranean termites (Isoptera : Rhinotermitidae) near structures; *J. Econ. Entomol.,* 88, 932-936.

Nelson, S.J., Robin, S., Rajaram, V. and Ramasamy, N., 1994. Foraging behaviuor and management of harvester termite, *Anacanthotermes viarum* (Konig) infesting rainfed rice; *J. Ent. Res.,* 18(2), 143-146.

Rajagopal, D., 2002. Economicaly important termite species in India; *Sociobiology,* 40(1), 19-21.

Rajavel, D.S., Premalatha, K. and Venugopal, M.S., 2007. A new bait system for monitoring and management of Subterranean termites, *Odontotermes*, *Microtermes* and *Macrotermes* (Termitidae : Isoptera); *Hexapoda,* 14(1), 20-23.

Evaluation of various organic components for the management of onion cutworm *Agrotis ipsilon* (Hufn.)

K. Suresh,* D. S. Rajavel,* R.K. Murali Baskaran* and B. Usha Rani*

Field experiment was conducted to evaluate the effect of various organic nutrients against onion cutworm *Agrotis ipsilon* (Hufn.) in the farmers holding at Therkutheru village of Madurai district during June – July 2005. The results revealed that basal application of FYM (12.5 t/ha) + biofertilizers (Phosphobacteria, *Azospirillum* and silica solubilising bacteria each @ 2 kg/ha) + neem cake (850 kg/ha) recorded less cutworm population of 1.68 larvae per plant as against 8.62 larvae per plant in inorganic NPK applied plots. Biochemical estimation, viz. phenol, OD phenol, pyruvic acid, sulphur, total chlorophyll and reducing sugars at 30 and 45 DAP revealed that high phenol, OD phenol, pyruvic acid and sulphur with low reducing sugars in the above promising treatment combination offered a better protection of onion from the cutworm and registered higher yield of 13.80 t/ha. Further prevalence of spiders was uniform in all the organic treatments and found to be safe.

Introduction

Onion is valued as condiment due to its flavour and is the most important bulbous vegetable grown in India. It is the second important vegetable crop next to tomato. Onion farmers face pest problems in two phases. In the first instance the cutworm *Agrotis ipsilon* (Hufn.) occurs sporadically, but now-a-days it is very serious in onion cultivated areas. Use of chemical sprays leaves toxic residues on onion. Being a readily consumable produce it is pertinent to develop alternate strategies such as use of organic farming. Organic farming management relies on developing biological diversity in the field to disrupt habitat of pest organisms and also the purposeful maintenance and replenishment of soil fertility.

Materials and Methods

A field experiment was conducted in farmers holding at Therkutheru village of Madurai district during June – July 2005 in a randomized block design (RBD) using test variety CO 4 with the following twelve treatments and replicated thrice to evaluate the various organic sources of nutrients and organic amendments against cutworm of onion. All the agronomic practices were followed uniformly in all the plots with plot size of 4 X 3 m.

Treatment details

T_1 FYM (12.5 t/ha) + Biofertilizers (2 kg/ha) + Neem cake (850 kg/ha) (FD)
T_2 FYM + Biofertilizers + Mahua cake (1500 kg/ha) (FD)
T_3 FYM + Biofertilizers + Pungam cake (1100 kg/ha) (FD)
T_4 FYM + Biofertilizers + NPK (HD) + Neem cake (HD) (425 kg/ha)
T_5 FYM + Biofertilizers + NPK (HD)+ Mahua cake (HD) (750 kg/ha)
T_6 FYM + Biofertilizers + NPK (HD)+ Pungam cake (HD) (550 kg/ha)

* Department of Agricultural Entomology, Agricultural College and Research Institute, Madurai - 625 104
Suresh_ento2006@yahoo.co.in

T_7	Vermicompost (1 t/ha) + Biofertilizers (2 kg/ha) + Neem cake (FD)
T_8	Vermicompost + Biofertilizers + Mahua cake (FD)
T_9	Vermicompost + Biofertilizers + Pungam cake (FD)
T_{10}	Vermicompost + Biofertilizers + NPK (HD) + Neem cake (HD)
T_{11}	NPK as inorganic form (60:60:30 kg/ha)
T_{12}	Untreated check

Farm yard manure (FYM) and vermicompost with computed quantity were applied basally at the time of main field preparation. The biofertilizers *viz., Azospirillum*, phosphobacteria and silica solubilizing bacteria each @ 2 kg/ha were incorporated in the soil in the respective treatments. Half of the dose of the total requirements of other organic amendments *viz.,* neem cake, mahua cake and pungam cakes were applied as basal and the remaining half of the dose was applied as top dressing in two equal splits at 20 days interval. Inorganic fertilizers in the form of urea, single super phosphate and muriate of potash with computed quantity were applied at recommended doses. Fifty per cent of total N and entire P and K were applied as basal and the rest of 50 per cent N was applied in two equal splits as top dressing at 20 days interval.

Ten plants were selected at random and the number of cutworm larvae present was recorded. The population was assessed from 30 to 51 DAT at 7days interval. The spider population was assessed from 15 to 50 DAP at seven days intervals on ten plants selected at random. Yield data were recorded at harvest and expressed as t/ha.

Results and Discussion

The mean cutworm population throughout the period of observation revealed that FYM + biofertilizers + neem cake full dose recorded less population of 1.68 per plant and was followed by FYM + biofertilizers + mahua cake full dose (3.17/plant) as against 8.62 per plant in NPK as inorganic form treated plots. The corresponding per cent reduction over NPK was 81.45 and 63.68 respectively. The next effective treatments in descending order were vermicompost + biofertilizers + neem cake full dose and FYM + biofertilizers + pungam cake. The population recorded in the above treatments was 4.23 per plant and 4.80 per plant with corresponding per cent reduction of 50.91 and 44.17 over NPK (Table 1).

In the present study the treatments with FYM + biofertilizers + neem cake in three splits as one basal and two top dressing at twenty days interval were found to be highly effective in reducing cutworm population than NPK as inorganic form throughout period of observation . This is in line with the findings of Godase and Patel (2001) who reported that the application of FYM significantly reduced the population of whitefly, leaf hopper and aphid in brinjal. This was again in consonance with the earlier findings of Godase and Patel (2002) who observed low incidence of leaf hopper in brinjal, when neem cake was applied.

Further, Krishnamoorthy *et al.* (2001) reported that the soil application of neem cake @ 250 kg/ha significantly reduced the leaf hopper population and this is in close agreement with the present investigation. This is again in consonance with the earlier observations, that the less incidence of sweet potato weevil, *Cylas formicarius* due to rational release of nitrogen when FYM was applied to the soil (Singh *et al.*, 1992). Suresh *et al.* (2006) also observed less incidence of shoot and fruit borer infestation in brinjal with FYM + biofertilizers + neem cake application than that of NPK as straight fertilizers Surekha and Arjuna Roa (2001) explained the reason that the inorganic fertilizers increase the plant growth by providing the nutrients to the plants in large quantity for shorter period of time. Thereby the plants are endowed with luxuriant growth which may offer adequate and highly preferred food to the insects leading to heavy insect population. Mohan *et al.* (1987) indicated that *Azospirillum* applied in soil as seed inoculant reduced the sorghum shootfly damage through the enhancement of phenolics in the plants. In the present investigation, biofertilizers (*Azospirillum*, phosphobacterium and silica solubilizing bacteria) applied as basal along with organic manures and amendments have been found effective in reducing the onion cutworm infestation through induced systemic resistance that can be attributed to the enhanced phenolic content in organically treated onion plants.

The differences in spider population was not significant during all the periods of observation. The population ranged from 0.13 to 0.36, 0.16 to 0.56, 0.36 to 0.62, 0.48 to 0.72, 0.62 to 0.91 and 0.40 to 0.64 per plant among treatments on 15, 22, 29, 36, 43 and 50 DAP respectively (Table 2). The present findings revealed that the prevalence of coccinellids and spiders was more or less uniform in all the treated plants. So, the results indicated that all the organic sources of nutrients along with amendments were found highly safe to the natural enemies.

In the present investigation, Biochemical contents, such as total phenol, OD phenol, pyruvic acid and sulphur were significantly more, while chlorophyll and reducing sugars were relatively less in onion treated with FYM + biofertilizers + full dose of neem cake (Table 3). In this study the organics treated plants showed higher phenol and OD phenol content which lowered the incidence of cutworm This might be due to the defensive mechanisms of phenols in plant and this is also in conformity with the findings of Chhabra *et al.* (1984) revealed that high content of total phenols in black gram cultivars served as defensive mechanism against whitefly, leaf hopper and pod borer complex. Further, Suresh (1988) reported that the phenols are often associated with feeding deterrence or growth inhibition. Further, Munakata (1997) reported that the phenols act as antifeedants and repellents to *Spodoptera litura* in several plants.

Table 1. Effect of organic sources of nutrients on cutworm in onion

S.No.	Treatments	Cut worm (No./plant)* Days after Planting 30	% reduction over NPK	37	% reduction over NPK	44	% reduction over NPK	Mean Cutworm (No./ Plant)	Mean % reduction over NPK
1.	FYM+BF+NC (FD)	2.53 (1.59)[a]	76.54	1.70 (1.30)[a]	79.92	0.80 (0.89)[a]	87.88	1.68	81.45
2.	FYM + BF + MC (FD)	4.07 (2.02)[b]	62.35	3.30 (1.82)[b]	61.02	2.13 (1.46)[b]	67.68	3.17	63.68
3.	FYM + BF + PC (FD)	6.03 (2.46)[d]	44.14	4.53 (2.13)[c]	46.46	3.83 (1.96)[d]	41.92	4.80	44.17
4.	FYM+BF+NPK (HD)+NC(HD)	7.20 (2.68)[ef]	33.33	6.03 (2.46)[de]	28.74	2.93 (1.71)[c]	55.56	5.39	39.21
5.	FYM+BF+NPK (HD)+MC(HD)	8.30 (2.88)[g]	23.15	7.07 (2.66)[f]	16.54	4.07 (2.02)[de]	38.38	6.48	26.02
6.	FYM+BF+NPK (HD)+ PC(HD)	9.17 (3.03)[h]	15.12	7.80 (2.79)[gh]	7.87	5.37 (2.32)[g]	18.69	7.44	13.89
7.	VC + BF + NC (FD)	5.30 (2.30)[c]	50.93	4.17 (2.04)[c]	50.79	3.23 (1.80)[c]	51.01	4.23	50.91
8.	VC + BF + MC (FD)	6.80 (2.61)[e]	37.04	5.57 (2.36)[d]	34.25	4.40 (2.10)[e]	33.33	5.59	34.87
9.	VC + BF + PC (FD)	7.57 (2.75)[f]	29.94	6.40 (2.53)[e]	24.41	4.93 (2.22)[f]	25.25	6.30	26.53
10.	VC+BF+NPK (HD)+ NC(HD)	8.60 (2.93)[gh]	20.37	7.33 (2.71)[fg]	13.39	5.50 (2.35)[g]	16.67	7.14	16.81
11.	NPK alone	10.80 (3.29)[i]	0.00	8.47 (2.91)[h]	0.00	6.60 (2.57)[h]	0.00	8.62	0.00
12.	Untreated check	8.57 (2.93)[gh]	20.68	7.40 (2.72)[fg]	12.60	5.47 (2.34)[g]	17.17	7.14	16.82

* Mean of three replications, Values in parentheses are square root transformations In a column, means followed by same letter(s) are not significantly different at P=0.05 as per DMRT.

Table 2. Influence of organic sources of nutrients on spider population in onion

S. No.	Treatments	Spiders (No./plant)* - Days after planting						
		15	22	29	36	43	50	Mean
1.	FYM+BF+NC (FD)	0.26	0.56	0.62	0.72	0.86	0.40	0.57
2.	FYM + BF + MC (FD)	0.23	0.50	0.60	0.70	0.80	0.42	0.54
3.	FYM + BF + PC (FD)	0.13	0.33	0.42	0.52	0.68	0.48	0.43
4.	FYM+BF+NPK(HD)+NC(HD)	0.16	0.26	0.52	0.58	0.76	0.44	0.45
5.	FYM+BF+NPK(HD)+MC(HD)	0.23	0.20	0.38	0.48	0.72	0.48	0.42
6.	FYM+BF+NPK(HD)+ PC(HD)	0.13	0.16	0.36	0.46	0.70	0.40	0.37
7.	VC + BF + NC (FD)	0.23	0.30	0.42	0.48	0.62	0.50	0.43
8.	VC + BF + MC (FD)	0.36	0.43	0.48	0.52	0.82	0.54	0.53
9.	VC + BF + PC (FD)	0.20	0.43	0.50	0.58	0.90	0.60	0.54
10.	VC+BF+NPK (HD) + NC(HD)	0.36	0.20	0.36	0.46	0.91	0.48	0.46
11.	NPK alone	0.36	0.50	0.56	0.60	0.81	0.62	0.58
12.	Untreated check	0.32	0.43	0.52	0.62	0.90	0.64	0.57

• Mean of three replications

In the present investigation, FYM + biofertilizers + neem cake recorded low reducing sugar content with less cutworm incidence where as NPK treated plants had high reducing sugar content with high pest incidence. This is in agreement with earlier workers (Knapp *et al.*, 1965 and Kalode and Pant, 1967) who observed the susceptibility of plants to insects at higher concentrations of sugars.

The results of the trial conducted to evaluate the effect of organic components indicated that FYM + biofertilizers + full dose of neem cake and NPK alone applied plots recorded more yield (Table 4). Yield in these treatments were 13.80 t/ha and 13.47 t/ha respectively which were statistically on par. Next to this FYM + biofertilizers + full dose of mahua cake recorded higher (12.47t/ha) yield. The yield recorded in untreated control was mere 7.30 t/ha. It is in conformity with the findings of Subbarao and Ravisankar (2001) who observed that the application of FYM together with all kinds of organic manures recorded the maximum fruit yield, while Ramachandran (1986) reported that organic amendments enhanced the growth of tomato plants with increased the fruit yields which are in corroboration with the present investigation.

Table 3. Effect of organic sources of nutrients on biochemical content in onion leaves

S.No.	Treatments	Biochemical constituents *				
		Phenol (mg/g)	OD phenol (mg/g)	Pyruvic acid (μmol/g)	Sulphur (mg/g)	Reducing sugars (%)
1.	FYM+BF+NC (FD)	4.58	3.02	5.67	0.61	2.03
2.	FYM + BF + MC (FD)	3.90	2.52	4.72	0.43	2.83
3.	FYM + BF + PC (FD)	3.63	1.85	3.23	0.40	2.88
4.	FYM+BF+NPK(HD)+NC(HD)	3.13	1.55	4.43	0.43	3.72
5.	FYM+BF+NPK(HD)+MC(HD)	2.75	1.40	3.25	0.40	4.03
6.	FYM+BF+NPK(HD)+ PC(HD)	2.55	1.23	3.23	0.32	4.28
7.	VC + BF + NC (FD)	2.95	1.42	4.57	0.35	4.10
8.	VC + BF + MC (FD)	2.93	1.18	3.38	0.32	4.27
9.	VC + BF + PC (FD)	2.53	1.02	3.08	0.30	4.33
10.	VC+BF+NPK (HD) + NC(HD)	2.30	1.10	3.38	0.43	4.32
11.	NPK alone	1.62	1.00	2.92	0.36	7.13
12.	Untreated check	1.72	0.68	2.38	0.25	5.07

* Mean of three replications

Table 4. Effect of organic sources of nutrients on onion yield

S. No.	Treatments	Yield (t/ha)*
1.	FYM+BF+NC (FD)	13.80[a]
2.	FYM + BF + MC (FD)	12.47[b]
3.	FYM + BF + PC (FD)	10.60[d]
4.	FYM+BF+NPK(HD)+NC(HD	11.77[c]
5.	FYM+BF+NPK(HD)+MC(HD)	11.67[c]
6.	FYM+BF+NPK(HD)+ PC(HD)	11.43[c]
7.	VC + BF + NC (FD)	10.47[d]
8.	VC + BF + MC (FD)	10.27[d]
9.	VC + BF + PC (FD)	9.60[e]
10.	VC+BF+NPK (HD) + NC(HD)	10.67[d]
11.	NPK alone	13.47[a]
12.	Untreated check	7.30[f]

* Mean of three replications; In a column, means followed by same letter(s) are not significantly different at P=0.05 as per DMRT.

References

Chhabra, K.S., Kooner, B.S., Saxena, A.K. and Sharma, A.K., 1984. Influence of biochemical components on incidence of insect pests and yellow mosaic virus in blackgram. *Indian J. Ent., 46(2),* 148-156.

Godase, S.K. and Patel, C.B., 2001. Studies on the influence of organic manures and fertilizer doses on the intensity of sucking pests, jassid (*Amrasca biguttula biguttula* Ishida) and aphid (*Aphis gossypii* Glover) infesting brinjal. *Pl. Prot. Bull.,* 53 (3-4), 10-12.

Godase, S.K. and Patel, C.B., 2002. Studies on the influence of organic manures and fertilizer doses on the intensity of whitefly, *Bemisia tabaci* Gennadius on brinjal. *Pl. Prot. Bull.*, 54 (1-2), 3-6.

Kalode, M.B. and Pant, N.C., 1967. Studies on the amino acids, nitrogen, and moisture content of maize and sorghum varieties and their relation to *Chilo zonellus* (Swin.) resistance. *Indian J. Ent.*, 29(2), 139-144.

Krishnamoorthy, P.M., Krishnakumar, N.K. and Edward Raja, M., 2001. Neem and pongamia cakes in the management of vegetable pests. In: Proceedings of the second National Symposium on Integrated Pest Management (IPM) in Horticultural Crops: New molecules, Biopesticides and Environment. Bangalore, p 74-75.

Knapp, J.L., Hedin, P.A. and Douglas, W.A., 1965. Amino acids and reducing sugars in silks of corn resistant or susceptible to corn earworm. *Ann. Ent. Soc. Amer.,* 58(3): 401-402.

Mohan, S., Jayaraj, S., Purusothaman, D. and Rangarajan, A.V., 1987. Can the use of *Azospirillum* biofertilizers control sorghum shootfly? *Curr. Sci.*, 72, 723- 725

Munakata, K., 1977. Insect antifeedants of *Spodoptera litura* in plants. In: Host Plant Resistance to Pests(ed. P.A. Hedin), American Chemical Society, Washington, pp 185-196.

Ramachandran, G., 1986. Studies on the management of the nematode (*Meloidogyne incognita*) and disease (*Rhizoctonia solani*) of tomato (*Lycopersicon escolentum* Mill). M.Sc.(Ag) Thesis. Tamil Nadu Agric. Univ. Coimbatore.

Singh, B., Yazdani, S.S. and Hameed, S.F., 1992. Influence of FYM or combination of synthetic fertilizers on weevil (*Cylas formicarius* Fab.) incidence in sweet potato (*Ipomoea batatas* Lam.). *Indian. J. Ent.,* 54, 376-370.

Subbarao, T.S.S. and Ravi Sankar, C., 2001. Effect of organic manures on growth and yield of brinjal, *South Indian Hort.* 49 (Spl.), 288-291.

Surekha, J. and P. Arjuna Rao, P., 2001. Managements of aphids on bhendi with organic sources of NPK and certain insecticides. *Andhra Agric. J.,* 48(1-2), 56-60.

Suresh, G., 1988. Plant natural products in relation to insect resistance-an analysis. In: Dynamics of Insect Plant Interactions.(Eds) T.N.Ananthakrishnan and A Raman) Oxford and IBH Publishing Co., New Delhi. 100 p.

Suresh, K., Rajendran, R., Manisegaran, S., Bhaskaran, R. and Usha Rani, B., 2006. Tritrophic relationship among plant nutrients major insect pests and natural enemies of brinjal in organic vegetable ecosystem. In : Biodiversity and Insect pest management, (eds.) S. Ignacimuthu and S. Jayaraj, Narosa Publishing House, Chennai, 399 p.

Transcriptional analysis of Defense related genes in rice plant against attack of Leaf folder and Plant hopper insects

R. Samiyappan,* K. Saveetha,* L. Karthiba,* T. Raguchander,*
M. Raveendran,** P. Balasubramanian** and D. Saravana Kumar***

Development of Rice varieties with durable resistance to major pests is the challenging task to the researchers. Among the major pests, leaf folder, *Cnaphalocrocis medinalis* and brown plant hopper, *Nilaparvata lugens* are the destructive insect pests in rice causing significant yield loss worldwide. Till now, no major resistance genes to those insect pests have been discovered in cultivated rice. In an attempt to find out effective biological means of controlling these pests, the bioagent *Pseudomonas fluorescens* strain TDK-1 (PGPR- Plant Growth Promoting Rhizobacteria) was found to exhibit higher degree of suppression of both pests by induced systemic resistance in rice (ISR). However, the molecular mechanisms behind the ISR mediated pest management by this PGPR are not well understood. In this investigation, we have attempted to study the PGPR mediated induced physiological, biochemical and molecular changes in rice. Transcript profiling through DD-RT-PCR strategy was done in rice plants, challenged with leaf folder as well as BPH in the presence and absence of PGPR. Results of this analysis showed that PGPR inoculation had significant effect on the increased expression of novel transcripts in rice, and it was twofold in both up and down regulation when compared without PGPR treatments. Leaf folder challenged plants induced the expression of key genes like aspartic proteinase, OsIAA21 - Auxin-responsive Aux/IAA gene family member, CIPK-like protein 1, TPR Domain containing protein and glucan endo-1,3-beta-glucosidase 3 precursor. The BPH challenged plants resulted in expression of transcripts like inositol-1-monophosphatase, putative, protease, NB-ARC domain containing protein, indole-3-acetate beta-glucosyltransferase, mitogen-activated protein kinase, LRR receptor-like protein kinase, esterase precursor, WRKY transcription factor protein in PGPR treated conditions. Our results also indicated that rice transcripts modulated by PGPR strain TDK-1 plays a significant role in induced resistance against both chewing and sucking insects. These findings shed lights on the understanding of molecular Plant- Pest- Microbe interactions and suggest novel and eco-friendly pest management strategy.

Introduction

Rice (*Oryza sativa* L.) is the most important cereal crop in the tropics and is the staple food for over half of the world's population by providing 23 per cent of total calories consumed worldwide (Brar and Khush, 2002). Asian countries *viz*., China, India, Japan and Korea produce and consume 90 per cent of the world rice and accounts for 60 per cent of daily calorie intake. Rice production is severely affected due to pest and disease incidence. The present study was initiated to know the gene(s) which are responsible for ISR mechanisms in rice plant against leaf folder and BPH. The literature pertaining to the present investigation has been reviewed under following headings: Importance of

* Centre for Plant Protection Studies, Tamil Nadu Agricultural University, Ciombatore, 641 003
**Centre for Plant Molecular biology, Tamil Nadu Agricultural University, Ciombatore, 641 003
*** AGROINNOVA, University of Turin[3], Turin, Italy.

Leaf folder and BPH in rice, PGPR and its role in ISR against major pests, feasibility of Differential display RT-PCR approach in identification of differentially expressed cDNA tags for comparison to identify putative novel genes.

Pest constraints to rice cultivation

The productivity of rice is threatened by insect pests attacking the crop from nursery to harvest causing enormous yield loss. Of these, leaffolder insect (*Cnaphalocrocis medinalis*) gained major importance because of their ability to defoliate or remove the chlorophyll content of the leaves leading to considerable reduction in yield and Brown Plant Hoper-BPH (*Nilaparvatha lugens)* by their ability to cause considerable physiological damage to the rice plants by removing nutrients and disrupting physiological processes and eventually affecting the growth and development of the rice plant (Watanabe and Kitagawa, 2000).

Rice genome information

The rice is an model crop next to *Arabidopsis*, sequencing of its genome was completed at the end of 2001 by the rice genome consortium, and more than 70 000 rice expressed sequenced tags (ESTs) are deposited in public databases (http://www.ncbi.nlm.nih.gov/dbEST/dbEST_summary.html). Its genome size is estimated to be 430 Mb in size (Eckardt, 2000) is the smallest of all cereal crops. High-density genetic and physical maps (Sasaki and Burr, 2000), and high-efficiency genetic transformation protocols are available and its genome presents high degrees of similarity among other cereal genomes (Gale and Devos, 1998). Thus, rice became a model organism for studying physiology, evolution, genetics and genomics of monocotyledons. Complete sequencing of its genome (Yu *et al.*, 2002) gives access to an unprecedented number of genes.

Economic importance of Leaf folder

India is one of the important rice growing countries in the world and it stands second in the production next to china. Rice leaffolder and brown plant hopper are the most pervasive insect pests through out the Asia (Torii, 1971). The lepidopteran insect pest *viz.,* leaffolder (*Cnaphalocrocis medinalis)* Guenee is a very serious insect pest of rice. Leaffolder has become a major threat to rice production in tropical and subtropical Asia (Heinrichs *et al.,* 1985). Khan *et al.* (1988) reported the serious menace of *C. medinalis* in tropical and subtropical Asia in last decade. The infestation of leaffolder was also heavy in China, Japan and Korea and in some parts of India, Sri Lanka, Bangladesh, Nepal and Malaysia as observed by (Khan *et al.* 1988). Losses due to these insects usually occur 5-10% and sometime reach up to 75 per cent in some areas (Balasubramanian *et al.*, 1973). In Tamil Nadu, the leaffolder complex has assumed a major pest status in the recent years (Bentur., *et al.* 1989; Pandi, 1997). Carbofuran and fenthion (Chandramohan and Jayaraj, 1976), bendiocarb, acephate and carbosulfan (Saroja and Raju, 1982), quinolphos, monocrotophos and phosphamidon (Raju *et al.*, 1990) and fenvalerate (Ramaraju and Natarajan, 1997) were the common insecticides used against the control of rice leaffolder. The frequent use of these insecticides leads to development of insecticide resistance (Endo *et al.*, 1987; Xiao *et al.*, 1994).

Economic importance of BPH

The brown planthopper (BPH), *Nilaparvata lugens* Stål (Homoptera: Delphacidae), is a monophagous pest and causes direct damage to the rice crops, and susceptible rice varieties often suffer severe yield losses annually from BPH infestations (Sogawa *et al.*, 2003). Feeding behaviour of these insects is highly specific to phloem sucking, as the phloem contains sucrose, potassium, and amino acids as its main constituents (Hayashi and Chino 1990). Such feeding activity causes considerable physiological damage to the rice plants by removing nutrients and disrupting physiological processes, and consequently affects the growth and development of the plant (Watanabe and Kitagawa 2000). Moreover, feeding by large numbers of BPH results in yellowing, drying of the leaves, stunting and wilting of the tillers, which is referred to as a hopper-burn condition (Mitchell and Maddison, 1983). BPH also causes indirect damage by transmitting viruses such as the ragged stunt and grassy stunt viruses, which results in severe damage (Heinrichs 1979). At this level, the loss is considered to be 100% (Pathak, 1994). BPH is a good example of pest resurgence and the intensive use of insecticide caused the outbreak of the brown plant hopper. Kenmore *et al.,* (1984) reported that the population of the brown plant hopper in the insecticide treated field was much higher than in the untreated field. The application of broad-spectrum insecticides can selectively destroy natural predators and increase BPH densities a thousand-fold compared with no pesticide application (Rombach and Gallagher, 1994). The damage and frequent outbreaks of BPH, along with the hazardous effects of using pesticides to protect rice crops, has now prompted researchers to identify BPH resistant germplasms, utilize the resistant lines in breeding programs, and attempt to isolate resistance genes.

Importance of PGPR on pest management

Biological control by antagonistic organisms has been studied extensively and rhizobacterial strains are emerged as potential biocontrol agents for the control of root and foliar diseases (Anuratha and Gnanamanickam, 1990; Raupach and Kloepper, 1998; Ramamoorthy *et al*., 2002b). PGPR are having the ability to protect above ground plant parts against viral (Harish), fungal and bacterial diseases by induced systemic resistance (ISR) (Kloepper *et al*., 1992). Among the PGPR, fluorescent pseudomonads are the most exploited bacteria for biological control of soil borne and foliar plant pathogens. In the past three decades numerous strains of fluorescent pseudomonads have been isolated from the soil and plant roots by several workers and their biocontrol activity against soil-borne and foliar pathogens was reported (Austin *et al*., 1977; Mew and Rosales, 1986; Rosales *et al*., 1993; Rabindran and Vidhyasekaran, 1996; Vidhyasekaran and Muthamilan, 1999; Nandakumar *et al*., 2001; Viswanathan and Samiyappan, 2001; Ramamoorthy *et al.,* 2002ab). Recently Saravana kumar (2006) experimentally proved that PGPR having ability to provide good resistance to plants against leaf folder insect. Sivasundaram (2006) reported PGPR in combination with entomopathogenic fungi *Beauveria bassiana* drastically reduced the incidence of leaf folder and BPH in rice under pot culture conditions. Through our *in-vitro* and *in-vivo* studies we have strongly proved that impact of PGPR strains on the pest control strategy. Since PGPR, *Pseudomonas fluorescens* strain TDK-1 has developed the ability to interact simultaneously both with plants and pest challenges, they can be used as model microorganisms to study complex and multiple-player plant–pests interactions.

Records on Interactive transcriptomic studies

Pathogenesis-related (PR) proteins, which are classified into 17 families, accumulate after pathogen infection or related situations in many plant species (van Loon *et al.*2006). Characteristic expression of twelve rice *PR1* family genes in response to pathogen infection, wounding, and defense-related signal compounds were studied in detail by Mitsuhara *et al*., 2008. In general, Pathogenesis-related (PR) proteins have been used as markers of plant defense responses. However, precise information on the majority members in specific PR families is still limited. Studies published so far on plant–pathogen interactions have been mainly focused on the molecular changes related to pathogen infection and/or plant response (Hammond-Kosack and Parker 2003; Ronald 1997; Suzuki *et al*. 2004). Several signal molecules and defence factors have been identified in plant (Cánovas *et al*. 2004; Ramonell and Somerville 2002; Rep *et al*. 2002), as well as virulence and avirulence factors in microbes (Kazemi-Pour *et al*. 2004; Smolka *et al*. 2003). Nevertheless, the molecular bases of plant pest interaction, multiple-player systems that may produce beneficial effects on plant health are largely unknown. Moreover, the influence of biocontrol agent on the interactions between a plant and pest has not yet been investigated by using transcriptomic approaches.

The responses of individual genes to salicylic acid, jasmonic acid, and ethylene induced defense signaling pathways in rice are likely to be different from those in dicot plants. Transcript levels in healthy leaves, roots, and flowers varied according to each gene. Analysis of the partially overlapping expression patterns of rice PR genes in healthy tissues and in response to pathogens, pests and other stresses would be useful to understand their possible functions and they can be used as characteristic markers for defense related studies in rice.

Genomic studies in plant-herbivore interactions

To date, 19 major BPH resistance genes have been identified (Yang *et al.*, 2004; Chen *et al.*, 2006; Jena *et al.*, 2006). The molecular mechanisms underlying plant resistance to herbivores have been mainly studied by examining interactions between host plants and chewing insects (Moran and Thompson, 2001). It has also been shown that both the feeding mode of a particular herbivore and the degree of plant tissue injury that results from this, exercise considerable influence on the gene expression patterns at the feeding site (Walling, 2000). Chewing insects usually cause extensive tissue damage and thus activate wound-signaling pathways in which jasmonic acid (JA) plays an important role (Rojo *et al.,* 1999). Reduced JA production in tomato appears to increase susceptibility to herbivores (Howe *et al.*, 1996). Consistently, the expression of genes associated with JA signaling was found to be up-regulated in *Nicotiana attenuata* injured by the herbivore *Manduca sexta* (Hui *et al.*, 2003). Furthermore, it has been reported that an attack by chewing herbivores leads to dramatic changes in the gene expression profiles of plants.

Examples of this are evident for both photosynthesis-related genes that are significantly down-regulated in response to stress, wounding and pathogens, and genes involved in carbon- and nitrogen-based defense pathways which are up-regulated under these conditions (Hermsmeier *et al.*, 2001). In contrast to studies of chewing herbivores, very little information is currently available regarding the plants response mechanisms to sucking insects. Although phloem feeding insects including BPH produce little damage to plant foliage, these minor injuries are still effectively perceived as pathogenic by the defense response pathways of the host plant (Moran *et al.*, 2002). Moreover, the plant defense mechanisms which are activated in response to herbivorous phloem-feeding insects have been found to principally involve the salicylic acid (SA) and the JA signalling pathways. For example, SA- and JA-dependent genes were shown to be induced in sorghum (*Sorghum bicolor*) in response to a greenbug (*Schizaphis graminum*) attack (Zhu-Salzman *et al.*, 2004). A JA-independent pathway may also be involved in such insect defense responses in sorghum as some transcripts which are exclusively activated by this invading greenbug were found to be independent of either the JA- or SA-regulated pathways (Zhu-Salzman *et al.* 2004). Consistent with this notion that JA-independent pathways are involved in plant defense against phloem feeding insects, no differences in the gene expression levels of lipoxygenase, a key enzyme in JA synthesis, were observed in a previous study of rice plants infested with BPH (Zhang *et al.*, 2004).

Genomic studies in Plant-BPH interactions

The interaction between rice and BPH has the potential to serve as an excellent model system for understanding the genetic basis of plant defense against phloem-feeding insects. To date, a number of BPH-responsive genes have been isolated in rice plants (Zhang *et al.*, 2004; Wang *et al.*, 2005) and most have been assigned to functional groups for signaling pathways, oxidative stress/apoptosis, wound-response, drought inducible and pathogen-related proteins. Moreover, the expression levels of many of these genes in insect susceptible varieties of rice were, lower than that in resistant varieties in most cases. It has also been found that genes involved in macromolecule degradation and plant defense are up-regulated in BPH susceptible rice varieties during an insect attack, whereas those involved in photosynthesis and cell growth are down-regulated, suggesting that leaf senescence is activated in these plants upon damage by insect feeding (Yuan *et al.*, 2005).

Recently the defense mechanisms in rice infested with BPH were elucidated, that it would be of great value to identify the genes that were specifically expressed in a resistant variety during insect feeding. In this regard, using the representational difference analysis (RDA) authors identified differentially expressed genes between two cDNA populations (Pastorian *et al.*, 2000). Upon completion of the RDA procedure, only the differentially expressed products was remained. In rice, RDA has been utilized previously to analyze genomic differentiation by isolating transposable elements (Panaud *et al.*, 2002), and to detect alien introgression into the rice genome (Vitte *et al.*, 2003) using genomic DNA. The present researcher employed RDA to identify BPH-responsive genes in a resistant rice variety using cDNAs. The use of the appropriate plant materials and the determination of the optimal subtraction strategy are key factors for success in these experiments. This information regarding their identified clones could therefore be useful for marker-assisted breeding of BPH resistant varieties and in the cloning of the genes that control BPH resistance (Park *et al.*, 2007)

BPH genome data

Hiroaki Noda (National Institute of Agrobiological Science, Japan) dealt in depth about research on the genome of BPH and at present over 31,000 EST sequences from different tissues of the insect have been made available for the public. He suggested development of different molecular markers for use in studies on pest migration, virulence composition, sex determination, wing polymorphism and insecticide resistance. Bo Seo (National Institute of Crop Science, Korea) explained the mechanism of plant resistance in terms of feeding physiology of the pest. He suggested development of different molecular markers for use in studies on pest migration, virulence composition, sex determination, wing polymorphism and insecticide resistance. Bo Seo (National Institute of Crop Science, Korea) explained the mechanism of plant resistance in terms of feeding physiology of the pest.

There is no previous record available on the interactive transcriptomic studies on rice-leaf folder interactions and the identification of rhizobacteria mediated expression of pest resistant genes in rice plants with response to the attack of both chewing as well as sucking insect pests, its utilization as molecular marker in resistant breeding programme and genetic engineering studies.

Materials and Methods

Plant material

Rice cultivar Co-43 (susceptible to major pests) was used throughout the experimental period. The seed material was obtained from Paddy Breeding Station, TNAU, Coimbatore.

Mass rearing of leaffolder

Live moths were collected from the Paddy Breeding Station, Tamil Nadu Agricultural University, Coimbatore and cultured in the greenhouse for further studies. The leaffolder adults were released into oviposition cages of 45x45x60 cm size containing 60 days old Co43 rice plants. After 4-5 days, these plants were transferred to rearing cage for emergence of larvae from eggs and used for conducting experiments. Mass culturing of uniform and continuous culture of leaffolder population was maintained. Ten per cent honey solution was provided in the cages as food to maintain the stock culture (Fujiyoshi *et al.*, 1980).

LEAFFOLDER INCIDENCE

For assessing the efficacy of *Pseudomonas* treatments against rice leaffolder, larvae of second instar were carefully collected from the cage by using the camel hair brush starved for 5 h and released on to the leaves of treated plants and allowed to feed for three days. The per cent damage was calculated using the formula

Per cent damage = (No. of damaged leaves / Total no. of leaves) x 100

Insect Bioassay

For testing the insect feeding preference under *in vitro*, detached leaves collected from the PGPR bioformulation treated plants were used. The rice leaves (1g) were cut and kept in Petri plates (9 cm diameter) lined with damp filter papers to maintain the humidity. The second instar larvae were collected from the mass-rearing cage, weighed their initial weight, starved the larvae for 5 h and then allowed for feeding the leaves. The plates were incubated at 25±2°C in the dark in 70% relative humidity for three days. After 72 h, reduction in leaf weight, larval mortality, leaf area consumed, larval and pupal weight and adult malformations were recorded for all the treatments (Maqbool *et al.*, 1998).

Plate.1 Mass rearing of leaf folder	Plate. 2 Mass rearing of BPH	Plate. 3 Plant material for DD-RT-PCR

Efficacy of fluorescent pseudomonad bioformulations against BPH insect

Mass rearing of brown plant hopper (N. lugens)

The method outlined by Medrano and Heinrichs (1985) was followed in mass culturing of BPH. Germinated rice seeds were sown in 40 x 30 x 5 cm trays and after 15 days, the seedlings were transplanted into earthen pots (12 cm dia) and kept partially immersed in plastic tray containing water. These trays were kept inside the insect rearing cages (3 x 2.5 x 2.75 m) with galvanized iron wire netting. Gravid females were collected from the field and released on individual potted plants kept inside the rearing cages for oviposition. Nymphs emerged a week after oviposition and the damaged plants due to feeding of BPH were periodically replaced with fresh potted plants. The plants were made free of spiders, mirid bugs and ants for effective culturing of BPH.

BPH INCIDENCE

For assessing the efficacy of *Pseudomonas* treatments on rice BPH, second instar nymphs were carefully collected from the cage by using aspirator starved for 5 h and released on to sheath of *Pseudomonas* treated rice plants covered with Mylar film and allowed to feed for seven days. Three replicates of ten nymphs were used in each case and 0.03 % Dimethoate served as chemical check. Three lots of plants sprayed with 10 ml of sterilized distilled water with 0.05 percent Tween 20 served as healthy control. The per cent damage was calculated using the formula

Per cent damage = (No. of damaged leaves / Total no. of leaves) x 100

Yield assessment

The talc-based formulation of individual and mixture of fluorescent pseudomonad strains were tested for their efficacy on growth promotion in rice under glass house conditions. The yield attributing parameters of rice *viz*., plant height, number of tillers and number of grains per panicle were recorded from individual treatments. Rice grain yield was recorded after harvesting the crop.

Assay of defense-related enzymes and compounds

Sample Collection

Samples were collected from individual treatments to study the induction of defense enzymes in response to pest attack in rice plants under glass house conditions. Leaves from fluorescent pseudomonads bioformulation treated plants inoculated with and without leaf folder/ BPH maintained under the same conditions were collected at 0 h, 24 h, 48 h, 72 h, 96 h, 120 h upto 7 days at 24 h interval.

Enzyme extraction

The leaf and sheath tissues were collected from bacterized and control rice plants and immediately homogenized with liquid nitrogen. One g of powdered sample was extracted with 2 ml of 0.1 M sodium citrate buffer (pH 5.0) at 4°C. The homogenate was centrifuged for 20 min at 10,000 rpm. Protein extracts prepared from rice tissues were used for estimation of defense enzymes. Sodium phosphate buffer 0.1 M (pH 7.0) was used for the extraction of polyphenol oxidase and phenylalanine ammonia lyase enzymes.

Determination of protein content

Protein was determined by the method of Bradford (1976). Ten mg of Coomassie brilliant blue G-250 was dissolved in 4.7 ml of absolute alcohol and 10 ml of concentrated phosphoric acid and the volume was made upto 100 ml with distilled water. A sample of 50 µl was added to 950 µl of dye solution and the mixture was incubated for 5 min at room temperature. The absorbance was recorded at 595 nm in GS5703 AT spectrophotometer. Bovine serum albumin was used as the standard.

Assay of phenylalanine ammonia-lyase (PAL)

One gram of plant sample was homogenized in 3 ml of ice cold 0.1 M sodium borate buffer, pH 7.0, containing 1.4 mM of 2-mercaptoethanol and 50 mg of insoluble polyvinylpyrrolidone (PVP). The resulting extract was filtered through cheese cloth and the filtrate was centrifuged at 20,000 g for 15 min at 4°C and the supernatant was used as the enzyme source. PAL activity (EC 4.3.1.5) was determined as the rate of conversion of L-phenylalanine to trans-cinnamic acid at 290 nm. Sample containing 0.4 ml of enzyme extract was incubated with 0.5 ml of 0.1 M borate buffer, pH 8.8 and 0.5 ml of 12 mM L-phenylalanine in the same buffer for 30 min at 30°C. The amount of trans-cinnamic acid synthesized was calculated using its extinction coefficient of 9630 $M^{-1}cm^{-1}$ (Dickerson *et al.*, 1984). Enzyme activity was expressed in fresh weight basis as nmol trans-cinnamic acid min^{-1} mg^{-1} of sample.

Assay of polyphenoloxidase (PPO)

One gram of sample was homogenized in 2 ml of 0.1 M sodium phosphate buffer (pH 6.5) at 4°C. The homogenate was centrifuged at 20,000 g for 15 min at 4°C. The supernatant served as enzyme source and polyphenoloxidase (EC 1.14.18.1) activity was determined as per the procedure given by Mayer *et al.* (1965). The reaction mixture consisted of 1.5 ml of 0.1 M sodium phosphate buffer (pH 6.5) and 200 μl of the enzyme extract. To start the reaction, 200 μl of 0.01 M catechol was added and the activity was expressed as change in absorbance $min^{-1}mg^{-1}$ of protein.

Chitinase

Assay of chitinase: The colorimetric assay of chitinase (EC 3.2.1.14) was carried out according to the procedure developed by Boller and Mauch (1988). One gram of rice tissue was extracted with 5 ml of 0.1 M sodium citrate buffer (pH 5.0). The homogenate was centrifuged for 10 min at 20,000 g at 4°C and the supernatant was used as enzyme source.

Preparation of reagents

Preparation of colloidal chitin: Colloidal chitin was prepared by treating 1 g of crab-shell chitin powder with acetone to form a paste, then slowly adding 20 ml of concentrated hydrochloric acid (HCl) while grinding in a mortar with the temperature maintained at 5°C. After several minutes, the syrupy liquid was filtered through glass wool and poured into vigorously stirred 50 per cent aqueous ethanol to precipitate the chitin in a highly dispersed state. The residue was sedimented and resuspended in distilled water several times to remove excess acid and alcohol and then dialysed against tap water. Chitin content of the suspension was determined by drying a sample *in vacuo* and adjusted with distilled water to a final concentration of 10 mg ml^{-1} (dry weight/volume) and stored at 5°C for further use (Berger and Reynolds, 1958).

Preparation of snail gut enzyme

Six hundred mg of the commercial lyophilized snail gut enzyme (Helicase, Sepracor, France) was dissolved in 10 ml of 20 mM potassium chloride (KCl) and chromatographed on a Sephadex G-25 column (38 x1.5 cm) using a 10 mM KCl solution, containing 1 mM EDTA and adjusted to pH 6.8 for equilibration and elution. The first 20 ml eluted after the void volume was collected (Boller and Mauch, 1988).

Preparation of *p*-dimethylaminobenzaldehyde (DMAB) reagent

The DMAB reagent was prepared by the procedure described by Reissig *et al.*, (1955). Stock solution of DMAB was prepared by mixing 8 g of DMAB in 70 ml of glacial acetic acid along with 10 ml of concentrated HCl. One volume of stock solution was mixed with 9 volumes of glacial acetic acid immediately before use.

Assay

The reaction mixture consisted of 10 μl of 0.1 M sodium acetate buffer (pH 4.0), 0.4 ml enzyme solution and 0.1 ml colloidal chitin. After incubation for 2 h at 37C, the reaction was stopped by centrifugation at 3,000 g for 3 min. An aliquot of the supernatant (0.3 ml) was pipetted into a glass reagent tube containing 30 μl of 1 M potassium phosphate buffer (pH 7.0) and incubated with 20 μl of 3% (w/v) snail gut enzyme for 1 h. After 1 h, the reaction mixture was brought to pH 8.9 by the addition of 70 μl of 0.1 M sodium borate buffer (pH 9.8). The mixture was incubated in a boiling water bath for 3 min and then rapidly cooled in an ice-water bath. After addition of 2 ml of DMAB, the mixture was incubated for 20 min at 37C. Immediately thereafter, the absorbance was measured at 585 nm. *N*-acetylglucosamine (GlcNAc) was used as a standard and the enzyme activity was expressed as n moles GlcNAc equivalents min^{-1} g^{-1} fresh weight.

Assay of lipoxygenase (LOX)

The spectrophotometeric assay of LOX was carried out through the procedure developed by Kermasha and Metche (1986). Rice leaf tissues were cooled in a dry ice bath (-30°C) and macerated to a fine powder and defatted with acetone and diethyl ether. 1.0 g of defatted material suspended in 3 volumes of Tris HCl buffer with mechanical stirring for 16 h and centrifuged at 4,800 g for 30 min. The supernatant was used for enzyme assay and stock solution was prepared using linoleic acid as substrate (A: 1% linoleic acid (w/v) in absolute ethanol; B: 7.1 ml of solution A with 0.25 ml of tween 20 and ethanol evaporated using rotary vacuum evaporator. Finally the residue dissolved in 100 ml of Na_2HPO_4 and pH adjusted to 9.0 using 1N NaOH). Stock solution was diluted ten times with 0.2 M citrate-phosphate buffer and used for further reactions. To 2.4 ml of substrate solution, 0.1 ml of sample was taken and absorbance was measured at 540 nm. Instead of sample, 0.1 ml of deionized water served as control. Rate of increase of optical density from the initial linear portion of the graph was taken and the enzyme activity was expressed as OD_{234}/min/mg of protein.

Assay of ascorbate peroxidases (APX)

APX activity (EC 1.11.1.11) was determined by following the decrease in A_{290} due to H_2O_2 dependent oxidation of ascorbate as described by Rao *et al.,* (1996). Rice leaves were ground to a fine powder in liquid N_2 and then homogenated in 10 mM potassium phosphate buffer (pH 7.0) containing 1mM ascorbate, 20% (w/v) sorbitol, 1 mm EDTA and 0.1% (w/v) phenylmethanesulfonyl fluoride using a mortar and pestle. The homogenate was squeezed through four layers of cheesecloth and then centrifuged at 10,000 rpm for 30 min. The supernatant (5 μL) was added to 2.0 mL of N_2 bubbling 50 mM potassium phosphate buffer (pH 7.0) containing 10 mM H_2O_2. The incubated mixture (1.98 mL) was sampled and mixed with 10 mL of 100mM ascorbate to terminate the inactivation at 1, 2, 3, and 5 min after the start of incubation. The residual oxidizing activity of ascorbate was then assayed by adding 10 ml of 20 mm H_2O_2. The oxidation of ascorbate was followed by a decrease in the A_{290} and the results were expressed in mmol/min/mg of protein.

Statistical analyses

The data were statistically analyzed using the IRRISTAT version 92 developed by the International Rice Research Institute Biometrics unit, the Philippines (Gomez and Gomez, 1984). Prior to statistical analysis of variance (ANOVA) the percentage values of the disease index were arcsine transformed. Data were subjected to analysis of variance (ANOVA) at two significant levels ($P < 0.05$ and $P < 0.01$) and means were compared by Duncan's Multiple Range Test (DMRT).

Gene profiling

Plant growth conditions and challenge inoculation

Rice plants (*Oryza sativa sp. Indica cv. Co 43*) were grown under green house conditions. Fifty day old plants were challenge inoculated with insect pests *viz*., leaf folder and BPH. One day before challenging, a set of plants to be inoculated with insects were treated with PGPR (*Pseudomonas fluorescens* strain TDK-1). Forty eight hours after inoculation, observations were made on pest/ disease development based on the symptoms. Leaf sheaths were collected in liquid nitrogen from the following treatments for RNA as well as protein isolation.

Treatment details:

1. Control	3. Leaf folder alone	5. BPH alone
2. PGPR alone	4. Leaf folder + PGPR	6. BPH + PGPR

RNA isolation and normalization

Total RNA was extracted from the leaf sheath tissues collected from the above treatments using the TRIzol reagent method (Invitrogen, San Diego, USA). The genomic DNA was removed from total RNA with DNase I. The quality of RNA was checked by 1.5% agarose gel electrophoresis and quantified by spectrophotometer. The A_{260}/A_{280} ratio was 1.9 which indicated purity of RNA devoid of DNA and protein contamination. Total RNA was normalized to a concentration of 0.1µg/µl for RT-PCR reactions and differential display studies.

Differential Display Reverse Transcription Polymerase Chain Reaction (DD-RT PCR)

First strand cDNA synthesis

Total RNA from rice leaf sheath tissues collected from control and treated plants was used for DD-RT-PCR analysis. All the protocols were followed as described in the DD-RT-PCR kit namely RNAimage kit (GenHunter- USA). Reverse transcription was done in the total volume of 20 µl containing 0.2mg of freshly diluted RNA, 125mM Tris - HCl, pH 8.3, 188mM KCl, 7.5mM MgC12, 25mM DTT, 250mM dNTP and 2mM of HT11M (where M may be G, A or C). RNA was heated at 65°C for 5min for denaturation and cooled to room temperature incubated at,37°C for 60min after adding 1 µl MMLV reverse transcriptase (100 U). After first strand cDNA synthesis the reaction mix was incubated at 75°C for 5min to destroy the RNA templates.

PCR amplification of cDNA

Each fraction of the single stranded cDNA was amplified by PCR with respective anchored primer and an arbitrary primer. Each 20 µl reaction contained 1x PCR buffer, 2 µl of first strand cDNA, 2.5 µM of each dNTPs, 0.2 µm each of anchored and arbitrary primer, 0.2 µCi α [^{33}P] dATP and 1 unit of *Taq* DNA polymerase, following the thermal program: an initial denaturation at 92° C for 5 min leading to 40 cycles of denaturation at 92° C for 30 sec, annealing at 40° C for 2 min, synthesis at 72° C for 30 sec, followed by a final extension for 5 min at 72° C. After the reaction, one fourth volume of reaction was used for electrophoresis on a 6% urea polyacrylamide gel.

Urea polyacrylamide gel electrophoresis of DD-RT-PCR fragments

The cDNA products of above DD-RT-PCR reaction were resolved on a 6% urea polyacrylamide gel (Sambrook *et al.*, 1989). The template was set by using a bind silane coated inner plate and repel silane coated outer plate and a pair of 0.4 mm thick spacers. The setup was made leak proof using clips and adhesive tapes. Required volume of 6% urea polyacrylamide gel mixture (120 ml of gel mixture = 50.4 g of ultra pure urea, 18 ml of acrylamide: bisacrylamide (19:1) solution, 12 ml of 10x TBE, 48 ml of distilled water, 50 µl of TEMED and 1.2 ml of 10% APS) was prepared and poured in the template. Immediately after pouring the gel mixture, an appropriate comb was inserted. The assembly was left undisturbed for 1-2 h. Gel was prerun for 1 h using 1x TBE buffer. Samples were loaded onto the gel (control and stress reaction products with same primer combinations were loaded adjacently) and electrophoresed at 1500 V for 3 h. Gel was autoradiographed using an x-ray cassette.

Elution and reamplification of cDNA fragments

Differentially expressed bands were identified in the autoradiogram. The cDNA bands showing differential expression were marked and carefully cut out and soaked in 10 µl of sterile water for 10 min. This was boiled for 15 min and cooled. Supernatant containing cDNA was collected after centrifugation and the cDNA was ethanol precipitated. After incubation (at -80° C for 1 h) DNA was spun down (at 12000 rpm for 20 min) and redissolved. The redissolved cDNA was used for further amplification under same conditions. Reamplified product was checked by agarose gel electrophoresis. Reamplified cDNA was purified by phenol: chloroform extraction and concentrated by ethanol precipitation.

Cloning of cDNA fragments

Purified cDNA fragments were cloned in a T/A cloning vector (pTZ5R/T) in a 10 µl reaction as per manufacturer's instruction (Fermentas Inc., USA).

Transformation of E. *coli*

Escherichia coli DH5α host cells were grown to mid log phase and the cells were made competent by treating with 100 mM calcium chloride solution (Sambrook *et al.,* 1989). About 2 µl of the ligation product was used for transformation. Ligation product was added to 100 µl of competent cell suspension and incubated on ice for 30 min. The cells were given heat shock at 42° C for 90 sec and again incubated on ice for 30 min. To this, 1 ml of LB medium was added and incubated at 37° C for 1 h. After incubation, 100 µl of the cell suspension was uniformly spread on LB agar (10 g/l tryptone, 10 g/l NaCl, 5 g/l yeast extract, 15 g/l agar, pH 7.2) medium containing X-gal and ampicillin (100 mg/ml) and incubated at 37°C for overnight, for the colonies to develop. Transformants were identified based on *lac z* selection (blue and white selection of *E. coli* cells harbouring recombinant plasmids grown as individual colonies which were checked by plasmid isolation and restriction digestion).

Sequencing of the partial cDNA clones

Cloned cDNAs were sequenced by Automated cycle sequencing and raw EST sequences were edited to remove adoptor and vector sequences using the Lucy software programme (Chou and Holmes, 2001).

Sequence analysis

To define the possible physiological function(s) of the cDNA inserts, the respective deduced amino acid sequences were compared with the previously described cDNAs in the database using NCBI – BLAST program and TIGR databases.

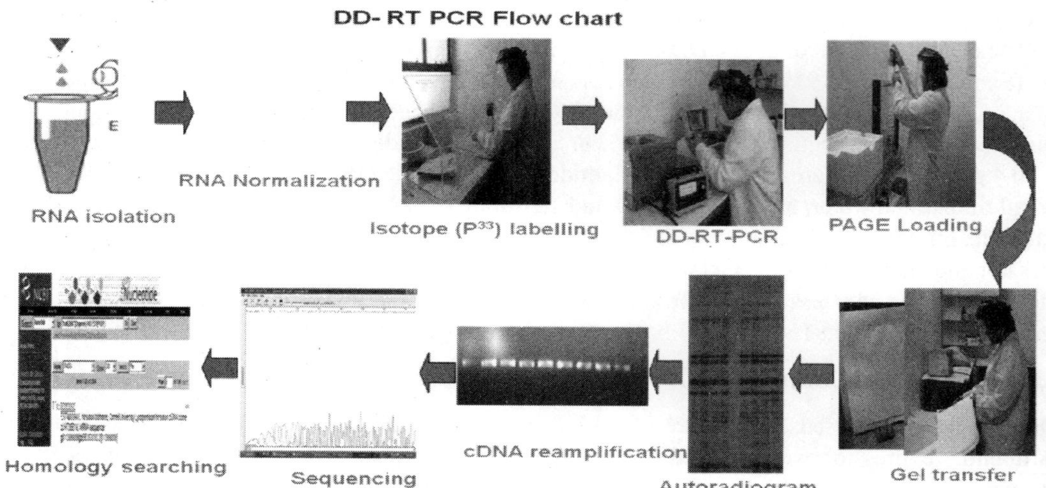

Plate. 4

Results

Glass house and field studies

Experiments on glass house and field studies revealed that the application of PGPR reduced the incidence of leaf folder and PBH drastically in rice plants, thereby it increased the yield considerably (Table. 1, 2; Plate 5).

Table 1. Effect of fluorescent pseudomonad strains on growth attributes of rice seedlings inoculated with Leaf folder pest under glass house conditions

S. No.	Treatments	Germination (%)	No. of tillers / hill	Plant height (cm) (75 DAP)	Leaf folder Pest incidence	Per cent reduction over control	1000 grain weight (g)	Grain Yield/ Plant (g)
1	PF-1	98.0	12.25^{d}	77.34	14.51^{j}(22.7)	50.4	24.61^{d}	11.08^{c}
2	TDK-1	99.5	12.95^{cd}	77.74	13.97^{I}(25.4)	43.8	24.71^{d}	11.15^{c}
3	KH-1	99.5	13.35^{cd}	79.94	12.62^{h}(21.1)	48.5	25.11^{c}	11.44^{bc}
4	PF-1+TDK-1	98.5	13.43^{c}	80.94	10.88^{g} (19.6)	37.8	25.31^{c}	11.58^{bc}
5	PF-1+KH-1	99.0	13.10^{cd}	79.14	10.39^{f}(18.5)	36.1	24.81^{d}	11.21^{c}
6	TDK-1+KH-1	100.0	15.56^{b}	84.54	7.23^{d} (18.6)	35.8	26.01^{b}	12.08^{ab}
7	PF-1+TDK-1+KH-1	100.0	16.00^{a}	84.94	5.82^{b} (15.4)	25.1	26.61^{a}	12.50^{a}
8	PF-1+PY-15+TDK-1	99.5	15.89^{a}	83.34	6.63^{c}(16.6)	20.2	25.51^{c}	11.73^{bc}
9	PF-1+PY-15+TDK-1+MDU2	97.5	12.80^{bc}	82.14	10.32^{e}(19.2)	23.0	25.41^{c}	11.65^{bc}
10	Chemical control	88.0	9.12^{d}	71.54	4.12^{a}(14.2)	14.3	20.8^{e}	08.30^{d}
11	Control	78.0	8.17^{e}	67.54	28.81^{k}(32.6)	-	17.1^{f}	07.43^{e}

Values are mean of two replications. Means in a column followed by same superscript letters are not significantly different according to Duncan's multiple range test at $P = 0.05$.

Table. 2 Effect of bioformulations containing mixture of fluorescent pseudomonad strains on the plant growth characters and pest incidence in rice under field conditions (Trial -Karur)

S. No.	Treatments	No. of tillers / hill	Plant height (cm) (75 DAP)	No. of panicles / hill	Leaf folder incidence (60 DAP)	BPH (No. / hill) (60 DAP)	1000 grain weight (g)	Grain Yield/ Plant (Kg/ha)
1	PF-1	18.33^{d}	105.2	12.80^{de}	13.5^{b} (21.55)	3.90^{b}	24.62^{cd}	7222^{d}
2	TDK-1	19.00^{cd}	109.47	12.87^{de}	12.7^{c} (20.89)	3.50^{c}	24.83^{cd}	7222^{d}
3	KH-1	19.60^{bc}	110.13	13.13^{de}	10.7^{e} (19.08)	2.52^{e}	26.08^{bc}	8056^{bc}
4	PF-1+TDK-1	19.73^{bc}	110.53	13.67^{cd}	11.4^{d} (19.72)	2.54^{d}	26.18^{bc}	8333^{bc}
5	PF-1+KH-1	19.27^{bc}	110.07	13.07^{de}	09.2^{f}(17.65)	2.52^{e}	25.58^{c}	7917^{c}
6	TDK-1+KH-1	20.13^{b}	113.07	15.07^{ab}	06.8^{i} (15.10)	1.51^{h}	28.73^{ab}	8611^{b}
7	PF-1+TDK-1+KH-1	22.27^{a}	113.93	15.33^{a}	05.7^{j} (13.80)	1.41^{i}	30.76^{a}	9444^{a}
8	PF-1+PY-15+TDK-1	20.13^{b}	111.4	14.20^{bc}	07.3^{h} (15.66)	1.74^{g}	27.12^{bc}	8611^{b}
9	PF-1+PY-15+TDK-1+MDU2	19.97^{b}	110.73	14.07^{c}	08.2^{g} (16.63)	1.90^{f}	26.78^{bc}	8472^{bc}
10	Chemical control	16.53^{e}	94.93	12.40^{e}	04.1^{k} (11.66)	0.57^{j}	22.71^{d}	6944^{de}
11	Control	13.27^{f}	88.27	9.60^{f}	32.8^{a}(34.93)	8.20^{a}	20.11^{e}	6389^{e}

Plate. 5a Leaf folder alone inoculated

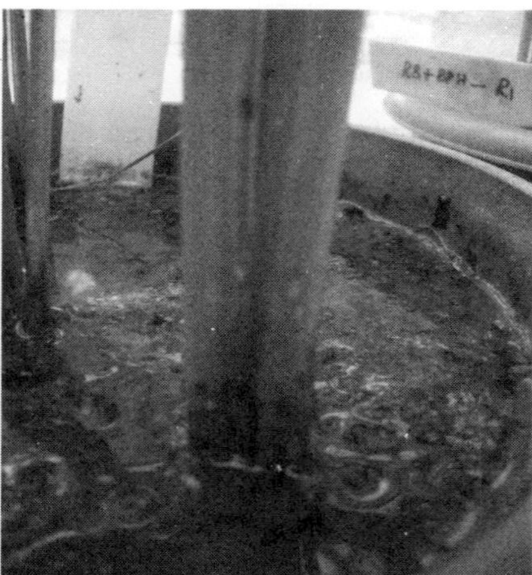

Plate. 5c BPH alone inoculated

Plate. 5b Leaf folder +PGPR

Plate. 5d BPH +PGPR

Enzyme assays and activity gel electrophoresis

Enzyme like Phenylalanine ammonia lyase (Fig. 1), Polyphenol oxidase (Fig. 2), chitinase (Fig. 3), Lipoxygenase (Fig. 4) and Ascorbate oxidase (Fig. 5) activity were enhanced when challenging with pest in the PGPR primed rice plants. The activity reached the peak generally after 96 hour challenge inoculation commonly in all the cases. The activity gel electrophoresis also proved the same (Plate 6a, 6b & 6c). With this background result we went for analysis of transcriptional regulation at the molecular level.

Transcriptional analysis

Total RNA isolation

Total RNA was isolated from rice leaf sheath tissues of 1) control, 2) PGPR primed, 3) Pests (leaf folder & BPH) challenged and from the combination both 2 & 3. The intactness of the isolated RNA

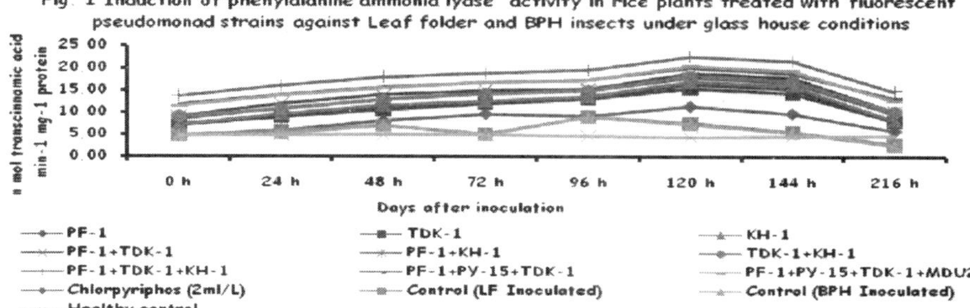

Fig. 1 Induction of phenylalanine ammonia lyase activity in rice plants treated with fluorescent pseudomonad strains against Leaf folder and BPH insects under glass house conditions

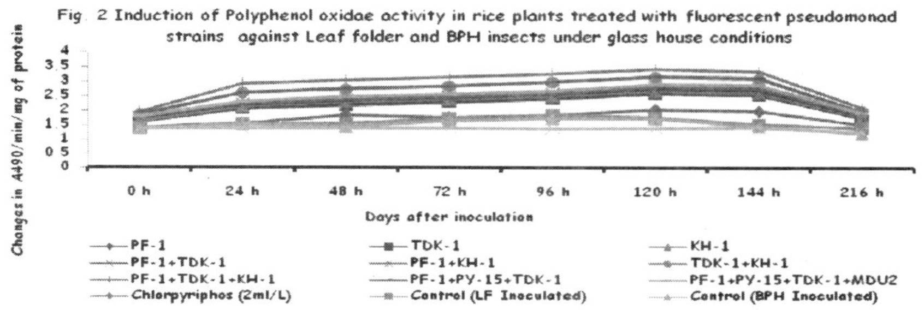

Fig. 2 Induction of Polyphenol oxidae activity in rice plants treated with fluorescent pseudomonad strains against Leaf folder and BPH insects under glass house conditions

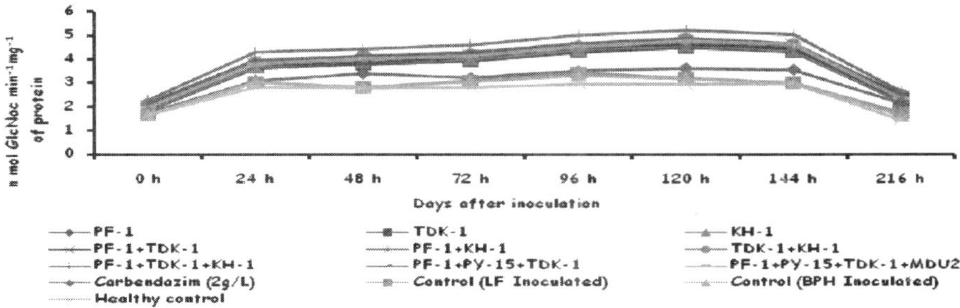

Fig. 3 Induction of Chitinase activity in rice plants treated with fluorescent pseudomonad strains against Leaf folder and BPH insects under glass house conditions

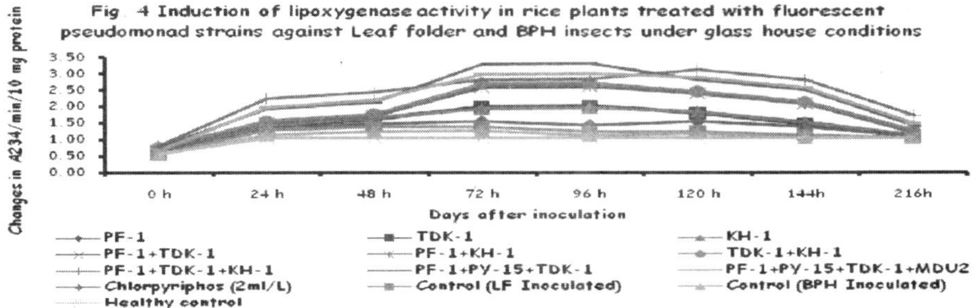

Fig. 4 Induction of lipoxygenase activity in rice plants treated with fluorescent pseudomonad strains against Leaf folder and BPH insects under glass house conditions

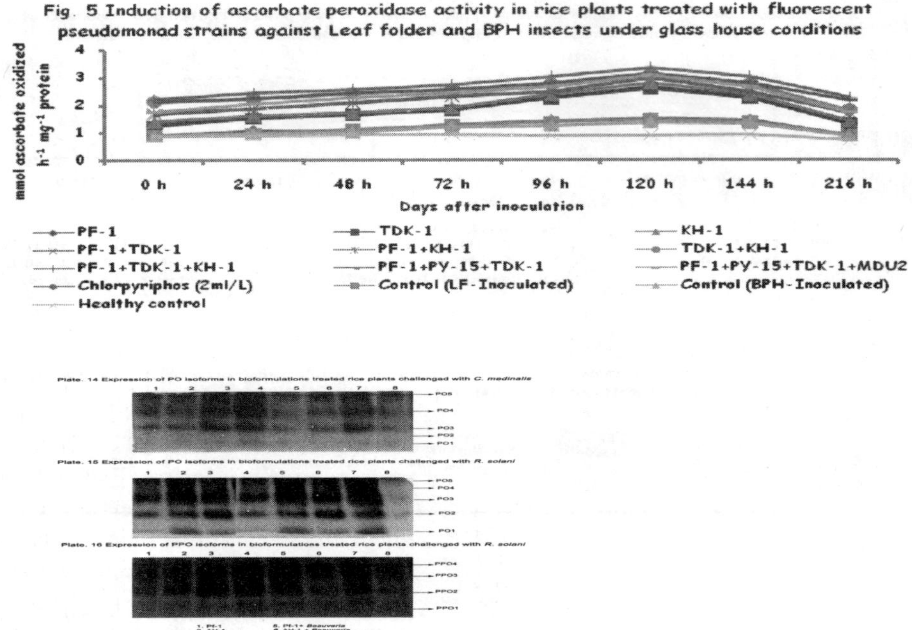

Plate. 6a Expression of PO isoforms in PGPR treated rice plants challenged with insect pests

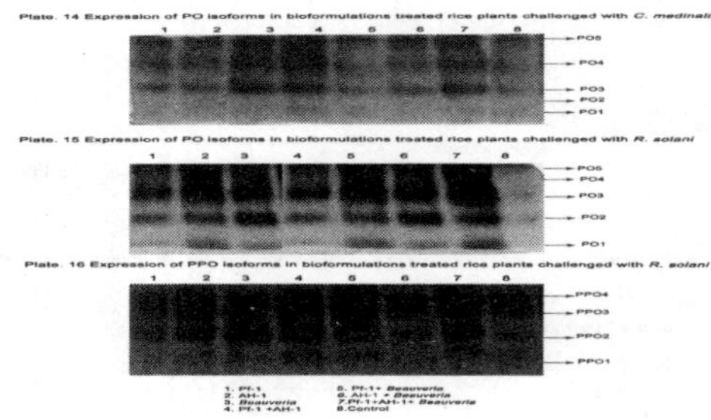

Plate. 6b Expression of PPO isoforms in PGPR treated rice plants challenged with insect pests

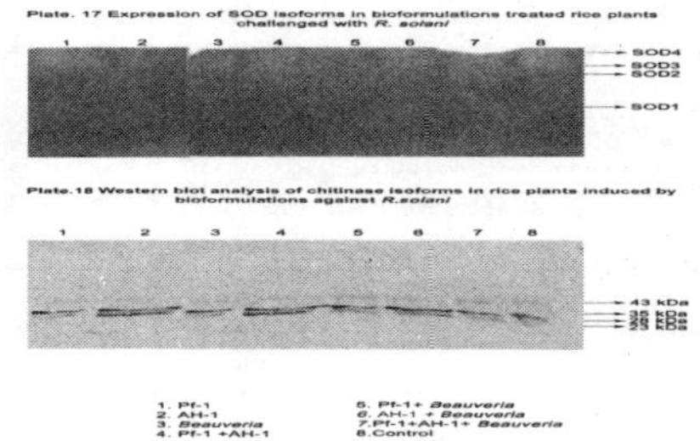

Plate. 6c Western blot analysis of Chitinase isoforms in PGPR treated rice plants challenged with insect pests

Treatments details

1. PF-1
2. TDK-1
3. KH-1
4. PF-1+TDK-1
5. PF-1+KH-1
6. TDK-1+KH-1
7. PF-1+TDK-1+KH-1
8. Control

was checked by a 1.2% agarose gel (Plate 7). The discrete bands of 25S and 18S rRNAs and their 2:1 ratio (a visual approximation) and the relatively clear interband zone indicated the intactness of the RNA. An optical density (OD) ratio (A_{260}/A_{280}) of >1.8 indicated the purity of the sample.

Treatment details:

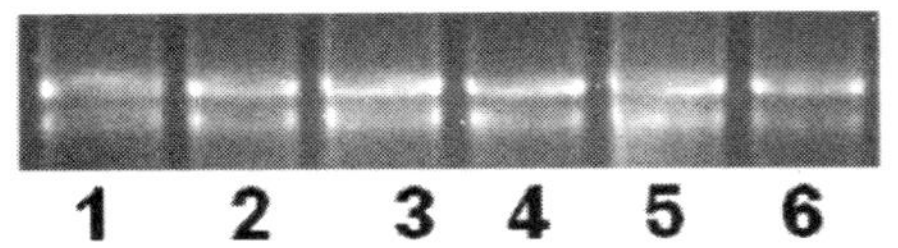

1. Control
2. PGPR alone
3. Leaf folder alone
4. Leaf folder + PGPR
5. BPH alone
6. BPH + PGPR

Plate. 7 Total normalized RNA

Identifying differentially expressed cDNAs and reamplification

About 2μg of total RNA from both control and treated samples were taken for DD-RT-PCR analysis. The first strand of cDNA was synthesized in both control and from different treatment imposed rice leaf sheath RNA using three different anchored primers. Each first strand cDNA product was used for the second strand amplification with 8 different arbitrary primers using α [^{33}P] dATP as one of the nucleotide. The product was resolved on 6% denaturing urea–polyacrylamide gels and it was exposed to X-ray film after drying. Plate 8 presents the autoradiogram showing differential display patterns of control and treatment imposed rice leaf sheath RNA. A comparison of amplification patterns between control, PGPR primed, Pest challenged with or without PGPR primed rice leaf sheath mRNA sample lanes using the anchored primers HT11(A/G/C) and the arbitrary primers HAP 1-8 indicated that many bands were common to all the samples and only a few bands appeared to exhibit a differential expression. Totally there were 96 cDNAs (including redundant ESTs) which showed differential expression due to various interactions. Out of the 96 cDNAs, 75% of the cDNAs were found to be up-regulated and 25% were found to be down-regulated (Plate. 8a).

Elution and reamplification of differentially expressed cDNAs

The differentially expressed cDNA fragments were retrieved from the gel and reamplified with the same set of primers as in the initial DD-RT-PCR reaction and analyzed electrophoretically (Plate 8b). The agarose gel electrophoresis showed that the cDNA fragments obtained were relatively short (100 – 600bp).

Cloning of cDNA fragments

Differentially expressed cDNA fragments were purified from the agarose gel and cloned into T/A cloning vector (pTZ5R/T). The ligated product was mobilized into *E. coli* and checked by blue (non-transformed) and white (transformed) colony selection and colony PCR analysis.

Sequencing of partial cDNA clones

Cloned cDNA fragments were sequenced at Agile life sciences (First base sequencing service Pvt. Ltd., Singapore). The sequencing results showed that the partial cDNA clones are of size between 600 bp and 150bp. The nucleotide sequence contains an oligo dT primer at the 3' end of the gene and a random arbitrary primer in the coding end.

Sequence analysis

To define the possible physiological function(s) of cDNAs, the nucleotide sequences of cDNA fragments were compared with the database using NCBI – BLAST, TIGR -BLAST program. The search of the SWISSPROT protein data-bases for the amino acid sequences similar to the partial cDNA clones revealed the homologous plant protein sequences. Analysis of 96 cDNA sequences (Table. 3) against various databases like TIGR, GRAMENE and NCBI revealed that 75% of the differential genes were from the treatments which received the PGPR priming. This includes both up and down regulation.

The differential cDNAs were classified into various functional categories (Fig. 6; Fig. 7; Fig. 8; Fig. 9; Fig. 10). In total from all the treatment combinations 16.5% of the cDNAs were found to be related to Signal transduction, 8.9% belongs to ISR followed by genes involved in Defense response (7.3%), Cellular organization and biogenesis (6.9%), Response to stress and biotic /abiotic stimulus (10.6%), Cellular transport (6.6%), Metabolism (9.6%), Energy metabolism (3.6%), Protein metabolism (10.9%), DNA/ RNA metabolism (8.6%), Unknown biological process (10.6%). Genes involved in various functional catogories were found to be down-regulated upon various treatments. The individual down regulated genes from various treatments were listed in control (Table.3). Of the total 96 cDNAs, 1% did not appear to exhibit any functional similarity to any of the sequence in the database. Another 10.6% were found to belong to the category of protein whose functions are not yet determined.

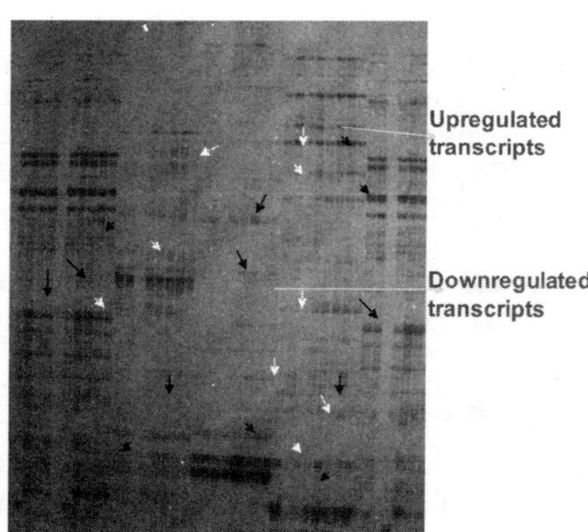

Plate. 8a Autoradiogram showing differentially expressed mRNAs in various treatments

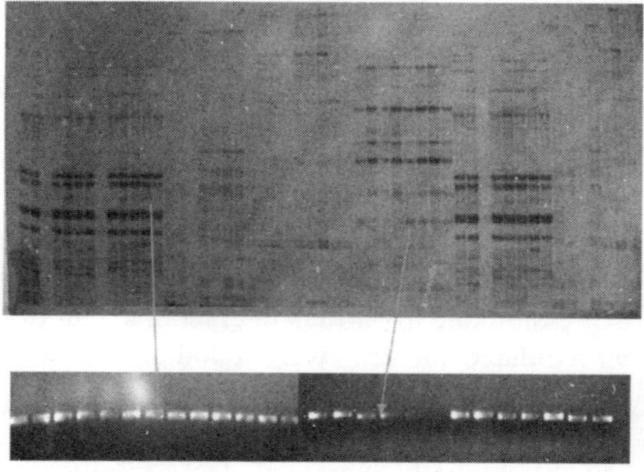
Plate. 8b Unique reamplified differential cDNA's specifically eluted from Autoradiogram

Discussion

Leaf folder and BPH are the major pest challenges for rice cultivation world wide and these insect pests lead to severe constraint in food production. These pest challenges urged the scientist to produce resistance cultivars through classical and molecular breeding approaches. So far there were no improved resistance cultivars available to eradicate the pest incidence, because the non availability of resistant genes resource and interactive functional gene informations. Our basic glasshouse and field studies on the effect of PGPR on pest management clearly indicated that, the PGPR application drastically reduced the pest incidence, thereby increased the yield at a significant level.

To find out the reason for above said plant growth and grain yield enhancement, we have collected the rice leaf sheath samples at different hours after the application of PGPR as well as pest challenge and determined the level of expression of defense related enzymes both qualitatively and quantitatively through spectrophotometer and native polyacrylamide gel electrophoresis respectively at various time intervals. The results of enzyme studies revealed the enhanced expression of defense related enzymes *viz*., Phenylalanine ammonia lyase, Poly phenol oxidase, Chitinase, Lipoxygenase

Table 3. ıdentified differential cDNA EST's are listed here with their specific functions. The sequences were submitted in GenBank (NCBI) The accession numbers of the each sequence with their source tissue, sequence length, blast match score and E-Value and the name of the protein identity were tabulated.

Sl.No	Lib -rary No.	Accesssion No	Seq -uence length	score	E-Value	Protein Name
			CONTROL (Genes which are down regulated in various treatments)			
1	I-1	FG346937	237	406	4.2e-13	unspliced-genomic expressed protein
2	I-2	FG346938	329	1361	9.0e-62	unspliced-genomic protein translation factor SUI1, putative, expressed
3	I-3	FG346939	241	1137	2.6e-45	unspliced-genomic vacuolar ATP synthase subunit E,putative
4	I-4	FG346940	209	359	4.4e-10	unspliced-genomic early response to drought 3, putative, expressed
5	I-5	FG346941	193	523	1.6e-17	unspliced-genomic cupin, RmlC-type, putative
6	I-6	FG346942	148	537	3.3e-18	unspliced-genomic zinc finger A20 and AN1domains-containing protein, putative, expressed
7	I-7	FG346943	142	580	3.6e-20	unspliced-genomic zinc finger A20 and AN1domains-containing protein, putative, expressed
8	I-8	FG346944	166	276	2.7e-06	unspliced-genomic zinc finger A20 and AN1domains-containing protein, putative, expressed
9	I-9	FG346945	127	526	1.3e-17	endoglucanase precursor, putative, expressed
10	I-10	FG346946	150	578	5.0e-20	receptor-like protein kinase 5 precursor, putative, expressed
11	I-11	FG346947	144	138	5.8	Phosphoenol pyruvate carboxylase 1, putative, expressed
12	I-12	FG346948	139	150	1.6	sialin, putative, expressed
13	I-13	FG346949	156	620	6.1e-23	unspliced-genomic expressed protein
14	I-14	FG346950	150	615	1.0e-22	unspliced-genomic expressed protein
15	I-15	FG346951	153	615	1.0e-22	unspliced-genomic expressed protein
16	I-16	FG346952	155	630	2.2e-23	unspliced-genomic expressed protein
17	I-17	FG346953	140	480	4.4e-16	unspliced-genomic expressed protein
18	I-18	FG346954	133	506	2.9e-17	unspliced-genomic hypothetical protein
19	I-19	FG346955	123	382	1.2e-11	unspliced-genomic hypothetical protein
				PGPR + LF		
20	II-1	FG347011	470	1439	2.6e-71	unknown
21	II-2	FG347012	241	165	0.34	DNA repair protein RAD54-like, putative
22	II-3	FG347013	357	210	0.0029	aspartic proteinase nepenthesin-2 precursor, putative, expressed
23	II-4	FG347014	154	277	2.7e-06	protein kinase APK1A, chloroplast precursor, putative, expressed
24	II-5	FG347015	258	1047	3.1e-41	farnesyl pyrophosphate synthetase, putative, expressed

25	II-6	FG347016	118	155	0.94	inositol-1-monophosphatase, putative, expressed
26	II-7	FG347017	177	163	0.43	unspliced-genomic oxidoreductase, FAD-binding,putative,
27	II-8	FG347018	212	163	0.39	unspliced-genomic hypothetical protein
28	II-9	FG347019	467	1271	3.7e-70	unspliced-genomic OsIAA21 - Auxin-responsive Aux/IAA gene family member, expressed
29	II-10	FG347020	360	1288	4.0e-52	CIPK-like protein 1, putative, expressed
30	II-11	FG347021	382	714	4.6e-26	TPR Domain containing protein
31	II-12	FG347022	389	1784	7.8e-76	unspliced-genomic zinc finger A20 and AN1 domains-containing protein, putative, expressed
32	II-13	FG347023	331	1361	1.4e-61	unspliced-genomic protein translation factor SUI1 putative
33	II-14	FG347024	122	308	1.1e-07	phosphatidylinositol transfer protein CSR1, putative, expressed
34	II-15	FG347025	157	635	1.3e-23	unspliced-genomic expressed protein
35	II-16	FG347026	156	645	4.6e-24	unspliced-genomic expressed protein
36	II-17	FG347027	200	361	4.6e-10	ferrochelatase-2, chloroplast precursor, putative, expressed
37	II-18	FG347028	152	630	2.2e-23	unspliced-genomic expressed protein
38	II-19	FG347029	155	630	2.2e-23	unspliced-genomic expressed protein
39	II-20	FG347030	119	357	5.9e-10	unspliced-genomic autophagy-related protein 8.precursor
40	II-21	FG347031	199	149	1.5	RNA binding protein
41	II-22	FG347032	196	163	0.41	glucan endo-1,3-beta-glucosidase 3 precursor, putative, expressed
42	II-23	FG347033	379	50	0.076	w11n Triticum aestivum cDNA clone
43	II-24	FG347034	141	135	7.9	ent-kaurene oxidase, putative, expressed
44	II-25	FG347035	211	160	0.76	2Fe-2S ferredoxin, putative, expressed
					LF Alone	
45	III-1	FG347036	470	1082	2.6e-52	OsIAA21 - Auxin-responsive Aux/IAA gene family member, expressed
46	III-2	FG347037	210	163	0.39	unspliced-genomic hypothetical protein
47	III-3	FG347038	466	721	6.4e-33	OsIAA21 - Auxin-responsive Aux/IAA gene family member, expressed
48	III-4	FG347039	175	769	3.7e-75	unspliced-genomic zinc finger A20 and AN1 domains-containing protein, putative, expressed
49	III-5	FG347040	292	875	1.3e-33	unspliced-genomic cupin, RmlC-type, putative, expressed
50	III-6	FG347041	173	624	3.4e-22	unspliced-genomic zinc finger A20 and AN1domains-containing protein, putative, expressed
51	III-7	FG347042	146	576	7.2e-20	unspliced-genomic golgi transport 1 protein B,putative, expressed

52	III-8	FG347043	141	491	5.6e-16	kinesin motor domain containing protein, expressed
53	III-9	FG347044	144	581	3.9e-21	unspliced-genomic expressed protein
54	III-10	FG347045	155	630	2.2e-23	unspliced-genomic expressed protein
55	III-11	FG347046	153	615	1.0e-22	unspliced-genomic expressed protein
56	III-12	FG347047	153	616	9.3e-23	unspliced-genomic expressed protein
57	III-13	FG347048	117	299	3.7e-08	unspliced-genomic conserved hypothetical protein
				PGPR+BPH		
58	IV-1	FG347050	260	1037	8.7e-41	farnesyl pyrophosphate synthetase, putative, expressed
59	IV-2	FG347051	145	171	0.18	inositol-1-monophosphatase, putative, expressed
60	IV-3	FG347052	250	163	0.42	protease 4, putative, expressed
61	IV-4	FG347053	150	159	0.63	succinyl-CoA ligase beta-chain, mitochondrial precursor, putative, expressed
62	IV-5	FG347054	202	511	7.1e-17	NB-ARC domain containing protein, expressed
63	IV-6	FG347055	264	1361	1.8e-60	protein translation factor SUI1, putative, expressed
64	IV-7	FG347056	213	640	.1e-22	indole-3-acetate beta-glucosyltransferase, putative
65	IV-8	FG347057	256	639	1.0e-22	phosphatidylserine decarboxylase, putative, expressed
66	IV-9	FG347058	207	851	2.3e-33,	unspliced-genomic 16kDa membrane protein
67	IV-10	FG347059	176	239	0.00014	spotted leaf protein 11, putative, expressed
68	IV-11	FG347060	185	760	2.4e-28	unspliced-genomic cupin, RmlC-type, putative,expressed
69	IV-12	FG347061	166	699	1.1e-25	60S ribosomal protein L7/L12 precursor, putative, expressed
70	IV-13	FG347062	140	490	6.2e-16	kinesin motor domain containing protein, expressed
71	IV-14	FG347063	140	536	4.7e-18	mitogen-activated protein kinase organizer 1, putative, expressed
72	IV-15	FG347064	148	390	2.0e-11	4Fe-4S ferredoxin, iron-sulfur binding protein
73	IV-16	FG347065	155	615	1.0e-22	expressed protein
74	IV-17	FG347066	152	637	1.0e-23	expressed protein
75	IV-18	FG347067	151	640	7.7e-24	expressed protein
76	IV-19	FG347068	154	316	4.8e-08	genomic nucleic acid binding protein
77	IV-20	FG347069	150	305	1.5e-07	genomic nucleic acid binding protein
78	IV-21	FG347070	331	1500	4.7e-63	unspliced-genomic expressed protein
79	IV-22	FG347071	183	139	5.1	LRR receptor-like protein kinase, putative, expressed
80	IV-23	FG347072	295	52	0.020	rootphos(-) Medicago truncatula cDNA clone MHRP-13D11
81	IV-24	FG347073	239	487	7.6e-17	expressed protein
82	IV-25	FG347074	207	299	3.0e-07	phosphatidylinositol-4-phosphate 5-Kinase family protein, expressed

83	IV-26	FG347075	166	623	5.0e-22	eukaryotic translation initiation factor 5A, putative, expressed
84	IV-27	FG347076	167	460	9.8e-	esterase precursor, putative, expressed
85	IV-28	FG347077	131	358	6.1e-10	WRKY transcription factor 46, putative, expressed
				BPH alone		
86	V-1	FG347078	170	135	7.7	unspliced-genomic retrotransposon protein,putative
87	V-2	FG347079	167	688	3.6e-25	NBS-LRR resistance protein, putative
88	V-3	FG347080	152	620	6.1e-23	unspliced-genomic expressed
89	V-4	FG347081	152	630	2.2e-23	unspliced-genomic expressed protein
90	V-5	FG347082	154	630	2.2e-23	unspliced-genomic expressed protein
91	V-6	FG347083	154	625	3.6e-23	unspliced-genomic expressed protein
92	V-7	FG347084	145	311	8.1e-08	unspliced-genomic nucleic acid binding protein,putative
93	V-8	FG347085	148	318	3.9e-08	unspliced-genomic expressed protein
94	V-9	FG347086	136	156	0.84	Unknown protein
95	V-10	FG347087	260	1135	3.3e-45	unspliced-genomic vacuolar ATP synthase subunit E
96	V-11	FG347088	215	892	3.2e-35	unspliced-genomic 16kDa membrane protein,

and Ascorbate oxidase. The activity attained a peak between 96 and 120 hours of PGPR application along with challenge inoculation of pests.

With this strong physiological background, we have taken a step to better understand the molecular mechanisms underlying the host plant defense, genes which are involved in this process, modulation in the expression level of genes in defense gene families, possible changes in the compounds and metabolites present in the signal transduction pathways. We employed DD-RTPCR to analyze cDNAs in a PGPR primed rice plant and its associated healthy and inoculated control plants. The damage and frequent outbreaks of BPH and incidence of leaf folder, along with the hazardous effects of using pesticides to protect rice crops, has now prompted researchers to identify pest resistant germplasms, to locate the source of resistance genes by the utilization of resistant lines in breeding programs, and attempt to isolate resistance genes. Through the molecular studies, we have noticed changes in about 96 genes in the expression level in both up and down regulation aspects. Among these, 75% of the genes belong to the treatments received PGPR priming with the maximum number of coding novelty proteins. These results clearly concluded that PGPR have great impact on the modulation R genes, Induction of systemic resistance, early accumulation secondary metabolites and need based regulation of members in R-gene families which belong to signal transduction pathways.

There are many evidences to show a gene-for-gene type of plant defense against sucking insects exists in a number of plant species (Brotman *et al.* 2002; Kaloshian 2004). Upon recognition of the invading pest, activation of plant defense is accompanied by an array of transcriptional reprogramming of pest responsive genes such as defense-related genes. Lot of studies were made on the rice, we are having complete genome sequence information, EST sequence data are also available for one rice-pathogen (Blast) interactions and one rice-pest (BPH) interaction. Due to changing nature of insect pest, most of the resistant varieties are vulnerable to the attack of insect pests. Searching the resistance gene, integration of the differential resistance gene which were obtained due to such type of interaction will not be a permanent solution for insect pest resistance. In this connection, utilization

Table 4: Percentage of EST in each library based on gene functional categories

Sl. No	Gene function categories	Treatment libraries				
		PGPR alone	PGPR + Leaf folder	Leaf folder	PGPR + BPH	BPH alone
1	Signal transduction	7	22	2	15	4
2	ISR	4	12	2	7	2
3	Defense response	3	6	1	9	3
4	Cellular organization and biogenesis	4	8	2	3	4
5	Response to stress and biotic /abiotic stimulus	5	14	2	7	4
6	Cellular transport	2	7	2	6	3
7	Metabolism	6	10	4	5	4
8	Energy metabolism	1	3	0	4	3
9	Protein metabolism	1	15	7	6	4
10	DNA/ RNA metabolism	4	8	2	7	5
11	Unknown biological process	8	12	5	4	3

Fig. 6 Functional categories of unique sequences expressed when rice plants Primed with PGPR

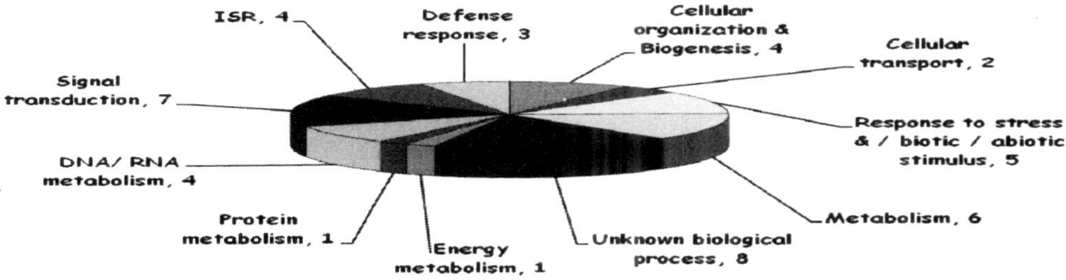

Fig. 7 Functional categories of unique sequences expressed when PGPR primed rice plants challenged with Leaf folder

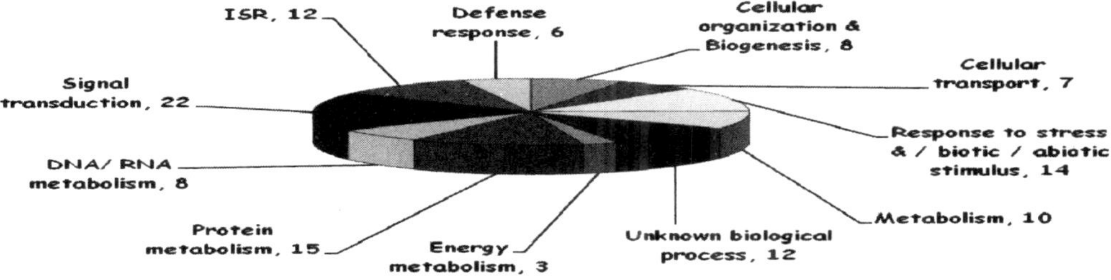

of beneficial microbes, *viz*., PGPR would result in ISR in plants thereby lead to creation of multiple pest, disease and stress resistance cultivars.

In the present study, rice plants were primed with PGPR as seed treatment, soil application and foliar spray for inducing resistance against insect pests both in laboratory and field conditions. Previous literatures were surveyed on the aspects of resistance gene information, their source, duration of resistance, the way it last their resistance capability. With an aim of identifying novel candidate genes conferring complete pest resistance, whole genome expression profiling was undertaken through DD-RT-PCR technique. The results obtained are discussed below.

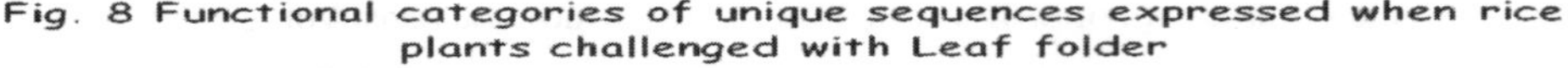
Fig. 8 Functional categories of unique sequences expressed when rice plants challenged with Leaf folder

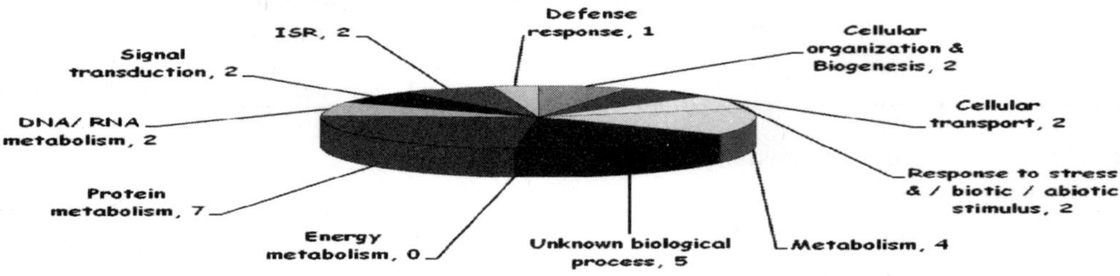

Fig. 9 Functional categories of unique sequences expressed when PGPR primed rice plants challenged with BPH

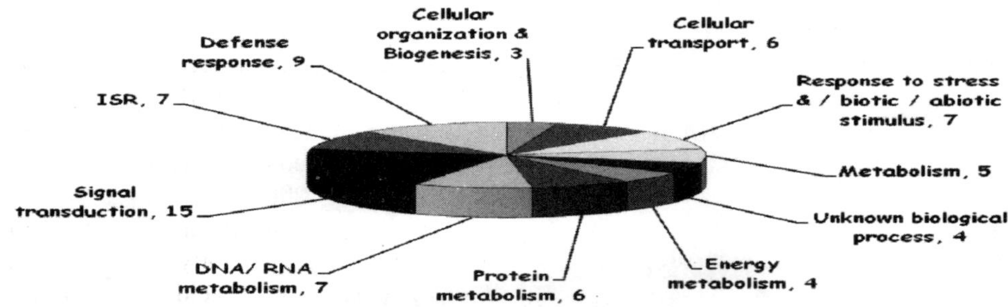

Fig. 10 Functional categories of unique sequences expressed when rice plants challenged with BPH

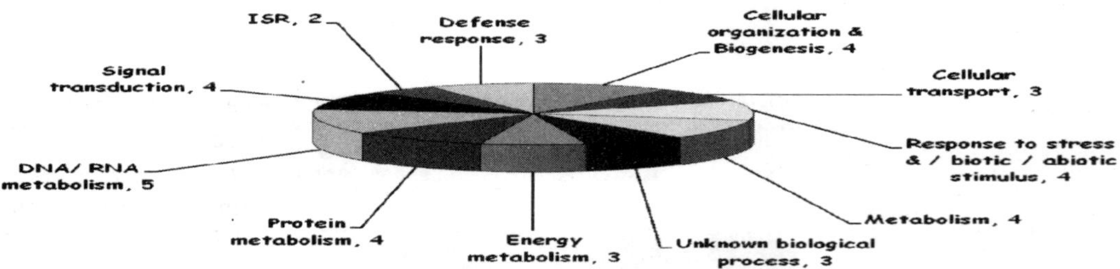

Ninety six genes were differentially expressed in the PGPR mediated pest resistant studies in the susceptible rice variety (Co-43). From the 96 genes we have subsequently identified number of well characterized novels response genes viz., aspartic proteinase, OsIAA21 - Auxin-responsive Aux/IAA gene family member, CIPK-like protein 1, TPR Domain containing protein and glucan endo-1,3-beta-glucosidase 3 precursor from leaf folder challenged along with PGPR priming, inositol-1-monophosphatase, putative, protease, NB-ARC domain containing protein, indole-3-acetate beta-glucosyltransferase, mitogen-activated protein kinase, LRR receptor-like protein kinase, esterase precursor, WRKY transcription factor were from the BPH challenged plants instead leaf folder and one novel protein coding gene namely, NBS-LRR resistance protein from BPH challenge alone. Ninety five percent of the expressed listed genes from the treatments received PGPR priming and five percentage from non PGPR primed pest alone challenge inoculated plants.

Our earlier studies in line with previous literatures clearly indicate the involvement of PGPR in pest resistance. Our current efforts will allow us to go for either QTL mapping or gene expression profiling between contrasting rice plants (Primed with/without PGPR) which will lead to isolate and characterize specific candidate genes controlling pest resistance. Identification of such candidate genes will help us to manipulate pest resistance trait in rice through genetic engineering.

Understanding the source of resistance

Unlike plant pathogen interaction studies, only few literatures are available in the aspects of plant pest interactions, that also in the model plant *Arabidopsis* and not in any cultivated crop plants. The following are the large-scale gene expression studies on *Arabidopsis*-herbivore interactions. Studies include the herbivores like 1) *pieris rapae* (larvae of cabbage white butterfly), 2) *Spodoptera littoralis* (larvae of Mediterranean brocade), 3) *Frankliniella occidentalis* (western flower thrip), 4) *Bemisis tabaci* (silver leaf whitefly nymphs), 5) *Brevicoryne brassicae* (cabbage aphid), and 6) *Myzus persicae* (green peach aphid) (De Vos *et al.*2005, Kempema *et al.* 2007, Little *et al* .2007, Reymond *et al.* 2000, Reymond *et al.* 2004, Kusnierczyk *et al.* 2007). These insects represented leaf-chewing larvae (*P. rapae* and *S. littoralis)* as well as cell-sucking (*F. occidentalis*) or phloem sap-feeding (*M. persicae, B. brassicae, B. tabaci)* adults with *P. rapae* and *B. brassicae* being specialist herbivores adapted to members of the *Brassicaceae* as their hosts. Recent literature complements previous literatures with an analysis of *Arabidopsis* rosette leaves fed upon by larvae of a different leaf-chewing specialist herbivore, *Plutella xylostella* (diamondback moth-DBM) (Ehlting, 2008). All the resistant host pest interaction studies indicated the differentially expressed genes which were associated with the signalling molecules jasmonate, auxin and cytokinin and they were significantly over-represented among genes up-regulated by insect feeding. Genes associated with gibberllic acid, brassinosteroids and abscisic acid were not over-represented among the differentially expressed genes unlike genes associated with jasmonate signalling, auxin and cytokinin metabolism. (Ehlting, 2008).

Identification of novel genes induced during Pest incidence

DD-RT-PCR

EST databases are a valid and reliable source of gene expression. Due to advances in computational molecular biology and biostatistics, it is possible to mine and analyze large-scale EST datasets efficiently and exhaustively (Ewing *et al.*, 1999; Ogihara *et al.*, 2003; Ronning *et al.*, 2003). Several recently developed bioinformatics and statistical tools have been used us to perform global expression analysis using EST data generated in various projects (Audic and Claverie, 1997; Ewing *et al.*, 1999; Greller and Tobin, 1999; Stekel *et al.*, 2000; Strausberg *et al.*, 2001). Generation of molecular information during progression of any biotic/abiotic stress, comparative gene expression profiling between varieties is proposed as a way of developing ESTs around specific traits for the candidate locus approach to mapping complex traits (Rudd, 2003).

In this study, DD-RT-PCR method was used to identify the PGPR mediated gene response in rice plants against pests. Differential display method has been used to identify novel genes induced during *Botrytis cinerea* infection in its host tomato (Benito *et al.*, 1996), symbiotic association between plant and arbuscular mycorrhiza (Lapopin *et al.*, 1999), Plant growth promoting rhizobacteria (*Paenibacillus polymyxa*) mediated gene expression in *Arabidopsis* (Timmusk and Wagener, 1999) and cyst nematode incidence in soybean (Hermsmeier *et al.*, 1998).At the physiological level, the association of the PGPR *Paenibacillus polymyxa* with *Arabidopsis thaliana* resulted in higher levels of plant resistance to the pathogen *Erwinia carotovora* (biotic stress) and in greater tolerance to drought (abiotic stress). DD-RT PCR identified an *Arabidopsis* gene, ERD15, which was induced by *P. polymyxa* , also induced by drought. This finding further demonstrate the activation of common biochemical pathways in response to different stimuli. A major advantage of DD-RT-PCR technique is that it permits the simultaneous identification of up and down regulated genes and hence it has been used to isolate mRNA species in plants induced in response to biotic and abiotic stresses. The usefulness of the method has been demonstrated for identification of novel genes induced during host-pathogen interaction in systems like *Cauliflower mosaic virus* (CaMV)-*Arabidopsis* (Geri *et al.*, 1999), *Phytophthora capsici*–pepper system (Munoz and Bailey, 1998) and soybean *Pseudomonas syringae* pv. *glycinea* interaction (Seehaus and Tenhaken., 1998).

In a recent study, BPH responsive genes in rice were identified through representational difference analysis similar to differential display (Park *et al.*, 2007). Their study demonstrated that these techniques can be used to identify genes that are altered in rice during BPH interaction. Depending on the type of input stimulus, defense associated genes respond with different expression timing and amplitudes (Eulgem, 2005). This is proved by the fact that a major portion of the cDNA clones obtained in the present investigation were related to stress and signaling molecules. Similarly in the present study also, DD-RT-PCR analysis revealed the induction/suppression of transcripts during pest incidence in Rice. Of the transcripts displayed by autoradiography, about 450 showed differential expression during various interactions. A total of 96 cDNA fragments were isolated, reamplified, cloned and sequenced.

Genes with altered expression in distal tissue included those which are involved in cellular housekeeping functions, suggesting modified resource allocation needed to respond to different stress conditions. Differences in local, systemic responses and the expression of several new transcripts played a role in insects attack and other signal transduction pathways (Sarosh and Meijer, 2007). Plant responses to pests are coordinated by multiple signalling systems, of which three have received much attention. One of these, the oxylipin-signaling pathway, which includes the hormone jasmonic acids (JA) and related compounds and it has been shown to influence the production of several metabolic defense mechanisms (Mikkelsen *et al.* 2003). Responses to some pathogens and insects involve a second pathway requiring an interaction between JA and ethylene (Penninckx *et al.* 1998). Plant responses to many microbes involve accumulation of a third signal, the phenylpropanoid derivative salicylic acid (SA), and activation of SA-responsive genes as part of systemic acquired resistance (Glazebrook *et al.* 2003).

Generally pathogen responsive genes were elicited at a later time, due to wounding or insect attack. Similarly most of the genes in disease resistance pathways were shown to be induced during wounding in *A. thaliana* (Cheong *et al.* 2002). Many of the genes during plant insect interaction system seem to be involved in intracellular signalling and transcription control. For example genes encoding zinc finger protein, leucine rich repeats, protein kinase, disease resistance protein, aconitase, GHMP kinase, phosphatidyl inositol kinase and GTP binding protein were induced by DBM larvae interactions in *B. napus* (Sarosh and Meijer, 2007).

Comparison studies between mechanical wounding using scissors or forceps and DBM feeding on *B. napus* revealed differences in local and systemic induction patterns and temporal changes in MBP (Myrosinase binding protein) transcript levels depending on insect or mechanical damage (Pontoppidan *et al.*, 2005). This finding was contrasted to cabbage aphids infestation of oilseed rape plants that caused other effects on MBP transcripts (Pontoppidan *et al.*, 2003). However, in certain cases mechanical wounding can be performed in a manner that mimics the insect damage (Mithofer *et al.*2005).

Functional Annotation of Rice ESTs

Sequence analysis of the differentially expressed cDNA fragments revealed the induction/suppression of genes involved in stress response, pathogenesis and signaling function. Many of the ESTs were assigned for their specific functions based on their sequence similarity to either proteins or genes of known functions in their own /other crops /organisms. The cDNA sequences were searched against the NCBI -nr nucleic acid and protein databases, TIGR and GRAMENE databases using BLAST algorithm. For each query sequence, top five matches were selected and based on the sequence identity and presence of conserved domains best matches were selected. By this way we were able to assign putative functions for around 80% of the cDNA fragments that showed differential expression in this study.

Approximately two third of the ESTs were similar to other publicly available sequences. The remaining sequences with no matches in the public databases at either the protein or nucleotide level may represent untranslated mRNAs, as well as interactive differential genes. Shoemaker at al. (2002) reported that 13% of the soybeans ESTs were having no matches against the GenBank nr database. The ESTs were categorized with respect to their predicted functions and based on the functional categorization they were grouped into eleven broad categories viz., 1) Signal transduction, 2) ISR, 3) Defense response, 4) Cellular organization and biogenesis, 5) Response to stress and biotic /abiotic stimulus, 6) Cellular transport, 7) Metabolism, 8) Energy metabolism, 9) Protein metabolism, 10) DNA/ RNA metabolism, 11) Unknown biological process. Sequence analysis indicated that the PGPR mediated rice interactive (Rice-PGPR-Pest (Leaf folder/BPH)) cDNA ESTs obtained in this study showed similarity towards the interactive genes expressed from the resistant varieties of following crops like Rice, *Arabidopsis*, Sugarcane, Sorghum, Barley, Wheat, *Brassica,* Chick pea, Maize, which were expressed during the interaction with various biotic / abiotic stress and the beneficial interactions like Legume-*Rhizobium*, *Arabidopsis*-PGPR, Maize-*Trichoderma* . This data lead us to speculate that expression of defense genes is independent of Crop-Microbe Interactions.

Role of differentially expressed cDNAs in Plant Resistance Mechanism

Signaling requires an initial signal transmitting molecule which is recognized as phytohormones like auxins, gibberellic acid, abscissic acid, ethylene and calcium channels (Walden and Ecker, 1998). This hormonal signaling can activate a large number of defense/resistance genes encoding proteins involved in diverse cellular functions including cell wall biosynthesis, hydrolysis of cell walls by chitinases and β-1,3- glucanases), PR proteins and synthesis of lignin, phytoalexins, and secondary signaling compounds (Heil and Bostock, 2002). Detailed knowledge of the signaling pathways mediating the interaction between rice and insect pest is lacking.

A strong and common theme of signaling networks in eukaryotic organisms is regulation by reversible protein phosphorylation. Protein kinases and protein phosphatases play a major role in the control of most of the pathways involved in growth and development as well as in responses to changes in the environment. It has been reported that the *Arabidopsis thaliana* genome encodes nearly 1000 genes belonging to the protein kinase superfamily and almost 300 genes encoding protein phosphatases (The *Arabidopsis* Genome Initiative, 2000). A large number (so far 80 clusters have been annotated) of rice EST clusters related to protein kinases were found in the TIGR database, including receptor-like protein kinases and two-component histidine kinases (Souza *et al.*, 2007). The MAP kinase (MAPK) cascade is the most studied phosphorylation pathway in plants and appears to transduce a vast array of signals (Valster *et al.*, 2000). Other signals transduced by the MAP kinase cascade include plant hormones and many environmental signals; however their complete pathways are still undefined.

In *Arabidopsis,* it was reported that a serine/threonine kinase OXI1 is required for downstream signaling during oxidative burst that occurs after pathogen attack. Knock-out of OXI1 resulted in enhanced plant susceptibility (Rental *et al.*, 2004) to oxidative damage. Recently it was demonstrated that the serine/threonine kinase ACIK1 is rapidly induced in tomato upon elicitation by the *Cladosporium fulvum* elicitor Avr9 and silencing of tomato ACIK1 resulted in decreased Cf-9-mediated resistance to *C. fulvum*. From the previous reports it was understood that a transcript showing homology to an *Arabidopsis thaliana* serine/threonine kinase mediates the interaction between sugarcane and *Puccinia melanocephala* (causal agent of brown rust) (Carmona *et al.*, 2004). In the present study, a number of kinases were found to be induced during pest challenge in PGPR primed rice plants including a transcript showing 60% homology to a sorghum serine/threonine kinase (resistant crop-pathogen interactions) which shows that there is active signaling that takes place during the process of pathogenesis in resistant crop. The main receptor proteins noticed in this study belongs to the serine/threonine receptor kinase family containing the leucine - rich repeat (LRR) which mediates the recognition of plant peptides and possibly pathogens (Gomez *et al.*, 2004). There is no specified reports are available in the aspect of plant-pest or plant-PGPR-pest interaction. Because our results also in line with the previous findings of plant-pathogen interactions, the same thing might have happened in our study also.

Auxins are one of the major plant growth hormones that control plant growth and development. It has also been described that auxins may participate in the defense responses against pathogen and microorganisms (Martínez-Noël *et al.*, 2001; Yamada, 1993). In several other pathosystems auxin-related genes are found to be induced upon pathogen challenge (Dowd *et al.*, 2004; Kuhlmann *et al.*, 2003). In *Arabidopsis* the auxin resistance locus AXR1 has been shown to be associated with pathogen resistance (Tiryaki and Staswick, 2002). Similarly a transcript was identified as Auxin-responsive protein IAA21 that is associated with the process of pathogenesis that correlates with the previous reports. In the present study Auxin responsive protein was not influenced by PGPR priming, since the expression was noticed in both PGPR primed and non-primed plants.

ACC induced ethylene-responsive element binding proteins (EREBP) transcription factors (one confirmed by qRT-PCR) as observed by Van Zhong and Burns (2003). EREBP are known to bind to GCC-box promoter elements that are found in many pathogen-responsive genes and, in *Arabidopsis* they have been implicated in the induction of defensins (Thatcher *et al.* 2005).

Transport of nucleotide sugars across the Golgi apparatus membrane is required for the luminal synthesis of a variety of plant cell surface components, such as cell wall polysaccharides (Baldwin *et al.*, 2001). In Maize- *Trichoderma*, three proteins were differentially identified as Golgi GDP Man transpoter. Two of these spots were up-regulated and one was down-regulated (Shoresh and Harman, 2008). But our study indicates that the expression of interactive Golgi transport 1 protein B is of pest origin.

Chitinase has been extensively studied in plants as a secondary hydrolase involved in defense. Its substrate chitin does not occur in higher plants but is present in the cell wall of many fungi and insect cuticle. Chitinase assays in cucumber–*Colletotricum lagenarium* system exhibited a strong increase in activity in resistant young leaves, which restricted the spread of infection (Boller, 1987). It has been reported that fungal infection can act as an independent elicitation of chitinase in bean, which exists as a small multigene family of approximately four members, at least two of which are differentially expressed during infection by pathogen (Broglie *et al.*, 1986). Similarly, differential expression and induction of chitinase gene was observed in the present study.

Recent studies on the interaction of Maize-*Trichoderma* were found to up-regulate following proteins which were similar to the previous findings of various interactions. Proteins *viz.*, Putative oxalate oxidase (rice), Type IIIa membrane protein cp-wap-13 (cowpea), DNA repair-recombination protein (RAD50) *Arabidopsis*, Putative RNA-binding protein (rice), Hypothetical protein (contain BTB/POZ domain) (rice), Hypothetical protein (At) contain 95% similar to Golgi GDP Man transpoter (GONST1) *Arabidopsis*, Putative 60s ribosomal protein L35 (*Arabidopsis*), ATP-binding HSP70 (*Arabidopsis*), PAL (*Vitis vinifera*), NBS/LRR resistance protein-like protein (*T.cacao*), Peroxidase 5 (*Triticum monococcum*), Putative TPR domain-containing protein (*Solanum demissum*), Hypothetical protein (rice) & Hypothetical protein At3g57440 (*Arabidopsis*) were upregulated and β-glucosidase (maize), Hypothetical protein (At) contain 95% similar to Golgi GDP Man transpoter (GONST1) *Arabidopsis* , β-glucosidase- chain B(Zmglu1) maize, Putative TPR domain-containing protein (*Solanum demissum*) & OSJNBa0057M08.27 probable nuclear protein similar to BRUSHY (rice) were down regulated (Shoresh and Harman, 2008).

Five up-regulated protein spots were identified as nucleotide site (NBS)/ Leu-rich repeat (LRR) resistance protein-like proteins in Maize- *Trichoderma* interactions. These proteins are known to have a major role in plant defense responses. Phenyl ammonium lyase (PAL; one spot) and another defense-related protein, were found to be up-regulated. The proteins oxalate oxidase, β-glucosidase, and Met synthase, which were described above, are also implicated in stress responses (Shoresh and Harman, 2008).

Ethylene is an essential signalling molecule in induced resistance responses (Bostock *et al.*, 2001). The jasmonate/ethylene pathway of induced resistance was shown to be induced in cucumbers inoculated by *Trichoderma asperellum* (Shoresh *et al.*, 2005). Evidence for the involvement of ethylene in plant systemic responses to *Trichoderma* inoculation was also demonstrated by Seggara *et al.* (2007) and by Djonovic *et al.*, (2007). The strong increase in Met synthase in maize whose roots are colonized by *Trichoderma* T22 and the induced systemic resistance that is generated in this system (Harman *et al*,. 2004b) are consistent with the concept that ethylene is involved in the response of maize to *Trichoderma* inoculation.

Numerous other proteins involved in stress and defence-related systems were found to be up-regulated in maize colonized by *Trichoderma*- T22. For example, forms of both PAL and peroxidise were up-regulated. The gene encoding for PAL is believed to be activated by the jasmonic acid/ethylene signalling pathway of induced plant resistance (Diallinas and Kanellis, 1994; Mitchell and Walters, 1995; Kato *et al.*, 2000). PAL is the first enzyme in the phenylpropanoid biosynthesis pathway, which provides precursors for lignin and phenols, as well as for salicylic acid (Mauch-Mani and Slusarenko, 1996). Other enzymes of the phenylpropanoid pathway, including peroxidises, are also induced in resistant reactions. Peroxidises are also known for their role in the production of phytoalexins, reactive oxygen species (ROS), and formation of structural barriers (Kawano, 2003; Passardi *et al.*, 2005).

Shoresh and Harman (2008) reported that chitinolytic enzymes were upregulated in Maize-*Trichoderma* interactions. Proteins with abilities to degrade chitin usually have acidic or basic isoelctric points and so were not detected in this study. These results are consistent with the observation that transcription of genes encoding these enzymes, and activity of these enzymes, also were reported to enhance in the cucumber-T. *asperellum* system (Yadidia *et al.*, 2000, 2003., Shoresh *et al.*, 2005)

In plants, the β-glucosidase is associated with a variety of functions that include chemical defense against pathogens and pests through the production of hydroxamic acid glucosides (Czjzek *et al.*, 2000).

To date, studies on lipoxygenase have indicated its role in seed germination, and in disease resistance, while its metabolism products have been found to have antimicrobial activities (Galliard and Chan, 1980). The differential expression of lipoxygenase in pathogenesis had been reported in wheat (Guss *et al.*, 1968) and rice (Ohta *et al.*, 1986) as that of the results of this study.

Several up-regulated protein spots were identified as NBS/LRR resistance protein –like proteins in the plant-*Trichoderma* interaction (Marra *et al.*, 2006). A recent study also showed that the level of NBS/LRR proteins increased in leaves interacting with *Trichoderma*. These disease resistance genes (R) are the specificity determinants of plant immune responses.

Protein with a role in plant growth and development, through a mechanism different from energy and sugar metabolism also were identified. The β-glucosidases identified in the interactions study were the gene products of ZmGLU1. In maize, ZmGLU1 is one of the β-glucosidases that has been suggested to hydrolyze the cytokinin-O-glucosides to liberate free cytokinins (Brzobohaty *et al.*,

1993). Although inactive cytokinin conjugate is abundant in plants, only a small amount of free cytokinins are available to stimulate and control plant growth (Sakakibara, 2006)

In tomato plants inoculated with *Trichoderma hamatum* during the expression of stress, cell wall, and RNA metabolism-related genes was also up-regulated, demonstrating similarities of plant responses to *Trichoderma* (Alfana *et al*., 2007). In this tomato- *Trichoderma* system, no positive growth response was recorded. Interestingly, in this system, there were also no carbohydrate metabolism-related genes up-regulated. This suggests that there may be a direct connection between the ability of *Trichoderma* to induce carbohydrate metabolism and its ability to induce growth response.

Detailed sequence analysis of novel transcripts

The RT-PCR analysis revealed the differential expression of the following novel genes in our PGPR mediated transcriptome analysis of rice transcripts in response to pest incidence in the present study.

Zinc finger A20 and AN1domains-containing protein

Park *et al*., (2007) identified the sequence of *OsBphi25*, from the BPH resistance rice varitey which encodes a putative zinc finger RING domain containing protein and they found out this is identical to OSIIEa09E16, an EST clone identified from a blast inoculated cDNA library (accession no. CB622502.1; Jantasuriyarat *et al.* 2005). Previous database searches have identified a total of 469 predicted RING domain-containing proteins in *Arabidopsis* (Stone *et al.* 2005), the majority of which are active in in vitro ubiquitination assays, suggesting that they function as ubiquitin E3 ligases. It has also been suggested that the presence of such a large and diverse number of zinc finger RING domain-containing proteins enables target- specific proteolysis to occur in response to complex and significant stress conditions (Stone *et al.* 2005). Earlier studies also suggested that ubiquitination may play an important role in plant disease resistance (Devoto *et al.* 2003).

We found that the enhanced expression of putative zinc finger A20 and AN1domains-containing protein in response to PGPR mediated leaf folder resistance studies. Moreover, no ubiquitin ligase targets that are associated with pest resistance have still been identified in plants, it is evident that the protein modification systems regulate plant resistance responses against pests. Of the differentially expressed transcripts, putative zinc finger A20 and AN1domains-containing protein was found to be expressed constitutively in response to Leaf folder challenge inoculation in both PGPR primed and nonprimed rice plants but there was a difference in the transcript size between them. In the PGPR primed plants, transcript size was around 389bp (NCBI Accession number, FG347022) whereas in the nonprimed plants it was around 173bp (NCBI Accession number, FG347041). Upon sequence matching, it was noticed that the transcript amplified from the nonprimed plants received higher score (1784) with doubled sequence length when compared to the cDNA from the plants received challenge inoculation alone. Almost the sequence length was matched with the *Arabidopsis* RING domain-containing proteins (Stone *et al.* 2005). The difference in the loss of sequence length in nonprime plants lead to susceptibility, wherein the difference may be due to deletions in the coding region which may lead to susceptibility.

Chitinases

Plant chitinases are to be considered pathogen-related (PR) proteins, since their activity is induced by fungal, bacterial, and viral infections, as well as by more general signals of stress such as wounding, salicylic acid, ethylene, and heavy metal salts (Graham and Sticklen, 1994). Moreover, chitinases also have important non-defensive functions in growth and development processes (Kasprzewska, 2003). Chitinases are involved in the regulation of nodulation signal molecules (Goormachtig *et al*., 1998), they affect embryogenesis (Helleboid *et al.*, 2000), and they might participate in programmed cell death (Passarinho *et al.*, 2001). Chitin and chitosan are major components of cell walls in insects and fungi. Incubation of concentrated Nepenthes trap liquid with colloidal chitin resulted in considerable digestion of chitin (Eilenberg *et al.*, 2006). However, this observed activity was partially attributed to the presence of chitinase-producing symbiotic micro-organisms in the open traps. Other than this, there were no strong records available on the involvement of beneficial microbes in enhancing the chitinase activity in crop plants against insects. In the present studies we observed enhanced chitinase activity in rice plants against the attack of insect pests. This activity was increased more than threefold due to PGPR priming compared to other treatments which received pest challenge alone and plants without any treatment. Eventhough we have quantified the enhanced chitinase activity in the varius time period upon constant interval we could not notice the differential gene in the autoradiogram, this may be because of the delayness in the aboundance of gene expression. This needs further confirmation, because of the lack of no previous records.

MAPKinase

The mitogen-activated protein kinase (MAP kinase) signal transduction cascades are routes through which eukaryotic cells deliver extracellular messages to the cytosol and nucleus ,but the downstream components of the MAPK have as yet not been elucidated (Innes, 2001). These signalling pathways direct cell division, cellular differentiation, metabolism, and both biotic and abiotic stress responses (Ichimura *et al.*, 2000). In plants, MAP kinases and the upstream components of the cascades are represented by multigene families, organized into different pathways which are stimulated and interact in complex ways. *Arabidopsis* genome sequence has revealed the presence of 23 MAPK genes in its genome, suggesting the complexity of MAPK in plant system. There is clear evidence for the involvement of MAP kinases in plant cell division and in the regulation of auxin signalling. Plant pathogens and pathogen-derived elicitors (from fungi, bacteria and viruses) and abiotic stresses including wounding, mechanical stimulation, cold, drought and ozone can elicit defence responses in plants through MAP kinase pathways. There are data suggesting that ABA signalling utilizes a MAP kinase pathway, and probably ethylene and perhaps cytokinins do so also (Morris, 2001). Fourteen kinase genes were differentially expressed in *Arabidopsis -Plutella xylostella* interactions at different time points. Most of the genes in this group codes for receptor-like kinases such as Leucin rich repeat (LRR) and peptodo-glucan (LysM) binding domain containing kinases. Other than these two mitogen activated protein kinase kinase kinases (MAPKKK) were reported as trancriptionally up-regulated due to specified interaction. To date some of the MAPKs have been identified and characterized in rice plants. Molecular and functional analysis of rice MAPK, BWMK1 revealed that this protein phosphorylates the rice transcription factor OsEREBP1 (rice ethylene-responsive element binding protein). (Song and Goodman, 2002; Wen *et al.*, 2002) Such EREBPs are known to bind to the GCC box DNA motif (AGCCGCC) that is located in the promoter of several PR genes and was associated with increase in resistance to pathogens and pests. These can be activated by SA / JA / ethylene associated plant defense signals (Cheong *et al*, 2003). In the present study PGPR mediated induction of MAPK was noticed when rice plant was challenged with BPH.

APX

Ascorbate peroxidase (APX) scavenges superoxide, hydroxyl radicals and singlet oxygen in the cytosol, chloroplast and mitochondria of higher plants. APX uses two molecules of ascorbate, the most important antioxidant substrate in plants, to reduce H_2O_2 to water (Asada, 1992). Decline of the SAPX protein level was detected in viral infected tobacco during programmed cell death (PCD), known as the hypersensitive response (HR) in plants (Mittler *et al.*, 1998). Suppression of

APX proteins supposedly contributes to PCD by allowing H_2O_2 accumulation in the cells (Mittler *et al*., 1998). In the present study APX was increased only in the PGPR primed lines. The same was reported by Lee *et al*. (2006) in resistance rice variety while interacting with *R. solani*. The proteins involved in defense of oxidative stress were up-regulated, such as glutathione S- transferase, ascobate peroxidase, and 20S proteasome. Production and removal of H_2O_2 must be strictly controlled during SAR because H_2O_2 acts not only as an antimicrobial compound and a signal molecule inducing defense proteins, but also as a toxic molecule to the host plants (Wang *et al*., 2002). Thus, in the response of the resistant and susceptible line in this study, the APX proteins were possibly involved in reducing H_2O_2 concentrations to protect cells from oxidative damage.

Ubiquitin

Targeted protein degradation via the ubiquitin/26S-proteasome pathway is one of the important regulatory processes in plant resistance system (Glickman and Ciechanover, 2002). Interaction studies on the *Arabidopsis*-DBM reported to the activation of 75 putative E3-ubiquitin-protein-ligases that were affected by DBM, in addition to other differentially expressed proteasome components, a single ubiquitin-like gene and ubiquitin activating enzyme. Large number of signal transduction components have been affected by DBM feeding in *Arabidopsis*. In particular, members of the AP2-EREBP family of transcription factors rapidly induced by herbivory suggesting roles for this family in plant-pest interaction induced signal transduction networks (Ehlting, 2008). These studies suggest that herbivory feeding can result in down-regulation of primary metabolic processes while at the same time activating defense related processes including secondary defence metabolism. In addition, massive reprogramming of primary and secondary metabolic processes as part of the insect-induced defence response involves rapid changes in signalling and other regulatory processes. In the same interaction studies, cytochrome P450 monooxygenase, which catalyzes the conversion of tryptophan to indole-3-acetaldoxime, the precursor of indole glucosinolates, camalexin, and auxin (Ljung *et al., .*2005, Mikkelsen *et al.* 2000). These secondary molecules are involved in lot of downstream signalling.

Functional genes in the octadecanoid pathway were found to be induced in comparative *Arabidobsis*-herbivory transcriptome meta-analysis. The role of octadecanoids in mediating herbivore-induced responses is well established, and it has been estimated that up to 80 percentage of all herbivore-induced *Arabidopsis* genes are octadecanoid regulated (Reymond *et al.* 2004). Most of our pest response genes in rice plants are also found to be present in the octadecanoid pathway.

To conclude, lot of genomic studies are available on the plant pest interaction aspects and reports are also available on the expression of novel gene. In the same time point, we are also provided with complete sequence information of model plant, *Arabidopsis* and our important food crop, Rice etc., While going through all the available records on interaction studies, we can notice the overlapping records on the expression of same gene due to various biotic/ abiotic stress and also lot of redundant genes in a particular type of induction itself. Lot of confusing and abundant records and the sequence information of temporal gene expression can not be used as a confident genetic marker resource for the production of complete, durable resistant rice cultivar. Use of plant growth promoting rhizobacteria (inducer of systemic resistance) to mediate the expression of novel genes can be the confident viable source. From our studies, we have opened the black box of PGPR mediated expression of resistant gene, this will be the stepping stone to the researchers in this emerging area of research in the coming years.

References

Alfana G, Ivey MLL, Cakir C, Bos JIB, Miller SA, Madden LV, Kamoun S, Hoitink HAJ, 2007. Systemic modulation of gene expression in tomato by *Trichoderma hamatum* 382. *Phytopathology,* 97, 429-437.

Anderson JP, Thatcher LF, Singh KB, 2005. Plant defence responses, conservation between models and crops. *Functional Plant Biology* 32,21-34.

Andreasson E, Taipalensuu J, Rask L, Meijer J, 1999. Age dependent wound induction of a myrosinase-associated protein from oilseed rape Brassica napus.. *Plant Mol Biol,* 41, 171-180.

Anuratha, C.S. and Gnanamanickam, S.S, 1990. Biological control of bacterial wilt caused by *Pseudomonas solanacearum* in India with antagonistic bacteria. *Plant Soil,* 124,109-116.

Audic S and Claverie J, 1997. The Significance of Digital Gene Expression Profiles *Genome Research,* 710, 986-995 .

Austin, B., Dickinson, C.H. and Goodfellow, M, 1977. Antagonistic interactions of phylloplane bacteria with *Drechslera dictyoides* Drechsler. Shoemaker, *Canadian J. Microbiol,* 23,710-715.

Balasubramanian G, Saravanabhavanandan M and Subramaniam, T.R., 1973. Control of rice leaffolder, *Cnaphalocrocis medinalis* Guenee. *Madras Agric. J,* 60, 25-427.

Baldwin TC, Hahdford MG, Yuseff M-I, Orellana A, Dupree P, 2001. Identification and characterization of GONSTI, a Golgi-localized GDP-mannose transporter in *Arabidopsis*. *Plant Cell* 13,2283-2295

Benito E.P, Have A, Klooster J.W and Kan J. A.L., 1998. Fungal and plant gene expression during synchronized infection of tomato leaves by *Botrytis cinerea*. European Journal of Plant Pathology, 104 2.

Bentur J.S, Chelliah S and Prakala Rao P.S, 1989. Approaches in rice pest management - Achievements and opportunities. *Oryza*, 26, 12-26.

Berger L.R. and Reynolds D.M, 1958. The chitinase system of a strain of *Streptomyces griseus*. *Biochem. Biophy. Acta,* 29, 522-534.

Boller T. and Mauch, F, 1988. Colorimetric assay for chitinase. *Meth. Enzymol,* 161, 430-435.

Bostock RM, Karban R, Thaler JS, Weyman PD, Gilchrist D, 2001. Signal interactions in induced resistance to pathogens and insect herbivores. *Eur J Plant Pathol,* 107, 103-111.

Bradford, M.M, 1976. A rapid and sensitive method for quantification of microgram quantities of protein utilizing the principle of protein-dye binding. *Anal. Biochem*, 72, 248-254.

Brar D.S and Khush G.S, 2002. Transferring genes from wild species into rice. In Kang, M.S. ed.. Quantitative genetics, genomics and plant breeding. Oxford, CABI.

Broglie K E, Gaynor J J and Broglie R M, 1986. Ethylene-regulated gene expression, molecular cloning of the genes encoding an endochitinase from Phaseolus vulgaris. *PNAS,* 8318., 6820-6824.

Brotman Y, Silberstein L, Kovalski I, Perin C, Dogimont C, Pitrat M, Klingler J, Thompson GA, Perl-Treves R, 2002. Resistance gene homologues in melon are linked to genetic loci conferring disease and pest resistance. *Theor Appl Genet,* 104, 1055–1063 .

Campo S, Carrascal M, Coca M, Abian J, San Segundo B, 2004. The defense response of germinating maize embryos against fungal infection, a proteomics approach. *Proteomics,* 4,383-396.

Cánovas FM, Dumas-Gaudot E, Recorbet G, Jorrin J, Mock HP, Rossignol M, 2004. Plant proteome analysis. *Proteomics*, 4, 285–298.

Carmona R, L Pérez-Alvarez , Muñoz M, Casado G, Delgado E, Sierra M, Thomson M, Vega Y, Parga E and Contreras G, 2004. Natural resistance-associated mutations to Enfuvirtide T20. and polymorphisms in the gp41 region of different HIV-1 genetic forms from T20 naive patients. *Journal of Clinical Virology* 323., 248 – 253.

Chandramohan, N and Jeyaraj S, 1976. Evaluation of certain granular and foliar insecticides against rice leaffolder, *Cnaphalocrosis medinalis* Guenee. *Madras Agric. J.,* 63, 364-366.

Chen JW, Wang L, Pang XF, Pan QH, 2006. Genetic analysis and fine mapping of a rice brown planthopper *Nilaparvata lugens* Stål. resistance gene *bpb19(t)*. *Theor Appl Genet,* 275,321–329.

Cheong YH, Chang HS, Gupta R, Wang X, Zhu T, Luan S, 2002. Transcriptional profiling reveals novel interactions between wounding, pathogen, abiotic stress, and hormonal responses in *Arabidopsis*. *Plant Physiol* 129, 661-677.

Chou H.H and Holmes M.H, 2001. DNA sequence quality trimming and Vector removal. *Bioinformatics*. 1712., 1093-1104.

Cipollini DF, Bergelson J, 2003. Plant density and nutrient availability constrain constitutive and wound-induced expression of trypsin inhibitors in *Brassica napus*. *J Chem Ecol,* 27,593-610.

Coram T.E, Pang E.DK, 2007. Transcriptional profiling of chickpea genes differentially regulated by salicyclic acid, methyljasmonate and aminocyclopropane carboxylic acid toreveal pathways of defense-related gene regulation.*Functional Plant Biology,* 34, 52-64.

Cowley T, Walters DR, 2005. Local and systemic changes in arginine decarboxylase activity, putrescine levels and putrescine catabolism in wounded oilseed rape. *New Phytol,* 165,807-811.

Czjzek M, Cicek M, Zamboni V, Burmeister WP, Bevan DR, Henrissat B, Esen A, 2000. Crystal structure of a monocotyledon maize ZMGlu1. beta-glucosidase and a model of its complex with p-nitrophenyl beta-D-thioglucoside. *Biochem J* , 354, 37-46.

Davuluri RV, Sun H, Palanisamy SK, Matthews N, Molina C, Kurtz M, Grotewold E, 2003. AGRIS, *Arabidopsis* gene gene regulatory information server, an information resource of *Arabidopsis* cis-regulatory elements and transcription factors. *BMC Bioinformatics,* 4, 25.

De Vos M, Van Oosten VR, Van Poecke RMP, Van Pelt JA, Pozo MJ, Mueller MJ, Buchala AJ, Metraux JP, van Loon LC, Dicke M, Pieterse CMJ 2005. Signal signature and transcriptome changes of *Arabidopsis* during pathogen and insect attack. *Mol Plant Microbe In* 18,923-937

Devoto A, Muskett PR, Shirasu K 2003. Role of ubiquitination in the regulation of plant defence against pathogens. *Curr Opin Plant Biol* 6,307–311

Diallinas G, Kanellis AK 1994. A phenylalanine ammonia-lyase gene from melon fruit, cDNA cloning, sequence and expression in response to development and wounding. Plant Mol Biol 26, 473-479

Dickerson D.P, Pascholati S.F, Hagerman A.E, Butler L.G and Nicholson R.L, 1984. Phenylalanine ammonia-lyase and hydroxy cinnamate CoA ligase in maize mesocotyls inoculated with *Helminthosporium maydis* or *Helminthosporium carbonum. Physiol. Plant Pathol,* 25, 111-123.

Djonovic S, Vargas WA, Kolomiets MV, Horndeski M, Weist A, Kenerley CM, 2007. A proteinaceous elicitor Sm1 from the beneficial fungus *Trichoderm virens* is required for systemic resistance in maize. *Plant Physiol,* 145, 875-889.

Park D.S, Lee S.K, LeeJ.H, Song M.Y, Song S.Y, Kwak D.Y, Yeo U.S, Jeon M.S, Park S.K, Yi G, Song Y.C, Nam M.H, Ku Y.C, Jeon J.S, 2007. The identification of candidate rice genes that confer resistance to the brown planthopper *Nilaparvata lugens.* through representational difference analysis. *Theor Appl Genet,* 115, 537–547

Dowd C, Wilson I.W and McFadden H, 2004. Gene expression profile changes in cotton root and hypocotyl tissues in response to infection with *Fusarium oxysporum* f. sp. *Vasinfectum. Mol. Plant Microbe Interact,* 17, 654–667.

Eckardt N. A 2000.. Sequencing the Rice Genome. *Plant Cell* 12, 2011–2017.

Ehlting J, Chowrira S.G, Mattheus N, Aeschliman D.S, Arimura G.I, and Bohlmann J., 2008. Comparative transcriptome analysis of *Arabidopsis thaliana* infested by biamond back moth Plutella xylostella. larvae reveals sifnatures of stress response, secondary metabolism, and signalling. *BMC Genomic,* 9, 154.

Endo S, Kozano H and Masuda T, 1987. Insecticide susceptibility of rice leaffolder larvae *Cnaphalocrosis medinalis* Guenee. *Appl. Entomol.Zool,,* 22, 145-192.

Eulgem, T, 2005. Transcriptional networks in plants Regulation of the *Arabidopsis* defense transcriptome. *Trend in Plant science,* 10, 71-78.

Ewing R, Poirot O and Claverie J.M, **1999.** Comparative analysis of the *Arabidopsis* and rice expressed sequence tag EST. **set**s. *Silico Biol,* 1, 18.

Ewing R.M, Kahla A.B, Poirot O, Lopez F, Audic S and Claverie J, 1999. Large-Scale Statistical Analyses of Rice ESTs Reveal Correlated Patterns of Gene Expression. *Genome,* 9(10)., 950-959.

Fujiyoshi, N., Noda, M. and Sakai, H, 1980. Simple mass rearing method of the grass leaf roller *Cnephalocrocis medinalis* Guenee of young rice seedlings. *Jap. J. Appl. Ent. Zool.,* 24, 194-196.

Gale M.D and Devos K.M, 1998. Comparative genetics in the grasses *Proc. Natl. Acad. Sci. USA* 95, 1971–1974.

Geri, Marmorstein A, Okamoto P and Vallee R, **1999. Altered patterns of gene expression in** *Arabidopsis* **elicited by cauliflower mosaic virus infection and by a CaMV gene VI transgenic,** *Mol. Plant-Microbe Interact.***, 12, 377-384.**

Glazebrook J, Chen W, Estes B, Chang HS, Nawrath C, Metraux JP, Zhu T and Katagiri F, 2003. Topology of the network integrating salicylate and jasmonate signal transduction derived from global expression phenotyping. *Plant J,* 34,217-228.

Glickman MH and Ciechanover A, 2002. The ubiquitin-proteasome proteolytic pathway, destruction for the sake of construction *Physiol Rev,* 82,373-428.

Gomez M.C, Pope C.E, Giraldo A, Lyons L.A, Harris R.F, King A.L, Cole A, Godke R.A and Dresser B.L, 2004. Birth of African Wildcat Cloned Kittens Born from Domestic Cats.*Cloning and Stem Cells,* 6, 247-258.

Gomez, K.A. and Gomez, A.A 1984. *Statistical Procedure for Agricultural Research.* John Wiley and Sons, New York.

Greller L.D and Tobi F.L, 1999. Detecting Selective Expression of Genes and Proteins. *Methods in Genome Res.,* 93., 282-296.

Gribskov M, Fana F, Harper J, Hope DA, Harmon AC, Smith DW, Tax FE and Zhang G, 2001. PlantsP, a functional genomic database for olant phosphorylation. *Nucl Acids Res,* 29,111-113.

Hamel F and Bellemare G, 1995. Characterization of a class I chitinase gene and of wound-inducible, root and flower-specific chitinase expression in *Brassica napus. Biochim Biophys Acta,* 1263, 212-220.

Hammond-Kosack KE and Parker JE, 2003. Deciphering plant pathogen communication, fresh erspectives for molecular resistance breeding. *Curr Opin Biotechnol,* 14 ,177–193.

Harman GE, Petzoldt R, Comis A and Chen J, 2004b. Interactions between *Trichoderma harzianum* strain T22 and maize inbred linr Mo17 and effects of this interaction on diseases caused by *Pythium ultim*um and *Colletotrihum graminicola. Phytopathology,* 94, 147-155.

Hayashi H and Chino M, 1990. Chemical composition of phloem sap from the uppermost internode of the rice plant. Plant Cell Physiol, 31,247–251.

Heinrichs EA, 1979. Control of leafhopper and planthopper vectors of rice viruses. In, Moramorosch K, Harris KF eds. Leafhopper vectors and plant disease agents. Academic Press, New York, pp 529–558.

Heinrichs, E., Camanag, A.E. and Romena, A., 1985. Evaluation of rice cultivars for resistance to *Cnaphalocrocis medinalis* Guenee Lepidoptera, Pyralidae.. *J.Econ.Entomol* 78(1), 274-278.

Hermsmeier D, Hart J.K, Byzova M, Rodermel S.R and Baum T.J, 2000. Changes in mRNA Abundance within *Heterodera schachtii*-Infected Roots of *Arabidopsis thaliana. Molecular plant microbe interactions,* 13, 309-315.

Hermsmeier D, Schittko U, Baldwin IT, 2001. Molecular interactions between the specialist herbivore *Manduca sexta* Lepidoptera, Sphingidae. and its natural host *Nicotiana attenuata.* I. Largescale changes in the accumulation of growth- and defense-related plant mRNAs. *Plant Physiol,,* 125,683–700.

Howe GA, Lightner J, Browse J, Ryan CA, 1996. An octadecanoid pathway mutant JL5. of tomato is compromised in signaling for defense against insect attack. *Plant Cel,l* 8, 2067–2077.

Hui D, Iqbal J, Lehmann K, Gase K, Saluz HP and Baldwin IT, 2003. Molecular interaction between the specialist herbivore Manduca sexta Lepidoptera, Sphingidae. and its natural host *Nicotiana attenuata,*V. Microarray analysis and further characterization of large-scale changes in herbivore-induced mRNAs. *Plant Physiol,* 131, 1877–1893.

Jalali B, Bhargava S and Kamble A, 2006. Signal transduction and transcriptional regulation of plant defence responses. *Journal of Phytopathology,* 154, 65-74.

Jantasuriyarat C, Gowda M, Haller K, Hatfield J, Lu G, Stahlberg E, Zhou B, Li H, Kim HR, Yu Y, Dean RA, Wing RA, Soderlund C,Wang GL and Wing R, 2005. Large-scale identification of expressedsequence tags involved in rice and rice blast fungus interaction. *Plant Physiol,* 138,105–115.

Jena KK, Jeung JU, Lee JH, Choi HC and Brar DS, 2006. High-resolution mapping of a new brown planthopper BPH. resistance gene, *Bph18*t., and marker-assisted selection for BPH resistance in rice *Oryza sativa* L... *Theor Appl Genet* 112,288–297.

Kaloshian I, 2004. Gene-for-gene disease resistance, bridging insect pest and pathogen defence. *J Chem Ecol.,* 30, 2419–2438.

Kato M, Hayakawa Y, Hyodo Y and Yano M, 2000. Wound-induced ethylene synthesi and expression and formation of 1-aminocyclopropane-1-carboxylase ACC. synthase, ACC oxidase, phenylalanine ammonia-lyase and peroxidise in wounded mesocarp tissue of Cucurbita maxima. *Plant Cell Physiol* ,41,440-447.

Kawano T, 2003. Roles of the reactive oxygen species-generating peroxidise reactions in plant defense and growth induction. *Plant Cell Rep.,* 21,829-837.

Kazemi-Pour N, Condemine G and Hugouvieux-Cotte-Pattat N, 2004. The secretome of the plant pathogenic bacterium *Erwiniachrysanthemi. Proteomics* 4,3177–3186.

Kempema LA, Cui XP, Holzer FM and Walling LL, 2007. *Arabidopsis* transcriptome changes in response to phloem-feeding silverleaf whitefly nymphs. Similarities and distinctions in responses to aphids. *Plant Physiol.,* 143, 849-865.

Kenmore P. E, Carifio F. O, Perez C. A, Dyck V. A and Gutierrez A. P, 1984. Population regulation of the rice brown plant hopper *(Nilaparvata lugens* Stal. within rice fields in the Phillippines. *J. Plant Prot. Tropics,* 1, 19-37.

Kermasha S and Metche M, 1986. Characterization of Seed Lipoxygenase of *Phaseolus vulgaris* Cv, Haricot. *J. Food Sci,* 51, 1224-1227.

Kessler A and Baldwin IT, 2002. Plant responses to insect herbivory; the emerging molecular analysis. *Annu Rev Plant Biol,* 53,299-328.

Khan Z.R, Barrion A.T, Litsinger J.A, Castilla N.P and Joshi R.C, 1988. A bibliography of rice leaffolders Lepidoptera , Pyralidae.. *Insect Sci. Applic,* 9,129-174.

Khasiana N, Eilenberg H, Pnini-Cohen S, Schuster S, Movtchan A and Zilberstein A, 2006. Isolation and characterization of chitinase genes from pitchers of the carnivorous plant. *Journal of Experimental Botany*, 57, 2775–2784.

Kloepper J.W, Tuzun S and Kuc J, 1992. Proposed definitions related to induced disease resistance. *Bio Control Sci. Technol* 2, 349-351.

Kuhlmann M, Horvay K, Strathmann A, Heinekamp T, Fischer U, Bottner S and Dröge-Laser W, 2003. The alpha-helical D1 domain of the tobacco bZIP transcription factor BZI-1 interacts with the ankyrin-repeat protein ANK1 and is important for BZI-1 function, both in auxin signaling and pathogen response. J. *Biol. Chem.,* 27, 8786–8794.

Kusnierczyk A, Winge P, Midelfart H, Armbruster WS, Rossiter JT and Bones AM, 2007. Transcriptional responses of *Arabidopsis* thaliana ecotypes with different glucosinolate profiles after attack by polyphagous *Myzus persicae* and oligophagous *Brevicoryne brassicae. J Exp Bot,* 58, 2537-2552.

Lapopin L, Gianinazzi-Pearson V and Franken P, 1999. Comparative differential RNA display analysis of arbuscular mycorrhiza in Pisum sativum wild type and a mutant defective in late stage development. *Plant Molecular Biology*, 415.

Little D, Gouhier-Darimont C, Bruessow F and Reymond P, 2007. Oviposition by pierid butterflies triggers defense responses in *Arabidopsis*. *Plant Physiol,* 143,784-800

Ljung K, Hull AK, Celenza J, Yamada M, Estelle M, Nonmanly J and Sandberg G, 2005. Sites and regulation of auxin biosynthesis in *Arabidopsis*. *Plant Cell,* 17, 1090-1104.

Maqbool, S.B., Husnain, T., Riazuddin, S., Masson, L. and Christou, P, 1998. Effective control of yellow stem borer and rice leaffolder in transgenic rice indica varieties Basmati 370 and M7 using the novel d-endotoxin cry2A *Bacillus thuringiensis* gene *Mol. Breed*, 4, 501- 507.

Marra R, Ambrosino P, Carbone V, Vinale F, Woo S, Ruocco M, Ciliento R, Lanzuise S, Ferraioli S, Soriente I, et al., 2006. Studymof the three-way interaction between *Trichoderma atroviride,* plant and fungal pathogens by using a proteomic approach. *Curr Genet* 50,307-321.

Martin H, Bostock and Richard M, 2002. Induced Systemic Resistance ISR. Against Pathogens in the Context of Induced Plant Defences. *Botanical Briefing Annals of Botany*, 89,503-512.

Martínez-Noël G, Tognetti J, Nagaraj V, Wiemken A and Pontis H, 2006. Calcium is essential for fructan synthesis induction mediated by sucrose in wheat. *Planta,* 225(1).

Mauch-Mani B and Slusarenko AJ, 1996. Production of salicylic acid precursors is a major function of phenylalanine ammonia-lyase in the resistance of *Arabidopsis* to *Peronospora parasitica. Plant Cell*, 8, 203-212.

Mayer A.M, Harel E and Shaul R.B, 1965. Assay of catechol oxidase a critical comparison of methods. *Phytochemistry,* 5, 783-789.

Medrano F.G and Heinrichs E.A 1985. In, *Handbook of insect rearing,* (P. Singh. and R.F. Moore, Eds.) Elseivier publishing co *Inc. New York*, 361-372.

Mew T.W. and Rosales A.M, 1986. Bacterization of rice plants for control of sheath blight caused by *Rhizoctonia solani. Phytopathol,* 76, 1260-1264.

Mikkelsen MD, Hansen CH, Wittstock U and Halkier BA, 2000. Cytochrome P450CYP79B2 from *Arabidopsis* catalyzes the conversion of tryptophan to indole-3-acetaldoxime, a precursor of indole glucosinolates and indole-3-acetic acid. *J Biol Chem*, 275, 33712-33717.

Mikkelsen MD, Larsen Petersen B, Glawisching E,Bogh Jensen A, Andreasson E, Halkier BA 2003. Modulation of CYP79 genes and glucosinolate profiles in *Arabidopsis* by defense signalling pathyways. *Plant Physiol,* 131, 298-308.

Mitchell A, Walters D, 1995. Systemic protection in barley against powdery mildew infection using methyl jasmonate. *Asp Appl Biol* 42,323-326.

Mitchell, W.C. and Maddison, P.A., 1983. Pests of taro. In, Wang, J-K., ed., Taro, *A Review of Colocasia esculenta and its Potentials*. University of Hawaii Press, 180–235.

Mithofer A, Wanner G and Boland W, 2005. Effect of feeding Spodoptera littoralis on lima bean leaves II. Continuous mechanical wounding resembling insects feeding is sufficient to elicit herbivory-related volatile emission. *Plant Physiol,* 137, 1160-1168.

Moran PJ, Cheng Y, Cassell JL and Thompson GA, 2002. Gene expression profiling of *Arabidopsis thaliana* in compatible plant–aphid interactions. *Arch Insect Biochem Physiol,* 51,182–203.

Moran PJ and Thompson GA, 2001. Molecular responses to aphid feeding in *Arabidopsis* in relation to plant defense pathways. *Plant Physiol,* 125,1074–1085.

Morris P.C, 2001. MAP kinase signal transduction pathways in plants. *New Phytologist,* 151, 67–89.

Nandakumar R, Babu S, Viswanathan R, Raguchander T and Samiyappan R, 2001. Induction of systemic resistance in rice against sheath blight disease by plant growth promoting rhizobacteria. *Soil Biol. Biochem,* 33, 603-612.

Odanaka S, Bennette AB and Kanayama Y, 2002. Distinct physiological roles of fructokinase isozymes revealed by gene-specific suppressin of Frk1 and Frk2 expression in tomato.*Plant Physiol.,* 129, 1119-1126.

Ogihara. Y., 2006. Biotechnology and Sustainable Agriculture 2006 and Beyond Proceedings of the 11th IAPTC&B Congress, August 31-18, Beijing, China Editors, Zhihong Xu, JiayangLi, Yongbiao Xue and Weicai Yang.

Panaud O, Vitte C, Hivert J, Muzlak S, Talag J, Brar D and Sarr A, 2002. Characterization of transposable elements in the genome of rice *Oryza sativa* L.. using Representational Difference Analysis RDA.. *Mol Genet Genomics,* 268, 113–121.

Pandi, V, 1997. Prediction of damage and yield loss caused by rice leaffolder in IR 50 by simulation model, *M.Sc. Thesis*, Tamil Nadu Agricultural University, Coimbatore, India, 96 p.

Passardi F, Cosio C, Penel C and Dunand C., 2005. Peroxidases have more functions than a Swiss army knife. *Plant Cell Rep*, 24, 255-265.

Pastorian K, Hawel III L and Byus CV, 2000. Optimization of cDNA representational difference analysis for the identification of differentially expressed mRNAs. *Anal Biochem,* 283, 89–98.

Paterson A, Bowers J, Kresovich S, Hash CT, Messing J, Peterson D, Schmutz J and Rokhsar D, 2007. Phytozome *Sorghum bicolar* [http,/www.phytozome.net/sorghum].

Pathak, M.D., 1994. *Insect pests of rice*. Manila Philippines., International Rice Research Institute. 89 p.

Penninckx I.A.M.A, Thomma B.P.H.J, Buchala A, Métraux J and Broekaert W.F . Concomitant Activation of Jasmonate and Ethylene Response Pathways Is Required for Induction of a Plant Defensin Gene in *Arabidopsis*. *Plant Cell* 10, 2103-2114

Pontoppidan B, Hopkins R, Rask L, Meijer J 2003. Infestation by cabbage aphid *Brevicoryne brassicae)* on oilseed rape *Brassica napus*. causes a long lasting induction of the myrosinase system. *Entom Exp Appl* 109,55-62

Pontoppidan B, Hopkins R, Rask L, Meijer J 2005. Differential wound induction of the myrosinase system in oilseed rape *Brassica napus*., contrasting insect damage with mechanical damage. *Plant Sci* 168,715-722

Rabindran R and Vidhyasekaran P 1996. Development of a formulation of *Pseudomonas fluorescens* PfALR2 for management of rice sheath blight. *Crop Protect* 15, 715-721.

Raju N, Gopalan M and Balasubiramanian G 1990. Ovicidal action of insecticides, mould inhibitors and fungicides in the eggs of rice leaffolder and stem borer, *Indian J. Pl. Prot.,* 18, 5-9.

Ramamoorthy V, Raguchander T and Samiyappan R 2002a. Enhancing resistance of tomato and hot pepper to *Pythium* diseases by seed treatment with fluorescent pseudomonads. *Euro. J. Plant Pathol* 108, 429-441

Ramamoorthy V, Raguchander T and Samiyappan R. 2002b. Induction of defense-related proteins in tomato roots treated with *Pseudomonas fluorescens* Pf1 and *Fusarium oxysporum* f.sp. *lycopersici*. *Plant and Soil.* 239, 55-68.

Ramaraju, K. and Natarajan, K., 1997.Control of leaffolder under extreme whether conditions. *Madras Agric. J.,* 66, 252-254.

Ramonell KM and Somerville S, 2002. The genomics parade of defense responses, to inWnity and beyond. *Curr Opin Plant Biol,* 5,1–4.

Rao M.V, Paliyath G and Ormrod D.P, 1996. Ultraviolet-B- and ozone-induced biochemical changes in antioxidant enzymes of *Arabidopsis thaliana. Plant Physiol,* 110, 125-136.

Raupach, G.S. and Kloepper, J.W, 1998. Mixtures of plant growth promoting rhizobacteria enhance biological control of multiple cucumber pathogens. *Phytopathology,* 88, 1158-1164.

Rentel M.C, Lecourieux D, Ouaked F, Usher S.L, Petersen L, Okamoto H, Knight H, Peck S.C, Grierson C.S, Hirt H and Knight M.R, 2004. OXI1 kinase is necessary for oxidative burst-mediated signalling in *Arabidopsis Nature,* 427, 858–861.

Rep M, Dekker HL, Vossen JH, de Boer AD, Houterman PM, Speijer D, Back JW, de Koster CG and Cornelissen BJC, 2002. Mass spectrometric identiWcation of isoforms of PR proteins in xylem sap of fungus-infected tomato. *Plant Physiol.,* 130, 904–917.

Reymond P, Remco N.B, Poecke M.P.V, Krishnamurthy V, Dicke M and Farmer E.E, 2004. A Conserved Transcript Pattern in Response to a Specialist and a Generalist Herbivore *The Plant Cell,* 16, 3132-3147.

Reymond P, Weber H and Farmer EE, 2000. Differential gene expression in response to mechanical wounding and insect feeding in *Arabidopsis. Plant Cell,* 12, 707-720.

Reymond P,Bodenhausen N, Van Poecke RMP, Krishnamurthy V, Dicke M and Farmer EE, 2004. A conserved transcript pattern in response to a specialist and a generalist herbivore. *Plant Cell,* 16, 3132-3147.

Rojo E, Leon J, Sanchez-Serrano JJ 1999. Cross-talk between wound signaling pathways determines local versus systemic gene expression in *Arabidopsis thaliana. Plant J* , 20, 135–142.

Rombach, M.C. and Gallagher, K.D, 1994. The brown plant hopper, promises, problems and prospects. In, Heinrichs, E.A., ed., *Biology and Management of Rice Insects*. New Delhi, India, Wiley Eastern/New Age International, 693–709.

Ronald PC, 1997. The molecular basis of disease resistance in rice. *Plant Mol Biol.,* 35,179–186.

Ronning C.M, Stegalkina S.S, Ascenzi R.A, Bougri O, Hart A.L, Utterbach T.R, Vanaken S.E, Riedmuller S.B, White J.A, Cho J, Pertea G.M, Lee Y, Karamycheva S, Sultana R, Tsai J, Quackenbush J, Griffiths H.M, Restrepo S, Smart C.D, Fry W.E, Rutger van der Hoeven, Tanksley S, Zhang P, Jin H, Yamamoto M.L, Baker B.J and Robin Buell C 2003. Comparative Analyses of Potato Expressed Sequence Tag Libraries *Plant Physiol,* 131, 419-429.

Rosales A.M, Vantomme R, Swings J, De Ley J and Mew T.W, 1993. Identification of some bacteria from paddy antagonistic to several fungal pathogens. *J. Phytopathol.,* 138,189-208.

Rudd S, 2003. Expressed sequence tags, alternative or complement to whole genome sequences? *Trends in Plant Science,* 8(7), 321-329.

Sakakibara H, 2006. Cytokinins, activity, biosynthesis, and translocation. *Annu Rev Plant Biol,* 57, 431-449.

Salzman R, Brady J, Finlayson S, Buchanan C, Summer E et al., 2005. Transcriptional profiling of sorghum induced by methyl jasmonate, salicyclic acid, and aminocyclopropane carboxylic acid reveals cooperative regulation and novel genes responses, *Plant Physiology,* 138,352-368.

Saravanakumar D, 2006. Molecular and biochemical marker asisted selection of fluorescent pseudomonad strains for ecofriendly mamagement of leaffolder insect pest and sheath rot disease in rice. *PhD, Thesis,* TNAU, Coimbatore-3, India. 238p.

Saroja R. and Raju, N., 1982. Effect of foliar insecticides on stem borer and leaffolder, *IRRN.,* 73., 14.

Sasaki T, Burr B 2000.. International Rice Genome Sequencing Project, the effort to completely sequence the rice genome. *Curr. Opin. Plant Biol* 3, 138–141.

Schenk P, Kazan K, Wilson I, Anderson J, Richmond T, Somerville S and Manners J, 2000. Coordinated plant defense responses in *Arabidopsis* revealed by microarray analysis. *Proceedings of the National Academy of Science* USA, 97, 11655-11660.

Seehaus K and Tenhaken R, 1998. Cloning of genes by mRNA differential display induced during the hypersensitive reaction of soybean after inoculation with *Pseudomonas syringae* pv. *glycinea. Plant Mol. Biol.,* 38, 1225–1234.

Seehaus, K and Tenhaken R, 1998. Cloning of genes by mRNA differential display induced during the hypersensitive reaction of soybean after inoculation with Pseudomonas syringae pv.glycinea. *Plant Mol. Biol*, 38, 1225-1234.

Seggara G, Casanova E, Bellido D, Odena Ma, Oliveria E and Trillas I, 2007. Proteome, salicyclic acid, and jasmonic acid changes in cucumber plants inoculated with *Trichoderma asperellum* atrain T34. *Proteomics,* 7, 3943-3952.

Shirsat AH, Wieczorek D and Kozbial P, 1996. A gene for *Brassica napus* extension is differentially expressed on wounding. *Plant Mol Biol,* 30,1291-1300.

Shoemaker R, Keim P, Vodkin L, Retzel E, Clifton S.W, Waterston R, Smoller D, Coryell V, Khanna A, Erpelding J, GaiJ X, Brendel V, Raph-Schmidt C,Shoop E.G, Vielweber C.J, Schmatz M, Pape D, Bowers Y,Theising B, Martin J, Dante M, Wylie T, and Granger C, 2002. A compilation of soybean ESTs, generation and Analysis. *Genome,* 45, 329–338 .

Shoresh B.R and Meijer J 2007. Transcriptional profiling by cDNA-AFLP reveals novel insights during methyl jasmonate, wounding and insect attack in *Brassica napus*. *Plant Mol Biol,* 64, 425-438.

Shoresh M and Harman GE, 2008. Genome-wide identification, expression and chromosomal location of the genes encoding chitinolytic enzymes in Zea mays. *Mol Genet Genomics* in press.

Shoresh M and Harman GE, 2008. The molecular basis of shoot responses of maize seedlings to *Trichoderma harzianum* T22 inoculation of the root, A proteomic approach. *Plant Physiol,* 147, 2147-2163.

Shoresh M, Yedidia I and Chet I, 2005. Involvement of jasmonic acid/ ethylene signalling pathway in the systemic resistance induced in cucumber by *Trichoderma asperellum* T203. *Phytopathology,* 95, 76-84.

Sivasundaram, V., 2006. Exploitation of biochemical and molecular tools for identification of effective *Beauveria* and *Metarhizium* pathogens against major insect pests in rice. *M.Sc.Ag., Thesis*, TNAU,Coimbatore-3 ,India.85p.

Smolka MB, Martins D, Winck FV, Santoro CE, Castellari RR, Ferrari F, Brum IJ, Galembeck E, Della Coletta Filho H, Machado MA, Marangoni S and Novello JC 2003. Proteome analysis of the plant pathogen *Xylella fastidiosa* reveals major cellular and extracellular proteins and a peculiar codonbias distribution. *Proteomics,* 3, 224–237.

Sogawa K, Liu GJ and Shen JH 2003. A review on the hyper-susceptibility of Chinese hybrid rice to insect pests. Chin J Rice Sci 17, 23–30.

Song F and Goodman RM, 2002. OsBIMK1, a rice MAP kinase gene involved in disease resistance response. *Planta*, 215, 997-1005.

Stekel D.J, Git Y and Falciani F, 2000. The Comparison of Gene Expression from Multiple cDNA Libraries. *Methods in Genome Res.,* 10(12), 2055-2061.

Stone SL, Hauksdottir H, Troy A, Herschleb J, Kraft E and Callis J, 2005. Functional analysis of the RING-type ubiquitin ligase family of *Arabidopsis*. *Plant Physiol.,* 137, 13–30.

Strausberg R.L and Levy S 2007. Promoting transcriptome diversity *Genome Res* 17,965-968

Suzuki H, Xia Y, Cameron R, Shadle G, Blount J, Lamb C and Dixon RA, 2004. Signals for local and systemic responses of plants to pathogen attack. *J Exp Bot,* 5539(5), 169–179.

Thatcher LF, Anderson JP and Singh KB, 2005. Plant defense responses, what have we learnt from *Arabidopsis*? *Functional Plant Pathology,* 55, 85-97.

Timmusk, S. and Wagener, E.G, 1999. The plant-growth-promoting rhizobacterium *Paenibacillus polymyxa* induces changes in *Arabidopsisthaliana* gene expression, a possible connection between biotic and abiotic stress responses. *Mol. Plant–Microbe Interact*, 12, 951–959.

Tiryaki I and Staswick P.E , 2002. An *Arabidopsis* mutant defective in jasmonate response is allelic to the auxin-signaling mutant *axr1*. *Plant Physiol* 130, 887–894.

Torii , T, 1971. The ecological studies of rice stem borer in Japan. *A review Mushi,* 451() , 1-49.

Valster A.H, Hepler P.K and Chernoff J, 2000. Plant GTPases, the Rhos in bloom. *Trends in Cell Biology*, 104., 141-146.

Van Zhong G and Burns J, 2003. Profiling ethylene-regulated gene expression in *Arabidopsis thaliana* by microarray analysis. *Plant Molecular Biology,* 53, 117-131.

Vidhyasekaran P and Muthamilan M, 1999. Evaluation of powder formulation of *Pseudomonas fluorescens* Pf1 for control of rice sheath blight. *Biocontrol Sci. Technol,* 9, 67-74.

Viswanathan R and Samiyappan R, 2001. Antifungal activity of chitinase produced by some fluorescent pseudomonads against *Colletotrichum falcatum* Went causing red rot disease in sugarcane. *Microbiol. Res.,* 155, 309-314.

Vitte C, Talag JD, Panaud O, Bernardo MA and Brar DS, 2003. Development of *O. minuta* specific clones using representational difference analysis RDA.. *Rice Genet Newsl,* 20, 121–122.

Walden R and Ecker J.R, 1998. Cell signalling and gene regulation, piecing the puzzle together. Current Opinion in Plant Biology. 1(5), 375-377.

Walling LL, 2000. The myriad plant responses to herbivores. *J Plant Growth Regul,* 19, 195–216.

Wang X.L, He R.F and He G.C, 2005. Construction of suppression subtractive hybridization libraries and identiWcation of brown planthopper-induced genes. *J Plant Physiol,* 162, 1254–1262.

Watanabe T and Kitagawa H, 2000. Photosynthesis and translocation of assimilates in rice plants following phloem feeding by the planthopper *Nilaparvata lugens* Homoptera, Delphacidae.. *J Econ Entomol.,* 93, 1192–1198.

Wen J-Q, Oono K, Imai R 2002. Two novel mitogen-activated proteinsignaling components, OsMEK1 and OsMAP1, are involved in a moderate low-temperature signalling pathway in rice. *Plant Physiol,* 129, 1880-1891.

Xiao, Z.Y., Huang, Z.X. and Huang, D.P., 1994. Resistance of rice leaffolder, *Cnaphalogrosis medinalis* to insecticides in Guangdong. *Guangdong Agric. Sci.,* 3, 35-37.

Yamada T, 1993. The role of auxin in plant-disease development. *Annu Rev Phytopathol.* 31, 253–273.

Yang H, You A, Yang Z, Zhang F, He R, Zhu and He G, 2004. Highresolution genetic mapping at the *Bph15* locus for brown planthopper resistance in rice *Oryza sativa* L... *Theor Appl Genet,* 110,182–191.

Yedidia I, Benhamou N, Kapulnik Y and Chet I, 2000. Induction and acculumation od PR proteins activity during early stages of root colonization by the mycoparasite, *Trichoderma harizianum* strain T-203. *Plant Physiol Biochem,* 38, 863-873.

Yedidia I, Shoresh M, Kerem Z, Benhamou N, Kapulnik Y and Chet I, 2003. Concomitant induction of systemic resistance to *Pseudomonas syringae pv.lachrymans* in cucumber by *Trichoderma asperellum* T-203. and acculumation of phytoalexins. *Appl Environ Microbiol,* 69, 7343-7353.

Yu J, Hu S, Wang J and Wong G.K, 2002. A Draft Sequence of the Rice Genome *Oryza sativa* L. ssp. *indica. Science,* 296, 79–92.

Yuan H, Chen X, Zhu L and He G, 2005. Identification of genes responsive to brown planthopper *Nilaparvata lugens* Stål (Homoptera: Delphacidae) feeding in rice. *Planta.* 221, 105–112.

Zhang F, Zhu L and He G, 2004. Differential gene expression in response to brown planthopper feeding in rice. *J Plant Physiol.* 161, 53–62.

Zhu-Salzman K, Salzman RA, Ahn JE and Koiwa H, 2004. Transcriptional regulation of *Sorghum* defense determinants against a phloem-feeding aphid. *Plant Physiol,* 134, 420–431

Fumigation Effect of essential oils and formaldehyde against stored product pests

G. Nattudurai,* M. Gabriel Paulraj* and S. Ignacimuthu*

Fumigation effect of camphor, eucalyptus oil and formaldehyde was tested individually and in different combination to check the toxic effect against *Callosobruchus maculatus*, *Sitophilus oryzae* and *Tribolium castaneum* adults in the laboratory. Among six different treatments, volatile oil + formaldehyde combination (1:1) and formaldehyde alone showed maximum adult mortality by fumigation effect. Both oils were less effective when used alone and in combination form at low concentrations. However eucalyptus and camphor oils killed 100 and 95 per cent *C. maculatus* respectively at a concentration of 250 µl/L in 24hr. *C. maculatus* was highly vulnerable and *T. castaneum* was the least susceptible one. Eventhough *T. castaneum* was resistant to pure oils, addition of formaldehyde (in 1:1 ratio) to eucalyptus or camphor oils drastically increased the mortality of *T. castaneum* to 100 per cent after 9 hr of treatment. The lethal effect of all the treatments was concentration dependent. Mortality also increased when the treatment time was increased.

Introduction

Stored products are destroyed by many insects and heavy loss is reported every year. Direct feeding by insects on stored products reduces grain weight and nutritional values. Infestation also leads to problems such as contamination, unpleasant odor and mold growth that reduce the quality of the grains and makes it unfit for processing food. Although many chemical fumigants have been abandoned in recent years, methyl bromide and phosphine are still used for stored product insect pest management in store houses (Rajendran and Sriranjini, 2007). The use of these two fumigants is being questioned because of the associated health and environmental problems and the possibility of development of resistance in insects (Leesch, 1995). It has already been reported that the number of cases of resistance to methyl bromide and phosphine has increased since the records from the FAO Global Survey in 1972 (Zettler, 1993).

These circumstances are leading to the search of safe fumigants like volatile oils for stored pest management. Many investigators have explored the efficacy of plant volatile oils in stored pest management (Grainge and Ahmed, 1988). Essential oils are potential sources of alternative compounds to currently used fumigants. Essential oils or volatile oils are commonly used for the flavoring of food cosmetics. Some essential oils are used in effective, preventive or curative treatments. Many studies conducted by various scientists clearly indicate that essential oils have anti-insect properties (Jacobson, 1989; Shaaya *et al.*, 1991, 1997; Regnault- Roger *et al.*, 1993). The volatile oils affect the stored product insects' growth, reproduction, physiology, and behavior (Nawrot and Harmatha, 1994; Isman, 2006). Camphor and eucalyptus oils are isolated from *Cinnamomum camphora* and *Eucalyptus globules* respectively. Formaldehyde is used in clinical laboratories as a disinfectant (Reuss *et al.*, 2005). While it is a satisfactory disinfectant or germicide, the value of the solution as an insecticide is small. Formaldehyde generated by the application of chloride of lime to liquid formalin appears to have been tried as an insecticide (Harvey and Hill, 1940).

* Entomology Research Institute, Loyola College, Chennai – 600 034

When used alone, volatile oils are not as effective as the synthetic pesticides. There is a need to enhance their activity for better stored product insect pest management. The present study was undertaken. To assess the efficacy of two volatile oils viz., eucalyptus and camphor oils along with formaldehyde three stored grain pests *viz., C. maculatus* (Coleopteran: Bruchidae), *S. oryzae* (L) (Coleopteran: Curculionidae) and *Tribolium castaneum* (Herbst) (Coleopteran: Tenebrionidae).

Materials and Methods

Insect: The test insects, *castaneum, S. oryzae* and *C. maculatus* were obtained from pure stock cultures maintained on wheat flour, whole rice and chick pea respectively at laboratory conditions (30±1°C; 60-65% R.H.; 11±0.5hr photoperiod) in the Entomology Research Institute. Three to seven days old adult insects were used in the experiments.

Volatile oils and formaldehyde: Volatile oils were purchased from local aromatic oil shop in Chennai and formaldehyde was purchased from Ranbaxy Company.

Treatments and Concentrations: Eucalyptus oil, camphor oil and formaldehyde were tested individually and also in combination. Following combinations were tested.

Treatment 1: Eucalyptus oil; Treatment 2: Camphor oil; Treatment 3: Formaldehyde; Treatment 4: Eucalyptus + Camphor oil; Treatment 5: Eucalyptus + Formaldehyde; Treatment 6: Camphor oil + Formaldehyde and Treatment 7: Eucalyptus + Camphor + Formaldehyde. Four different concentrations viz., 80, 180, 250 and 400µl/L were used.

Testing procedure: The fumigant toxicity of essential oils and formaldehyde was tested by cotton dip method (Negahban *et al*., 2007). Cotton balls were impregnated with treatment solutions and fixed inside the screw caps of containers (500 ml). Ten adults of each species irrespective of sex were introduced in to the containers. Mortality was recorded up to 4 days and corrected by Abbott's corrected formula (Abbott, 1925).

$$\text{Abbot's corrected mortality} = \left(1 - \frac{\text{n in T after treatment}}{\text{n in Co after treatment}}\right) \times 100$$

n = Insect population; T = treated; Co = control

Median lethal concentration: The median lethal concentration (LC_{50}) was calculated for all the treatments by using probit analysis software package.

Results

The toxic effect of volatile oils and formaldehyde on three insect species is presented in figures 1 to 3. Among the three insects tested, *C. macultus* was the most susceptible insect to the treatments. The combinations of volatile oils and formaldehyde were more effective than individual oils and formaldehyde. Eucalyptus oil with formaldehyde (1:1), camphor with formaldehyde (1:1) and eucalyptus + camphor + formaldehyde (1:1:1) combinations killed 100 per cent insects after 9 hr exposure at 180 µl/L concentration. When the concentration was increased to 250 µl/L, both eucalyptus oil and formaldehyde treatments gave 100 per cent mortality at 12 hr exposure period. At this concentration level, other treatments gave more than 90 per cent mortality at 12 hr period.

T. castaneum was highly resistant to the treatments at lower concentration. High concentrations and increased treatment time were necessary to kill 100 per cent population of *T. castaneum* as evidenced from figure 3. At 250 µl/L concentration eucalyptus + camphor + formaldehyde combination presented 100 per cent mortality after 96 hr period. This same treatment killed 100 per cent *T. castaneum* after 48hr when the concentration was increased to 400 µl/L.

Sitophilus oryzae was moderately resistant to the volatile oils and formaldehyde treatments. At 80 and 180 µl/L concentrations *T. castaneum* was found to be more resistant than *S. oryzae*. But at higher concentrations *viz*., 250 and 400 µl/L, *T. castaneum* was at high risk for survival especially when the exposure period was extended. At lower concentrations, formaldehyde alone and eucalyptus + formaldehyde combination were found to be the effective adulticides for *S. oryzae* at 24, 48 and 96 hr exposure period. At 250 µl/L concentration level, camphor + formaldehyde and eucalyptus + camphor + formaldehyde combinations killed maximum *S. oryzae* adults and 100 per cent mortality was recorded in T3, T5, T6 and T7 after 96 hr. When the two oils were compared eucalyptus was more effective than camphor oil. It gave 100 per cent mortality in *C. maculatus* at 250 µl/L concentration after 12 hr. In the case of *S. oryzae* and *T. castaneum*, eucalyptus was effective only when it was mixed with formaldehyde.

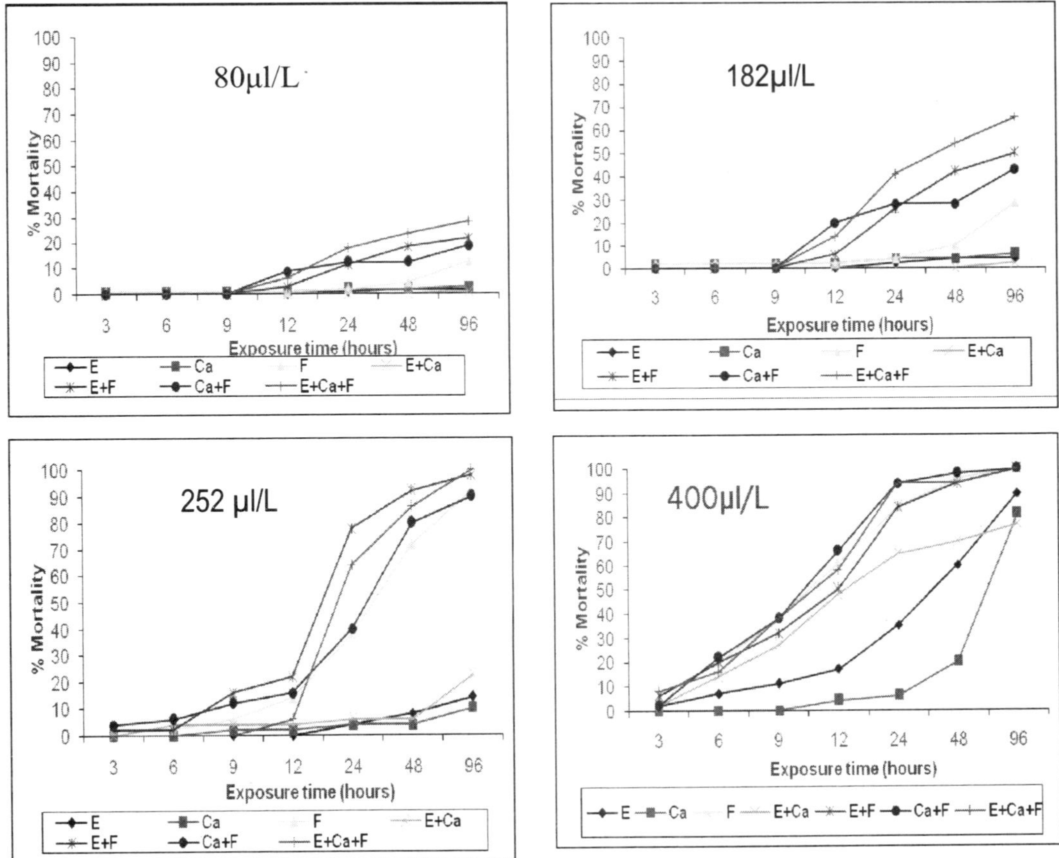

Fig.1. Percent mortality of *Tribolium castaneum* exposed to different combinations of essential oil and formaldehyde at different time duration.

Median lethal concentration: Median lethal concentration of eucalyptus oil, camphor oil, formaldehyde and different combinations of them were calculated using the mortality data recorded for the four different concentrations tested in this study (Table 1). The LC_{50} of different treatments for *C. maculatus* and *S. oryzae* was calculated (Table 1). But in the case of *T. castaneum* LC_{50} of some treatments viz., T1, T2 and T4 could not be obtained (Table 1) since the mortality of *T. castaneum* in these three treatments was below 50 per cent even at the highest concentration used. The least LC_{50} was calculated for T5 against *C. maculatus*, which demonstrated that this treatment could be an

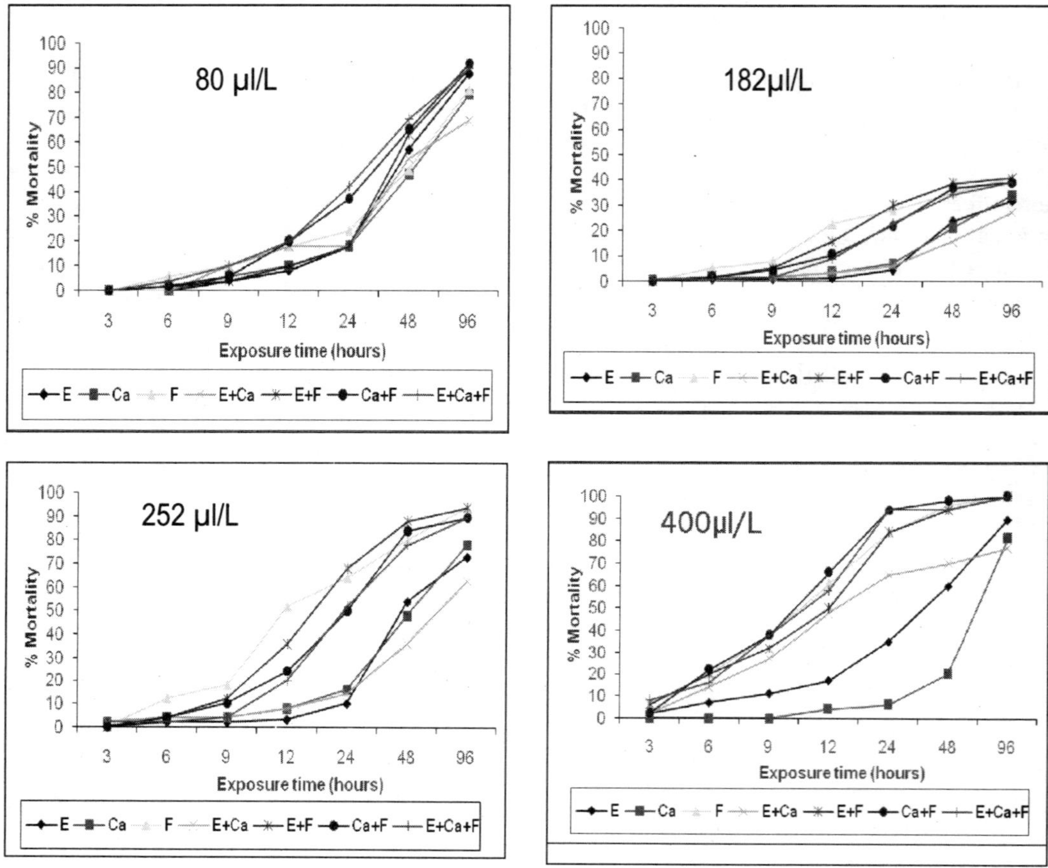

Fig.2. Effect of volatile oil + formaldehyde combinations on the survival of *Sitophilus oryzae* at different concentrations and different exposure period.

effective one for this species. Higher LC_{50} values were obtained for eucalyptus oil (402.5 µl/L) and camphor oil (404.3 µl/L) against *S. oryzae*.

Discussion

The volatile oils, camphor and eucalyptus, were less toxic to the insects when tested individually. Formaldehyde also presented low toxic effect when used individually. But activity increased when formaldehyde and volatile oils were combined together. This clearly indicated the synergistic activity of volatile oils and formaldehyde. Volatile compounds of many plant extracts and essential oils consist of alkanes, alcohols, aldehydes and terpenoids, especially monoterpenoids, and exhibit fumigant activity (Coats *et al.* 1991, Ahn *et al.* 1998, Kim and Ahn 2001, Kim *et al.*, 2003). EL-Nahal *et al.* (1989) and Shaaya *et al.* (1997) have also reported the fumigant activity of plant essential oils against some stored insects. Paulraj and Ignacimuthu (2007) reported that Eucalyptus and Camphor oil killed 100 and 40 per cent adult *Lasioderma serricorne* (Fab) (Coleoptera: Anobiidae) respectively at 200 µl/L concentration in 24 hr.

Negahban *et al.* (2007) have tested the toxic effect of *Artemisia sieberi* oil against *C. maculatus*, *S. oryzae* and *T. castaneum* adults. They reported that 444 µl/L concentration was needed to kill 100 per cent *C. maculatus*, *S. oryzae* and *T. castaneum* in 6, 12 and 12 hr respectively. They found that

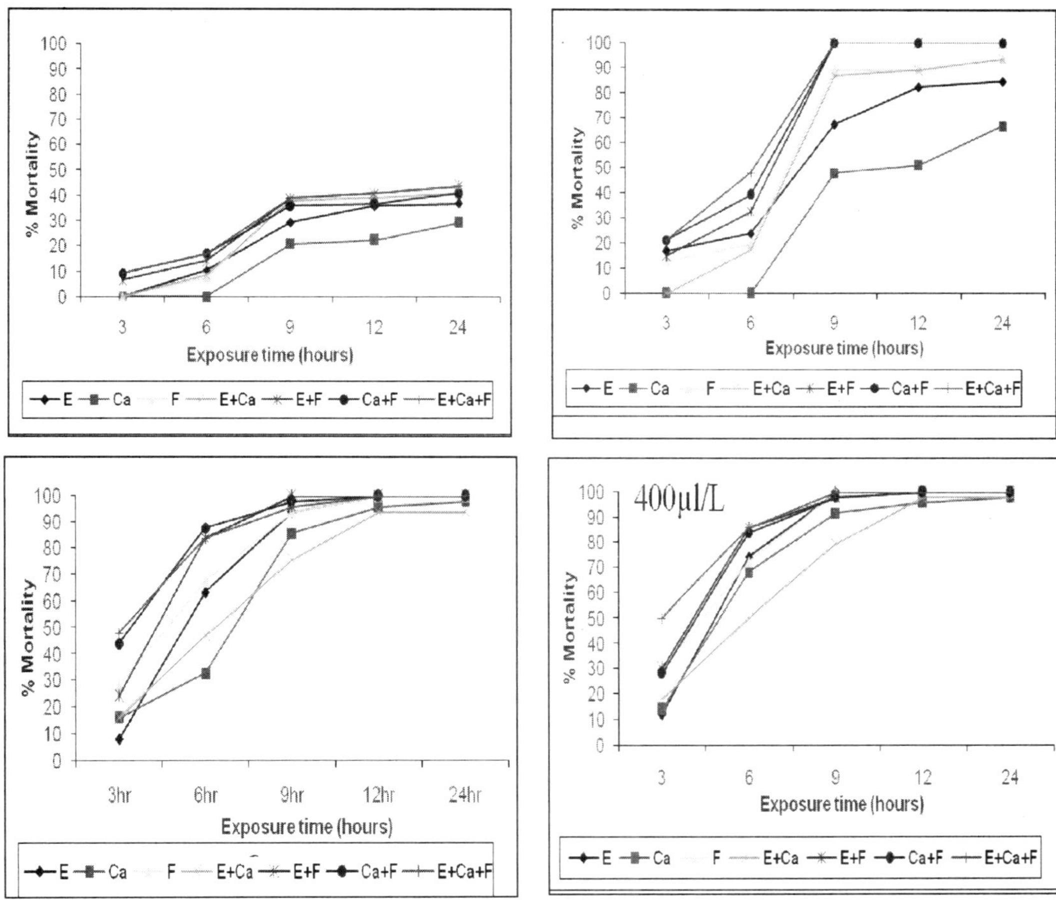

Fig.3. Fumigation toxicity of volatile oils + formaldehyde combinations against *Callosobruchus maculatus* adults at different concentration and different exposure period.

C. maculatus was the most susceptible and *T. castaneum* the least susceptible and this finding coincides with the present finding. Shaaya *et al.* (1997) have reported that the essential oil from Labiatae species (ZP51) killed 85 to 100 per cent adults of *T. castaneum, S. oryzae, Rhyzoertha dominica* (F.) and *Oryzaephilus surinamensis* (L.) within 4 days of exposure at 70 µl/L air. In the present study 180 to 400 µl/L concentrations of volatile oils and formaldehyde was required to kill 80 to 100 per cent insects in four days.

The present study indicated that camphor and eucalyptus oils had the potential of killing stored product insects and the effect could be increased by adding formaldehyde to them. Formaldehyde is commonly used in clinical laboratories as disinfectant. When compared to synthetic chemicals formaldehyde is far safer to use in store houses as fumigant. However formaldehyde may cause irritation in the respiratory system when we are exposed to high concentrations of formaldehyde vapour. The present study had the prime objective of minimizing the quantity of formaldehyde in treatments by substituting volatile oils. The results obtained were encouraging. In future the activity of effective treatments can be improved by replacing formaldehyde with some other less toxic volatile compounds.

Table 1. Median lethal concentration of different combinations of volatile oils and formaldehyde against *C. maculates*, *S. oryzae* and *T. castaneum*

Treatment	LC_{50} (µl/ L)	Lower(µl/ L) fiducial limit	Upper(µl/ L) fiducial limit	Slope ± SE	Chi square (x2)
			C. maculatus		
T1	102.46	82.84	119.77	4.77 ± 0.70	4.035
T2	129.72	102.86	152.53	4.27 ± 0.64	3.283
T3	97.06	80.45	112.18	5.48 ± 0.83	0.897
T4	99.31	75.79	119.60	3.64 ± 0.56	5.578
T5	95.42	76.84	111.71	4.77 ± 0.72	2.144
T6	100.59	81.89	117.18	4.69 ± 0.68	3.344
T7	<80	-	-	-	-
			S. oryzae		
T1	402.53	319.39	701.68	2.86 ± 0.89	0.692
T2	404.35	299.31	784.78	2.03 ± 0.59	0.944
T3	125.34	81.79	161.85	2.08 ± 0.42	2.435
T4	125.47	77.05	165.57	1.88 ± 0.41	3.274
T5	121.56	82.19	154.63	2.17 ± 0.42	1.607
T6	137.39	86.29	181.18	1.88 ± 0.41	3.274
T7	127.20	85.25	162.68	2.17 ± 0.43	1.482
			T. castaneum		
T1	>400	-	-	-	3.023
T2	>400	-	-	1.82 ± 1.05	0.287
T3	343.01	283.41	462.92	2.78 ± 0.61	1.809
T4	>400	-	-	-	-
T5	158.42	127.81	187.78	2.76 ± 0.43	4.131
T6	150.62	121.47	178.02	2.86 ± 0.43	3.448
T7	124.09	71.71	166.18	1.54 ± 0.37	0.795

REFERENCE

Abbott, W.S., 1925. A method of computing the effectiveness of an insecticide. *Journal of Economic Entomology*, 18, 265-267.

Ahn, Y.J., Lee, S.B., Lee, H.S. and Kim, G.H., 1998. Insecticidal and acaricidal activity of carvacrol and thujaplicine derived from *Thujopsis dolabrata* var. hondai sawdust, *J. Chem. Ecol.*, 24, 81 – 90.

Coats, J.R., Karr, L.L and Drewes, C.D., 1991. Toxicity and neurotoxicity effects of monoterpenoids in insects and earthworms, pp. 305- 316. In P. A. Hedin [ed.], *Naturally occurring pest bioregulators. Am. Chem.. Soc. Symp.* Ser. 499.

EI – Nahal, A.K.M., Schmidt, G.H., 1989. Vapours of Acorus calamus oil-a space treatment for stored product insect. *J. Stored Prod. Res.*, 25, 211-216.

Grainge M. and Ahmed S., 1988. Handbook of Plant with Pest – *Control Properties. Wiley – Interscience,* New york.

Harvey. W.M. and Hill, H., 1940. Insect pests. *H. K. lewis & co. Ltd.* London.

Isman M.B., 2006. Botanical Insecticides, Deterrents, and Repellents in modern agriculture and an increasingly regulated world. *Annual Review of Entomology.,* 51, 45 – 66.

Isman, M.B., 2000. Plant essential oils for pest and disease management. *Crop Prot.*, 19, 603-608.

Jacobson, M., 1989. Botanical pesticides: past, present, future. In: Arnason, J.T., Philogene, B.J., Morland,p.(Eds.), *Insecticides of Plant Origin. American Chemical Society*, Washington, dc, pp. 1-10.

Kim. D. H. and Ahn Y.J., 2001. Contact and fumigant activities of constituents of foeniculum vulgare fruit against three coleopteran stored- product insects. *Pest Management Science*, 57(3), 301 – 306.

Kim. S.I., Park, C., Ohh, M.H., Cho, H.C, and Ahn, Y.J., 2003. Contact and fumigant activity of aromatic plants extracts and essential oils against *Lasioderma serricorne* (Coleoptera: anobiidae). *J. Stored Prod. Res*, 39, 11 – 19.

Leesch J.G., 1995. Fumigant action of acrolein on stored- product insects. *journal of economic entomology*, 88, 326-330.

Negahban, M., Moharramipour, S. and Sefidkon. F., 2007. Fumigant toxicity of essential oil from *Artemisia sieberi* Besser against three stored- product insects. *Journal of Stored Product Research*, 43, 123-128.

Nawrot, J. and Harmatha, J., 1994. Natural products as antifeedents against stored product insects. *Post Harvest News and Information*. 5, 17N – 21N.

Paulraj, M.G. and Ignacimuthu, S., 2007. Effect of some volatile oils on life stages and orientation behaviour of *Lasidoderma serricorne* (Fab.) (Coleoptera: Anobiidae). In: *Biotechnology and Insect Pest Management* eds. 171-176.

Rajandren .S. and Sriranjini, V., 2007. plant products as fumigants for stored –product insect control. *Journal of Stored Product Research* .

Regnault – Roger, C., Hamaroui, A., Holeman, M., Theron, E. and Pinel, R., 1993. Insecticidal effect of essential oils from Mediterranean plants upon *Acanthoscelides obtetus* (Say) (Coleopter: Bruchidae), a pest of kidney bean (*Phaseolus vulgaris* L.). *Journal of Chemical Ecology,* 14, 1965-1975.

Reuss. G., Disteldorf . w., Gamer. A.O. and Hilt. A., 2005. "Formaldehyde" In *Ullmann's Encyclopedia of Industrial Chemistry Wiley-VCH,* Weinheim.

Shaaya, E., Kostjukovski, M., Eilberg, J. and Sukprakarn, C., 1997. Plant oils as fumigants and contact insecticides for the control of stored product insects. *Journal of Stored Product Research*, 33, 7 – 15.

Shaaya, E., Ravid , U., Paster, N., Juven, B., Zisman, U. and Pisarrev, V., 1991. Fumigant toxicity of essential oils against four major stored – product insects. *Journal of Chemical Ecology,* 17, 499 – 504.

Zettler. J. L., 1993. Phosphine resistance in stored product insects. In *proceeding of an international conference on controlled atmosphere and fumigation in grain storage* (Edited by Navarro and Donahaye E.) pp. 449-460. Winnipeg, Canada, 1992.

Prospects of Utilizing Sea Weeds in Mosquito Control

A. Jebanesan*

Mosquito problem has become acute in recent years and death of millions of people every year due to mosquito borne diseases has resulted in the loss of socio economic wealth. Chemical is not favoured at present because of insecticide resistance among vectors and environmental imbalance created. Therefore alternative control methods are needed. Sea weed possess a wide range of pharmaceuticals and insecticidal properties. In the present study, thirteen different sea weeds viz. *Ulva lactuca, Caulerpa scalpelliformis, Caulerpa racemosa, Dictyota dichotoma, Enteromorpha clathrata, Sargassum wightii, Turbinaria conoides, Turbinaria ornata, Padina gymnospora, Gracilaria edulis, Hypnea musciformis, Acanthophora spicifera* and *Amphiroa anceps* were collected from Mandapam and Rameshwaram, South east coastal areas of India. The bioactivity of methanol extract of thirteen sea weeds were studied under laboratory conditions for larvicidal ovicidal, oviposition and repellent activity against the filarial vector mosquito, *Culex quinquefasciatus*. Among the sea weeds tested highest larvicidal and repellent activity was observed with *D. dichotoma*. *G. edulis* possesses potential oviposition attractant activity and *E. clathrata* showed highest ovicidal activity against *Culex. quinquefasciatus.*

Introduction

Culex quinquefasciatus is a major vector in India as well as in tropical regions of the world and has been shown to be directly responsible for 80 million annual lymphatic filariasis of which 30 million cases exist in chronic infection. There are 45 million cases of lymphatic filariasis in India alone (Bowers et al., 1995). The present proliferation of this disease is not only due to higher number of breeding places in urban area, but also due to increasing resistance of mosquitoes to current commercial insecticides such as organochlorides organophosphates, pyrethroid, and carbamates (Das and Amalraj, 1997); secondly the synthetic insecticides are toxic and adversely affect the environment by contaminating soil, water and air. There is a need to find alternatives to these synthetic insecticides. Identification of botanical insecticides are alternative method in that they are effective, environment – friendly, easily biodegradable, inexpensive and also multiple effects in a variety of insect species (Su and Mulla, 1999). These effects include larvicidal, ovicidal, antifeedancy, growth regulation, fecundity suppression, male sterility and changes in oviposition activity mostly as repellency. Sea weeds are marine algae growing in the sea and brackish water. They are different from sea grass (which are flowering plants). Sea weeds not only supply oxygen to the biosphere but also serve as sources of food, feed, manures and as sources of drugs, energy and industrial chemicals. The present study was undertaken to assess the bioactivity of seaweeds against *Cx. quinquefasciatus* in a search for effective natural products to be used in the control of filariasis vector.

Materials and Methods

Seaweed collection and extraction

Generally alive, healthy and well grown algae at a depth of about four meters were collected from shells, coral reefs and rocks which are submerged under sea water at low tide, they were exposed with

* Division of Vector Biology, Department of Zoology, Annamalai University, Annamalainagar – 608 002

the heavy growth of sea weeds. The algae were cleaned on the spot with sea water to remove epiphytes, shells and sand particles and other extraneous matter. Nearly 13 algal species of Clorophyceae, Paeophyceae and Rhodophyceae were collected from Rameshwaram and Mandapam coastal areas, southeast coast of India.

Table 1. Seaweeds used in the present study

S. No.	Family Name	Name of the algal specimens (seaweeds)
	Chlorophyceae	
1.	*Ulvaceae*	*Ulva lactuca* (Linnaeus)
2.	*Caulerpaceae*	*Caulerpa racemosa* (Porskal) J. Agardh
3.	*Caulerpaceae*	*Caulerpa scalpelliformis* (Turn)Ag.
4.	*Dictyotaceae*	*Dictyota dichotoma* (Hudson) Lamouroun
5.	*Ulvaceae*	*Enteromorpha clathrata* (Roth) J. Agardh
	Phaeophyceae	
6.	*Sargassaceae*	*Sargassum wightii* (Greville) J. Agardh
7.	*Sargassaceae*	*Turbinaria conoides* (J. Agardh) Kuetzing
8.	*Sargassaceae*	*Turbinaria ornata* (Turner) J. Agardh
9.	*Dictyotaceae*	*Padina gymnospora* (Kuetzing) Vickers
	Rhodophyceae	
10.	*Gracilariaceae*	*Gracilaria edulis* (Gmell) Silva
11.	*Hypneaceae*	*Hypnea musciformis* (J. Agardh) (Wulf.) Lamour
12.	*Rhodomelaceae*	*Acanthophora spicifera* (Vahl.) Boergs
13.	*Corallinaceae*	*Amphiroa anceps* (Lamk.) Decsne

The collected species were identified in the Department of Botany and Centre for Advanced Studies in Marine Biology, Annamalai University. Voucher specimens were labelled and presented in Department of Zoology, Annamalai University. A list of algal specimens collected is presented in Table 1. Sea weeds were dried and powdered using a mechanical grinder. From each sample 100 g of powdered material was loaded in Soxhlet apparatus and extracted in 300 ml methanol for 8 h and the extract was concentrated in a rotary vacuum evaporator to yield the sea weed extract or residue, which was used for the bioactivity.

Mosquito culture

Adult *Cx. quinquefasciatus* were obtained from a laboratory colony maintained at 27±2°C, 70-80% relative humidity and 14:10 light and dark photo period cycle. Larvae were fed with dog biscuits and yeast powder in the ratio of 3:1. The adults mosquitoes were reared in the glass cages and provided with 10% sucrose solution and were periodically blood fed on restrained 5-7 week old chicks.

Larvicidal activity

The larval bioassay test was carried out by following the standard procedure (WHO, 1996). The sea weed extracts was volumetrically diluted separately in 250 ml filtered tap water to obtain the test concentrations of 20, 40, 80, 100 ppm. The methanol was served as control. For each concentration five replicates were run at a time. The larval mortality was recorded after 24 h of exposure, during which no food was given to larvae. The lethal concentration LC_{50} was calculated by probit analysis (Finney, 1971).

Ovicidal activity

Ovicidal activity of the extract was determined by using the following procedure (Su and Mulla 1998).

The extract was volumetrically diluted in filtered tap water to obtain the test concentration ranging from 50 to 250 ppm. Control was set up with the 1 ml of methanol in 99 ml of water. The egg rafts of two separate age groups 0-9 and 9-18 h interval after oviposition are used. Ten egg rafts of each age group was individually transferred to different concentrations of extract for 3 h treatment. After treatment, 5 egg rafts selected at random from each concentration and control individually transferred to distilled water for hatching assessment after counting the eggs in each egg raft under microscope. Each treatment was done with five replicates. The percentage hatching was assessed 120 h after treatment by the following formula

$$\frac{\text{Number of hatched larvae}}{\text{Total no. of eggs in all 5 egg rafts}} \times 100$$

The oviposition behaviour test was determined by using the following protocol (Xue et al. 2001). Twenty gravid female *Cx. quinquefasciatus* (10 days old, 5 days after blood feeding) transferred to each mosquito cage (45 x 38 x 38 cm) covered with plastic screen, with a glass top and muslin sleece for access. Concentrations of 25, 75, 125, 175 and 225 ppm were made in 100 ml filtered tap water. Two enamel bowls holding 100 ml of water were placed in opposite corners of each cage. One treated with the test material and other with a methanol control. The positions of the bowls were alternated between the different replicates so as to nullify and effect of position on oviposition. Five replicates for each concentration were run, with cages placed side by side for each bioassay. After 24 h, the number of egg rafts laid in treated and control bowls were recorded. The percent repellency for each concentration of the extract was calculated by the following formula.

$$ER(\%) = \frac{NC - NT}{N} \times 100$$

where, NC is the number of egg rafts laid in control and NT is the number of egg rafts laid in test solution.

Repellency test

The percentage of protection in relation to dose method was used (WHO, 1996) and 3-4 days old blood – starved female *Cx. quinquefasciatus* (100) were kept in a net cage (43 x 38 x 38 cm). The arms of the volunteer ware washed and cleaned with ethanol and ethanol served as a control. After air drying the arms of the volunteer, only 25 cm^2 dorsal side of the skin on the each arm was exposed and the remaining area being covered by rubber gloves. The sea weed extract of 1.0 mg/cm was applied. The control and treated arm were introduced simultaneously into the cage. The number of bites were counted over 5 minutes every 30 minutes from 18.00 h to 6.00 h. In the event of no bites, the initial 5 minutes the test arm was exposed after every 30 min for a duration of 5 min until a confirmed bite was received. The test was over after the confirmation of mosquito bite in extract to be tested. The mosquito repellency was measured on the basis of the protection time (minutes). The experiment was replicated five times. It was observed that no skin irritation occurred from seed weed extract tested.

Results

Results on the larvicidal ovicidal, oviposition effects, and repellent activity of selected sea weed extracts observed in the study confirm their potential for control of the mosquito populations. Based on the LC_{50} toxicity values the sea weed samples were classified into four groups *viz*., highly effective, effective, less effective and ineffective. The highly effective samples were those having LC_{50} values less than 100 ppm (< 100 ppm). The effective sea weed samples were having LC_{50} values between 100 and 500 ppm. The less effective seaweed samples were having LC_{50} between 500 and 1000 ppm.

Table 2. Larvicidal activity of the seaweed extracts against the mosquito *Culex quinquefasciatus*

Species	% of larval mortality Concentration (ppm)					LC_{50} (LC_{90}) ppm	95% Confidence limit LCL LC_{50} (LC_{90}) ppm	UCL LC_{50} (LC_{90}) ppm	Regression equation	χ^2 value
	20	40	60	80	100					
Caulerpa scalpelliformis	10	28	58	94	100	48.38 (96.55)	28.78 (34.65)	81.37 (269.03)	Y = -2.1971 + 4.272X	9.524
Enteromorpha clathrata	7	4	28	48	82	73.66 (173.20)	48.79 (64.16)	111.20 (467.61)	Y = -1.4461 + 3.4521X	9.638
Dictyota dichotoma	18	38	62	88	100	43.70 (100.22)	24.337 (37.79)	78.48 (382.04)	Y= -0.2344 +3.19070X	10.976
	Concentration (ppm)									
	100	200	300	400	500					
Caulerpa racemosa	5	13	26	48	95	347.08 (768.33)	216.71 (268.97)	555.89 (2056.36)	Y = -6.0844 + 4.3632X	5.455
Ulva lactuca	2	3	9	18	33	804.08 (2320.01)	594.40 (1216.25)	1087.73 (4425.45)	Y = -3.09244+ 2.7854X	6.133
Hypnea musciformis	8	23	38	58	96	301.32 (657.49)	196.13 (273.38)	462.91 (1581.28	Y = -4.37727+ 3.782X	5.106
Sargassum wightii	1	4	8	22	40	643.17 (1499.31)	529.79 (971.47)	780.82 (2313.90)	Y = -4.79365+ 3.4873X	6.373
	Concentration (ppm)									
	600	700	800	900	1000					
Turbinaria ornate	2	3	7	20	40	1087.00 (1478.99)	1015.36 (1278.59)	1163.70 (1710.80)	Y = -24.1014+ 9.5847X	4.386
Turbinaria conoides	2	6	12	32	60	973.92 (1262.31)	938.54 (1161.54)	1010.63 (1371.88)	Y = -29.0070+ 11.379X	4.534
Acanthophora spicifera	4	9	20	39	76	915.03 (1179.60)	837.06 (967.54)	1000.25 (1438.25)	Y = -29.41169+11.6199X	10.364
Padina gymnospora	-	2	4	8	18	1385.29 (2114.18)	1127.60 (1439.03)	1701.86 (3106.07)	Y = -16.93282+ 6.98154X	1.024
Gracilaria edulis	-	2	4	7	21	132.68 (1943.21)	1111.73 (1403.09)	1568.91 (2691.24)	Y = -18.85096+ 7.64258X	2.657
Amphiroa anceps	-	2	4	10	24	1089.79 (1369.21)	1420.15 (2288.34)	1244.05 (1770.09)	Y = -20.901 + 8.3693X	2.094

X^2 values are significant at P<0.05 level; '-' no activity; LCL – Lower confidence limit. UCL – Upper confidence limit.

The ineffective seaweed samples were having LC_{50} values above 1000 ppm (>1000 ppm) (Table 2). Among the thirteen seaweeds tested only three sea weeds, viz., *D. dichotoma, E. clathrata* and *C. scalpelliformis* showed LC_{50} toxicity values below 100 ppm, three sea weeds viz., *C. racemosa, H. musciformis* and *S. wightii* exerted LC_{50} values between 100 and 500 ppm, five seaweeds viz. *T. ornate, U. lactuca, T. conoides, A. spicifera* and *G. edulis* possessed LC_{50} values between 500 ppm upto 1000 ppm. Only two sea weeds *P. gymnospora* and *A. anceps* showed LC_{50} values above 1000 ppm.

The data on the ovicidal activity are presented in Table 3. Egg rafts with ages of 0-9 and 9-18 h were exposed to 50, 100, 150, 200 and 250 ppm sea weed extracts for ovicidal activity. Ovicidal activity of the sea weed extracts was influenced by concentration of the extract and age of the egg rafts. It is also noted that early stage age groups of egg rafts showed poor hatchability rate when exposed to higher concentrations of extracts and older age group of egg rafts showed higher hatchability rate when exposed to lower concentration of extract. Among the sea weeds *E. clathrata* showed lowest egg hatchability (100 percent egg mortality) at 150 ppm concentration for 0-9 h old egg rafts. *D. dichotoma* and *C. scalpelliformis* showed 100 percent egg mortality at 250 ppm for 0-9 h old egg rafts.

Table 3. Ovicidal activity of seaweed extracts against the mosquito Culex quinquefasciatus Say

Seaweed name	Age of the egg rafts (h)	Egg hatchability (%) Concentration (ppm)					Control
		50	100	150	200	250	
D. dichotoma	0-9	26.4 ± 0.6	21.5 ± 0.9	15.2 ± 1.7	10.3 ± 0.3	NH	92.4 ± 1.2
	9-18	89.8 ± 0.5	80.1 ± 0.9	53.8 ± 2.2	35.9 ± 0.3	17.1 ± 0.2	100.0 ± 0.0
S. wightii	0-9	54.9 ± 1.2	46.5 ± 072	28.5 ± 0.4	21.9 ± 0.4	8.2 ± 0.4	94.4 ± 1.3
	9-18	100.0 ± 0.0	100.0 ± 0.0	64.1 ± 0.3	46.2 ± 0.3	18.4 ± 0.3	100.0 ± 0.0
E. clathrata	0-9	16.3 ± 0.3	8.7 ± 0.4	NH	NH	NH	94.4 ± 1.4
	9-18	46.6 ± 0.5	29.3 ± 0.4	19.1 ± 0.7	11.2 ± 0.	NH	100.0 ± 0.0
P. gymnospora	0-9	83.7 ± 0.6	64.7 ± 0.4	50.9 ± 1.7	35.3 ± 0.3	24.6 ± 0.5	34.4 ± 1.4
	9-18	100.0 ± 0.00	100.0 ± 0.00	100.0 ± 0.00	77.9 ± 0.4	52.6 ± 0.3	100.0 ± 0.00
C. scalpelliformis	0-9	58.5 ± 0.4	26.1 ± 0.3	16.3 ± 0.6	13.4 ± 0.5	NH	89.9 ± 1.6
	9-18	100.0 ± 0.0	88.8 ± 0.3	55.9 ± 0.9	36.4 ± 0.6	29.1 ± 0.8	100.0 ± 0.0
C. racemosa	0-9	81.3 ± 0.6	60.7 ± 1.2	44.3 ± 0.4	31.4 ± 0.3	22.9 ± 0.9	93.7 ± 1.4
	9-18	100.0 ± 0.0	100.0 ± 0.0	94.0 ± 1.1	70.5 ± 0.9	51.4 ± 1.1	100.0 ± 0.0
T. conoides	0-9	58.5 ± 0.8	42.3 ± 0.8	25.6 ± 0.8	18.7 ± 1.1	10.8 ± 0.6	89.7 ± 0.8
	9-18	100.0 ± 0.0	94.5 ± 1.2	55.8 ± 0.7	40.7 ± 0.3	22.6 ± 0.7	100.0 ± 0.0
T. ornate	0-9	100.0 ± 0.0	70.5 ± 0.8	61.6 ± 1.2	53.4 ± 0.5	37.2 ± 0.6	100.0 ± 0.0
	9-18	100.0 ± 0.0	100.0 ± 0.0	100.0 ± 0.0	84.4 ± 0.7	69.7 ± 0.5	100.0 ± 0.0
H. musciformis	0-9	88.4 ± 0.6	56.3 ± 0.4	49.1 ± 0.6	31.6 ± 0.3	27.3 ± 0.9	92.3 ± 1.1
	9-18	100.0 ± 0.0	100.0 ± 0.0	89.4 ± 0.4	67.2 ± 0.9	49.1 ± 0.5	100.0 ± 0.0
U. lactuca	0-9	96.3 ± 0.3	89.4 ± 0.4	79.9 ± 0.4	61.4 ± 0.8	49.2 ± 1.2	98.8 ± 0.6
	9-18	100.0 ± 0.0	100.0 ± 0.0	100.0 ± 0.0	94.5 ± 0.9	89.6 ± 0.6	100.0 ± 0.0
G. edulis	0-9	98.3 ± 0.8	73.8 ± 0.6	62.4 ± 0.3	53.3 ± 0.5	45.5 ± 0.3	98.3 ± 0.17
	9-18	100.0 ± 0.0	93.0 ± 0.9	79.7 ± 0.6	67.9 ± 0.4	54.9 ± 0.6	100.0 ± 0.0
A. spicifera	0-9	100.0 ± 0.0	100.0 ± 0.0	94.7 ± 0.7	78.4 ± 0.6	56.8 ± 0.7	100.0 ± 0.0
	9-18	100.0 ± 0.0	100.0 ± 0.0	100.0 ± 0.0	100.0 ± 0.0	89.9 ± 0.5	100.0 ± 0.0
A. anceps	0-9	100.0 ± 0.0	100.0 ± 0.0	87.7 ± 0.6	71.6 ± 0.4	65.4 ± 0.3	100.0 ± 0.0
	9-18	100.0 ± 0.0	100.0 ± 0.0	100.0 ± 0.0	100.0 ± 0.0	100.0 ± 0.0	100.0 ± 0.0

NH – No hatchability (100% egg mortality); Each value X ± S.E) represents mean of six values

The oviposition activity of the sea weed extracts are shown in Table 4. The number of egg rafts laid by *Cx. quinquefasciatus* seaweed extracts progressively increased with concentration 25 to 225 ppm. Oviposition index also increased with increasing concentrations of sea weed extracts. But in some cases the OAI values revealed that the sea weed extracts had repelling activity at higher dosages resulting in failure of oviposition or oviposition of very few eggs. Out of thirteen sea weeds tested six sea weeds namely *G. edulis*, *A. anceps*, *P. gymnospora*, *A. spicifera*, *T. conoides* and *T. ornata* showed their O.A.I. values of + 0.1. Seven sea weeds such as *S. wightii, H. musciformis, U. lactuca, C. racemosa, C. scalpelliformis E. clathrata and D. dichotoma* were showed values between - 0.1 to + 0.1. As suggested by Kramer and Mulla (1979), compound with an O.A.I. of +0.1 and above are considered as oviposition attractants, while those with –0.1 and below are considered as repellents.

Sea weed extracts were tested for their repellent activity against *Cx. quinquefasciatus* and the results are presented in Table 5. *D. dichotoma, E. clathrata* and *C. scalpelliformis* showed above 5.00 h protection. Two sea weeds namely *C. racemosa* and *U. lactuca* showed three to five hours protection. Three seaweeds *H. musciformis, S. wightii* and *T. ornate* showed two to three hours protection. The remaining five sea weeds such as *T. conoides, A. spicifera, P. gymnospora, G. edulis* and *A. anceps* showed less than one hour protection.

Table 4. Ovipositional activity of the seaweed extracts, against the mosquito *Culex quinquefasciatus* (Say)

	Concentration (ppm)	Number of egg rafts laid		Effective attractancy (EA %)	O.A.I
		Treated ($\bar{X}$ ± S.E)	Control ($\bar{X}$ ± S.E)		
	25	4.4 ± 0.3[a]	11.4 ± 0.2[a]	-175.0	-0.46
	75	6.3 ± 0.3[b]	8.3± 0.6[b]	-33.3	-0.14
A anceps	125	11.2 ± 0.4[c]	5.1 ± 0.4[cd]	54.54	0.37
	175	14.2 ± 0.4[d]	6.2 ± 0.3[bc]	56.64	0.40
	225	32.4 ± 0.4[e]	4.1 ± 0.3[d]	87.03	0.83
	25	5.0 ± 0.4[a]	10.0 ± 0.3[a]	-100.0	-0.33
	75	6.3 ± 0.3[a]	8.7 ± 0.4[a]	-50.00	-0.2
G. edulis	125	9.1 ± 0.4[b]	7.3 ± 0.7[b]	22.22	0.125
	175	14.1 ± 0.4[c]	7.5 ± 0.4[b]	50.00	0.333
	225	21.6 ± 0.7[d]	8.1 ± 0.4[a]	63.63	0.466
	25	5.0 ± 0.3[a]	12.1 ± 0.2[a]	-142.0	-0.41
	75	4.5 ± 0.3[a]	11.6 ± 0.7[a]	-157.7	-0.44
P. gymnospora	125	6.3 ± 0.4[a]	9.5 ± 0.3[a]	-50.79	-0.20
	175	11.4 ± 0.6[b]	7.4 ± 0.3[b]	35.087	0.21
	225	16.8 ± 0.4[c]	5.2 ± 0.7[b]	69.04	0.52
	25	4.4± 0.422[a]	10.6 ±0.428[a]	-140.90	-0.657
	75	4.6± 0.494[a]	8.7 ±0.224[b]	-89.13	-0.308
A. spicifera	125	5.9 ±0.307[a]	5.9 ±0.428[c]	0.00	0.00
	175	9.1 ±0.307[b]	6.4 ±0.428[cd]	29.67	0.174
	225	14.8±0.428[c]	4.5 ±0.307[d]	69.59	0.533
	25	4.0±0.0316[a]	12.2±0.428[a]	-205.0	-0.506
	75	5.3±1.0428[ab]	9.8 ±0.333[b]	-84.90	-0.298
T. conoides	125	6.5 ±0.333[b]	8.0 ± 0.365[c]	-23.07	-0.103
	175	8.7 ±0.333[c]	7.3±0.428[cd]	16.09	0.087
	225	13.4 ±0.422[d]	6.4 ±0.422[d]	52.23	0.353

	25	3.0 ±0.258^{a}	9.1±0.307^{ab}	-203.3	-0.504
	75	3.9 ±0.307^{a}	10.2±0.428^{a}	-210.0	-0.446
T. ornata	125	5.4 ±0.333^{b}	8.3± 0.333^{bc}	-53.70	-0.211
	175	9.8 ±0.333^{c}	7.4± 0.422^{cd}	24.48	0.139
	225	11.6 ±0.333^{d}	6.6 ± 0.428^{d}	43.10	0.2724
	25	3.3±0.244^{a}	11.1± 0.422^{a}	-236.36	-0.582
	75	3.9 ±0.307^{a}	10.8± 0.333^{a}	-176.92	-0.469
S. wightii	125	4.0 ±0.365^{a}	9.4 ± 0.211^{b}	-135.0	-0.402
	175	5.8 ±0.428^{b}	7.3 ± 0.211^{c}	-25.86	-0.114
	225	7.6±0.333^{c}	8.5± 0.224^{bc}	-11.84	-0.055
H. musciformis	25	4.5 ± 0.333^{bc}	11.17±0.365^{a}	-146.66	-0.602
	75	0.0 ±0.000^{a}	10.3 ±0.494^{a}	-1.0	-1.0
	125	3.5 ±0.428^{b}	8.4±0.428^{b}	-140.0	-0.411
	175	4.8 ± 0.401^{bc}	9.5±0.428^{ab}	-97.9	-0.328
	225	5.0 ± 0.365^{c}	7.9±0.365^{b}	-58.00	-0.224
U. lactuca	25	0.00± 0.000^{a}	9.2 ±0.307^{a}	-100.0	-1.0
	75	0.00 ±0.00^{a}	7.0 ±0.333^{b}	-100.0	-1.0
	125	3.25± 0.211^{b}	6.4 ± 0.333^{b}	-100.0	-0.326
	175	4.8 ±0.307^{c}	5.6 ± 0.333^{bc}	-16.66	-0.076
	225	4.9 ±0.307^{c}	4.8 ±0.307^{c}	2.040	0.010
C. scalpelliformis	25	0.00± 0.00^{a}	11.1±0.477^{ab}	-100.0	-1.00
	75	0.00 ±0.00^{a}	12.1±0.477^{a}	-100.0	-1.00
	125	0.00 ±0.00^{a}	9.6 ± 0.333^{b}	-100.0	-1.00
	175	5.3 ±0.494^{c}	7.2 ± 0.477^{c}	-35.84	-0.152
	225	4.0 ±0.258^{b}	5.1 ±0.719^{c}	-27.5	-0.120
C. racemosa	25	0.00± 0.000^{a}	11.2±0.601^{a}	-100.0	-1.00
	75	0.00 ±0.00^{a}	8.3 ±0.437^{b}	-100.0	-1.00
	125	4.3± 0.307^{b}	5.9± 0.4307^{b}	-37.20	-0.156
	175	3.7 ±0.333^{b}	6.2 ± 0.601^{c}	-67.56	-0.252
	225	4.5 ±0.422^{b}	4.6 ±0.494^{c}	-2.22	-0.010
E.clathrata	25	3.7± 0.333^{b}	10.3±0.307^{a}	-178.3	-0.471
	75	0.00 ±0.00^{a}	8.7 ± 0.307^{bc}	-100.00	-1.00
	125	0.00± 0.00^{a}	9.4 ± 0.211^{ab}	-100.00	-1.0
	175	4.5 ±0.224^{b}	7.8 ±0.307^{b}	-73.33	-0.418
	225	4.2 ±0.307^{b}	6.0 ± 03657^{c}	-42.85	-0.176
D. dichotoma	25	0.00± 0.00^{a}	11.3±0.224^{a}	-100.00	-1.00
	75	0.00 ±0.00^{a}	10.6±0.333^{a}	-100.00	-1.00
	125	0.00 ±0.00^{a}	6.7 ± 0.307^{c}	-100.00	-1.00
	175	4.9 ±0.307^{c}	8.6 ± 1.49^{b}	-75.51	0.274
	225	3.8 ±0.333^{b}	5.4 ±0.422^{c}	-42.10	-0.173

OAI – Oviposition Active Index; Each value ($\overline{X} \pm$ S.E) represents mean of six replicates; Mean values followed by the different letter are significantly different at $p<0.05$ level (Turkey's test).

Table 5. Repellent activity of the seaweed extracts against the mosquito Culex quinquefasciatus Say.

	Mean number of mosquito bites		Total % of protection for 6 h	Mean number of 100% protection (h)	p-value
	Control	Treated			
	$\overline{X}$ ± S.E	$\overline{X}$ ± S.E			
Dictyota dichotoma	26.3 ± 0.91	4.66 ± 0.19	100.00	6.0	0.0002
Enteromorpha clathrata	35.03 ± 0.57	2.0 ± 0.00	94.28	5.5	0.0003
Caulerpa scalpelliformis	28.0 ± 0.52	9.0 ± 0.33	67.85	5.0	0.0000
Caulerpa racemosa	31.16± 0.50	11.5 ± 0.36	64.50	4.0	0.0000
Ulva lactuca	26.0 ± 0.78	10.3 ± 0.56	60.26	3.5	0.0005
Hypnea musciformis	21.0 ± 0.52	10.0 ± 0.53	52.53	2.5	0.0019
Sargassum wightii	44.3 ± 0.2	23.0 ± 0.90	48.08	2.5	0.0005
Turbinaria ornata	40.16± 0.58	25.0 ± 0.81	37.77	1.5	0.0005
Turbinaria conoides	45.33± 0.82	32.0 ± 0.52	29.40	58.0 minutes	0.0012
Acanthophora spicifera	38.0 ± 0.82	30.0 ± 0.638	21.05	45.0 minutes	0.0013
Padina gymnospora	47.0 ± 0.73	34.8 ± 0.28	25.95	60.0 minutes	0.0021
Gracilaria edulis	50.6 ± 0.53	46.33± 0.45	8.51	25.0 minutes	0.0116
Amphiroa anceps	36.83± 0.78	32.0 ± 0.521	13.11	30.0 minutes	0.0073

Each values ($\overline{X}$ ± S.E) represent mean of six values; p-Values are significant at $p<0.01$ level

Discussion

The larvicidal results of this present study are similar to our earlier report of mosquitocidal compound octacosane which was isolated from leaves of *Moschosma polystachyum* which exhibited remarkable larvicidal activity against *Cx. quinquefasciatus* (Rajkumar and Jebanesan, 2004a). The bioactivity of phytochemicals against mosquito larvae vary widely from promising to very unpractical doses depending on plant species, plant part used, age of plant part, solvent used in extraction, method of extraction and mosquito species (Sukumar *et al.*, 1991). The results of ovicidal activity obtained in this present work indicate that egg rafts of 0-9 h old were more susceptible than 9-18 h older age of egg rafts and also increasing concentration of the extract found to increase the ovicidal activity. The preset ovicidal results are comparable with our previous study of *Solanum trilobatum* leaf extract against egg rafts of *Cx. quinquefasciatus* and *Cx. tritaeniorhynchus* (Rajkumar and Jebanesan, 2004b).

The results of oviposition activity indicate that mosquitoes are acutely sensitive to chemical stimuli, with significant amount of oviposition deterrence occurring in response to different concentrations of extract. Mosquitoes are known to select or reject their specific hosts and oviposition sites by sensing chemical signals, that are detected by sensory receptors on the antennae (Davs and Bowen 1994). The oviposition deterrent activity study is comparable to previously screened leaf extract of *Solanum trilobatum* against eggs of *An. stephensi* (Rajkumar and Jebanesan, 2005a). And also comparable to well established synthetic DEET which exhibit deterrent activity against *Ae. albopictus* (Xue *et al.*, 2001).

In the present study repellent activity was very high at the initial stage (h) of exposure. Increase in the exposure period showed reduction in repellent activity. The repellent activity depends upon the concentration of extract and density of mosquito. These results were compared with the previous reports. The skin repellent activity of *Solanum trilobatum* leaf extract against *An. stephensi* with higher concentration provided over 100 minutes protection against mosquito bites. (Rajkumar and Jebanesan, 2005a). The volatile oils of *Moschosma polystachyum* and *S. xanthocarpum* possessed effective skin repellent activity against *Cx. quinquefasciatus*. The biological activity of the volatile oil might be due to the various compounds and these compounds may jointly or independently contribute to produce skin repellant activity (Rajkumar and Jebanesan, 2005b).

The screening of locally available plants for mosquitocidal activity may eventually lead to their use in natural- product based mosquito abatement practices. Such practices would generate local employment, reduce dependence on expensive imported products and stimulate local efforts to enhance public health (Bowers et. al., 1995). In this way the results obtained in this study suggest that the sea weed extracts of *D. dichotoma, E. clathrata, C. scallpeliformis* and *G. edulis* are promising natural insecticides against *Cx. quinquefasciatus*. Moreover, these results could be useful in the search for newer, more selective, and biodegradable mosquitocidal natural compounds.

References

Bowers WS., Sener, B., Evans, Ph., Bingol, F. and Erdogan, I., 1995. Activity of Turkish medicinal plant against mosquitoes *Aedes aegypti* and *Anopheles gambiae*. *Insect Sci. Appl.,* 16, 339-342.

Das, PK. And Amalraj, D., 1997. Biological control of malaria vectors. *Ind. J. Med. Res.,* 106, 174-197.

Davis, EE. And Bowen, FM., 1994. Sensory physiological basis for attraction in mosquitoes. *J. Amer. Mosq. Cont. Ass.,* 10, 316-325.

Finney, D.J., 1971. Probit analysis, 3rd ed. Cambridge University Press, Londong, UK. 38.

Mordue, AJ. And Blackwell, A., 1993. Azadirachtin an update. *Insect Physiol.* 39, 903-924.

Rajkumar, S. and Jebanesan, A., 2004a. Mosquitocidal activities of octacosane from *Moschosma polystachyum* Linn. (lamiaceae). *J. Ethnopharmacol.* 90, 87-89.

Rajkumar, S. and Jebanesan, A., 2004b. Ovicidal activity of *Solanum triblobatum* Linn (Solanaceae) leaf extract against *Culex quinquefasciatus* Say and *Culex tritoeniorhynchus* Gile (Diptera: Culicidae). *International J. Trop. Insect Sci.,* 24, 340-342.

Rajkumar, S. and Jebanesan, A., 2005a. Oviposition deterrent and skin repellent activities of *Solanum trilobatum* leaf extract against the malarial vector *Anopheles stephensi. J. Insect Sci.,* 5(15), 1-3.

Rajkumar, S. and Jebanesan, A., 2005b. Repellency of volatile oils from *Moschosma polystachyum* and *Culex quinquefasciatus*. *Trop. Biomed*., 22(2), 130-142.

Shaalan, E.A.S., Canyon, D., Younes, M.W.F., Wahab, H.A. And Mansour, A.H., 2005. A review of botanical phytochemicals with mosquitocidal potential. *Environ. Int.* 31: 1149-1166.

Su, T. and Mulla, M.R., 1999. Oviposition bioassay responses of *Culex tarsalis* and *Culex quinquefasciatus* to neem products containing azadirachtin. *Entomologia Experimentalis et Applicata*., 91, 337-345.

Su, T. and Mulla, M.S., 1998. Ovicidal activity of neem products (Azadirachtin) against *Culex tarsalis* and *Culex quinquefasciatus* (Diptera: Culicidae). *J. Amer. Mosq. Cont. Assoc*., 14, 204-209.

Sukumar, K., Perich, M.J. and Boobar, L.R., 1991. Botanical derivatives in mosquito control. A review. *J. Amer. Mosq. Cont. Asso*., 7, 210-217.

WHO, 1996. Report of the WHO Informal consultation on the evaluation and testing of insecticides. CTD/WHOPES/IC/96.1, pp.69.

Xue, R.D., Barnard, D.R. and Ali, A., 2001. Laboratory and field evaluation of insect repellents as oviposition deterrents against the mosquito *Aedes albopictus*. *Med. and Vet. Entomol*., 15, 126-131.

Chikungunya fever outbreak investigations conducted in the Lakshadweep islands, India

P. Philip Samuel* and R. Krishnamoorthi*

Due to the emergence of Chikungunya in Lakshadweep islands early in November 2006, an entomological investigation was conducted to find out the vector mosquitoes involved in this Chikungunya epidemic. Of all the islands, Andrott and Kalpeni had recorded maximum number of suspected Chikungunya cases.

Aedes albopictus was found to be the predominant vector species in these islands. Mostly the breeding of the mosquitoes is recorded in the outdoor habitats. The emergence records showed *Ae. albopictus* breeding in the cement tanks, cement cistern, metal containers, areca nut soaked jars and plastic containers. *Armigerus subalbatus* was found breeding in areca nut soaked jars and coconut shells. Characteristically, *Aedes aegypti* breeding was not found during this entomological survey. In the outdoors landing collections mosquitoes collected were mainly belonging to *Aedes albopictus, Cx. quinquefasciatus* and *Armigeres subalbatus*.

Studies have suggested that the main breeding sources of the *Ae. albopictus* mosquitoes are the rodent devoured coconuts and coconut shells, areca nut soaking mud and plastic pots, plastic tea-cups, discarded containers, grinding stones, metal containers and plastic containers. *Aedes albopictus* is pre-dominantly present in the Lakshadweep islands and possibly play a major role as the vector mosquitoes responsible for the chikungunya transmission. More awareness camps are to be conducted periodically after fixing more educative banners in the main areas to create awareness among the public to protect them from this dreaded disease.

Introduction

Mosquito borne diseases have caused havoc for millions of people in developing countries both among urban and rural population and the loss in terms of human lives is irrevocable. Malaria, Filaria, Japanese encephalitis, Dengue and Chikungunya are the major mosquito-borne diseases in India. Chikungunya fever is a viral disease transmitted to human beings by *Aedes* genus mosquitoes. In Asia, epidemics have occurred in urban areas where *Ae. aegypti* and *Ae. albopictus* are vectors (Tesh *et al.,* 976, Banedee *et al.,* 1988, Turell *et al.,* 1992). Since the beginning of 2006, a crippling mosquito-borne disease has shown an explosive emergence in nations in the Indian Ocean area. By March 7, 2006, more people had been infected in the French island La Réunion, and the disease had spread to the islands of Seychelles, Mauritius, and Mayotte (French). Subsequently, the disease appeared in India, China, and European countries. The World Health Organization is taking measures to assist in fighting the epidemic (Ligon 2006). Epidemic of chikungunya emerged early in November 2006 in the Lakshadweep Islands. Of all the islands, Andrott (population of 12,000) and Kalpeni (population of 5000) had recorded maximum number of suspected Chikungunya cases. Epidemic started in Andrott from 20th November onwards and about 200 cases were recorded daily during the peak time. Similar survey was conducted in the neighboring island (Kalpeni), which was also badly affected with more than 90% population with the suspected Chikungunya fever.

* Centre for Research in Medical Entomology (ICMR), Madurai- 625 002

Materials and methods

Entomological investigations were carried out in these two islands. Larval survey was carried out in the Andrott and Kalpeni islands where stored waters were kept in the containers like cement cistern, cement tank and plastic drum and the peri-domestic containers like mud tubs, grinding stones, metal container, tyre, unused well etc. All immature samples were kept for emergence and the emerged adults were identified and pooled separately from adult mosquito samples in the field. Adult collections were also made from the indoor resting /human landing and out door resting/ human landing collections. A KAP survey was also carried out to understand the awareness on Chikungunya and its vector among residents of these islands. A total 74 households were included in this survey. Salient findings of the survey are briefed under the results section.

Results

Larval survey and adult landing collections both were conducted extensively. Larval survey showed several breeding habitats, viz., rodent-devoured coconuts and their shells, arecanut soaking mud and plastic pots, discarded containers, grinding stones, metal containers, plastic containers and wells all of which were found breeding profusely. The larvae collected were emerged for identification. *Aedes albopictus* immatures were found in different types of water storage receptacles. *Aedes albopictus* was found to be the predominant vector species in this island. The emergence records showed *Ae. albopictus* bred in cement tanks, cement cistern, metal containers, arecanut soaked jars and plastic containers. *A. subalbatus* was found breeding in areca nut soaked jars and coconut shells. Characteristically, *Aedes aegypti* breeding was not at all found during this entomological survey. *Culex quinquefasciatus* breeding was, however, also observed in less intensity. The emerged adults were subsequently screened for trans-ovarial transmission. During the resting collections from the indoor only *Cx. quinquefasciatus* was sampled. In the landing collections outdoors, however, mosquitoes collected were mainly belonging to *Aedes albopictus, Cx. quinquefasciatus* and *A. subalbatus.*

Most of the people (97.3%) from these islands were aware of this Chikungunya fever. The basic knowledge about the main causative agent was known only by 17 % of the total people surveyed. Mode of Chikungunya fever transmission by mosquito bite was known by 52% of the people. Even though the awareness was there about the vector mosquito, only one 32% of the people were aware of the day biting behavior of these vector mosquitoes. It is interesting to note that even the illiterate people (97%) are aware of the Chikungunya fever. People's perception was only 20.6% regarding the breeding sources of the Chikungunya vectors.

Discussion

Studies have suggested that the main breeding sources of the mosquitoes are the rodent devoured bitten coconuts and coconut shells, areca nut soaking mud and plastic pots, plastic tea-cups, discarded containers, grinding stones, metal containers and plastic containers. Local community and school students should be encouraged to prevent accumulations of such discarded containers viz. plastic tea cups & coconut shells which are among the man made breeding sites. Until an efficient Chikungunya vaccine is available, the vector control is the only way to diminish the Chikungunya transmission. Regular educational campaigns are need to be organized within the community involving the women groups present there about the potential breeding grounds of mosquitoes and their control aspects. More awareness camps are to be conducted periodically. Various educative banners about this fever are to be fixed up in the main areas to educate the public about the significance of this disease and mosquito control. Mosquito control has to be an ongoing continuous programme that never ends as long as the threat of epidemic Chikungunya transmission exists in this area.

Acknowledgement

We are thankful The Director General, Indian Council of Medical Research, New Delhi, Officer in-Charge, Centre for Research in Medical Entomology, Indian Council of Medical Research, Madurai, Director, National Vector Borne Diseases Control Programme, Delhi, Director of Medical and Health Services, UT of Lakshadweep, Kavarati-673555 and The Lakshadweep Administration for the facility and encouragement and help in pooling the resources needed to complete this study. We appreciate the excellent help rendered by Shri. Victor Jerald Leo and Shri K. Venkatasubramani, CRME, Madurai in preparation of this manuscript.

References

Tesh, R.B., Gubler, D.J. and Rosen, L., 1976. Variation among geographic strains of *Ae. albopictus* in susceptibility to infection with chikungunya virus. *Am. J. Trop. Med. Hyg.,* 33, 176–181.

Banedee, K., Mourya, D.T. and Malunjkar, A.S., 1988. Susceptibility and transmissibility of different geographical strains of *Ae.albopictus* mosquitoes to chikungunya virus. *Indian J. Med. Res.,* 87, 134–138.

Turell, M.J., Beaman, J.R. and Tammariello, R.F., 1992. Susceptibility of selected strains of *Ae. aegypti* and *Ae. albopictus* (Diptera: Culicidae) to Chikungunya virus. *J Med Entomol* 29, 49–53.

Ligon, B.L., 2006. "Reemergence of an unusual disease: the chikungunya epidemic." *Semin. Pediatr. Infect. Dis.,* 17(2), 99 - 104.

Development of Granular Formulation for an Indigenously Isolated Mosquitocidal Strain (ISPC-8) of *Bacillus sphaericus*

A.B. HADAPAD,* N. VIJAYALAKSHMI,* R. S. HIRE* AND T. K. DONGRE*

In our laboratory, we have isolated three strains of *Bacillus sphaericus* and designated as ISPC-5, ISPC-6 and ISPC-8. Among these isolates, ISPC-8 was found to be the most potent and effective against different mosquito species. Attempts have been made to develop suitable formulation based on this potent isolate, ISPC-8. In order to identify the desirable formulation components, different binding agents, UV protectants and floating agents were screened under laboratory conditions. Sodium alginate (3%) and gelatin (15%) were found to be the most suitable binding agents and congo red (0.1%) as a UV protectant. Incorporation of cork powder (20 mesh) was the best carrier which floats uniformly and sustains for a longer period in water bodies. The experimental formulations based on above components were prepared and tested against third instar larvae of *Culex quinquefasciatus*. It was found that, a concentration of 3 mg/l (10^6 spores) was sufficient to kill all the mosquito larvae within 48 h of feeding. When unformulated *B. sphaericus* spores were exposed to sunlight for 6 h, complete loss of spore viability (1.1%) and larvicidal activity (14.42%) was observed whereas, polymer based formulations protected more than 90% of spore viability as well as larvicidal activity after the same period of exposure. There was no significant reduction in larvicidal activity of gelatin and sodium alginate based formulations even after 240 days of storage at room temperature. The polymer based formulations exhibited considerable shelf-life and good protection from the sunlight. Therefore, these formulations would have a great potential to use under field conditions for mosquito control.

INTRODUCTION

Bacillus sphaericus Neide (*Bs*) is an aerobic, endospore forming bacteria having larvicidal activity against different mosquito species (Regis *et al.*, 2001). Recently attempts have been made to utilize this organism as an effective biopesticide agent. The success of using *B. sphaericus* as an effective biopesticide mainly depends upon the development of desirable formulations having considerable persistence in the treated habitats and available for longer time in larval feeding zone. Although, available formulations based on *B. sphaericus* have been proved to be fairly efficacious in aquatic environment (Karch *et al.*, 1991; Skovmand and Sanogo, 1999; Medeiros *et al.*, 2005), their problems of inactivation by UV radiation (Hadapad *et al.*, 2008) and subsequent rapid settling needs to be resolved. In present study, attempts have been made to develop the sustained-release biopolymer based formulations and their efficacy has been tested under laboratory conditions by using third-instar larvae of *Culex quinquefasciatus* Say.

MATERIALS AND METHODS

Mosquito

The culture of *C. quinquefasciatus* was maintained in our laboratory with an optimal temperature of 28 ± 2^{0}C and 85% RH. Eggs were allowed to hatch in plastic bowls containing 1 lit tap water

* Nuclear Agriculture and Biotechnology Division, BARC, Trombay, Mumbai – 400 085

supplemented with 0.125 g sterilized larval food (13:6:1 of wheat flour, chickpea flour and yeast extract). For all the assays, third instar larvae of the same age and size were used.

Bacterial strains and their growth conditions

The strains ISPC-5, ISPC-6 and ISPC-8 of *B. sphaericus* were isolated in our laboratory from diseased mosquito larvae. The standard strain, 1593 of *B. sphaericus* was obtained from the Pasteur Institute, Paris (France). All the bacterial cultures were maintained on nutrient broth (NBM) medium (5 g peptone, 3 g meat extract and 5 g Nacl, in 1000 ml distilled water) supplemented with 0.3% molasses (3 g). The cultures were grown as method described in Hadapad *et al.* (2008). The final spore-crystal mixture of lyophilized powder was stored at 4°C.

Larvicidal activity of different strains of *B. sphaericus*

For preliminary screening of the larvicidal activity stock suspensions of all *B. sphaericus* strains were prepared in sterile distilled water and tested against third-instar larvae of *C. quinquefasciatus*. Different concentrations of all these strains were tested in 100 ml tap water containing 25 larvae in each beaker (150 ml) with four replications for each concentration. The total larval mortality was scored after 48 h of treatment.

Screening of floating agents, UV protectants and biopolymers

Various carriers such as cork powder (Coarse, 20 and 40 mesh), saw dust, corn cob circles and coconut fiber powder were screened for their prolonged floating characteristics on water bodies. These floating materials were released on water bodies and their floating ability was observed for 60 days.

For UV protectant property, various synthetic and organic compounds were screened under laboratory condition by exposing mixture of active ingredient and UV protectant compounds to UV B radiation for 6 h (Hadapad *et al.*, 2008). The larvicidal activity of exposed and unexposed samples was tested against third-instar larvae of *C. quinquefasciatus*.

Various biopolymers were screened for their sustained-releasing property by embedding active ingredient with different polymers and testing their persistence in 100 ml water. The persistence was assessed by releasing fresh batch of 25 *Culex* larvae at weekly interval for the period of 42 days.

Preparation of formulations

In order to prepare the slow-release formulations, three biopolymers such as gelatin (Sigma Chemicals Co., St Louis MO), acacia gum (domestic use), and sodium alginate (Sigma Chemicals Co., St Louis MO) were used. Essentially all formulations contain same concentration of active ingredient (10% w/w) and 0.1% (w/w) solar protectant, congo red (Merck Products, Darmstadt, Germany). Initially each polymer (15% w/w) was dissolved in adequate amount of distilled water and then active ingredient, congo red were added and agitated to get the homogenized product. Finally cork powder passed through 20 mesh sieve (U.S. Standard Sieve series, Dual Manufacturing Co. Chicago, Il) was added in all formulations. These mixtures were dried at room temperature and moisture content was assessed in final product.

Effect of sunlight on viability and larvicidal activity of different formulations

To assess the effect of sunlight on spore viability and larvicidal activity of formulated and unformulated *Bs* spore-crystal mixture, a single dose was selected from the previous bioassays studies (Hadapad *et al.*, 2008). Formulated and unformulated mixture with active ingredient were dispensed in 5 mL water and exposed to sunlight in an open Petri dish (3 cm dia.) for 6 h.

The viability was assessed after the active ingredient (*Bs* extract) released from the polymer matrix by using different procedures (Murat-Elcin 1995; Hadapad *et al.*, 2008a). Spore viability was determined by plating desired dilutions on nutrient agar plates before and after sunlight exposure. The number of colony forming unit (CFU/ml) was determined after 24 h of incubation at 27 ± 2°C. Sunlight exposed samples were tested for mosquitocidal activity against third-instar larvae of *C. quinquefasciatus* with 4 replications containing 25 larvae each. The total mortality was scored after 48 h of treatment. The original larvicidal activity remaining (OAR) was calculated according to Patel *et al.*, (1996).

Determination of shelf-life

The formulations prepared in Perti dishes as mentioned above were sealed with paraffin film, and stored at room temperature. In order to determine shelf-life of formulated products, larvicidal activity was studied at the interval of 30 days for 8 months and LC_{50} values were calculated after 48 h of treatment.

Data analysis

For all experiments, percentage of spore viability and mean larval mortality were calculated and data were normalized using arcsine transformation before subjecting to analysis of variance (ANOVA, $p<0.05$). Statistical significance between means of different treatments was calculated by using Tukey's HSD multiple comparison test (SPSS 12.0 for Windows). Larvicidal activity data of shelf-life studies were analyzed by using probit analysis and the LC_{50} values were calculated at 95% confidential limit (SPSS 12.0 for Windows).

Results and Discussion

Selection of virulent entomopathogenic bacterium is a prerequisite for the development of suitable formulation for effective control of insect pests. In present study, we have screened various isolates of *B. sphaericus* against third-instar larvae of *C. quinquefasciatus* and results showed a varying virulence patterns. The isolate, ISPC-8 was found to be highly toxic to *Culex* larvae compared with other strains ($F = 6.48$; $df = 3$; $p < 0.016$). In all isolates, the concentration dependent mortality was observed (Table 1). At concentrations of 10^3 and 10^4 spores/ml of ISPC-5 and ISPC-6 exhibited low mortality as compared with ISPC-8 and standard strain, 1593. Moreover, there was no significant difference in the total mortality of standard strain 1593 and ISPC-8 (Table 1).

Table 1. Total percent mortality of indigenous isolates of *Bacillus sphaericus* against third-instar larvae of *Culex quinquefasciatus*.

Isolates	Percent mortality at different concentrations (spores/ml) ± SE		
	10^3	10^4	10^5
ISPC-5	3.3 ± 0.6b	6.7 ± 0.7c	69.3 ± 1.6c
ISPC-6	3.3 ± 0.6b	3.3 ± 0.6c	80 ± 2b
ISPC-8	36.7 ± 0.4a	93.3 ± 0.6a	100 a
1593	33.4 ± 0.3a	76.7 ± 0.3b	98.7 ± 0.8a

SE: Standard error of four replications; Means with the same letter are not significantly different ($p<0.05$)

The sustained-release of active ingredient, floating ability of the carrier and protection against UV radiation are the important properties in a development of suitable formulation for surface feeding mosquito larvae. In present study, various floating agents were screened. Among them, cork powder

(20 mesh) was found to be the most suitable floating carrier exhibiting floating characteristics on the water bodies for a longer period. However, powders from cob corn circles, coconut fiber and saw dust showed fast sinking characteristics. These floating carriers are known to faster absorb and this lead to rapid settling of the carriers in water bodies. Therefore, cork powder was selected and used in the preparation of sustained-release formulations for ISPC-8 strain.

Table 2. Effect of sunlight on spore viability and larvicidal activity of unformulated and biopolymer based formulations (ISPC-8) after 6 h exposure.

Formulation type	Spore viability (%) ± SE[b]	OAR (%)[a] ± SE[b]
Gelatin	90.12 ± 5.0a	86.65 2.4a
Acacia gum	84.88 ± 5.6ab	79.57 3.4ab
Sodium alginate	91.06 ± 5.9a	88.85 3.9a
Control (*Bs*)	0.46 ± 1.1b	14.45 2.8b

[a]Original activity remaining after sunlight exposure; [b]Standard error of four replications; Means with the same letter are not significantly different ($p<0.05$)

Among the different synthetic and organic compounds screened for UV protection property, congo red was offered good protection of spores viability as well as larvicidal activity even after 168 h of exposure to UV-B radiation (Hadapad *et al.*, 2008b). The gelatin, acacia gum and sodium alginate based formulations showed sustained-release of larvicidal entities for a longer period of time. Sustained-release formulations prepared with these three polymers and a solar protectant showed more than 90% mortality in *C. quinquefasciatus* larvae (Fig. 1). Significantly higher mortality was obtained with sodium alginate followed by gelatin and acacia gum ($F = 10.81$; $df = 3$; $p<0.05$) based formulations.

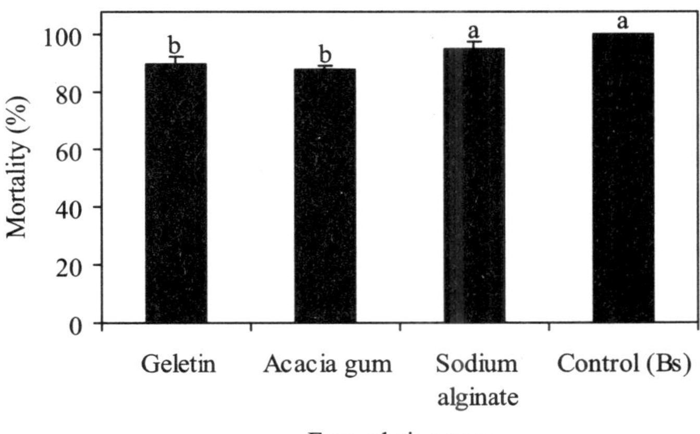

Fig. 1. Bioefficacy of polymer based formulations tested against third-instar larvae of *Culex quinquefasciatus*. Means with the same letter are not significantly different ($p<0.05$).

Similarly, the larvicidal activity (14.42%) of spores was also drastically reduced when tested against third instar *C. quinquefasciatus* larvae after 6 h of exposure to sunlight (Table 2). Similar results were observed with the use of aqueous suspensions of *B. sphaericus* and *B. thuringiensis* subsp. *israelensis* (Araújo *et al.*, 2007). Thus, the addition of substances that protect the activity of the *B. thuringiensis* subsp. *israelensis* toxin from sunlight could extend the residual activity of formulations in the field (Maldonado-Blanco *et al.*, 2002).

In the present study, bioassays were carried out to determine the effect of storage on unformulated and formulated products with different polymers at room temperature. The results showed that larvicidal activity was not affected even up to 8 months storage under regular environmental conditions (temperature 25-30°C; 60-80% RH). After 90 days of storage, maximum larvicidal activity retained in sodium alginate (0.13 mg/l) based formulation followed by gelatin (0.15 mg/l) and acacia gum (0.28 mg/l) (Fig. 2). However, after 240 days of storage, formulation with acacia gum showed increased LC_{50} values compared with the other mixtures. Similarly, Nayar *et al.*, (1999) have reported that storage of the *Bti* formulation under normal environmental conditions for 8 months did not decrease the potency significantly. The larvicidal activity was retained for a longer time when spores were preserved in sodium alginate matrix (Prabakaran and Hoti, 2007). These observations support the present findings on stability of polymer based formulated *B. sphaericus* at room temperature.

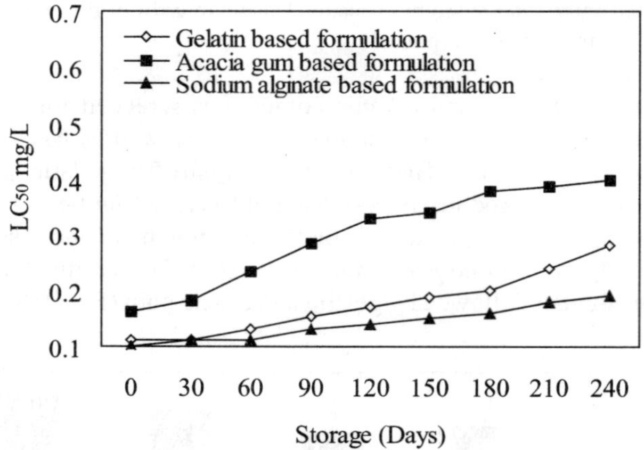

Fig. 2. Shelf-life of biopolymer based formulations based on ISPC-8 strain of *B. sphaericus* stored at room temperature.

Biopolymer based formulation of ISPC-8 exhibited higher toxicity to *Culex* larvae, showed considerable shelf-life and also got protected from the sunlight. Therefore, these formulations will have great potential to use under field conditions for mosquito control.

References

Araujo, A.P., Melo-Santos, M.A.V., Carlos, S.O., Rios, E.M.M.M. and Regis, L., 2007. Evaluation of an experimental product based on *Bacillus thuringiensis* sorovar. *israelensis* against *Aedes aegypti* larvae (Diptera: Culicidae). *Biol. Control.*, 41, 339-347.

Hadapad, A.B., Hire, R.S., Vijayalakshmi, N. and Dongre T.K., 2008. UV protectants for the biopesticide based on *Bacillus sphaericus* Neide and their role in protecting the binary toxins from UV radiation. *J. Invertebr. Pathol. (In press).*

Hadapad, A.B., Vijayalakshmi, N., Hire, R.S. and Dongre, T.K. 2008a. Sustained-release polymer based formulations for an indigenous strain (ISPC-8) of mosquitocidal organism, *Bacillus sphaericus*. *Biocontr. Sci. Technol. (In Press).*

Hadapad, A. B., Vijayalakshmi, N., Hire, R.S., Dongre, T.K., 2008b. Effect of ultraviolet radiation on spore viability and mosquitocidal activity of an indigenous ISPC-8 *Bacillus sphaericus* Neide strain. *Acta Trop.*, 107, 113-116.

Karch, S., Manzambi, Z.A. and Salaun, J.J., 1991. Field trials with Vectolex (*Bacillus sphaericus*) and Vectobac (*Bacillus thuringiensis* (H-14) against *Anopheles gambiae* and *Culex quinquefasciatus* breeding in Zaire. *J. Am. Mosq. Control Assoc.,* **7,** 176–179.

Maldonado-Blanco, M.G., Galan-Wong, L.J., Padilla, C.R. and Matinez, H.Q., 2002. Evaluation of polymer-based formulations of *Bacillus thuringiensis israelensis* against larval *Aedes aegypti* in the laboratory. *J. Am. Mosq. Control Assoc.,* 18, 352-358.

Medeiros, F.P., Santos, M.A., Regis, L., Rios, E.M., Rolim, P.J., 2005. Development of a *Bacillus sphaericus* tablet formulation and its evaluation as a larvicide in the biological control of *C. quinquefasciatus*. *Mem. Inst. Oswaldo Cruz.,* 100, 431-434.

Murat-Elcin, Y., 1995. *Bacillus sphaericus* 2362-calcium alginate microcapsules for mosquito control. *Enz. Microbial Technol.* 17, 587-591.

Nayar, J. K., Knight, J. W., Ali, A., Carlson, D. B, O'Bryan, P.D., 1999. Laboratory evaluation of biotic and abiotic factors that may influence larvicidal activity of *Bacillus thuringiensis* serovar. *israelensis* against two Florida mosquito species. *J. Am. Mosq. Control Assoc.,* 15, 32-42.

Patel, K. R., Wyman, J.A., Patel, K.A., Burden, B.J., 1996. A mutant of *Bacillus thuringiensis* producing a dark brown pigment with increased UV resistance and insecticidal activity. *J. Invertebr. Pathol.,* 67, 120-124.

Prabhakaran, G. and Hoti, S. L. 2007, Immobilization of alginate as a technique for the preservation of *Bacillus thuringiensis* var. *israelensis* for long term preservation. *J. Microbiol. Methods.*, 72, 91-94.

Regis, L., Silva-Filha, M.H., Nielsen-LeRoux, C., Charles, J.F., 2001. Bacteriological larvicides of dipteran disease vectors. *Trends Parasitol.,* 8, 377-380.

Skovmand, O. and Sanogo, E., 1999. Experimental formulations of *Bacillus sphaericus* and *B. thuringiensis israelensis* against *Culex quinquefasciatus* and *Anopheles gambiae* (Diptera: Culicidae) in Burkina Faso. *J. Med. Entomol.,* **36**, 62–67.

Effects of variation in quality of leaf detritus on production of three vector species of mosquitoes

S. Dinakaran* and S. Anbalagan*

Massive works have been published on chemical control of vector species of mosquitoes whereas little formal attention has been received towards influence of leaf litter on production dynamics on vector species of mosquitoes in tropical Asia. The present study was conducted to examine the growth and production of three vector species namely *Aedes aegypti*, *Culex tritaeniorhynchus* and *Culex quinquefasciatus* with three natural leaf litter microcosms (*Pongamia glabra*, *Cassia siamea* and soil microcosm). *C. siamea* contains large quantities of nitrogen and phosphorus than *P. glabra* and soil microcosms. The pupal development of *A aegypti* was rapid (5th day) than *C. tritaeniorhynchus* (8th day) and *C. quinquefasciatus* (10th day). The growth rate of *A. aegypti* and *C. tritaeniorhynchus* was higher in C. *siamea* than *P. glabra*. *C. quinquefasciatus* growth rate was lofty in *P. glabra* than *C. siamea*. In contrast, bacterial growth was also abundant in *C. siamea* than *P. glabra* microcosm. Consequently, the larval and pupal development of three vector species was absent in soil microcosm.

Introduction

Aquatic macroinvertebrate productivity and distribution are affected by the quality and quantity of detritus (Egglishaw, 1964; Stout *et al.,* 1985; Sweeney and Vannote, 1986; Corkum, 1992), which limit the growth of some taxa (Gee, 1988; Richardson, 1991; Dobson and Hildrew, 1992). On land and in water, leaves tend to degrade at species-specific rates (Stout, 1980; Webster and Benfield, 1986; Cornejo *et al.,* 1994; Cornelissen, 1996), and the degradation rate of detritus and its nutritional quality are often correlated (Swift *et al.,* 1979; Golladay *et al.,* 1983). Thus, different species of leaves may not be equal in nutritional value to the organisms that consume them (Sweeney and Vannote, 1986; Canhoto, and Graca, 1995; Walker *et al.,* 1997). These limited data are inconclusive but suggest that understanding of the relationship between macroinvertebrate production and leaf litter. In tropical areas, the influence of leaf litter species on macroinvertebrate growth and production has received less attention. However, these litter microcosm studies will be needed to improve our understanding of the role of detritus in the food webs of tropical areas and it would be very useful for vector control in aquatic ecosystems. The objective of the present study was to examine the effects of variation in quality of leaf detritus on growth of three vector species of mosquitoes (*Aedes aegypti*, *Culex tritaeniorhynchus* and *Culex quinquefasciatus*).

Materials and methods

Experimental set up

A. aegypti, *C. tritaeniorhynchus* and *C. quinquefasciatus* are found in wetlands throughout much of south India. Wetlands of south India have many species of riparian trees. The dominant herbaceous cover includes *Pongamia glabra*, *Cassia siamea*, *Azadirachta indica,* etc. However, the two dominant plant leaves of *Pongamia glabra*, *Cassia siamea* and soil from man-made pond (which is located in our college campus) were taken for this study. Three cement tanks (62 cm in diameter and 40 cm in

* PG Department of Zoology, The Madura College, Madurai – 625 011

height) were taken. In each tank, the above represented of three samples were added and they were closed with fine nylon mesh (1 mm). This set up was kept outside of the laboratory in open areas and they were maintained for 7 days for developing microcosm.

Laboratory studies

Eggs of *A. aegypti*, *C. tritaeniorhynchus* and *C. quinquefasciatus* were brought out from Centre for Research in Medical Entomology (Indian Council for Medical Research), Madurai and they were floated on de-chlorinated water for 6 hours for hatching the larvae. Four 1litre beakers were taken and 500ml of microcosm sample were taken from *P. glabra*, *C. siamea* and mud water tank and another one as control (distilled water). Twenty larvae of each species were added in each beaker. In control yeast-biscuit powder was added until pupae formation. All those experiments were kept in room temperature and they were maintained until pupae formation. During this study the larval growth, biomass and survival rate were noted. Physico-chemical parameters such as temperature, dissolved oxygen, pH, Total dissolved solids; conductivity, nitrogen and phosphorus were estimated from four samples according to standard methods (APHA, 1995).

Aliquot of fluid (0.25-ml) taken daily from each pitcher was used to census bacteria. Cell growth on agar plates was used to estimate the relative abundance of bacteria by counting colony-forming units (CFUs). Prior research has demonstrated that CFUs from plate counts do increase proportionally as prey are introduced into pitchers (Kneitel and Miller, 2002; Miller *et al.*, 2002) and that bacterial abundances are significantly correlated with changes in the number of ants captured by pitchers through time (Miller and Kneitel, 2005). A 0.05-ml sub-sample was used to create a 10^{-4} dilution, then 0.1 ml of the dilution was spread on a half-strength Luria agar plate (Cochran-Stafira and von Ende, 1998; Kneitel and Miller, 2002). The plates were incubated at room temperature (31°C) for 48 h, after which the numbers of CFUs were determined by direct count.

Three vector species of mosquitoes in all treatments were observed daily for emergence. At the first sign of emergence as indicated by either the presence of adults or pupal-exuviae, all pupae present in the microcosms were sacrificed and converted to individual biomass (mg dry mass) using subfamily- specific length-to-biomass equations (Benke *et al.*, 1984). Adult biomass was assumed to be equivalent to that of the largest larva present in a given treatment at the time of emergence. The daily instantaneous growth rate (g) for each genus was calculated using the following equation:

$$g = \ln(M_t/M_o)/\Delta t$$

where Mt is the mean final dry mass, M_o is the mean initial dry mass and t is the number of days from hatching to first emergence (Benke, 1984). The growth rates calculated for each one and they were compared between chamber treatments using General Linear Model ANOVA (PAST 1.42 version). Instantaneous growth rate curves were derived for all treatments using combined data from microcosms.

Results

Physico-chemical parameters

The physico-chemical parameters of three experimental samples are given in Table 1. Dissolved oxygen was high in soil microcosm whereas consequently absent in *C. siamea* microcosm. pH was high in *P. glabra* and low in *C. siamea* whereas conductivity was high in *C. siamea* and low in soil microcosm. Total dissolved solids were greater in *C. siamea,* which indicate high nutrients present rather than others. The microcosm of *C. siamea* consist greater proportion of nitrogen and phosphorus than *P. glabra* and soil microcosms.

Survival rate

Larvae of three mosquito species were studied until their pupae formation on three different microcosms. These results showed that the maximum pupae of *A aegypti* were developed on 5th days in both *P. glabra* and *C. siamea* microcosms. The greatest number of pupae of *C. tritaeniorhynchus* was developed on 5th and 8th day in *C. siamea* in *P. glabra* microcosms respectively. In *C. quinquefasciatus*, the higher rate of development of pupae was on 10th days in *P. glabra* and there was no significant development in *C. siamea* microcosms. For control, the pupae of three mosquito species were developed between 5th and 10th day. Interestingly there is no larval and pupal development of three vector species of mosquitoes in soil microcosm because it contains microfauna (*Paraplea*; Aquatic Hemiptera), which killed the all larvae of mosquito species (Table 2).

Growth rate and biomass

Instantaneous daily growth rates were significantly different between three microcosms treatments ($p > 0.0238$). Maximum growth rates of *A aegypti* (0.68 d^{-1}) and *C. tritaeniorhynchus* (0.71 d^{-1}) were found in the microcosm of C. *siamea* whereas the higher growth rate of *C. quinquefasciatus* (0.54 d^{-1}) were observed in *P. glabra*. Minimum developmental times, measured as the time in days from the hatching of an egg mass to emergence of the first adult, ranged from 5 to 11 days (Table 3).

Table 1. The mean physico-chemical parameters of three different microcosms.

	P. glabra	*C. siamea*	Soil microcosm
Water temperature (°C)	28	28	28
Dissolved oxygen (mgL^{-1})	2.5	0.0	10.08
pH	7.24	6.85	7.11
Total dissolved solids (mgL^{-1})	1500	1530	1450
Conductivity (µmhos)	2.39	2.45	2.29
Nitrogen (%)	2.7	2.55	2.55
Phosphorus (%)	3.11	4.84	0.75

Table 2. Pupal development of three mosquito species (in days) on three different microcosms ('-' indicates that no emergence after 2 weeks).

Days	*P. glabra*	*C. siamea*	Soil microcosm	Control
A. aegypti				
0	20	20	20	20
1	0	0	0	0
2	0	0	0	0
3	0	0	0	0
4	0	0	0	0
5	9	7	0	0
6	5	1	0	1
7	0	1	0	19
8	1	3	0	0
9	0	2	0	0
10	2	2	0	0
11	0	1	0	0
12	-	-	0	0

C. tritaeniorhynchus				
0	20	20	20	20
1	0	0	0	0
2	0	0	0	0
3	0	0	0	0
4	0	0	0	0
5	3	16	0	0
6	0	0	0	6
7	2	0	0	4
8	6	0	0	0
9	1	0	0	5
10	0	0	0	2
11	0	0	0	0
12	-	-	-	-
C. quinquefasciatus				
0	20	20	20	20
1	0	0	0	0
2	0	0	0	0
3	0	0	0	0
4	0	0	0	0
5	0	0	0	2
6	0	0	0	4
7	3	2	0	3
8	1	1	0	6
9	3	1	0	1
10	6	0	0	2
11	0	0	0	0
12	-	-	0	0

Table 3. Daily instantaneous growth rates (g) ($\mu g\ \mu g^{-1}\ d^{-1}$) and average biomass (initial/final, μg) by three different microcosms on three mosquito species.

		A. aegypti	*C. tritaeniorhynchus*	*C. quinquefasciatus*
P. glabra	Total Numbers	17	12	13
	Growth rate	0.5	0.49	0.53
	Biomass	0.13	0.11	0.11
C. siamea	Total Numbers	17	17	4
	Growth rate	0.68	0.71	0.14
	Biomass	0.30	0.28	0.009
Soil microcosm	Total Numbers	0	0	0
	Growth rate	0.0	0.0	0.0
	Biomass	0.0	0.0	0.0
Control	Total Numbers	20.0	20	18
	Growth rate	0.37	0.31	0.54
	Biomass	0.14	0.07	0.14

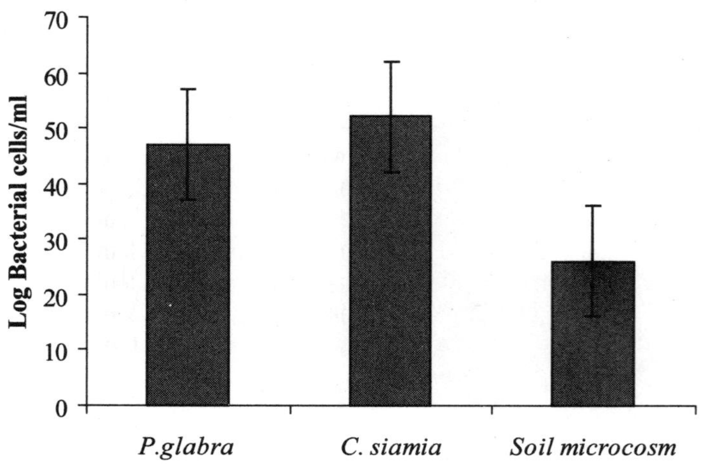

Fig. 1. Dynamics of bacterial colonies associated with three different microcosms

Discussion

The current study suggests a role of leaf litter microcosm in structuring both feeding and production of larval *A. aegypti*, *C. tritaeniorhynchus* and *C. quinquefasciatus*. This may have important consequences in shaping the ecological and evolutionary patterns of three vector mosquitoes' species populations in the tropical mosquito breeding sites. Thus, in tropical mosquito-breeding sites, the selection indirectly exerted by the riparian vegetation on larval mosquito fauna through dietary leaf-litter polyphenols may have generated only postingestive adaptative responses. Because of the predominance of a polyphenolic-rich leaf litter in a given breeding site, the larvae may have no possible diet choice. Thus, the acquisition of metabolic detoxification processes may have appeared as the main way for larval tolerance to those ingested leaf-litter toxic polyphenols. Such a metabolic adaptation has been considered as the key innovation (Berenbaum *et al.,* 1996) that may have promoted ecological diversification (David *et al.,* 2000) and shaped evolutionary history (Rey *et al.,* 2001) of mosquito taxa throughout the tropical breeding sites surrounded by vegetation with different polyphenol richness. The pupae of three mosquito species were rapidly developed in both leaf litter microcosms. This result reflects the fact that the findings of leaf litter in a tropical stream may serve as food or substrate for macroinvertebrates (Dudgeon and Wu, 1999). In tropical countries where many leaf types of varying palatability and diverse defensive compounds are present (including a greater proportion of species with high levels of condensed tannins: Stout, 1989), the patch – specific response of faunal densities to changes in the total amounts of these mixture can be expected to be rather weak, and macroinvertebrate abundance is unlikely to correlate closely with litter biomass. The larva and pupa growth was conspicuously absent in soil microcosm. This may have growth of microfauna (*Paraplea*: Aquatic Hemiptera) present in this microcosm, which may be killed all the larvae of three species of mosquitoes. However, microfauna abundance in soil microcosm, which remove the mosquito larva than in litter microcosm and soil microcosm suggests that increased predator efficiency or nutrient limitation caused local extinctions.

Although different microcosms demonstrated significantly direct effects of resources on bacteria and mosquitoes, almost direct effects through intermediate trophic levels were obvious. For example, nitrogen richness increased bacterial abundance, but this increase in bacteria had no effect on the abundance of any of the bacteriovores. Similarly, mosquitoes had strong effects on bacteriovores, but this suppression of bacteriovore abundances did not cascade down to cause increases in bacteria. The bacterial and bacteriovore responses to nutrient additions should be interpreted with some caution. The use of plate counts to enumerate relative changes in bacterial abundance assumes that the proportion of total number of bacterial cells that will grow on the media is independent of treatment. If for example, the addition of carbon disproportionately increases the abundance of culturable bacteria, then we may have overestimated the potential effects of carbon addition on the bacteriovores (Bakken, 1997). Rate of three species of mosquitoes' growth was high in leaf litter microcosm rather than soil microcosm; it may also be influenced by many factors including food quality/availability and temperature (Ward and Cummins, 1979; Hauer and Benke, 1991; Benke, 1998). The lower growth rates found in soil microcosm, may be attributable to a less favorable food source, either in quantity or quality.

Acknowledgement

We thank Ms. K. Viveka, Ms. S. Roseline, Ms. R. Kavitha and Mr. C. Balachandran for laboratory assistance. We also thank Secretary, Principal and Vice-Principal of Madura College for permitting us to carry out this study.

References

APHA, 1995. Standard methods for the examination of water and wastewater, 16th edition. American Public Health Association (APHA), Washington, D.C.

Bakken, L.R., 1997. Culturable and unculturable bacteria in soil. In van Elsas, J.D., Trevors, J.T. and Wellington, E.M.H. (eds), Modern Soil Microbiology. Marcel Decker, New York, 47–61.

Benke, A.R., 1984. Secondary production of aquatic insects. pp. 289-322. *In* Ecology of Aquatic Insects. Resh, V.H. and Rosenberg, D.W. (eds.) Prager. New York, NY.

Benke, A.C., 1998. Production dynamics of riverine chironomids: extremely high biomass turnover rates of primary consumers. *Ecology* 79, 899–910.

Berenbaum, M.R., Favret, C. and Schuler, M., 1996. On defining "key innovations" in an adaptive radiation: Cytochrome P 450 and Papilionidae. *Am. Nat.,* 148, S139–S155.

Canhoto, C. and Graca, M.A.S., 1995. Food value of introduced eucalypt leaves for a Mediterranean stream detritivores: *Tipula lateralis*. *Freshwater Biol* 34, 209-214

Cochran-Stafira, D.L. and von Ende, C.N., 1998. Integrating bacteria into food webs: studies with *Sarracenia purpurea* inquilines. *Ecology* 79, 880–898.

Corkum, L.D., 1992. Relationships between density of macroinvertebrates and detritus in rivers. *Arch Hydrobiol* 125, 149-166.

Cornejo, F.H., Varela, A. and Wright, S.J., 1994. Tropical forest litter decomposition under seasonal drought: nutrient release, fungi and bacteria. *Oikos,* 70, 183-190.

Cornelissen, J.H.C., 1996. An experimental comparison of leaf decomposition rates in a wide range of temperate plant species and types. *J. Ecol.*, 84, 573-582

David, J.P., Rey, D., Cuany, A., Amichot, M. and Meyran, J.C., 2000. Comparative ability to detoxify alder leaf litter in field larval mosquito collections. *Arch. Insect Biochem. Physiol.*, 44, 143–150.

Dobson, M. and Hildrew, A.G., 1992. A test of resource limitation among shredding detritivores in low order streams in southern England. *J. Anim. Ecol.*, 61, 69-77.

Dudgeon, D. and Wu, K.K.Y., 1999. Leaf litter in tropical stream: food or substrate for macroinvertebrates? *Arch. Hydrobiol.*, 146(1), 65-82.

Egglishaw, H.J., 1964. The distributional relationship between the bottom fauna and plant detritus in streams. *J. Anim. Ecol.*, 33, 463-476.

Gee, J.H.R., 1988. Population dynamics and morphometrics of *Gammarus pulex* L.: evidence of seasonal food limitation in a freshwater detritivore. *Freshwater Biol.,* 19, 333–343.

Golladay, S.W., Webster, J.R. and Benfield, E.F., 1983. Factors affecting food utilization by a leaf shredding aquatic insect: leaf species and conditioning time. *Holarctic Ecol.* 6, 157-162.

Hauer, F.R. and Benke, A.C., 1991. Rapid growth of snagdwelling chironomids in a blackwater river: the influence of temperature and discharge. *J. North Am. Benthol. Soc.,* 10, 154–164.

Kneitel, J. and Miller, T.E., 2002. Resource and top-predator in the pitcher plant (Sarracenia purpurea) inquiline community. *Ecology,* 83, 680–688.

Miller, T.E. and Kneitel, J., 2005. Inquiline Communities in Pitcher Plants as a Prototypical Metacommunity. In Holyoak, M. Leibold, M. and Holt, R. (eds), Metacommunities: spatial dynamics and ecological communities. University of Chicago Press, Chicago, 122–145.

Miller, T.E., Horth, L. and Reeves, R. 2002. Trophic interactions in the phytotelmata communities of the Pitcher Plant, Sarracenia purpurea. Community *Ecology,* 3, 109–116.

Rey, D., Despres, L., Schaffner, F. and Meyran, J.C., 2001. Mapping of resistance to vegetable polyphenols among *Aedes* taxa (Diptera: Culicidae) on a molecular phylogeny. *Mol. Phyl. Evol.,* 19, 317–325.

Richardson, J.S., 1991. Seasonal food limitation of detritivores in a montane stream: an experimental test. *Ecology,* 72, 873-887.

Stout, J., 1980. Leaf decomposition rates in Costa Rican lowland tropical rainforest streams. *Biotropica,* 12, 264-272.

Stout, R.J., Taft, W.H. and Merritt, R.W., 1985. Patterns of macroinvertebrate colonization on fresh and senescent alder leaves in two Michigan streams. *Freshwater Biol.,* 15, 573-580.

Stout, R.J., 1989. Effects of condensed tannins on leaf processing in mid-latitude and tropical streams: a theoretical approach. *Can. J. Fish. Aquat. Sci.,* 46, 1097-1106.

Sweeney, B.W. and Vannote, R.L., 1986. Growth and production of a stream stonefly: influences of diet and temperature. *Ecology,* 67, 1396-1410.

Swift, M.J. and Heal, O.W. and Anderson, J.M., 1979. Decomposition in terrestrial ecosystems. University of California Press, Berkeley.

Walker, E.D., Kaufman, M.G., Ayres, M.P., Riedel, M.H. and Merritt, R.W., 1997. Effects of variation in quality of leaf detritus on growth of the eastern tree-hole mosquito, *Aedes triseriatus* (Diptera: Culicidae). *Can. J. Zool.,* 75, 707-718.

Ward, G.M. and Cummins, K.W., 1979. Effects on food quality on growth of a stream detritivore, *Paratendipes albimanus* (Meigen) (Diptera: Chironomidae). *Ecology,* 60, 57–64.

Webster, J.R. and Benfield, E.F., 1986. Vascular plant breakdown in freshwater ecosystems. *Ann. Rev. Ecol. Syst.*, 17, 567-594.

Bioefficacy of some essential oils against the larvae of vector mosquitoes *Culex quinquefasciatus* and *Aedes aegypti* (Diptera : Culicidae)

R. Maheswaran,* K. Baskar,* S. Kingsley* and S. Ignacimuthu*

Five essential oils *viz.,* pine, clove, orange, rosemary and cedarwood were tested at different concentrations for their larvicidal effects against the laboratory-reared mosquito species of *Culex quinquefasciatus* and *Aedes aegypti* under laboratory conditions. *Cx. quinquefasciatus*, the major filarial vector in India and *A. aegypti,* the main vector of chikungunya, dengue and dengue hemorrhagic fever in urban areas, were used. All essential oils showed some larvicidal activity against fourth instar larvae of the two mosquito species after 24 hr exposure. Pine oil and clove oil showed 100% mortality at 1000 ppm concentration on both species. Pine oil and clove oil, could be used as a potential mosquito control agents.

Introduction

Mosquitoes, one of the major groups of arthropods carry and spread several diseases like dengue fever malaria filariasis, yellow fever etc. It is estimated that every year, at least 600 million people suffer from one of these infections (WHO, 1996; Ravi Kiran, 2007). The mosquito, *Culex quinquefasciatus* is one of the potential vectors of *Wuchereria bancrofti*, the causative agent of human lymphatic filariasis (HLF) all over the world (Birley, 1993; Ahamed, 1994; Pailey *et al.,* 1995). Recent estimates suggest that in 73 countries some 120 million people are infected with HLF (WHO, 1997). *Aedes aegypti* is one of the most important vectors of arbovirus infections, especially dengue and yellow fever, throughout the tropics.

Repeated use of synthetic insecticides for mosquito control has disrupted natural biological control systems and led to resurgences in mosquito populations. It has also resulted in the development of resistance Recent studies have also stimulated the investigation of insecticidal properties of chemicals derived from plant material and many medicinally important plant extracts were studied for their efficacy to kill the larvae of different species of mosquito (Kumar and Dutta, 1987; Tare *et al.*, 2004).

Studies of the essential oils obtained from the plants such as *Cymbopogan citrates* (Sukumar *et al.,* 1991), *Mentha piperata* (Ansari *et al.,* 2000), *Lippia sidoides* (Carvalho, 2003), *Hyptis martiusii* (Araujo *et al.,* 2003) have demonstrated promising larvicidal activities. A large number of plant essential oils may be potential sources of mosquito larvicides, because they constitute a rich source of bioactive components (Cetin *et al.,* 2006). Many plant oils such as basil, cinnamon, citronella and thymus are promising mosquito larvicides (Cavalcanti *et al.,* 2004).

Investigation of plant oil-derived larvicides used for mosquito control is, therefore, a vital point of the present investigation. The selection of plants used in this study focused on those belonging to similar or associated plant families reported to have potential against mosquitoes (Sukumar *et al.,* 1991). Commercially, essential oils are used in four primary ways: pharmaceuticals, flavour enhancers in many food products, odorants in fragrances, and insecticides particular emphasis has been placed

* Entomology Research Institute, Loyola College, Chennai – 600 034, India.

on their antibacterial (Chang *et al.*, 2000a), antifungal (Chang *et al.*, 2000b), antimite (Chang *et al.*, 2001a), antitermite and insecticidal activities (Chang *et al.*, 2001b).

Materials and Methods

Essential oils: Pine, clove, orange, rosemary and cedarwood oils were procured from the Government recognized aromatic oil store, Chennai, Tamil Nadu, India. The oils were selected based on medicinal and insecticidal properties.

Mosquito culture: Cx. quinquefasciatus and *Ae. aegypti* larvae collected from various clean stagnant water bodies in and around Chennai, Tamil Nadu, India, were colonized and maintained continuously for generations since 2005 in the laboratory, free of exposure to pathogens, insecticides and repellents. They were maintained at 27±2°C, 75-85% RH with 14:10 L/D photoperiod cycle.

The larvae were fed with dog biscuits, yeast extract in the ratio of 3:1. Pupae were transferred from the trays to a cup, containing tap water and placed in screened cages (45x45x45 cm in dimension), where adults emerged. The adult mosquitoes were reared in the screened cages of 45x45x45 cm in dimension. Adults were continuously provided with 10% sucrose solution in a petridish with cotton wick. On the day of post-emergence, the adult females were deprived of sucrose for 12 hours and then provided with a mouse placed in resting cages overnight for blood feeding. After three days, ovitrap was kept into the cages, and the eggs were collected and transferred to the enamel trays. They were maintained at the same condition for experiment.

Preparation of essential oils: One per cent stock solution was prepared by dissolving 1 ml of plant essential oil in 99 ml of distilled water in a standard flask, from the stock solution 1000, 500, 250, 125 and 62.5 ppm concentrations were prepared.

Larvicidal Bioassay: Larvicidal activity was evaluated by WHO method (1975) with some modifications. Five batches of early fourth instar larvae with 20 numbers in each batch, were taken in 99 ml of tap water and 1.0 ml of the desired essential oil concentration. Tween 80 was used as an emulsifier. Control was set up with Tween 80. Mortality was recorded after 24 hours of the exposure period. Dead larvae were identified when they failed to move after probing with a needle in the siphon or cervical region. Moribund larvae were those incapable of rising to the surface (within a reasonable period of time) or showing the characteristic diving reaction when the water was disturbed. Larvae were also observed for discolouration, unnatural positions and uncoordination movement.

$$\text{Percentage of Mortality} = \frac{\text{No. of Dead Larvae}}{\text{No. of Larvare introduced}} \times 100$$

Results and Discussion

In the present investigation the larvicidal activities of tested oils against fourth instar larvae of *Cx. quinquefasciatus* and *Ae. aegypti* is shown in Tables 1&2. The larval mortality of *Cx. quinquefasciatus* and *Ae. aegypti* was dose dependent. Increasing the concentrations from 125 to 1000 ppm, increased the larval mortality range from 25-100, 29-100, 19-82, 18-72 and 16-83% on *Cx. quinquefasciatus* in pine, clove, orange, rosemary and cedarwood oil. Against *Ae. aegypti* the larval mortality from 125 to 1000 ppm concentrations was 21 -100, 25-100 15-84, 20-77 and 16-75% in pine, clove, orange, rosemary and cedarwood oil, respectively. Pine oil and clove oil showed 100% mortality at 1000 ppm and 25% and 29% mortality at 125 ppm on *Cx. quinquefasciatus*. On *Ae. aegypti*, the larval mortality of pine and clove oil was 100% at 1000 ppm concentration and at 125 ppm, the mortality was 21% and 25% respectively. Maximum larval mortality of 88% and 96% was also observed by Ansari *et al.*, (2005) by Pine oil at 200 ppm concentration against *Cx. quinquefasciatus* and *Ae. aegypti* respectively. In the present investigation, the highest potential larvicidal activity was observed in pine and clove oil at 1000 ppm concentration followed by orange, rosemary and cedarwood oil on

Cx. quinquefasciatus and *Ae. aegypti*. From the results of the present investigation it was observed, the larval mortality increased significantly with an increase the concentration of the test oils. This corroborates with earlier findings of Ansari *et al.* (2005), Cavalcanti *et al.* (2004) and Cheng (2003) on *Cx. quinquefasciatus* and *Ae. aegypti*. In the present study cedarwood oil showed a maximum of 75% of larval mortality at 1000 ppm concentration. In the present work, after exposure to the essential oils, the treated larvae exhibited restlessness, sluggishness, tremors, and convulsions followed by paralysis at the bottom of the bowl. The same abnormal and irregular movement of larvae was also observed earlier by Senthil Nathan (2007). The larvicidal mode of action of essential oils was investigated by Corbet *et al.* (1995), who noted the susceptibility of mosquito larvae and pupae to surface materials entering this tracheal flooding and chemical toxicity.

The results obtained from the present investigation suggest that the pine and clove oil are potential larvicide against *Ae. aegypti* and *Cx. quinquefasciatus*. Moreover, these results could be useful in the search for newer, more selective, and biodegradable larvicidal natural compounds.

Table 1. Larvicidal activity of essential oils against *Culex quinquefasciatus*

S. No	Name of the Oil	% of mortality				
		Control withT-80	1000ppm	500 ppm	250 ppm	125 ppm
1	Pine oil	00±00	100±00	78±0.89	56±0.83	25±1.00
2	Clove	00±00	100±00	72±1.14	51±1.09	29±0.83
3	Orange	00±00	82±1.14	64±0.83	47±0.89	19±0.83
4	Rosemary	00±00	72±0.89	57±1.14	39±0.83	18±1.14
5	Cedarwood	00±00	83±1.14	39±0.83	32±0.89	16±0.83

Table 2. Larvicidal activity of essential oils against *Aedes aegypti*

S. No	Name of the Oil	% of mortality				
		Control withT-80	1000 ppm	500 ppm	250 ppm	125 ppm
1	Pine oil	00±00	100±00	67±0.89	52±1.14	21±1.14
2	Clove oil	00±00	100±00	74±1.92	56±0.83	25±1.00
3	Orange oil	00±00	84±0.83	56±1.30	41±0.83	15±0.70
4	Rosemary	00±00	77±1.14	48±1.14	36±0.83	20±0.70
5	Cedarwood	00±00	75±0.70	51±0.83	35±1.00	16±0.83

Reference

Ahamed, S.S., 1994, Human filariasis in South and South-East Asia. A brief appraisal of our present knowledge. *Proc. Pakistan congr. Zool.,* 14, 1-23.

Ansari, M.A., Vasudevan, P., Tandon, M. and Razdan, R.K., 2000, Larvicidal and mosquito repellent action of peppermint (*Mentha piperita*) oil. *Bioresource Technol.*, 71, 267-71.

Araujo, E.C.C., Silveira, E.R. Lima, M.A.S, Neto, M.A., Andrade, I. and Lima, M.A.A., 2003. Insecticidal activity and chemical composition of volatile oils from *Hyptis martiusii* benth. *J. Agric. Food Chem.*, 51, 3760-3762.

Birley, M.H., 1993, A historical review of malaria, kala-azar and filariasis in Bangladesh in relation to the flood action plan. *Ann. Trop. Med. Parasitol.*, 87, 319-334.

Brown A.W.A., 1986, Insecticide resistance in mosquitoes: pragmatic review. *J. Am. Mosq. Control Assoc. 2:* 123–40.

Carvalho, A.F.U., Melo, V.M.M. Craveiro, A.A. Machado, M.I.L., Bantim, M.B. and Rabelo, E.F., 2003, Larvicidal activity of the essential from *Lippia sidoides* Cham. against *Aedes aegypti* L. *Mem. Inst. Oswaldo Cruz,* 98, 569 – 571.

Cavalcanti, E.S.B., de Morais, S.M., Lima, M.A.A. and Santana, E.W.P., 2004, Larvicidal activity of the essential oil from brazilian plants against *Aedes aegypti*. *Mem Inst. Oswaldo Cruz.,* 99(5), 541-544.

Cetin, H. and Yanikoglu, A., 2006, A study of the larvicidal activity of *Origanum* (Labiatae) species from southwest Turkey. *Journal of Vector Ecology,* 31(1), 118 – 122.

Chang, S.T., Chen, P.F. and Chang, S.C., 2000b, Antibacterial activity of essential oils and extracts from Taiwania (*Taiwania cryptomerioides* Hayata). *Quart. J. Chin. For*., 33(1), 119 – 125.

Chang, S.T., Chen, P.F., Wang, S.Y. and Wu, C.L., 2001a, Antimite activity of essential oils and their constituents from *Taiwania cryptomerioides. J. Med. Entomol.,* 38(3), 455-457.

Chang, S.T., Cheng, S.S. and Wang, S.Y., 2001b, Antitermitic activity of essential oils and their components from (Taiwania) *Taiwania cryptomerioides*. *J. Chem. Ecol*., 27(4), 717-724.

Chang, S.T., Wang, S.Y., Wu, C.L., Chen, P.F. and Kur, Y.H., 2000a, Comparisons of the antifungal activities of cadinane skeletal sesquiterpeniods from Taiwania (*Taiwania cryptomerioides* Hayata) heartwood. *Holzforschung.,* 54(3), 241-245.

Cheng, S.S. Chang, H.T., Chang, S.T, Tsai, K.H. and Chen, W.J., 2003, Bioactivity of selected plant essential oils against the yellow fever mosquito *Aedes aegypti* larvae. *Bioresource Technol,* 89, 99 - 102.

Kumar, A and Dutta, G.P., 1987, Indigenous plant oils as larvicidal agent against *Anopheles* mosquitoes. *Curr. Sci.,* 56, 959 – 960.

Pailey, K.P., Hoti, S.L., Manonmani, A.M. and Balaraman, K., 1995, Longevity and migration of *Wuchereria bancrofti* infective larvae and their distribution pattern in relation to the resting and feeding behaviour of the vector mosquito, *Cx. quinquefasciatus*. *Ann. Trop. Med. Parasitol*., 89, 39-47.

Ravi Kiran, S. and Sita Devi, P., 2007, Evaluation of mosquitocidal activity of essential oil and sesquiterpenes from leaves of *Chloroxylon swietenia* DC. *Parasitol Res*.,101,413-418.

Sukumar, K., Perich, M.J. and Booba, L.R., 1991, Botanical derivatives in mosquito control: a review. *J. Am. Mosq. control Assoc.* 7, 210-237.

Tare, V., Deshpande, S., Tueni, M., Taoubi, K., Nadar, N.A. and Mrad, A., 2004. Suscepbility of two different strains of Aedes aegypti (Dipetera :Culicidae) to plant oils. J. Econ. Entomol. 97, 1734 - 1736.

World Health Organisation, 1975, Instruction for determining the susceptibility or resistance of mosquito larvae to insecticides, mimeographed document. WHO/VBC/1975:583.

World Health Organisation, 1996, Report of the WHO informal consultation of the evaluation and testing of insecticides. WHO, Geneva, pp 9, 32 – 36, 50 – 52.

World Health Organisation, 1997, UNDP/World Bank/ WHO Special Programme for Research and Training in Tropical Diseases (TDR), Tropical Disease Research: Progress 1995-96. Thirteenth Programme Report of the UNDP/World Bank/WHO Special Programme for Research and Training in Tropical Diseases, Geneva, Switzerland. 141pp.

Bacterial mosquito biolarvicides as an alternative to chemical mosquito control

P. R. Sharma,* R. Parshad*, U. Jamwal* and I. Baruah**

The synthetic chemical insecticides for mosquito vector control in recent decades have caused environmental, health and aqueous ecosystems pollution and insecticide resistance in many mosquito species. Alternative biological larvicides are becoming more popular in mosquito management programme the world over because of safety to environment, human beings, animals and other non-target organisms. *Bacillus sphaericus* (Bs) and *Bacillus thuringiensis* serovar *israelensis* (Bti) have been widely used in mosquito control programmes. An indigenous microbial strain has been isolated and identified as *Bacillus sphaericus*. The LC_{50} and LC_{90} of this microbial strain in *Culex quinquefasciatus* are 0.022 and 0.045mg/l and in *Aedes aegypti* are 0.707 and 1.414mg/l. Simulated field trials using indigenous strain at 0.05mg/l showed 100% larval mortality in *C. quinquefasciatus* after 24 to 48 h. The field trials using our indigenous Bs strain showed 100% larval mortality in *C. quinquefasciatus* after 24 to 48 h. The liquid formulation of above strain when stored at 4 - 25°C for six months resulted in no loss in bioefficacy. *Bacillus thuringiensis* serovar *israelensis* (1884) and *B. sphaericus* (1593) were also grown and taken as control. *Bacillus thuringiensis* serovar *israelensis* (1884) bioefficacy in *Aedes aegypti* and in *A. albopictus* was LC_{50} and LC_{90} 0.015 and 0.040mg/l and in *C. quinquefasciatus was* 0.016 and 0.040mg/l . Bioefficacy of *B. sphaericus* (1593) showed almost similar LC_{50} and LC_{90} values in *C. quinquefasciatus* and in *A. aegypti* as our indigenous strain. The morphology and ultrastructure of spores of both *B. sphaericus* and *B. thuringiensis* serovar *israelensis* has been studied by light and electron microscopy. All the above strains of *Bacilli* sp. are being grown on a large scale at IIIM Jammu for mosquito larvae control.

Introduction

Mosquito-borne diseases, such as malaria, filariasis, dengue, Japanese encephalitis, etc. cause millions of deaths every year. Repeated use of synthetic insecticides for mosquito control has disrupted natural biological control systems and led to resurgence in mosquito populations. The drug resistance (Trape 2001) and increasing insecticide resistance (Wattal *et al.,* 1981; Hargreaves *et al.,* 2000; Lima *et al.,* 2003) have stimulated the use of alternative larvicides. The use of synthetic insecticides has also caused environmental pollution resulting in bioamplification of food chain contamination and harmful effects on beneficial insects. In recent years, the search for new insecticides that are easily biodegradable and do not have any ill effect on non target organisms remains the top priority (Redwane *et al.,* 2002). With the discovery of the mosquitocidal strains of *Bacillus thuringiensis* var. *israelensis* de Barjac (Bti) and *Bacillus sphaericus* Neide (Bs) during the mid-1970s, larvicides have become available that are highly effective, yet selective in action (Charles and Nielsen-LeRoux, 2000), and therefore environmentally safe to non-target organisms, as well as for human exposure (WHO, 1999). Larviciding and source reduction have a major advantage in that they control mosquitoes before they disperse and transmit disease (Killeen *et al.,* 2002). Malaria was successfully eradicated in the United States, Europe and the Middle East by rigorous larval control measures, integrated with any other

* Indian Institute of Integrative Medicines (CSIR), Canal Road Jammu Tawi (J&K) State- 180001

** Defense Research Laboratory (DRDO), Tezpur Assam, India-784001 available tool (Hays, 2000). The microbial insecticides are selective in their action on mosquitoes and blackflies. The *Bacillus* products are characterized by the ease of handling, cost-effectiveness and capability of being produced locally. Furthermore, application of larvicides does not require expensive equipment, and can be organized locally (Becker 1992). Bs is a aerobic, spore forming bacterium that produces a highly mosquitocidal binary toxin (Kalfon *et al.,* 1983) and is also ideal for commercial mosquito control, where its residual activity is important for controlling mosquito larvae that breed in polluted water. However, the target spectrum of Bs is strictly limited to mosquitoes, primarily *Culex* and certain *Anopheles* species (Delécluse *et al.*, 2000). The first pathogenic strains of Bs were isolated from *Culiseta incidens* larvae (Kellen *et al.,* 1965). Currently many toxic strains are known and many studies have been performed on strains 1593 and 2362 isolated respectively in Indonesia (Singer, 1973) and Nigeria (Weiser, 1984). This organism does not infect non-target invertebrates (including bees) or cold-blooded vertebrates and it is also innocuous to mammals in laboratory tests (Davidson, 1985). The World Health Organization (WHO) recommends the utilization of this bacterium in public health programs (WHO, 1985). Almost all mosquito species from the genus *Culex*, some of them vectors of filariasis, are susceptible to it. *B. sphaericus* activity is because of a presence of different kinds of protein toxins that differ both in their composition and time of synthesis. The parasporal crystal of Bs, which is produced during the sporulation phase is a binary toxin composed of two components designated as P51 and P42 on the basis of their molecular masses (Baumann *et al.,*, 1987), and now termed Bin B and Bin A, respectively. Bs is a very promising microorganism and several laboratories around the world are looking for new strains that may be able to produce novel toxins or which may be more adapted to local environmental conditions, in order to have better effect in the field and which could be used in resistance management. *Bti* a spore forming bacterium has proved to be a potent control for mosquito larvae since its discovery by Goldberg and Margalit (1977). The great advantage of Bti over all other larvicides is the low probability of developing resistance (Charles and Nielsen-LeRoux 2000). In the past decades, considerable research had been carried out on development and field-testing of both of these biological control agents against disease spreading mosquitoes

The present work describes the larvicidal activity of indigenous strain of Bs both in the laboratory as well as in field conditions. The larvicidal activity of strain 1593 of Bs and 1884 of BTi has also been studied. The morphology of spores of all the three strains has been studied by light and scanning electron microscopy (SEM).

Materials and methods

Phase contrast and scanning electron microscopy

The lypholised powder of spores of Bs and Bti was dissolved in distilled water and studied under the phase contrast microscope . The photography was done using Olympus digital Camera C-4000 attached to the Olympus research microscope (Vanox). For SEM studies the spore-crystal suspension in distilled water of all the three strains were air-dried on cover slips, mounted on stubs, coated first with carbon in a JEOL-JEE-4X vacuum evaporator, and then with gold in a Polaron sputter coater. Finally, the samples were observed in a JEOL-100CXII electron microscope with ASID operating at 40 kV. The photography was done using black and white 120 roll film (Nova).

Maintenance of mosquito cultures

All laboratory experiments were carried out with third instar larvae of laboratory-reared, *Culex quinquefasciatus, Aedes aegypti and A. albopictus. C. quinquefasciatus and A. aegypti* were collected from Jammu and colonies were established. *A. albopictus* was procured from DRDO Tejpur, Assam. The cultures were maintained at 26 ± 2°C with 70 to 75% relative humidity and with a photoperiod consisting of 14 h of light and 10 h of darkness at IIIM, Jammu. Larvae were reared in enamel trays filled with dechlorinated tap water and fed ground dog biscuits and yeast at a ratio of 3 : 2 in water as a nutrient source. The adults mosquitoes were fed with a 10% honey solution through cotton pads. The female mosquitoes were fed on mice to achieve oviposition.

Bacillus sp. cultures

The indigenous culture of *Bacillus sphaericus* (Bs) was isolated at DRDO, Tejpur Assam and identified as Bs by 16S ribosomal RNA technique at IIIM, Jammu. The cultures of Bs (1593) and Bti (1884) were kindly provided by Dr Christina Nielsen Le-Roux, Institute Pasteur, Paris, France.

Bioassays

Bs and Bti *spores* lyophilized powders were used as the toxin source. A liquid formulation prepared from indigenous strain of Bs, was also evaluated both in laboratory as well as in field trials. The preparation of stock solution from these bacterial samples and bioassays were conducted as described earlier (WHO, 2005). Test media were prepared by adding appropriate volumes of Bs or Bti in 200 ml of water in glass beakers in triplicate and 10 early third instar larvae from all the three mosquito species were introduced separately in each of the test concentrations. For each strain, preliminary tests were conducted in three replicates of six different concentrations, and controls in order to determine their respective active ranges. While food supplement was provided for larvae under Bs treatment, no food was provided for Bti treated larvae (WHO, 2005). The bioassays were repeated three times on different days. The lethal concentrations were expressed in miligram of spore toxins per litre. The LC_{50} and LC_{90} values were calculated using regression analysis.

Simulated field trials

Prior to open field trials, the simulated field trials with indigenous strain Bs were conducted between March –April 2007 to determine the minimum effective dosages, according to standard testing procedures (WHO 2005). For comparison, strain 1593 of Bs was evaluated simultaneously. Following the laboratory studies, the needed test aliquots were added to 8 l of dechlorinated tap water dispensed in 10 l plastic containers in triplicate, which contained 100 larvae each. Moribund larvae were considered dead and included in the analyses. The bioassays were run in two different concentrations of 0.05 and 0.1mg/l and controls and replicated thrice, on different occasions. All trials were conducted at ambient temperatures that ranged from 23 to 30°C. Data from all replicates were pooled and % mortality was calculated.

Open field trials

Open field trials with the indigenous strain of Bs were conducted between May-June 2007 (dry season) on *C. quinquefasciatus* at three different locations in Jammu. The field trials were done on the basis of total water volume in pond/field followed by spore application at 0.1 to 0.2 mg/l (solution made in water).

Results

General description of spores from the phase contrast and SEM study

Preliminary observations with phase contrast microscopy indicated that, spores of Bs appeared spherical to ovoid in shape and that of Bti were square and flat (Figure 1). This morphology was corroborated by SEM (Figure 2), by which the surface ultrastructure of spores and crystals was clearly seen. The spores of Bs are 1.25 to 1.85 μm in length and 1.1 to 1.5 μm in width whereas those of Bti are 1.55 to 2 μm in length and 0.85 to 1.16 μm in width.

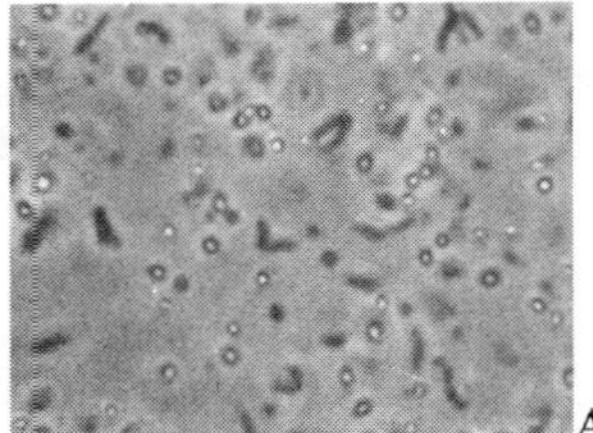

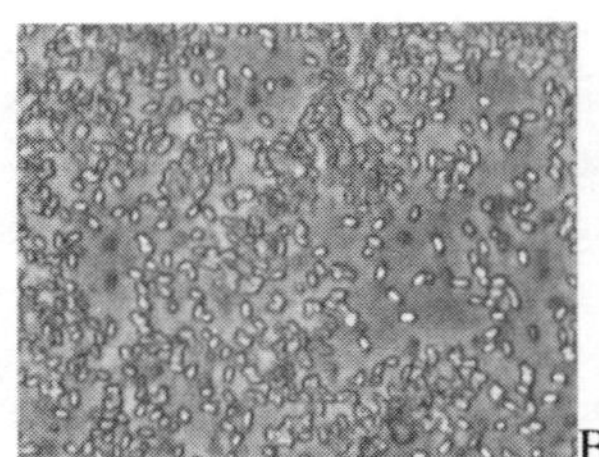

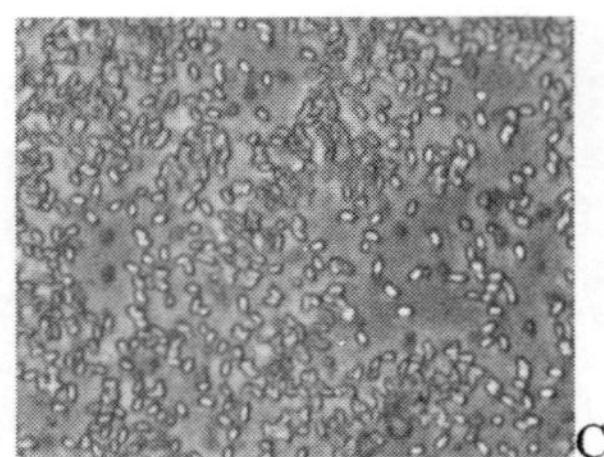

Fig. 1. Phase contrast microscopy of spores of *Bacillus sphaericus* (Bs) and *B. thuringiensis* serovar *israelensis*(Bti) . A- Indigenous strain of Bs, B- strain 1593 of Bs and C- Bti. Original magnification, 270X

Maintenance of mosquito cultures

The different stages of *C. quinquefasciatus* and *A. aegypti* were collected from Jammu and their cultures were raised in the laboratory. *C. quinquefasciatus* is very common in Jammu from February to October whereas *A. aegypti* is very common during July to October especially during rainy season. *A. albopictus* was procured from DRDO, Tejpur, Assam.

Laboratory studies

The comparative toxicities of *Bs* samples both indigenous and 1593 and *Bti* are shown in Table 1. The LC_{50} and LC_{90} values for Bs indigenous strain and Bs strain 1593 against *C. quinquefasciatus* larvae were respectively 0.022 and 0.045 mg/l and 0.022 and 0.044mg/l. The liquid formulation developed from indigenous strain also showed similar values in laboratory studies. Both the strains however showed LC_{50} and LC_{90} of 0.707 and 1.414 mg/l respectively against *A. aegypti.* The *Bti* showed LC_{50} and LC_{90} of 0.016 and 0.040 mg/l in *C. quinquefasciatus.* In *A. aegypti* and *A. albopictus* the LC_{50} and LC_{90} values of *Bti* were observed to be and 0.015 and 0.040 mg/l.

Table 1. LC_{50} and LC_{90} of indigenous stain of *B. sphaericus* and two commercial strains (*B. sphaericus* 1593 and *B. thuringiensis* servopar *israelensis* strain 1884) against mosquito larvae. Exposure time: 48 h in case of *B. sphaericus* and 24 h in case of *B. thuringiensis* servopar *israelensis*. Control mortality: Nil

S. No.	Bacterial strain	Mosquito species	LC_{50} (mg/l)	LC_{90} (mg/l)
1	*B. sphaericus* Indigenous strain	*Cx. quinquefasciatus*	0.022	0.045
		A. aegypti	0.707	1.414
2	*B. sphaericus* strain 1593	*Cx. quinquefasciatus*	0.022	0.044
		A. aegypti	0.707	1.414
3	*B. thuringiensis* servopar *israelensis* strain 1884	*Cx. quinquefasciatus*	0.016	0.040
		A. xaegypti	0.015	0.040
		A. albopictus		

Simulated field trials

Simulated field trials were conducted on 3rd instar larvae of *C. quinquefasciatus.* After 24 h treatment, the percentage mortality at concentrations of 0.05 and 0.1mg/l, was respectively 81.66, 83.33; 98.33, 95, in indigenous and 1593 strains of *Bacillus sphaericus.* After 48 h treatment, the percentage mortality at concentrations of 0.05 and 0.1mg/l, was recorded as 100 in both the strains of Bs (Table 2).

Open field trials

Open field trials on *C. quinquefasciatus* using liquid formulation of Bs at 0.2 mg/l of the indigenous strain showed 100% mortality after 48 h.

Table 2. Simulated field trial

Bacterial strain	Dose mg/l	No. of Larvae Treated	% of Larvae Dead after 24 h (no.)	% of Larvae Dead after 48h (no.)
B. sphaericus, Indigenous strain	0.05	300	81.66(245)	100(300)
	0.1	300	98.33(295)	100(300)
B. sphaericus, strain 1593	0.05	300	83.33(250)	100(300)
	0.1	300	95(285)	100(300)
Control	0	300	Nil	Nil

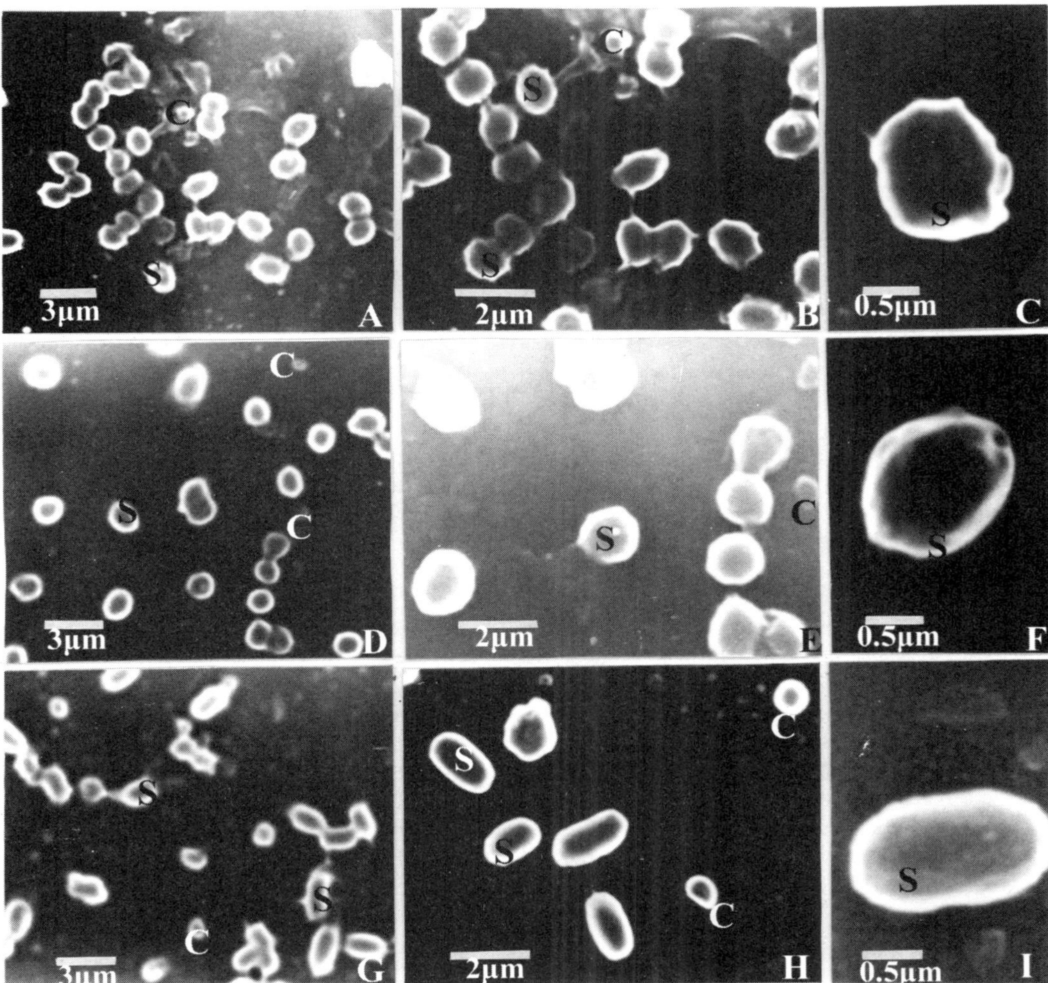

Fig. 2. Scanning electron microscopy of spores (S) and crystals (C) of *Bacillus* sp. A-C, Indigenous strain of *B. sphaericus*, D-F, strain 1593 of *B. sphaericus*, G-I, strain 1884 of *B. thuringiensis* serovar *israelensis* (Bti).

Discussion

In recent years, environmental friendly and easily biodegradable insecticides have gained renewed importance. Microbial larvicides have several advantages over other mosquito control agents, not only in high efficacy but also in environmental safety and safety for human consumption, for instance when applied in drinking water (WHO 1999), which makes them powerful vector control tools that are gaining more ground for disease control in Africa and other parts of the tropics. It is imperative therefore that new formulations, which are suitable for use in tropical settings, are developed (Barbazan *et al.,* 1998) and evaluated. Bs is an important microbial agent for mosquito control, yet its practical field life is limited by the propensity for mosquitoes to evolve resistance. Various strategies have been proposed to manage this problem. Recent evidence also suggests that increased toxin diversity may be the more effective approach. Toxin diversity could be increased by formulating insecticides to contain both Bs and *Bti* or through recombinant bacteria engineered to express toxins from both microbes (Federici *et al.,*, 2003; Zahiri and Mulla, 2003). Although a cautious approach to adopting novel insecticides is needed, they may provide the effective tools needed for controlling mosquito populations.

This article is the first report of *Bs* and *Bti* studies on the larvae of *C. quinquefasciatus* and *A. aegypti* from our area and the data obtained may contribute for better control measures using these bioinsecticides against mosquito larvae.

References

Barbazan, P., Baldet, T., Darriet, F., Escaffre, H., Djoda, D.H. and Hougard, J.M., 1998. Impact of treatments with *Bacillus sphaericus* on *Anopheles* populations and the transmission of malaria in Maroua, a large city in a savannah region of Cameroon*; J. Am. Mosquito Cont. Ass.*, 14, 33–39.

Baumann, P., Baumann, L., Bowditch, L. and Broadwell, A.H., 1987. Cloning of the gene for the larvicidal toxin of *B. sphaericus* 2362, evidence for a family of related sequences; *J. Bacteriol.,* 163, 738–747.

Becker, N., 1992. Community participation in the operational use of microbial control agents in mosquito control programmes; *Bull. Soc. Vec. Ecol*., 17, 114–118.

Charles, J.F. and Nielsen-Le Roux, C., 2000. Mosquitocidal bacterial toxins: diversity, mode of action and resistance phenomena; *Memorias Do Instituto Oswaldo Cruz,* 95, 201–206

Davidson, E.W., 1985. *Bacillus sphaericus* as a microbial control agent for mosquito larvae; in *Integrated Mosquito Control Methodologies* (ed) M Laird and J Miles (USA: Academic Press) vol. 2, pp 213–226.

Delécluse, A., Juárez-Pérez, V. and Berry, C., 2000. Vector-active toxins: Structure and diversity. In: Charles, J.-F., Delécluse, A., Nielsen-LeRoux, C. (Eds.), Entomopathogenic Bacteria: From Laboratory to Field Application.Kluwer Academic Publishers, Dordrecht, The Netherlands.

Federici, B.A., Park, H.W., Bideshi, D.K., Wirth, M.C. and Johnson, J.J., 2003. Recombinant bacteria for mosquito control. *J. Exp. Biol*., 206, 3877–3885.

Goldberg, L.J. and Margalit, J., 1977. A bacterial spore demonstrating rapid larvicidal activity against *Anopheles sergentii, Uranotaenia unguiculata* , *Culex univitattus* , *Aedes aegypti* and *Culex pipiens;*. *Mosq. News,* 37, 355- 358.

Hargreaves, K., Koekemoer, L.L., Brooke, B.D., Hunt, R.H., Mthembu, J. and Coetzee, M., 2000. *Anopheles funestus* resistant to pyrethroid insecticides in South Africa; *Med. & Vet. Ent*., 14, 181–189.

Hays, C.W., 2000. The United States Army and malaria control in World War II; *Parasitologia,* 42, 47–52.

Kalfon, A., Larget-Thiery, I., Charles, J.F. and de Barjac, H., 1983. Growth, sporulation and larvicidal activity of *Bacillus sphaericus; Euro. J. Appl. Microbiol. Biotechnol.,* 18, 168–173.

Kellen, W.R., Clark, T.B., Lindegren, J.E., Ho, B.C., Rogoff, M.H. and Singer, S., 1965. *Bacillus sphaericus* Neide as pathogen of mosquitoes; *J. Invertebr. Pathol*., 7, 442–448.

Killeen, G.F., Fillinger, U., Kiche, I., Gouagna, L.C. and Knols, B.G.J., 2002. Eradication of *Anopheles gambiae* from Brazil: lessons for malaria control in Africa; *Lancet Infectious Diseases,* 2, 618–627.

Lima, J.B.P., Pereira da Cunha, M., Silva-Jr, R.C.S. et a!., 2003. Resistance of *Aedes aegypti* to organophosphates in several municipalities in the states of Rio de Janeiro and Espírito Santo, Brazil; *Am J Trop Med Hyg.,* 68, 329-333.

Redwane, A., Lazrek, H.B., Bouallam, S., Markouk, M., Amarouch, H. and Jana, M. 2002. Larvicidal activity of extracts from *Querus lusitania* var. infectoria galls (oliv); *J. Ethnopharm.*, 79, 261-263.

Singer, S., 1973. Insecticidal activity of recent bacterial isolates and their toxins against mosquito larvae; *Nature,* 244 110.

Trape, J.F., 2001. The public health impact of chloroquine resistance in Africa; *Am. J. Trop. Med. & Hyg.*, 64, 12–17.

Wattal, B.L., Joshi, G.C. and Das, M., 1981. Role of agricultural insecticides in precipitation vector resistance; *J Comm. Dis.,* 13, 71-73.

Weiser, J.A., 1984. A mosquito-virulent *Bacillus sphaericus* in adult *Simulium damnosum* from northern Nigeria; *Zentralbatt Mikrobiologie,* 139, 57–60.

WHO, 1985. Informal Consultation on the Development of *Bacillus sphaericus* as a Microbial Larvicide; *Special Programme for Research and Training in tropical Diseases*, TDR/BCV/ PHAERICUS/85.3. Geneva.

WHO, 1999. International programme on chemical safety (IPCS): microbial pest control agent *Bacillus thuringiensis*. *Environ. Hea. Crit,* 217, 1–105.

WHO, 2005. Guidelines for laboratory and field testing of mosquito larvicides. *WHO/CDS/WHOPES/ CDPP/2005,*13.

Zahiri, N.S. and Mulla, M.S., 2003. Susceptibility profile of *Culex quinquefasciatus* (Diptera: Culicidae) to *Bacillus sphaericus* on selection with rotation and mixture of *Bacillus sphaericus* and *Bacills thuringiensis israelensis*. *J. Med. Entomol.*, 40, 672–677.

Population Dynamics of the Oriental Fruit Fly, *Bactrocera dorsalis* Hendel in Relation to Hosts and Abiotic Factors

S. Manisegaran,* S. Balsubramaniyan,* S. Natarajan* and S. Sithanandam**

Studies conducted on the population dynamics of the oriental fruit fly, *Bactrocera dorsalis* Hendel, in two mango orchards of Horticultural College and Research Institute, Periyakulam from January 2006 to December 2007 using methyl eugenol traps revealed that three distinct population peaks in March/ April, May /June and September during both years. These peaks coincided with ripening of guava and mango fruits which are the major hosts or *B. dorsalis*. Among abiotic factors, day degrees, the maximum and minimum temperature showed a significant positive correlation while the rain fall and relative humidity had a significant negative correlation with trap catches of adult flies. This indicates that abiotic factor played a major role in the population dynamics of *B. dorsalis*. The influence of placement and position of traps in the orchards are also discussed in this paper.

Introduction

Historical records show that several invasive fruit flies have been introduced into countries other than their place of origin, mainly through infested commodities (fruits) Kapoor (1993). For example, some *Bactrocera* spp. have established in Hawwaii, French Guiana and Surinam; *Ceratitis capitata* has spread from Africa to many parts of the World through fruit movements (White and Elson-Harris 1994). Recently, Varghese *et al.* (2006) and Drew *et.al.* (2005) reported introduction of *B. invadens* on the West coast of South India from Goa to Kanyakumari and Eastern Africa respectively. The oriental fruit fly, *Bactrocera dorsalis* Hendel, is one of the most destructive pests of mango (Verghese *et.al.*, 2004). The fruit fly adults cause damage to the mango and guava fruits when they lay their eggs causing blemishes and discolouration in the fruits. The damaged fruits become unfit for human consumption. This fly is active throughout the year in South India. It has been suggested to use methyl eugenol traps for estimating the male fly population. Information on seasonal population fluctuations and peak period of *B. dorsalis* activities are essential for development of suitable control strategies. The present investigation reports population fluctuations of *B. dorsalis* in relation to hosts and various abiotic factors using methyl eugenol traps.

Materials and methods

Investigations were carried out on fruit fly populations in two mango orchards at the Horticultural College and Research Institute, Periyakulam from January 2006 to December 2007. A total of 13 traps supplied by Sun Agro Industries India, Chennai were hung on mango trees in different directions *viz.,* North, South, East, West, Centre, Northeast, Southeast, Northwest, Southwest, Northeast edge, Southeast edge, Northwest edge, Southwest edge and in the four corners of both the orchards. Fruit fly catches were recoded weekly and the solution changed once in three months for two years. Weekly average of maximum and minimum temperature relative humidity and rainfall were recorded The

* Horticultural College and Research Institute Tamil Nadu Agricultural University, Periyakulam
** Sun Agro Research Centre, Porur, Chennai – 600 116

daily mean temperature figures were converted to day degree or thermal unit (T.U.) using a base temperature of 5°C.(Aliniazee 1976). The correlation coefficients (r) were then calculated between weekly trap catches and weekly average of day degrees.

$$T.U = \frac{\text{Maximum} + \text{minimum temperature}}{2} - 5$$

Results and discussion

Population fluctuations in relation to hosts

During 2006, the population fluctuations of *B. dorsalis* started increasing in the fourth week of February and peaked during the fourth week of March (509 flies in orchard 1) and in the first week of April 535 in orchard 2). The second population peak was noticed in the second week of May (420 flies) in orchard 1 and in the second of week June (433 flies) in orchard 2 respectively .The third peak was observed during the first week of September in orchard 1(408 flies) and in the first week of October in orchard 2 (376 flies). In 2007, the first peak in the population of *B. dorsalis* was noticed during the first week of March (638 in orchard 1) and in the first week of April (653 in orchard 2). Whereas, the second population peak was observed in the second week of May (480 flies) and June (456) flies in orchard 1 and orchard 2 respectively. A third peak was noticed in the first week of September (559 flies in orchard 1 and in the first week of October (385 in orchard 2).

From the 2006- 2007 data, it is evident that there are three distinct peaks in *B. dorsalis,* the first in March/ April, the second in May/ June and the Third in September/October. These peaks coincide with ripening period of guava in March/ April and September/October, and mango in May/June. These findings are in agreement with the reports of Narayanan and Batra (1960), Prasad and Bagle (1978) and Yao and Lee (1978) who reported that mango and guava as the major hosts of *B. dorsalis.*

Population fluctuations in relation to abiotic factors

The influence of temperature, relative humidity rainfall and day degrees on the number of *B. dorsalis* caught in traps is presented in table 1.

Temperature (Thermal units): There was a significant positive correlation between maximum temperature and number of *B. dorsalis* captured. The weekly trap catches of flies were significantly and positively correlated with day degrees (thermal units). The maximum number of flies emerged (and were trapped) during the fourth week of March 2006 when mean day degrees were 23.09° C Similarly highest numbers of flies were trapped during third week of March 2007 when average day degree were 22. 89 – 25 93 ° C. Generally, higher numbers of flies were caught during weeks of highest maximum temperature. Similar trend was noticed with regard to minimum temperature. These results suggest that temperature plays an important role in regulating *B. dorsalis* abundance. Aliniazee (1976) has reported that peak abundance of B. dorsalis at thermal units of 22- 24 ° C and emergence of various levels of Western cherry fruit fly, *Rhadoletis indifferens* Curran.

Relative humidity: Relative humidity affected the abundance of *B. dorsalis.* Weekly catches of flies had a low and negative and no significant correlation with the weekly average of relative humidity (Table 1). In generally, maximum numbers of *B. dorsalis* were captured during the weeks of minimum relative humidity in both the years. This clearly indicates that relative humidity also plays an important role in regulating the *B. dorsalis* population.

Rainfall: There was a significant negative correlation between weekly rainfall and numbers of *B. dorsalis* except orchard 2 during 2006(Table1). The fly population was high during weeks of low rainfall. This is perhaps rainfall may prevent both the emergence and flight of adult flies. Christenson and Foote (1960) observed that *B. dorsalis* become inactive during moderate to heavy rainfall.

Table 1.Coefficient of correlation between weekly trap catches of *B. dorsalis* and abiotic factors

Abiotic factors	Coefficient of correlation (r)			
	2006		2007	
	Orchard 1	Orchard 2	Orchard 1	Orchard 2
Day degrees(thermal units)	0.6772**	0.61360	0.6122	0.5491
Maximum temperature	0.5321	0.5926	0.6237	0.5704
Minimum temperature	0.3157	0.3490	0.3640	0.3351
Relative humidity	-0.0758 Ns	-0.0627	-0.0641	-0.0145
Rainfall	-0.2641*	-0.2728	-0.2946*	-0.4841**

**Significant at 1% level *Significant at 5 % level; Ns Non significant

Influence of trap placement on capture of *B. dorsalis*

The table 2 revealed that the numbers of *B. dorsalis* captured during 2006-2007 in the traps placed at different location had a significant positive influence on the capture of *B. dorsalis* in orchard 1 and 2 in both the years. However, traps placed on the edges of orchards caught higher number of flies than traps placed in the centre of the orchards (Table 3).This may be attributed due easy recognition of tarp colour in the edges than at the centre. It is relevant to quote at this juncture that Jiji *et al* (2005) and Jalaluddin *et al*, (1996) observed that flies were attracted to an increasing scale of " redness" with a peak at Yellow /orange rather than red itself. There was a highly significant positive correlation between monthly trap catches in the centre verse the edges of the orchards in both the years. This suggests that it is essential to fix the traps in edges of the orchard for efficient trapping of *B. dorsalis.*

Table 2.Effect of trap placement in different locations of the orchard on capture of *B. dorsalis*

Locations	Number of fruit flies trapped per year per trap	
	2006	2007
North	815 (1215) *	1602 (917) *
South	1480 (1525)	1507 (1336)
East	1417 (1302)	1120 (1312)
West	1020 (992)	1268 (998)
Centre	839 (817)	984 (829)
Northeast	768 (929)	1213 (1332)
Southeast	1125 (1019)	1321 (1123)
Northwest	802 (1009)	1262 (913)
Southwest	1066 (1141)	1302 (1207)
Northeast edge	1488 (1796)	1365 (1574)
Southeast edge	1541 (1473)	1506 (1402)
North west edge	1332 (1758)	1423 (1231)
Southwest edge	982 (1432)	1578 (989)

* Figures in parentheses are the number of *B. dorsalis* captured in orchard 2 while other figures are for orchard 1

Table 3. Effect of trap placement in centre and edges of orchards on capture of *B. dorsalis*

Months	Number of *B. dorsalis* per trap									
	2006					2007				
	Centre	North east edge	South east edge	North west edge	South west edge	Centre	North east edge	South east edge	North west edge	South west edge
January	12	49	30	40	26	16	16	39	19	30
	(15)*	(52)*	(19)*	(46)*	(26)*	(21)*	(31)*	(42)*	(38)*	(41)*
February	81	110	142	102	107	95	49	62	41	46
	(23)	(96)	(85)	(91)	(40)	(31)	(52)	(58)	(46)	(38)
March	186	276	318	432	215	248	319	261	323	385
	(191)	(452)	(285)	(456)	(234)	(184)	(312)	(339)	(482)	(210)
April	162	310	407	360	203	139	148	253	123	267
	(158)	(251)	(263)	(249)	(199)	(113)	(242)	(207)	(135)	(92)
May	147	305	216	35	141	148	200	250	207	203
	(134)	(268)	(248)	(168)	(293)	(123)	(264)	(198)	(192)	(149)
June	121	204	168	72	120	102	119	152	207	181
	(102	(183)	(192)	(121)	(301)	(49)	(121)	(92)	(182)	(83)
July	12	68	29	68	31	21	102	89	146	76
	(19)	(118)	(95)	(130)	(28)	(9)	(23)	(36)	(93)	(33)
August	18	52	43	62	30	14	57	61	70	36
	(26)	(81)	(69)	(117)	(52)	(7)	(32)	(23)	(36)	(27)
Septem.	42	44	65	36	26	92	112	81	42	73
	(36)	(52)	(81)	(92)	(68)	(102)	(103)	(158)	(21)	(48)
October	33	22	34	32	28	52	143	97	28	116
	(43)	(56)	(52)	(81)	(45)	(64)	(201)	(167)	(72)	(108)
Novem.	15	20	48	43	30	32	42	41	62	87
	(28)	(87)	(50)	(99)	(83)	(12)	(128)	(38)	(69)	(18)
Decem.	10	28	41	50	25	28	58	38	55	78
	(42)	(100)	(39)	(102)	(67)	(14)	(92)	(42)	(47)	(52)
Total	839	1588	1541	1332	982	984	1365	1506	1423	1527
	(817)	(1796)	(1473)	(1758)	(1438)	(829)	(1574)	(1420)	(1231)	(989)
	'r'	0.93	0.94	0.74	0.97	'r'	0.93	0.94	0.74	0.97
	Value	(0.93)	(0.96)	(0.75)	(0.90)	Value	(0.93)	(0.96)	(0.75)	(0.90)

* Figures in parentheses are the number of *B. dorsalis* captured in orchard 2 while other figures are for orchard 1

Acknowledgements

The senior author is thankful Dr. V.V Ramamurthy and Dr. Debjani Dey, senior Scientist of Indian Council Agricultural Research, New Delhi for identification of fruit flies.

References

Aliniazee, M.T., 1976. Thermal unit requirements for determining adult emergence of the Western cherry fruit fly (Diptera: Tephritidae) in the Willamette Valley of Oregon. *Environmental Entomology*, 5, 397-404.

Christenson, L.D. and Foote, R.H., 1960. Biology of fruit fly Annual review of Entomology 5, 171-192.

Drew, R.A.I., Tsuruta, K and White. I.M., 2005. A new species of pest fruit fly (Diptera : Tephritidae: Dacinae) from Sri Lanka and Africa. *African Entomology*, 13(1), 149-154.

Jalaluddin, S.M., Natarajan, K., Sadakathulla, S. and Rajukkannu, K., !998. Effect of colour, height and dispenser on the catches of guava

Jiji, T., Thomas, J., Singh, H.S., Jhala, R.C., Patel, R.K., Napoleon, A., Senthilkumar, S., Vidya, C.V., Mohantha, A., Sisodiya, D.B., Joshi,B.K., Stonehouse, J.M., Verghese, A and Mumford, J.D., 2005. *Pest Management in Horticultural Ecosystem* 11(2), 88-90.

Kapoor, V.C., 1993. Indian Fruit flies (Insecta: Diptera: Tephritidae). Oxford and IBH Publishing Company, New Delhi. pp 228.

Lee, H., 1976. Investigation on the oriental fruit fly, Dacus dorsalis Hendel, damaging papaya. *J. Agrl. Res. China.* 25(2), 156-162.

Narayanan, E.S. and Batra, H.N., 1960. Fruit flies and their control. Monograph Indian Council of Agricultural Research, New Delhi pp 68.

Prasad, V.D. and Bagel, B.G., 1978. Population dynamics of oriental fruit fly, *Dacus dorsalis* Hendel, by male annihilation technique. *Bull. Entomo.,* 19, 103-105.

Steiner, L.F ., Mitchell, W.C ., Harria, E.J., Kozuma, T.T and Fuzimoto, M.S., 1965. Oriental fruit fly eradication by male annihilation technique. *J. Econ. Entomol.,,* 58, 961.

Steiner, L.F., 1965. A method of estimating the size of native populations of oriental melon and Mediterranean fruit flies to establish the over flooding rates required for sterile male release. *Journal of Economic Entomology,* 62, 4-7

Verghese, A., Tandon, P.L. and Stonehouse, J., 2004. Economic evaluation of the integrated management of oriental fruit fly, *Bactocrea dorsalis* in mango in India. *Crop Protect. (U.K),* 23, 61-63.

Verghese, A., Sreedevi, K., Nagaraj, D.K and Jayanthimala, B.R. A farmer friendly trap for management of fruit fly *Bactrocera* spp.(Tephritidae: Diptera). *Pest Managt. Horticult. Ecosyst.,* 12(2), 75-84.

White, I.M and Elson-Harris, M.M., 1994. Fruit flies of economic importance: Their identification and Bionomics. CAB International, Australian Centers for International Agricultural Research, pp 601.

Yao, A.L and Lee, W.Y., 1978. A population study of the Oriental fruit fly, *Dacus dorsalis* Hendel (Diptera : Tephritidae) in guava, citrus fruits and wax apple fruits in Northern Taiwan. *Bull. Inst. Zool. Acad. Sinica,* 17(2), 103-108.

Influence of Mealybug infestation on mulberry leaves on the silkworm, *Bombyx mori* (L).

VITTHALRAO B. KHYADE,* SUNANDA V. KHYADE* AND VIVEKANAND V. KHYADE*

The infestation of mulberry crop with mealybugs (*Maconellicoccus hirsutus*) is generally called as Tukra disease. Healthy and Tukra affected mulberry leaves were assessed for their biochemical contents and used for rearing of polyvoltine (PM), bivoltine (NB_4D_2) and crossbreed (PM x NB_4D_2) races of silkworm larvae. Tukra affected mulberry leaves were found with increased proteins, soluble sugars, reducing sugar and amino acids. Cocoons harvested from the group of larvae reared on Tukra affected mulberry leaves were found no more inferior to that of the cocoons of larvae fed with healthy leaves. The silk filament parameters also appeared parallel with that of the cocoons. The study suggests that mealybug affected leaves should not be wasted/ discarded, but utilized for rearing of silkworm larvae.

INTRODUCTION

Biochemical components present in the leaves of mulberry play an important role for successful cocooning (Radha *et al.*, 1978). The quality of nutrition in the larval stage will significantly influence the resulting pupae, adult and silk production (Aftab Ahmed, 1998). Depletion in the bio-chemical components in response to infection/infestation is quite common in mulberry leading to economic losses.

Tukra caused by the pest *Maconellicoccus hirsutus (*Green) (Pseudoeoccidae: Hemiptera) is known to affect the leaf quality and cocoon production (Thangamani and Vivekanandan, 1983; Shree and Umashankar, 1989; Pradipkumar *et al.,* 1992). Mealy bug infestation causes stunted growth of plant and leaf curling. This results in the reduction of leaf yield by about 21-28 percent (Kumar, *et al.,*, 1989). It is believed that the infested leaves are nutritionally inferior and not fit for feeding silkworm as they affect cocoon production related parameters.

Although, few workers have analysed the bio-chemical components of "tukra" leaves and made bio-assay studies (Thangamani and Vivekanandan, 1983), their results are not similar to those obtained by Pradip kumar *et al.,* (1992). There is a concern as to whether "Tukra" leaves can be used for feeding silkworm. Hence an effort was made to analyse the biochemical components of infested leaves and to make bioassay studies with the infested leaves by rearing popular pure and cross breeds of *Bombyx mori (*L).

MATERIAL AND METHOD

Healthy and "Tukra" affected tender leaves of M-5 mulberry variety were collected from the mulberry garden at the farm of Agriculture Development Trust, Malegaon and used for biochemical and bioassay studies. The leaves were dried at 55^{0}C for about 75 hours and powdered by using mortor and pestle. This was used for bio-chemical analysis. Standard methods were followed for estimation of proteins, soluble sugars, reducing sugars, DNS total phenols, amino acids and micronutrients like Fe, Mn an Zn by Di acid digestion method (Shree and Umashankar, 1989).

---

* 'Shrikrupa', Malegaon Colony, Tal, Baramati, Dist. Pune. 413115

Bioassay studies were conducted by feeding healthy and mealy Bug infested leaves separately to bivoltine (NB_4D_2), multivoltine Pure Mysore (PM) and their cross breed PM x NB_4D_2 from the beginning of fourth instar to spinning in four replications of hundred larvae each. Standard rearing practices were followed (Krishnaswami, 1978). The harvested cocoons were subjected for studying parameters like single cocoon weight, shell weight, filament length, denier, renditta, etc. The data were statistically analyzed using student's 't-test'.

Results and Discussion

Biochemical estimation of assay mixture of mealy bug infested mulberry leaves was found deviated from the constituents of healthy leaves. There was increase in some biochemical constituents of mealy bug infested leaves like soluble proteins (143.5 mg/g) and amino acids (103.5 m mole/g) compared to healthy laves (11.5 mg/g and 29.;25 m mole/g respectively). Soluble sugars (18.7 mg/gm) and reducing sugar (22.2 mg/gm) recorded significant increase in the mealy bug infested leaf (5 mg/gm and 1.6 mg/gm respectively) compared to healthy leaf. There was no significant change in the phenolic content of mealy bug infested leaf. Micronutrients like Fe showed non significant increase, but Zn (42.6 ppm) and Mn (40.5 ppm) contents of mealy bug infested leaf were significantly high i.e., 10.17 and 8.13 ppm, respectively (Table-1).

The increased nutritive value of the mealy bug infested leaf may be due to diversion of the above mentioned constituents to the site of infestation by the stimulus activity (Sadasivan and Subramaniam, 1960) or the overall increase of the above constituents may be by the stunted growth in the apical portion of mealy bug infested plant leading to the translocation and accumulation of the nutrients from the roots to the apical zone. Increased nutritive value may lead to the healthy growth of silkworms and yield good quality cocoons. The higher nutritive value of the "tukra" affected mulberry leaves is known to result in better cocoon characters (Thangamani and Vivekanandan, 1983).

Table 1. Biochemical components of healthy and Tukra infested mulberry leaves.

Sr. No.	Biochemical constituents	Healthy mulberry leaves	Tukra infested mulberry leaves
1.	Moisture (%)	72 ± 5.263	74.4* ± 6.258
2.	Moisture retention		
	6 hours	72.80 ± 3.898	75.5* ± 4.896
	12 hours	56.40 ± 4.253	61.16* ± 5.341
	18 hours	44.50 ± 2.510	50.21* ± 4.389
	24 hours	31.50 ± 1.562	32.542* ± 3.418
3.	Proteins		
	Soluble (mg/gm)	132.60 ± 5.631	143.5* ± 9.27
	Total (mg/gm)	548.320 ± 17.473	441.94* ± 18.628
4.	Sugars		
	Reducing (mg/gm)	34.7 ± 3.487	48.7* ± 2.513
	Soluble (mg/gm)	20.713 ± 1.637	22.2* ± 3.562
5.	Amino acids (m moles/gm)	74.25 ± 2.598	103.5* ± 5.192
6.	Phenols (mg/gm)	31.326 ± 3.117	28.726 ± 3.896
7.	Micro nutrients		
	Fe (ppm)	372.27 ± 5.313	214.43* ± 9.628
	Zn (ppm)	32.50 ± 2.687	42.67 ± 7.23
	Mn (ppm)	31.87 ± 1.312	40.52 ± 6.321

Each figure is the mean of three replications; Figure with ± sign is the standard deviation; * = significant at five percent (t-test).

Table 2. Economic parameters of silkworm, *Bombyx mori* (L). fed with healthy and Tukra infested mulberry leaves.

Race	PM		NB_4D_2		PM x NB_4D_2	
Group	Control	Expt.	Control	Expt.	Control	Expt.
Parameters						
Larval duration (hrs.)	365±1.26	365*±0. .98	311±0.82	304±0.82	311±0.95	304*± 0.96
Larval weight (gm)	2.668 ±0.33	2.736*±0.42	3.649±0.27	4.033*±0.33	3.631±0.54	3.718*±0.96
Cocoon weight (gm)	1.07±0.53	1.073*±0.63	1.86±0.43	1.968*±0.57	1.51±0.39	1.69* ±0.45
Shell weight (gm)	0.134±0.09	0.47*±0.04	0.376±0.81	0.439*±0.09	0.228±0.01	0.247*±0.03
Shell ration	13.137	13.699*	20.215	21.9*	15.10	14.615*
Silk filament length (m)	372±9.33	376*±16.71	1019±5.49	1138*±19.67	767± 2.88	873* ±21.71
Silk filament weight (gm)	0.123±0.01	0.127*±0.02	0.244±0.01	0.279*±0.08	0.189±0.00	0.205*±0.05
Denier scale	2.9754	3.093*	2.1546	2.2059*	2.2176	2.1132*
Renditta (kg)	13.79					

Each figure is the mean of three replications; Figure with ± sign is the standard deviation; * = significant at five per cent (t-test).

The moisture content (74.4%) and its retention of the mealy bug infested mulberry leaf recorded significant increase after 4,8,12 and 24th (2.7, 4.7, 5.7, and 13% respectively) (Table-1) related to the shorter larval duration and quality cocoons (Thangamani and Vivekanandan, 1983; Umashankar *et al.,* 1992). The increase in the moisture content and its retention may be because of the phytotoxic effect or due to formation of unstried layer as a result of the curling of laminar area.

Parameters like larval duration of fourth and fifth instars (311 h in both NB_4D_2 and PM x NB_4D_2) was more in healthy leaf fed worms compared to mealy bug infested leaf fed worms (304 h in both NB_4D_2 & PM x NB_4D_2) (Table-2) but there was no change in the larval duration of the Pure Mysore. This may be due to less sensitivity of PM race to the nutritional quality compared to NB_4D_2 and PM x NB_4D_2. Larval weight was higher in all the verities of silkworm fed with infested leaves (NB_4D_2 : 10.523%; PM : 2.54%, PM x NB_4D_2 : 2.39%). There was no significant effect observed in the ERR (%) of the silkworms studied. Cocoon weight differences (NB_4D_2 : 0.108 gm. PM : 0.053g, PM x NB_4D_2 : 0.45 gm) and CSR% (NB_4D_2 : 1.685%, PM : 0.562%, PM x NB_4D_2 -0.484%) were significantly high in the mealy bug infested leaf fed worms. There was no effect on the pupation rate, fecundity, hatching in all breeds studied. Difference in filament length of NB_4D_2 (119 m) and PM x NB_4D_2 (106 m) was more in the mealybug infested leaves fed silkworms but not significant in PM (4 g.). There was non-significant change in the filament denier in all the races of study. Renditta was decreased in NB_4D_2 (0.19g); and PM (1.18 gm) PM x NB_4D_2 (0.21g).

The above results clearly indicate that the nutritional quality and bioassay of the "Tukra" affected leaves are in-no way inferior to the healthy leaves. Preventive measures for Tukra should be followed in the mulberry garden. If at all there is infestation of Tukra, the mulberry leaves should not be rejected and wasted, instead they can safely and effectively be used as feed for silkworms as they accept it. Efficient use of available system help to improve the quality of silk and the life of silkworm, *Bombyx mori*.

References

Aftab Ahmed, C.A. Chandarakala, M.V. Shivakumar, C. and Raghuraman, R., 1998. Food and water utilization pattern under restricted feeding durations in *B. mori* of PM race. *J. Exp. Zool. India,* **1**, 29-34.

Vitthalrao B. Khayde, 2004. Influence of Juvenoids on Silkworm, *Bombyx mori* (L), Thesis, Shivaji University, Kolhapur.

Krishnaswami, S., 1978. *New Technology of Silkworm Rearing.* Bull No. 2 (C.S.R.T.I. Mysore) p. 21.

Kumar, P. Kishore, R., Noamani, M.K.R. and Sengupta, K., 1989. Effect of feeding tukra affected mulberry leaves on silkworm rearing performances. *Oriental Entomology Congress,* Agra press.

Pradipkumar, Ramkishore, Noamani, M.K.R. and Sengupta, K., 1992. Effect of feeding tukra affected mulberry leaves on silkworm rearing performances, *Indian, J. Seric..* **31,** 27-29.

Radha, N.V., Athchoumanance, I., Rajeswari and Obliswami, G., 1978. Effect of feeding with leaves of different mulberry varieties on the races of silkworm, *All India Symposium on Seric. Sciences.* UAS, Bangalore (Abst) pp. 52.

Sadasivan and Subramanian, 1960. *Plant Pathology* (an advanced treatise). Horsefal and A.f. Diamond, Academic Press **1**, pp 306.

Shree, M.P. and Umashankar, N.N., 1989. Bio-chemical changes in tukra affected exotic mulberry plants *Current Science,* **58**(22), 1251-1253.

Thangamani, R. and Vivekanandan, M., 1983. Effective utilization of tukra diseased mulberry leaves as silkworm, *B. mori* feed *Sericologia,* **23**, 211-223.

Umashankar N.N., Shree, M.P. and Pavankumar R., 1989. of chemical constituents of mulberry variety Vietnam, *Geobios,* **16**(6), 283-284.

Index